Bacterial and Eukaryotic Porins

Edited by
Roland Benz

Further Titles of Interest

M. Futai, Y. Wada, J. Kaplan (eds.)
Handbook of ATPases – Biochemistry, Cell Biology, Pathophysiology

2004
ISBN 3-527-30689-7

Y. Yawata
Cell Membrane – The Red Blood Cell as a Model

2003
ISBN 3-527-30463-0

G. Krauss
Biochemistry of Signal Transduction and Regulation (3rd Ed.)

2003
ISBN 3-527-30591-2

M. Hoppert
Microscopic Techniques in Biotechnology

2003
ISBN 3-527-30198-4

G. Winkelmann (ed.)
Microbial Transport Systems

2001
ISBN 3-527-30304-9

Bacterial and Eukaryotic Porins

Structure, Function, Mechanism

Edited by
Roland Benz

WILEY-VCH

WILEY-VCH Verlag GmbH & Co. KGaA

Edited by

Prof. Dr. Roland Benz
Lehrstuhl für Biotechnologie
Biozentrum der Universität Würzburg
Am Hubland
97074 Würzburg
Germany

Cover Illustration
Top view of a trimeric LamB maltoporin
with bound oligosaccharides
(courtesy of R. Benz)

Library of Congress Card No.: applied for
British Library Cataloguing-in-Publication Data
A catalogue record for this book is available from the British Library.

Bibliographic information published by Die Deutsche Bibliothek
Die Deutsche Bibliothek lists this publication in the Deutsche Nationalbibliografie; detailed bibliographic data is available in the Internet at http://dnb.ddb.de.

Printed in the Federal Republic of Germany.
Printed on acid-free paper.

Typesetting Hagedorn Kommunikation GmbH, Viernheim
Printing betz-druck gmbh, Darmstadt
Bookbinding J. Schäffer GmbH & Co. KG, Grünstadt

ISBN 3-527-30775-3

Contents

Bacterial and Eukaryotic Porins. Edited by Roland Benz

ISBN 3-527-30775-3

Preface

Cell wall containing eubacteria have traditionally been divided into gram-positive and gram-negative bacteria; according to their staining properties with crystal violet, also know as Gram stain. Electron microscopic analysis into the structure of the so-called gram-negative bacteria soon suggested that their cell wall contained an additional membrane, called outer membrane [1]. The lipid composition of the outer membrane composed of lipids and lipopolysaccharides is well known for a long time. Starting with the early seventies of the last century knowledge accumulated on the rather simple protein composition of the outer membrane: only a few bands were observed on SDS-PAGE of outer membrane proteins. Some of them are heat-modifiable because they change the position on SDS-PAGE when heated to 100°C. Electron microscopic analysis of the surface of *Escherichia coli* outer membrane revealed the presence of a protein with a regular structure [2]. This protein termed matrix protein' was considered the permeability pathway for hydrophilic solutes through the outer membrane. After identification of the pore-forming unit in the outer membrane of *Salmonella typhimurium* [3] and *E. coli* [4] the name porin for the outer membrane channels came into use.

The study of the amino acid composition of porins and the first amino acid sequence of a porin of *E. coli* deduced from the mature protein represented a big surprise [5]. The porins had an amino acid composition similar to those of water-soluble proteins, containing more than 50% hydrophilic amino acids. The primary sequences of the first and all other outer membrane porins sequenced to date do not show any indication for the presence of α-helical structures, which were considered since Kyte and Doolittle [6] to represent the typical structural elements of membrane proteins. Vogel and Jähnig [7] suggested that outer membrane proteins are arranged in amphipathic β-strands and form a β-barrel cylinder. The cylindrical structure implies that on average every second amino acid in membrane-spanning β-sheets is hydrophobic because it faces the hydrocarbon core of the membrane or it is hydrophilic and points to the channel interior. From the functional standpoint porins were divided in general diffusion pores and substrate-specific porins. Besides these more classical porins, the outer membrane contains also energy-coupled transporters and channel-tunnels (see below). General diffusion pores represent more or less structured holes in the outer membrane but they have a defined exclusion limit for the passage of hydrophilic solutes (see chapters

Bacterial and Eukaryotic Porins. Edited by Roland Benz

ISBN 3-527-30775-3

2 and 6). Specific porins contain binding sites for substrates such as sugars or nucleosides (see chapters 9 and 10). Luckily enough it was possible to crystallize both, general diffusion pores and specific porins of bacterial outer membranes. Chapter 2 describes structural properties of general diffusion porin and chapter 9 those of specific porins. Surprisingly, the carbohydrate-specific LamB and ScrY have two β-strands more than the general diffusion pores OmpF, OmpC and PhoE.

The expression of general diffusion pores and specific porins can be regulated dependent on the requirements of the growth media. A interesting example for porin regulation represents the regulation of OmpF/OmpC porins of *E. coli* by the two-component regulatory system EnvZ/OmpR dependent on osmolarity, pH and other environmental parameters (see chapter 1). This means that bacteria are able to respond to environmental stress. Chapter 3 describes the role of bacterial porins in antibiotic susceptibility. This chapter describes the relationships between porins and antibiotic molecules and its impact on the development of resistance against certain antibiotics. Besides bacterial response against stress created by antibiotics there exist also other mechanisms of outer membrane permeability control. This type of control is described in chapter 5 of this book and deals with rapid modulation of porin function. Other gram-negative bacteria have a certain "natural" resistance against many antibiotics. A prominent example of this intrinsic antibiotic resistance is the opportunistic human pathogen *Pseudomonas aeruginosa*. Chapter 4 deals with the properties of the porins of this organism and explains the high intrinsic antibiotic resistance of this organism, which is in part the result of the control of outer membrane permeability for hydrophilic solutes. Part of this control is also OprF of *P. aeruginosa* outer membrane, which has not the "classical" trimeric form of outer membrane porins of the OmpF type as it is described in chapter 7. OprF and also OmpA of *E. coli* and other enteric bacteria exist in two different configurations. The majority of these outer membrane proteins have 8 β-strands and a very low permeability, whereas a small fraction of OprF and OmpA allow diffusion of large solutes that cannot penetrate OmpF of *E. coli* (chapter 7).

Reconstituted systems allow a meaningful study of porin function. As already mentioned, chapter 5 deals with the study of porin modulation in model membranes using the patch-clamp technique. Chapter 6 describes general diffusion porin reconstitution in lipid bilayer membranes and chapter 10 the study of specific diffusion porins containing binding-sites in the same system. Porin trimers are very stable and the lipid bilayer technique can be a useful tool in the area of development of biosensors.

Besides the classical Omp pores also several outer membrane transporters for iron chelates and one for vitamin B12 were crystallized (chapters 11 and 12). These transporters are of special interest because the outer membrane is not energized. Energy is provided through proteins of the cytoplasmic membrane and their coupling to the outer membrane transporter is of special interest. The transporter have a novel structure of outer membrane proteins. In particular, they are formed by a β-barrel cylinder of 22 β-strands. The hole in the cylinder is plugged by a cork or a hatch as it is described for FhuA (siderophore receptor, chapter 11) and BtuB (vitamin B12 receptor, chapter 12) of *E. coli*. Another class of outer membrane permeability pathway is formed by the channel tunnel prototype TolC of *E coli*, which has also been crystallized (chapter 8). The outer membrane channels of the TolC-type are formed by homotrimers. However, differently to the porins, the trimer contains only one channel. The channel-tunnels can be divided into different parts. The 4 nm long β-barrel cylinder of 12 β-barrels is connected to the 10 nm long á helical tunnel domain of 12 α-helices, which presumably spans the periplasmic space. The channel-tunnels are important parts of efflux pumps exports systems as discussed in chapter 8.

Because of the homology of the electron transport chains and the existence of two membranes, it has been hypothesized that mitochondria are descendants of certain strictly aerobic bacteria. In any case, the permeability properties of the mitochondrial outer membranes show some resemblance to those of bacterial outer membranes as discussed in chapter 13. A considerable part of the permeability properties of mitochondrial outer membranes is caused by the presence of a general diffusion pore, called mitochondrial porin or VDAC [8], which forms voltage-dependent channels in reconstituted systems. Mitochondrial porins have a secondary structure highly homologous to that bacterial outer membrane proteins, which means that the channel formed by a porin monomer is essentially a β-barrel cylinder (chapter 13). Several different isoforms of not well understood function exist in many organisms as discussed in chapters 14 and 15. Unfortunately, mitochondrial porin or VDAC could not be crystallized to date in order to obtain any useful 3D-structure, hence the exact 3D-structure is still not known and currently a matter of debate. Chapter 16 describes that mitochondrial porins play an important role in the physiology of these cell organelles, which means that they have a communicative function. Mitochondrial porins/VDACs regulate the movement of mitochondrial metabolites between the cytosol and the mitochondrial compartments. They are possible components of the mitochondrial permeability transition pore and may participate in very interesting mitochondrial functions such as apoptosis (see chapter 14 and 16).

I am very grateful to all authors who made valuable contributions in this book and who produced their manuscript so promptly. Many thanks go also to Christian Andersen who provided the picture of the front cover and Frank Orlik who helped me with the index.

Würzburg, June 2004 Roland Benz

References

1 J.-M. Ghuisen, R. Hakenbeck (eds.), *Bacterial Cell Wall*, Elsevier, Amsterdam, **1994**.

2 A. C. Steven, B. ten Heggeler, R. Môller, J. Kistler, J. P. Rosenbusch, *J. Cell. Biol.* **1977**, *72*, 292–301.

3 T. Nakae, *Biochem. Biophys. Res. Commun.* **1975**, *64*, 1224–1230.

4 T. Nakae, *Biochem. Biophys. Res. Commun.* **1976**, *71*, 877–884.

5 R. Chen, C. Krämer, W. Schmidmayr, U. Henning, *Proc. Natl. Acad. Sci. USA.* **1979**, *76*, 5014–5017.

6 J. Kyte, R. F. Doolittle, *J. Mol. Biol.* **1982**, *157*, 105–132.

7 H. Vogel, F. Jähnig, *J. Mol. Biol.* **1986**, *190*, 191–199.

8 S. J. Schein, M. Colombini, A. Finkelstein, *J. Membrane Biol.* **1976**, *30*, 99–120.

List of Contributors

Christian Andersen
Department of Biotechnology
Würzburg University Biocenter
Am Hubland
97074 Würzburg
Germany
andersen@biozentrum.
uni-wuerzburg.de

Keltoum Anflous
Department of Molecular and
Human Genetics
Baylor College of Medicine
One Baylor Plaza
Houston, TX 77030
USA

Arnaud Baslé
Department of Biology and
Biochemistry
University of Houston
369 Science & Research Bldg.
Houston, TX 77204-5001
USA

Roland Benz
Lehrstuhl für Biotechnologie
Biozentrum der Universität Würzburg
Am Hubland
97074 Würzburg
Germany
roland.benz@mail.uni-wuerzburg.de

Michael Braun
Fakultät für Biologie
Universität Tübingen
Auf der Morgenstelle 28
72076 Tübingen
Germany

Volkmar Braun
Fakultät für Biologie
Universität Tübingen
Auf der Morgenstelle 28
72076 Tübingen
Germany
volkmar.braun@mikrobio.
uni-tuebingen.de

Dieter Brdiczka
Department of Biology
University of Konstanz
Universitätsstr. 10
78457 Konstanz
Germany
dieter.brdiczka@uni-konstanz.de

Nathalie Cadieux
Department of Microbiology
University of Virginia
1300 Jefferson Park Avenue
Charlottesville, VA 22908-0734
USA

Bacterial and Eukaryotic Porins. Edited by Roland Benz

ISBN 3-527-30775-3

David P. Chimento
Department of Microbiology
University of Virginia
1300 Jefferson Park Avenue
Charlottesville, VA 22908-0734
USA

William J. Craigen
Department of Molecular and
Human Genetics
Baylor College of Medicine
One Baylor Plaza
Houston, TX 77030
USA
wcraigen@bcm.tmc.edu

Christophe Danelon
Biophysique Membranaire
Institut de Pharmacologie et de
Biologie Structurale
205, route de Narbonne
31077 Toulouse
France

Anne H. Delcour
Department of Biology and
Biochemistry
University of Houston
369 Science & Research Bldg.
Houston, TX 77204-5001
USA
adelcour@uh.edu

Vito De Pinto
Laboratory of Biochemistry and
Molecular Biology
Department of Chemical Sciences
University of Catania
Viale A. Doria, 6
95125 Catania
Italy
vdpbiofa@mbox.unict.it

Alvin Go
Department of Microbiology and
Immunology
University of Illinois at Chicago
835 S. Wolcott Avenue
Chicago, IL 60612
USA

Robert E. W. Hancock
Department of Microbiology and
Immunology
University of British Columbia
6174 University Blvd
Vancouver, BC V6T 1Z3
Canada
bob@cmdr.ubc.ca

Robert J. Kadner
Department of Microbiology
University of Virginia
1300 Jefferson Park Avenue
Charlottesville, VA 22908-0734
USA
rjk@virginia.edu

Linda J. Kenney
Department of Microbiology and
Immunology
University of Illinois at Chicago
835 S. Wolcott Avenue
Chicago, IL 60612
USA
kenneyl@uic.edu

Angela Messina
Laboratory of Biochemistry and
Molecular Biology
Department of Chemical Sciences
University of Catania
Viale A. Doria, 6
95125 Catania
Italy

Hiroshi Nikaido
Department of Molecular & Cell Biology
University of California, Berkeley
426 Barker Hall
Berkeley, CA 94720-3202
USA
nhiroshi@uclink4.berkeley.edu

Frank Orlik
Lehrstuhl für Biotechnologie
Biozentrum der Universität Würzburg
Am Hubland
97074 Würzburg
Germany

Jean-Marie Pagès
Université de la Méditerranée
Faculté de Médecine
27 Bd. Jean Moulin
13385 Marseille cedex 05
France
jean-marie.pages@medecine.univ-mrs.fr

Tilman Schirmer
Division of Structural Biology
Biozentrum, University of Basel
Klingelbergstrasse 70
4056 Basel
Switzerland
tilman.schirmer@unibas.ch

Georg E. Schulz
Institut für Organische Chemie und Biochemie
Universität Freiburg
Albertstrasse 21
79104 Freiburg
Germany
schulz@bio.chemie.uni-freiburg.de

Etsuko Sugawara
Department of Molecular & Cell Biology
University of California, Berkeley
426 Barker Hall
Berkeley, CA 94720-3202
USA

Sandeep Tamber
Department of Microbiology and Immunology
University of British Columbia
6174 University Blvd
Vancouver, BC V6T 1Z3
Canada

Mikhail Vyssokikh
A. N. Belozersky Institute of Physico-Chemical Biology
Moscow State University, Building A
Moscow, 119899
Russia

Don Walthers
Department of Microbiology and Immunology
University of Illinois at Chicago
835 S. Wolcott Avenue
Chicago, IL 60612
USA

Mathias Winterhalter
International University Bremen
School of Engineering and Science
Campus Ring 1
28759 Bremen
Germany
m.winterhalter@iu-bremen.de

1
Regulation of Porin Gene Expression by the Two-component Regulatory System EnvZ/OmpR

Don Walthers, Alvin Go and Linda J. Kenney

1.1
Introduction

The major paradigm for signal transduction in bacteria is the two-component regulatory system. The first component is a sensor kinase, most often a membrane protein, which senses an environmental signal and is phosphorylated by ATP on a conserved histidine residue. The second component is the response regulator, which catalyzes the phosphotransfer of the phosphoryl group onto a conserved aspartic acid residue (see [1, 2] for recent reviews). Most response regulators are two-domain proteins and phosphorylation of the receiver domain alters the output of the effector domain, which is usually a stimulation of DNA binding. In some systems, the histidine kinase alters the level of the phosphoresponse regulator by stimulating its dephosphorylation, rather than by stimulating its phosphorylation.

The sensor kinase EnvZ and the response regulator OmpR comprise the two-component system that is responsible for the regulation of expression of the outer membrane proteins OmpF and OmpC (see Figure 1.1). Porin levels are influenced by a wide variety of environmental conditions, including osmolality, temperature, pH and growth phase. Although the total amount of OmpF and OmpC remains constant, the relative level of the two proteins fluctuates with respect to the osmolality of the growth medium. At low osmolality, the major porin present is OmpF, while at high osmolality, the expression of *ompF* is repressed and OmpC becomes the predominant porin [3]. The two porins differ from one another by the size of their pores and their flow rates, with OmpC having the smaller pore and slower flux [4]. Sensing the osmolality of its surroundings is one strategy by which *Escherichia coli* senses its environment and this ability is crucial for its survival. It is proposed that osmosensing enables *E. coli* to determine whether or not it is in a host environment (high osmolality) or a dilute environment (low osmolality).

Bacterial and Eukaryotic Porins. Edited by Roland Benz

ISBN 3-527-30775-3

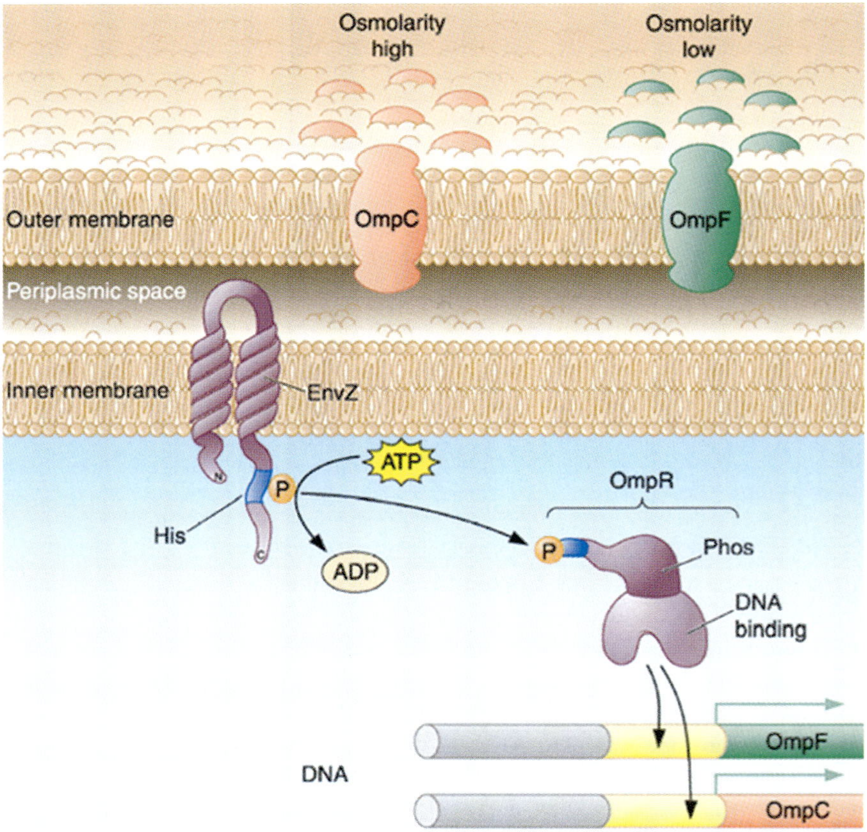

Figure 1.1 Regulation of the porin genes by EnvZ/OmpR. When the osmolality of the growth medium is low, OmpF is the predominant porin in the outer membrane. When the osmolality increases, *ompF* is repressed and OmpC becomes the major porin in the outer membrane. Regulation is mediated by the two-component regulatory system that consists of the two proteins EnvZ and OmpR. EnvZ is a sensor kinase, located in the inner membrane. ATP phosphorylates EnvZ on a conserved histidine residue and it transfers the phosphoryl group to OmpR. OmpR is a two-domain response regulator. Phosphorylation in the N-terminal receiver domain at a conserved aspartic acid residue alters the conformation of the C-terminal DNA binding domain. Phospho-OmpR (OmpR-P) binds to the regulatory regions of the porin genes *ompF* and *ompC* and alters their expression.

1.2
The Structure of EnvZ

EnvZ is a 450-amino-acid protein, located in the inner membrane (see Figure 1.2). EnvZ is comprised of two transmembrane domains flanking a 117-amino-acid periplasmic region at the N-terminus and a kinase/phosphatase catalytic domain at the C-terminus (EnvZc). EnvZc can be further separated into two functionally

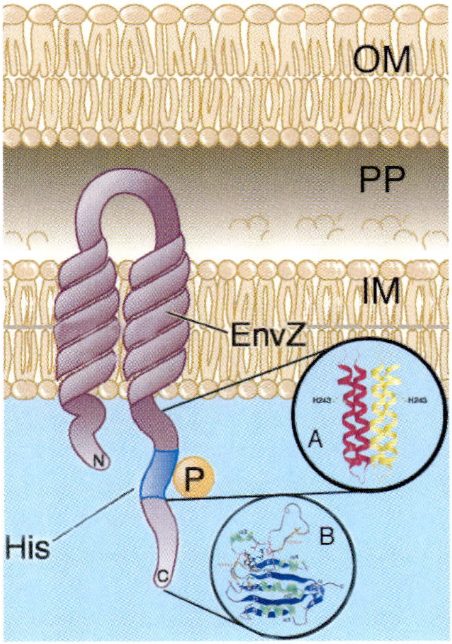

Figure 1.2 EnvZ topology. EnvZ is located in the inner membrane, with both N- and C-termini in the cytoplasm. It has two transmembrane domains and a periplasmic loop. Whether or not there is a role for the periplasmic domain in osmoregulation has not been clearly established. Structures of two domains of the cytoplasmic portion of EnvZ have been determined by nuclear magnetic resonance spectroscopy (see insets). Domain A consists of a four-helix bundle and is a dimer in solution. The histidine that is phosphorylated is located in domain A. Domain B contains the ATP binding site and is composed of a conserved α/β fold that is also found in other ATP binding proteins (see text for details). OM = outer membrane, PP = periplasm, IM = inner membrane.

distinct subdomains. Domain A (amino acids 223–289) is the phosphorylation and dimerization domain, and contains the site of autophosphorylation at His-243. Domain B (amino acids 290–450) contains the ATP-binding site as well as several regions conserved amongst all members of the histidine kinase family [5–7].

Domain A of EnvZc is located in the cytosol and is separated from the second transmembrane domain by a linker of approximately 43 amino acids. This HAMP linker (histidine kinase, adenylyl cyclase, methyl-accepting chemotaxis protein and phosphatase) likely consists of two amphipathic helices, is structurally conserved among many sensor proteins and may play a role in signal transduction [8]. When expressed separately, domain A forms a stable homodimer in solution with an apparent molecular weight of 19 kDa [9]. The homodimer consists of a four-helical bundle with 2-fold symmetry. Each monomer of domain A contains two α-helices, $\alpha 1$ (residues 235–255) and $\alpha 2$ (residues 265–286), separated by a 9-amino-acid loop. The helices of the subunits pack in the dimer such that each $\alpha 1$ is surrounded by and aligns antiparallel to an $\alpha 2$ of each subunit. The core of the bundle is hydrophobic and is composed of a number of methyl-containing residues, consistent with other histidine kinases. His-243, which lies in $\alpha 1$, is oriented opposite its counterpart in $\alpha 1'$ and protrudes away from the helical bundle, where it is solvent-accessible for phosphorylation by ATP. The phosphotransfer domains of the histidine kinases CheA and ArcB also contain four-helix bundles. In each of these structures, the active histidine (His-48 in CheA and His-717 in ArcB) lies in the center of their respective helix and points outward [10, 11]. The structure of domain A also resembles the cytoplasmic domains of the *E. coli* chemoreceptors Tar and Tsr, which are the sensory components of the chemotaxis system [12]. Like EnvZc, the functional chemoreceptor is a homodimer in which the cytoplasmic domains of each monomer form a four-helical bundle.

The structure of domain B consists of an α/β sandwich composed of a five-stranded β-sheet (strands B: residues 319–323; D: 356–362; E: 367–373; F: 420–423; and G: 431–436) on one fold and three α-helices (α1: 301–311; α2: 334–343; and α3: 410–414) on the other [13]. This structure resembles the ATP-binding proteins Hsp90 and DNA gyrase B. Between these two folds is a hydrophobic core containing many structural hydrophobic residues conserved amongst other histidine kinases. Between α3 and α4, a long polypeptide loop extends, termed the "central loop", which has no defined structure and may be mobile in solution. Binding of ATP occurs at α3 and the central loop, and also involves contacts with β-strands F and G. This central loop is near the ATP-binding pocket formed by Asp-347, Asp-373, Ile-378 and Phe-387, and may interact with His-243 in domain A, possibly stabilizing phosphorylation. The triphosphate chain of the ATP molecule is exposed on the surface of the protein, to allow the transfer of the phosphate to His-243 in domain A. Several conserved glycines, forming the G1 and G2 boxes, previously shown to be essential for kinase activity, are also located in the catalytic core.

1.3
Biochemical Activities of EnvZ underlie Signaling

The EnvZ kinase has the following enzymatic activities:

(1) EnvZ + ATP	$\rightarrow$	EnvZ–P + ADP (autophosphorylation)
(2) EnvZ–P + OmpR	$\rightarrow$	EnvZ + OmpR-P (phosphotransfer)
(3) EnvZ + OmpR-P	$\rightarrow$	EnvZ + OmpR + P_i (phosphatase)

EnvZ could potentially modulate the level of OmpR-P by adjusting the activity of its autokinase (1), the phosphotransferase activity (2) or the OmpR-P phosphatase activity (3) separately, or in various combinations (see Figure 1.3).

It has been proposed that domain A contains the phosphatase activity of EnvZ [14]. The half-life of OmpR-P alone was reported to be approximately 90 min, whereas in the presence of domain A, the half-life of OmpR-P decreased to 8.7 min. This result led to the interpretation that the A domain was the source of the phosphatase activity. However, if domains A and B were intact (i. e. EnvZc), the half-life of OmpR-P further decreased to 2.5 min. It is evident that stimulation of OmpR-P turnover by EnvZc is most efficient in the presence of the intact cytoplasmic domain. Either both A and B domains contribute to OmpR-P dephosphorylation or the A domain must be in a preferred conformation that requires tethering to the B domain in order for the A domain to fully function. An important remaining question is how the A and B domains of EnvZc are organized with respect to one another in the intact protein.

1.4
What is the EnvZ Activity Regulated by the Stimulus?

In a recent attempt to elucidate the stimulus to which EnvZ responds, the kinase was overproduced, purified and reconstituted into proteoliposomes [15]. The EnvZ autokinase activity was stimulated by addition of potassium (activity 1), but neither phosphotransfer to OmpR (activity 2) nor the phosphatase activity of EnvZ (activity 3) were affected by the presence of potassium. However, the activities measured were extremely low, making interpretation of the experiments difficult. An osmotic upshift imposed by various sugars, glycine betaine, proline or Tris–MES was without effect. Since potassium accumulation is an early response to osmotic upshift by *E. coli*, it may be that the autokinase of EnvZ is sensitive to this step, arguing that the kinase activity (1) is the osmosensitive reaction that is regulated. A previous study also reported that potassium stimulated the level of OmpR-P, but the autokinase and phosphotransferase activities of the kinase were not separated [16]. Interestingly, the phosphorylation of an OmpR mutant (OmpR3, phenotype F^-C^+) was constitutively high at low KCl concentrations and was not stimulated by further addition of KCl [16]. This result implies that the OmpR3 mutant has altered interactions with EnvZ that lead to high levels of OmpR-P at low potassium concentrations, or that the phosphotransfer activity is the step altered by high potassium (activity 2), in contrast to the results of the proteoliposome study [15].

An earlier study by Jin and Inouye [17] proposed that at high osmolality, OmpR-P levels increase as a result of a decrease in the phosphatase of EnvZ (activity 3). This hypothesis is based on experiments with a chimeric kinase Taz, which contains the periplasmic domain of the aspartate chemoreceptor Tar fused to the cytoplasmic domain of EnvZ [18]. This construct activates *ompC* in response to aspartate. However, this construct has several serious limitations, which cast doubt on whether conclusions based on this construct are physiologically meaningful. For example, Taz requires 1–5 mM aspartate to activate *ompC* compared to Tar, which binds aspartate with a K_d of 1.2 μM [19] and the addition of maltose, which also binds to Tar, did not enhance *ompC* expression. Furthermore, aspartate did not affect *ompF* transcription. In any case, the results from several Tar and Trg chimeras support a view that a common transmembrane signal transduction mechanism exists [20, 21].

Activation of EnvZ, by an as yet undetermined signal, leads to phosphorylation at His-243 from ATP and subsequent phosphorylation of OmpR at Asp-55. Phosphorylation of OmpR increases its affinity by at least 10-fold for the regulatory regions upstream of the *ompF* and *ompC* genes [22]. More recently, it was shown that the presence of DNA stimulates OmpR phosphorylation [23], i. e. the communication between OmpR domains is bidirectional. These experiments led to the proposal that OmpR might be activated while bound to its target DNA. This series of activation events would require that a complex exists between the membrane-embedded sensor kinase EnvZ with OmpR while complexed to the regulatory regions of *ompF* and *ompC* DNA. Genetic evidence for such a kinase/response regulator/DNA com-

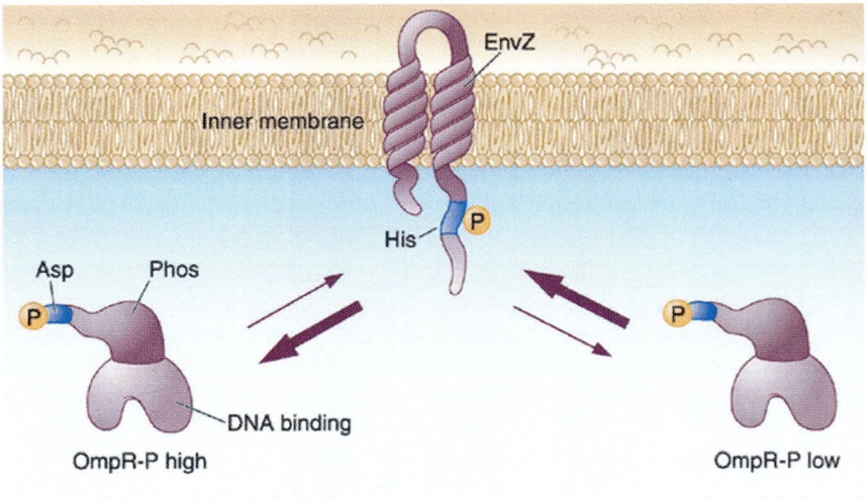

Figure 1.3 Biochemical activities of EnvZ control the concentration of OmpR-P. At low osmolality, the OmpR-P concentration is low, either because the EnvZ kinase activity is low or the phosphatase activity is high (right arrows). OmpR-P levels increase at high osmolality – this is either due to stimulation in the EnvZ kinase or a reduction in the EnvZ phosphatase activity (left arrows).

plex has been reported by Silhavy et al. in the homologous Cpx system that senses envelope stress (P. DiGiuseppi and T. J. Silhavy, personal communication). A four-state model can be described (Figure 1.4) in which OmpR exists as an equilibrium mixture between the unphosphorylated form (A), OmpR-P (B), the unphosphorylated, DNA-bound form (C) and the phosphorylated form bound to DNA (D). The reaction step that is most affected by the presence of DNA depends upon the phosphodonor employed. When phosphorylating with the small molecule phosphodonor, acetyl-phosphate, DNA binding dramatically stimulates the rate of phosphorylation with little effect on the dephosphorylation rate of OmpR-P. Estimates of initial rates indicate that phosphorylation by acetyl-phosphate is at least 25-fold faster in the presence of DNA than in its absence (i. e. C to D is much faster than A to B, Figure 1.4 [23]). Furthermore, DNA binding slows dephosphorylation about 2-fold (D to C is slightly slower than B to A, Figure 1.4 [23]). In contrast, when phosphorylating with the phosphokinase (EnvZ-P), the step most affected by DNA binding is the rate of EnvZ-stimulated OmpR-P dephosphorylation (i. e. D to C is much slower than B to A, Figure 1.4 [24]). In either case, the overall effect of DNA is to increase the net rate of OmpR-P formation on the order of 50-fold. Based on their findings, Qin *et al.* proposed that when OmpR-P binds to DNA, it is effectively made inaccessible to EnvZ and thus DNA binding inhibits EnvZ stimulation of OmpR-P breakdown [24]. If this proposal were true, it is difficult to imagine how the phosphatase activity (activity 3) could be the important physiologically regulated step [17], since OmpR-P bound to DNA would then be inaccessible to EnvZ. In order to address this question, OmpR was labeled with a fluorescent probe and equilibrium binding was measured using fluorescence anisotropy. The K_d for EnvZ binding to OmpR was 425 nM and the presence of *ompF* or *ompC* DNA did not affect the interaction [25]. However, when OmpR was phosphorylated, the affinity of interaction with EnvZ was so low (at least 10-fold lower) that it was not measurable. These results are in conflict with the proposed role of DNA in the OmpR/OmpR-P equilibrium mediated by EnvZ, in which DNA prevents the interaction of EnvZ with OmpR [24]. However, the results are consistent with previous measurements in the chemotaxis system in which phosphorylation of the OmpR-homolog CheY reduced its affinity for the kinase CheA [26]. The lower affinity for CheA of CheY-P also favors binding to the switch proteins of the flagellar motor. In keeping with this analogy, if OmpR were phosphorylated by EnvZ-P while bound to DNA, the reduced affinity of OmpR-P for EnvZ would favor the release of EnvZ, enabling OmpR to interact with RNA polymerase and activate transcription. With a cellular concentration of OmpR of 3.5 μM and an EnvZ concentration of 180 nM [27], an apparent $K_d > 5$ μM for EnvZ binding to OmpR-P indicates that these two partners would only rarely be associated. Our favored interpretation is that OmpR-P dephosphorylation is sufficiently rapid *in vivo* to promote turnover and that the more likely osmosensitive reaction of EnvZ is the autokinase activity (activity 1). This view is consistent with the observation of Jung *et al.* that potassium stimulates autophosphorylation [15].

The issue of bifunctionality of the sensor kinase (i. e. phosphorylation and dephosphorylation functions) was addressed and a system with these features was

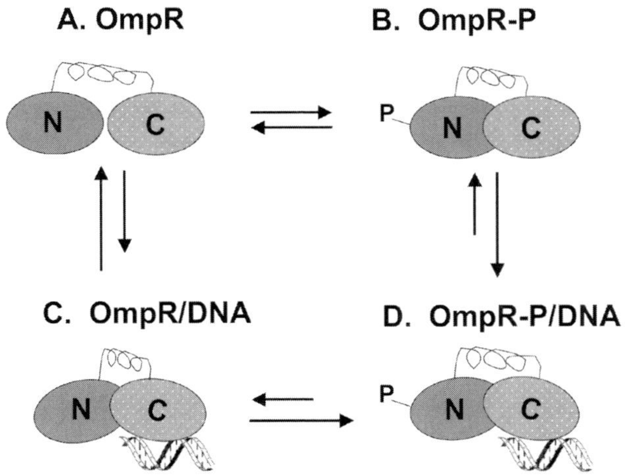

Figure 1.4 Model for OmpR phosphorylation and DNA binding. OmpR is depicted as a two-domain protein with the N-terminus joined to the C-terminus by a flexible linker region. The protein is shown alone (A), phosphorylated (B), bound to DNA (C), and phosphorylated and bound to DNA (D). The arrows depict transitions between these states. Note that the conformation of the linker changes when OmpR is phosphorylated (B), bound to DNA (C) or both (D).

compared to one in which spontaneous dephosphorylation of the response regulator was the sole means of turnover [28]. The concentrations of EnvZ and OmpR were varied independently, and the effect on *ompF* and *ompC* transcription was determined at low and high osmolality. The authors tested whether or not the system was robust with respect to the components EnvZ and OmpR. Changes in EnvZ levels had little effect on *ompF* and *ompC* transcription, whereas increasing the concentration of OmpR dramatically increased *ompC* transcription. This effect was not observed until OmpR levels had risen at least 10-fold and was especially pronounced at high osmolality. Presumably, the over-expression of OmpR enables unphosphorylated protein to occupy the low-affinity sites, and stimulate *ompC* transcription and repress *ompF*.

1.5 How is the Signal Propagated?

Reports differ as to the domains of EnvZ that are essential for signal transduction. An early study engineered large deletions (24–40 amino acids) in the periplasmic domain and examined porin phenotypes [16]. The resulting EnvZ constructs produced constitutive expression of a high osmotic phenotype, (i. e. $OmpF^-$, $OmpC^c$), regardless of the osmolality of the growth medium. Although the construction of the mutants resulted in the addition of a few extra amino acids and the mutants were over-expressed, the study suggested that in a low osmotic environment, there was an interaction with the periplasmic domain of EnvZ that was removed or not present at high osmolality. The authors further claimed that the mutants were defective in the EnvZ-stimulated dephosphorylation of OmpR-P (activity 3), although the autokinase activity and phosphotransferase activities were not examined in detail [16]. An interesting observation was that the EnvZ mutants produced pleiotropic phenotypes that were $PhoA^-$, $LamB^-$ and Mal^-. The explanation for the effects on multiple pathways outside of the normal porin repertoire was that the accumulation of OmpR-P as a result of the altered dephosphorylation by EnvZ enabled OmpR to act on genes that it normally does not regulate. This hypothesis has not been adequately tested to determine whether or not OmpR-P levels are actually higher in these EnvZ backgrounds or whether OmpR-P directly affects these additional genes, but it remains an intriguing hypothesis.

A more recent study compared EnvZ molecules from two different organisms [29] and noted the absence of a periplasmic domain in the EnvZ from *Xenorhabdus nematophilus*. Interestingly, *envZ* from *X. nematophilus* was able to complement an *envZ*-null strain of *E. coli* and restore osmoregulation of the porin genes [29]. Replacement of the periplasmic domain of EnvZ with the non-homologous domain of PhoR (a sensor kinase not involved in porin gene expression) produced a chimera capable of osmoregulation of *ompF* and *ompC* [30]. However, a 91-amino-acid periplasmic deletion showed a similar phenotype (F^-C^c) reported in the Tokoshita study [16]. It would be of interest to compare a randomized amino acid sequence in the periplasmic domain, rather than the replacement with the sensing domain from another kinase. The result of Leonardo and Forst implies that spacing between the transmembrane domains is essential and that specific amino acid contacts are not important for signaling [30]. It argues against a specific ligand-binding site for EnvZ, which is not surprising, given that both sucrose and NaCl (i. e. unrelated ligands) can induce a high osmotic response. However, neither study rules out a role for periplasmic contacts during signal transduction. One explanation of the previous results of Tokoshita *et al.* with the periplasmic deletions [16] is that mutant proteins are produced that lock the transmembrane domains in the high osmotic form of EnvZ and the periplasmic domain is not directly involved in coupling the stimulus to the output. Their results also point to the somewhat counterintuitive view that the stimulus is present at low osmolality rather than high osmolality.

1.6
Is there a Role for Acetyl-phosphate in OmpR-P Production?

A number of reports exist of EnvZ-independent osmoregulation of the porin genes. It is clear that the porin genes are regulated in response to medium osmolality and that phosphorylated OmpR is required for expression of *ompF* and *ompC* [31]. It is presumed that OmpR-P levels increase at high osmolality and one study measured higher OmpR-P levels *in vivo* [32]. However, this result was also obtained in the absence of *envZ*, suggesting that multiple pathways can affect OmpR-P levels. The results do not exclude the possibility that EnvZ is the major factor involved in regulation of OmpR-P levels, but a second, redundant pathway might contribute to regulation in the absence of *envZ*. A likely candidate is acetyl-phosphate, as it can phosphorylate response regulators [33, 34]. Outer membrane protein profiles were examined to determine the effects of deletions in *envZ* or *ackA–pta* (required for acetyl-phosphate synthesis) on OmpF and OmpC expression. When a wild-type copy of *envZ* was present, deletion and subsequent complementation of *ackA–pta* did not significantly affect porin levels at either low or high osmolality [35]. In an *envZ* deletion strain, OmpF levels were reduced slightly and did not change at high osmolality, whereas OmpC was absent at both high and low osmolality. When both *envZ* and *ackA–pta* were deleted, OmpF production was abolished. Thus, OmpF production in the absence of *envZ* is dependent on production of acetyl-phosphate. Furthermore, over-expression of the *ack–pta* genes produces OmpR-P levels sufficient for OmpC expression and OmpF repression at high osmolality.

1.7
The OmpR Subfamily

OmpR is an extensively studied member of a subfamily of response regulators that includes PhoB, VirG, ResD and CpxR among others. This subfamily has 14 homologs in *E. coli* alone [36]. The crystal structure of the C-terminus of OmpR (OmpRc) was determined independently by two groups [37, 38]. The structures revealed the presence of a winged helix-turn-helix motif (wHTH) in the DNA binding domain of OmpR (see Figure 1.5). The recognition helix is unusually long and appears to be the major determinant of DNA binding specificity. The turn of the HTH is also unusually long; most canonical HTH proteins have turns composed of 3–4 amino acids, whereas the OmpR subfamily members have turns of 10 amino acids [39]. The turn is dynamic, based on the high-temperature factors reported and substitutions in this region were reported to affect interactions with RNA polymerase (RNAP) [40–42]. Although there is strong evidence that the turn of the HTH of PhoB interacts with the σ subunit of RNAP, recent evidence suggests that in OmpR, the turn affects DNA binding [43].

The only detailed description of DNA binding by this group of response regulators has been determined for the OmpR subfamily member PhoB [44]. PhoB was co-crystallized with a 23-bp DNA fragment containing a *pho* box and the structure was determined to 2.5-Å resolution. The structure reveals a novel tandem arrangement in which several monomers bind head to tail to successive 11-bp direct-repeat sequences, coating one face of a smoothly bent double helix. The single wing of PhoB makes minor groove contacts and a non-specific contact with the DNA backbone. One of the unique features of OmpR subfamily members is the presence of a

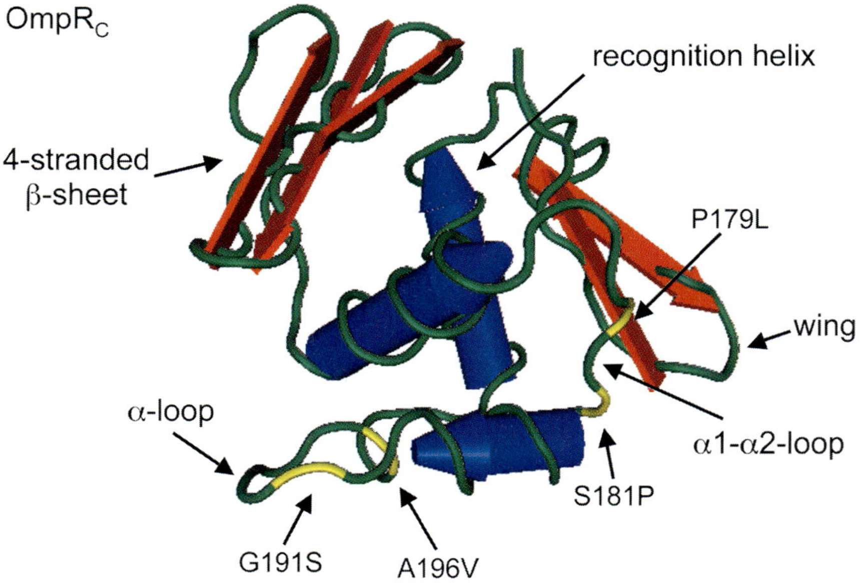

Figure 1.5 The structure of OmpRc. The worm diagram of the C-terminus of OmpR was generated with the Cn3D software package version 4.1 (available at http://www.ncbi.nlm.nih.gov). Structural information was obtained from the Protein Data Bank ID 1ODD and the determination was described in [37, 38]. The peptide backbone is depicted in green, and the α-helix and β-sheet structures are colored red and blue, respectively. Structural features and substitutions described in the text are also indicated.

four-stranded β-sheet at the N-terminal end of the effector domain (see Figure 1.5). A model whereby phosphorylation of the receiver domain results in subtle conformational changes that are transmitted to the β-sheet of the C-terminus was proposed from the co-crystal structure [44]. Cysteine-scanning mutagenesis through this region in OmpR suggests that the β-sheet may be involved in oligomeric interactions [45], indicating that the role of the β-sheet may vary among OmpR subfamily members.

Recent structural determinations have enhanced our view of OmpR subfamily members. Two full-length structures of members DrrD and DrrB from the thermophile *Thermotoga maritima* have been determined [46, 47]. Although essentially nothing is known about the function of these proteins, their structures reveal some important features. The first is that even though DrrD and DrrB are both in the OmpR subfamily, their interdomain linkers vary considerably in length (5 and 3 amino acids, respectively, compared to 15 for OmpR) and their interdomain interfaces are vastly different from one another. In the case of DrrD, there is essentially no interdomain interface, suggesting the lack of a fixed orientation with respect to the N-terminal phosphorylation domain and the C-terminal DNA binding domain [46]. In contrast, DrrB has an extensive interdomain interface of 751 $Å^2$ [47]. Earlier studies have highlighted the importance of the interdomain linker in communicating between domains [23, 48, 50]. The variation observed in the interdomain interface in the structures implies that different signaling mechanisms exist within the subfamily, and that different strategies for communication between the N- and C-terminal domains are employed [23, 47–50].

1.8 OmpR Binding Sites

OmpR binds to three tandem sites immediately upstream of the –35 element (see Figure 1.6) between approximately –100 and –40 from the transcriptional start site at both the *ompF* and *ompC* promoters [51–54]. At *ompF*, there is an additional upstream site F4 that is required for repression [55, 56]. In order to determine precisely which OmpR binding site(s) participates in the recruitment of RNAP, a genetic approach was employed [57]. DNA fragments containing the binding sites were used in a gel retardation assay with purified OmpR, EnvZ and ATP. Only the DNA fragments containing the site between –100 and –80 of either promoter were shifted in the presence of protein and ATP (F1 and C1 sites; see Figure 1.6). The interpretation of these results was that F1 and C1 each comprise high-affinity OmpR binding sites. The F1 and C1 fragments were cloned in either orientation upstream of the *ompC* –35 element. Both fragments were able to direct transcriptional activation when present in either direction, but, in both cases, the forward facing F1 or C1 fragments were more active. Insertion of a 4-bp spacer between the binding site and the –35 element abolished transcriptional activation. Interestingly, insertion of a 10-bp fragment at either promoter reduced activation, but the level of transcription was still considerably higher than the control strain [57]. Ap-

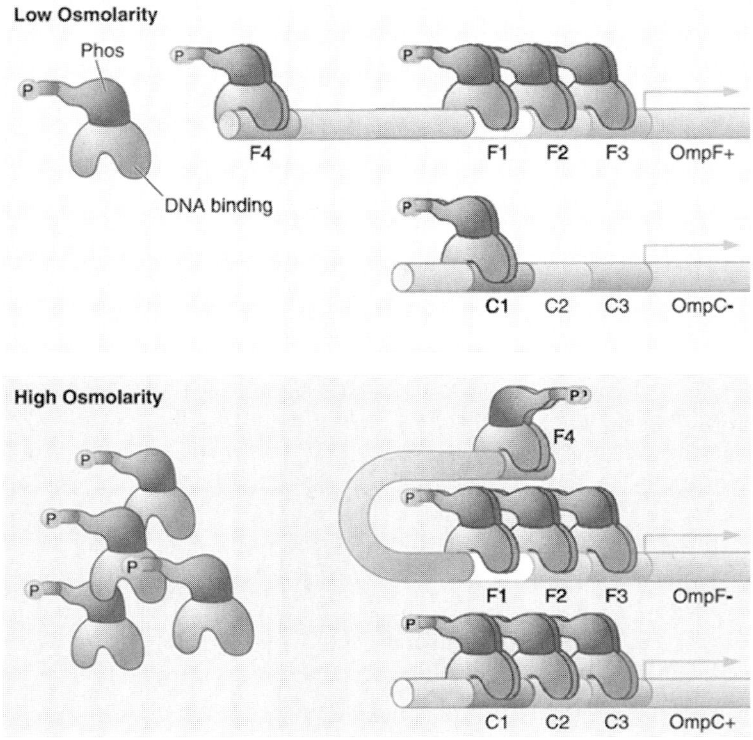

Figure 1.6 Osmoregulatory scheme for porin gene expression. At low osmolality, there is a low concentration of OmpR-P. Based on our previous studies, we think it likely that EnvZ phosphorylates OmpR while it is bound to the regulatory regions. OmpR-P bound to the *ompF* regulatory region activates *ompF*. OmpR-P is also bound to C1. Because OmpR-P has similar affinity for the low-affinity sites C2 and C3 as it does for low-affinity sites F2 and F3, the observation that *ompC* is not expressed at low osmolality suggests that OmpR-P may be prevented from binding at these sites at low osmolality. At high osmolality, the concentration of OmpR-P increases, a loop forms (facilitated by IHF binding) and OmpR-P molecules bound at the upstream site F4 interact with OmpR-P molecules bound at the low-affinity sites F2 or F3. The result is to repress *ompF* expression. At the *ompC* regulatory region, a change in accessibility of the low-affinity sites C2 and C3 enables OmpR-P to bind, activating *ompC*.

parently, the interaction between OmpR and RNAP is flexible, provided that the proper helical phasing of each protein bound to DNA is maintained. Thus, OmpR bound immediately upstream of the –35 element or at one helical turn upstream is sufficient to activate transcription, indicating that at the wild-type *ompF* or *ompC* promoters, occupancy of the F3 or C3 sites is required for activation [58]. A prediction from these studies is that OmpR bound to DNA exists as a head-to-head or tail-to-tail dimer [45]. This is intriguing, because the OmpR binding sites are asymmetric and protein dimers in head-to-head or tail-to-tail binding modes typically bind to inverted repeat elements. Furthermore, a previous study reported a head-to-tail mode of binding based on Cu-phenanthroline cleavage of OmpR bound to DNA [59]. These conflicting results need to be reconciled and further experiments will be required to do so. OmpR exists as a monomer in solution and dimerization occurs on the DNA [45]. Thus, the sequence requirements for binding to the downstream half site may be less stringent than for initial binding to the upstream half site, implying that the binding sites may actually contain a poorly defined inverted repeat.

1.9
Recruitment of RNA Polymerase to OmpR-dependent Promoters

OmpR-mediated transcription occurs through a recruitment mechanism whereby OmpR facilitates binding of RNAP at OmpR-dependent promoters. The *ompF* and *ompC* promoters contain –10 elements that deviate from the *E. coli* consensus sequence (see below). A genetic study examined the role of each –10 element in OmpR-dependent transcriptional activation.

	*
Consensus	TATAAT
ompF	AAAGAT
ompC	GAGAAT

When the first base of either –10 element was converted to a consensus T (labeled with an asterisk above), the promoters were highly active in an *ompR* mutant background [60]. In an *ompR*$^+$ background, the level of transcription was enhanced further. Apparently, the non-consensus –10 promoter element necessitates the requirement for OmpR in transcriptional activation. The addition of OmpR and/or mutation towards the consensus sequence results in stimulation of transcription. These results are consistent with a recruitment mechanism for RNAP binding to promoter DNA. OmpR most likely recruits RNAP to promoter DNA via a direct protein/protein interaction. Alternatively, OmpR binding could drive a structural change in the promoter that then facilitates RNAP binding. Since a single nucleotide change can stimulate OmpR-independent transcription, the former possibility seems more likely. A point mutation is more likely to result in the formation of a favorable nucleotide/amino acid interaction rather than mimicking a structural change in the DNA due to binding of an upstream activator.

An elegant study from the Ishihama laboratory demonstrated a direct requirement for the C-terminal domain of the α subunit of RNAP (αCTD) in activation at the *ompC* promoter [61]. RNAP was reconstituted with a full-length α subunit or a deletion lacking the αCTD for *in vitro* transcription assays at various promoters. It was shown that the location of the transcription factor binding site determined the requirement for αCTD at the promoters under examination. OmpR binds to sites upstream of the –35 region of the *ompC* promoter and requires the reconstituted RNAP containing the full-length αCTD to activate transcription, suggesting that a direct interaction occurs between OmpR and αCTD. In contrast, the OmpR family member PhoB binds to a site at the *pstS* promoter in the position normally occupied by the −35 element and does not require αCTD. It was later shown that PhoB recruits RNAP to promoters by directly interacting with the σ^{70} subunit of RNAP holoenzyme [40, 41]. Thus, OmpR subfamily members can activate transcription via interactions with different subunits of RNAP and their mechanism of activation is not conserved throughout the subfamily.

Subsequent genetic studies identified substitutions in the α subunit that affected expression uniquely at the porin genes, but not at other test promoters [62, 63]. However, it was not determined whether the substitutions affected the interaction between OmpR and α or the interaction between α and promoter DNA. It is well established that α stimulates transcription at certain promoters by directly binding to high-affinity sites termed UP elements (reviewed in [64]), but most activator-dependent promoters do not contain UP elements. It has been proposed that at some promoters that lack UP elements, non-specific or low-affinity interactions between α and DNA are stabilized by interactions with an activator and α or with σ^{70} and α [65–71]. Thus, it remains possible that mutations in α that specifically affect transcription of OmpR-dependent promoters alter the interaction of α with DNA, and not with α and OmpR. More direct evidence for an interaction between OmpR and α was provided by the identification of a suppressor mutation in *rpoA* (encoding the α subunit) of the OmpR mutant P179L [72]. The suppressor resulted in a glycine substitution for valine at position 264 of α [72]. Surprisingly, the co-crystal structure of a CAP/αCTD/DNA complex identified residue 264 of αCTD as making direct contacts with the DNA backbone [65]. It thus seems more likely that the V264G substitution enhances the affinity of αCTD for DNA rather than suppressing the OmpR positive control defect. From further analysis of the OmpR P179L mutant, it is evident that the mutant is severely defective in binding to DNA. This defect can be overcome by increasing the concentration of the mutant in an *in vitro* transcription assay, leading to a stimulation of transcription [43]. Thus, it is still unresolved whether the requirement for α at OmpR-dependent promoters is the result of an interaction with OmpR and α or the promoter DNA and α. Two studies reported inhibitory effects of over-expression of α on the *ompF* and *ompC* promoters, suggestive of a direct OmpR/α interaction [62, 73]. High concentrations of α inhibit OmpR-dependent transcription of *ompF*, presumably by sequestering OmpR from the promoter. However, it is possible that α bound to promoter DNA prevents RNAP holoenzyme from interacting with OmpR and/or the promoter DNA, a scenario that would not require a direct interaction between OmpR and α.

1.10 OmpR–RNAP Interaction Surface

The precise interaction surfaces on OmpR and α remain undetermined. Genetic screens for positive control mutants, which are defined as mutants defective in their interaction with RNAP, but not with DNA, identified three potential surfaces on OmpR for the interaction with RNAP [42, 74, 75]. These surfaces include the N-terminal receiver domain and two regions in the C-terminus, the α loop and the α1–α2 loop (see Figure 1.6). In all of the previous studies, the lack of a DNA binding defect was confirmed only with the high-affinity site C1 [42]. It is noteworthy that in two of these cases, a partial defect in binding to C1 DNA was apparent using a direct *in vitro* approach [74, 75].

Subsequent experiments with the OmpR mutant T83I indicated that altered binding can occur at the low-affinity sites without a simultaneous defect at the high-affinity C1 site [58], suggesting that high-affinity binding is not a stringent enough requirement for a positive control mutant. This result raises the possibility that the positive control mutants previously identified could be defective in DNA binding at the low-affinity sites, rather than having an impaired interaction with RNAP. Because OmpR binding to C3 is required for activation of the wild-type *ompC* promoter, defects in binding to this region would confer a transcriptional activation defect [58].

A subset of these mutants were expressed and purified to examine their DNA binding and transcriptional activation properties. OmpR R42H is a substitution in the N-terminal domain [42]; OmpR G191S and A196V are substitutions in the so-called α loop [42], which is the turn of the HTH motif [38]. OmpR P179L [42, 74] and S181P [75] are substitutions in the α1–α2 loop, immediately upstream of the HTH motif. Four of the five mutants had DNA binding defects to varying degrees at both promoters [43]. S181P was defective in binding to *ompC*, but protected *ompF* from DNase I cleavage similarly to wild-type OmpR. All of the mutants, with the exception of S181P, were able to significantly activate transcription *in vitro*. The results indicate that the substitutions in the N-terminus and the α loop are actually DNA binding mutants and not positive control mutants. When the DNA binding defect is overcome by increasing the concentration of the mutant proteins, they are able to activate transcription. Based on these results, it is proposed that a location in the N-terminus containing R42 (α1) and the α loop are potential candidates for regions that determine DNA binding specificity [43]. The α1–α2 loop remains a likely candidate as an RNAP interaction surface. The crystal structure of the OmpR C-terminus (OmpRc) reveals that residues in this region interact with residues within the recognition helix [38, 39, 76]. Thus, it is not unexpected that substitutions in the α1–α2 loop could affect DNA binding. Clearly, additional studies are required to identify the regions of OmpR required for an interaction with RNAP.

1.11
Affinity Model of Porin Gene Regulation

The predominant model that attempted to explain the differential regulation of *ompF* and *ompC* is referred to as the affinity hypothesis. The model was based on genetic analysis of a series of mutants that indicated that (i) OmpR-P was the form that regulated the porin genes and OmpR did not play a role [31] and (ii) EnvZ controls the concentration of OmpR-P via its biochemical activities (described in Section 1.3) [77]. The hypothesis is outlined briefly as follows: at low osmolality, there is a low level of OmpR-P present, either due to a low activity of the EnvZ kinase or a high activity of the EnvZ phosphatase (see Figure 1.3). OmpR-P binds to high-affinity sites on *ompF* that activate its expression. At high osmolality, the kinase activity of EnvZ increases or the phosphatase activity decreases, thereby increasing the OmpR-P concentration. OmpR-P binds to low-affinity sites on *ompF* that repress OmpF expression and it also binds to low-affinity sites on *ompC* that activate expression of OmpC. The hypothesis had two basic predictions – that phosphorylation of OmpR increases its affinity for the porin regulatory regions and that the upstream sites have different affinities for OmpR-P. The *ompF* activating sites should be high affinity, the repressing sites should be low affinity and the *ompC* activating sites should be low affinity. Thus, at low OmpR-P (low osmolality), the high-affinity site on *ompF* would be occupied, leading to OmpF production. At high OmpR-P (high osmolality), the low-affinity sites at *ompF* would be bound, repressing *ompF* transcription and the low-affinity sites at *ompC* would also be occupied, activating transcription of *ompC*. It was proposed that the differences in affinity would need to be on the order of 20-fold in order to account for the observed differences in OmpF and OmpC expression levels [77].

1.12 A Test of the Affinity Model

A quantitative analysis of the binding affinities of OmpR and OmpR-P for individual OmpR binding sites and the composite *ompF* and *ompC* promoters was performed using fluorescence anisotropy [22]. OmpR was phosphorylated by acetylphosphate and OmpR-P concentrations were measured on a C4 column using reverse-phase HPLC. The binding assays revealed that OmpR-P has similar affinities for both the isolated F1 and C1 sites. This result was surprising, as the affinity hypothesis had predicted that the F1 site should be considerably higher affinity than the C1 site. When OmpR-P binding at the composite F1F2F3 and C1C2C3 sites was examined, the affinity for *ompF* DNA (K_d = 15 μM) was only 2-fold higher than for *ompC* DNA (K_d = 31 μM). The binding assays indicate that the affinity of OmpR-P for each promoter is not sufficiently dissimilar to account for osmoregulation of porin gene expression on the basis of differences in binding site affinity. More interestingly, the binding data indicated that the affinity of unphosphorylated OmpR is greater for *ompC* DNA (K_d = 87 μM) than for *ompF* DNA (K_d = 194 μM). Thus, even at low OmpR-P concentrations, it is predicted that occupancy of *ompC* and *ompF* DNA by either OmpR or OmpR-P would be quite similar. This conclusion is supported by recent DNase I protection assays using similar conditions at the *ompF* and *ompC* promoters [43].

Considering that the affinities of OmpR-P for *ompF* and *ompC* are quite similar, the simplest explanation for the lack of *ompC* activation at low osmolality is a type of gene silencing mechanism. Such a mechanism would prevent OmpR and/or RNAP binding or otherwise negatively regulate steps downstream of RNAP recruitment [78]. Occlusion of OmpR binding to *ompC* DNA at low osmolality by another bound factor or an altered DNA structure is not in disagreement with one aspect of the affinity hypothesis. That is, differences in OmpR-P levels in response to changes in medium osmolality correlate with differential regulation of *ompF* and *ompC*. Although the affinity of OmpR-P for each promoter is similar, displacement of another bound factor or binding to an unfavorable structure *in vivo* at high osmolality may require higher concentrations of OmpR-P.

1.13 Conformational Changes in OmpR Contribute to Differential Regulation of the Porin Genes

The affinity hypothesis was attractive because it was elegant and simple. If the hypothesis were no longer valid, how can osmoregulation of the porin genes be accounted for? Recent studies on the OmpR mutant T83I has led to an alternate model for differential porin gene expression [58]. A threonine in the phosphorylation site is a highly conserved residue among response regulators (position 83 in OmpR). It participates in a phosphorylation-dependent conformational change in the N-terminal receiver domain and transmission to the C-terminal DNA binding domain of OmpR. Upon phosphorylation at D55, the side-chain hydroxyl of T83 hydrogen bonds to the phosphoryl group. This creates a pocket that enables Y102 to rotate from an outward, solvent-exposed to an inward, buried position [79–82]. This "aromatic switch" is likely a general mechanism to transduce the phosphorylation event into a conformational change in the protein. Such a conformational change would then modulate protein function, for example, by increasing affinity for DNA. When over-expressed *in vivo*, OmpR T83I is unable to activate transcription of either *ompF* or *ompC*. This defect is not due to a defect in phosphorylation by EnvZ. DNase I footprinting assays indicated that the OmpRT83I mutant was able to bind to *ompF* DNA identically to wild-type OmpR, yet displayed altered binding at *ompC*. Thus, OmpR must adopt different conformations when bound to *ompF* versus *ompC*. In the case of T83I, the mutant appears to be locked into a low osmolality conformation that enables normal binding at *ompF*, but not at *ompC*. It is especially intriguing that this substitution is in the phosphorylation site, not in the DNA binding domain of OmpR. Binding at *ompC* would require that OmpR adopt a different conformation and this switch is prevented by the substitution. The mechanism that facilitates the conformational change of OmpR when bound to one promoter versus the other remains elusive. It is possible that OmpR somehow responds to intracellular changes in osmolality. Another possibility is that the different conformation required for DNA binding at *ompC* occurs post-DNA binding. In other words, once OmpR is bound to *ompC* DNA, a conformational change might occur to facilitate oligomerization, RNAP interaction or some other event required for transcriptional activation. A shift to high osmolality may directly or indirectly alter the structure of the *ompC* promoter that facilitates DNA binding. For example, a previous report demonstrated that integration host factor (IHF) protects approximately 40 bp upstream of the OmpR binding sites at the *ompC* promoter [83]. If IHF binding at *ompC* depends on environmental osmolality, binding or release at high osmolality could lead to a structural change in the downstream DNA that favors OmpR binding. In agreement with this prediction, a recent report demonstrated that IHF binding to an upstream activating sequence of the *ilv*P_G promoter mediates duplex destabilization in the –10 element that favors RNAP binding [84]. In other words, a structural change in the DNA transmitted by distally bound IHF favors RNAP binding in the absence of any direct interaction between the two proteins.

The characterization of OmpR T83I revealed an additional property of OmpR-dependent transcriptional activation. T83I binds to *ompF* DNA similarly to wild-type OmpR, yet it is unable to activate transcription. Since EnvZ can phosphorylate T83I *in vitro*, the substitution likely confers a defect in signaling that is subsequent to phosphorylation and is required for transcriptional activation. Likely possibilities are OmpR oligomerization or a conformational change that is required for a productive interaction with RNAP.

1.14 Other Factors that Regulate *ompF* and *ompC*

Additional factors participate in the regulation of *ompF* and *ompC* (see [85] for a review). For example, the *micF* gene, which is divergently transcribed from upstream of the *ompC* gene, produces two mRNA species, both of which contain sequences complementary to the ribosome binding site and translational initiation sequence of *ompF* [86, 87]. Induction of *micF* results in an mRNA that is capable of interfering with *ompF* translation. MicF appears to be more important in regulating *ompF* in response to high growth temperature rather than to changes in osmolality.

The ubiquitous, nucleoid-associated protein IHF has also been implicated in the regulation of not only *ompF*, but the *ompC* and *ompR–envZ* loci as well. At the *ompF* promoter, IHF is believed to facilitate formation of a DNA loop that represses *ompF* at high osmolality [88]. *In vivo* expression studies indicated elevated *ompF* levels in an IHF mutant strain, even when grown at high osmolality [89]. A separate study examined the interaction between purified IHF and *ompF* DNA *in vitro* [88]. DNase I protection assays combined with electrophoretic mobility shift assays identified two protected regions – one of high affinity between –200 and –160 relative to the transcriptional start site and a lower-affinity site between –80 and –40 [88]. The more promoter-distal binding site is likely involved in DNA looping, as it lies between the F4 site and the promoter-proximal sites. However, DNA looping at the *ompF* promoter has yet to be directly demonstrated. The downstream site may also participate in repression of *ompF*, as it overlaps the OmpR binding sites F2 and F3. Addition of IHF to an *in vitro* transcription reaction inhibited OmpR-dependent transcription and an increase in OmpR concentration was sufficient to overcome this effect [88].

IHF has also been shown to affect transcription of *ompC*. DNase I footprinting identified an IHF binding site upstream of the *ompC* promoter at –200 to –160 and addition of IHF to an *in vitro* transcription system resulted in reduced OmpR-dependent transcription of *ompC* [83]. Suppressor analysis suggests that IHF-mediated repression of *ompC* may be physiologically relevant. The OmpR mutant V203M constitutively expresses *ompF*, but is defective at activating *ompC*. V203M has a reduced affinity for *ompC* DNA when compared to wild-type OmpR [90]. In a *himA* null background (encoding one of the two IHF subunits), the *ompC* expression defect was suppressed [83]. These results suggest that IHF inhibits OmpR binding at the *ompC* promoter, although this possibility has not been tested directly. It is unclear whether or not IHF levels are coordinated to changes in medium osmolality and under what conditions IHF binds at *ompF* and *ompC* (i.e. at low versus high osmolality or both). Thus, the precise contributions made by IHF to differential porin gene expression and the signals that feed into this pathway remain elusive. A further complication is that IHF has also been shown to negatively regulate transcription of the *ompR–envZ* locus in both *in vitro* and *in vivo* assays [91]. An intriguing question is whether or not changes in IHF levels and/or function in response to changes in medium osmolality are important for differential regulation of the three loci. However, IHF is a pleiotropic, global regulatory protein, and any models concerning IHF-dependent regulation will likely be difficult to test *in vivo*.

1.15 OmpR is a Global Regulator

In addition to modulating the ratio of OmpF to OmpC in the cell in response to changes in osmolality, OmpR plays a wider role in regulating gene expression. Recent transcriptome analysis reported that an *ompR–envZ* deletion resulted in changes in the expression of 125 different genes [92]. What remains to be determined is how many of these genes are directly regulated by OmpR. Although some of these changes may be the result of defects in porin synthesis, they include a diverse array of genes, such as those necessary for flagellar-mediated motility, amino acid biosynthesis (cysteine and leucine) and enterochelin synthesis. The lack of both EnvZ and OmpR results in a variety of phenotypic changes, including increased hexose utilization (allose, fructose, mannitol, *N*-acetyl-D-glucosamine and glucose), increased resistance to antibiotics (cephalosporins, β-lactams, topoisomerase inhibitors and folate antagonists) and a decreased resistance to some metal ions (cobalt, selenite, chromate, lithium and dichromate), but, interestingly, little change in sensitivity to osmolality [93]. More recently, a deletion in *ompR* was shown to increase expression of a variety of drug exporters [94]. Previous work from many groups has shown that OmpR plays a role in the regulation of a wide variety of genes, including the flagellar master regulon *flhDC* [95], expression of *csgDEFG* operon and curli fimbriae production [96–100], expression of the cryptic porins OmpS1 and OmpS2 in *Salmonella typhi* [99, 100], the stationary phase acid tolerance response in *S. typhimurium* [101, 102], and expression of many virulence genes [103–106].

1.15.1 Flagellar Biosynthesis

The flagellar master regulator FlhDC is encoded by the *flhDC* operon and is responsible for activating the expression of motility and chemotaxis genes, including the flagellar components, the motor apparatus, the chemoreceptors and *che* genes (for review, see [12, 107]). Several studies report that expression of the *flhDC* operon is partially affected by acetyl-phosphate concentration [108–110]. Shin and Park demonstrated that OmpR-P binds to the promoter and represses *flhDC* [95]. Mutations in *pta* (phosphotransacetylase) or *ompR* increase *flhDC* expression, whereas mutations in *ack*A (acetate kinase), *envZ11* or increasing osmolality decrease *flhDC* expression. On soft-agar motility plates, Δ(*ompR–envZ*) mutations have a stronger swarm phenotype than the wild-type [92]. Therefore, acetyl-phosphate may regulate flagellar biosynthesis through phosphorylation of OmpR, which binds to the *flhDC* promoter region and inhibits expression of the FlhDC master regulator. Electrophoretic mobility shift assays and DNase I footprinting identify two binding sites in the *flhDC* promoter region [95]. Phosphorylation of OmpR leads to a 10-fold increase in DNA binding affinity, as also observed for binding at *ompF* and *ompC* [22] and results in protection at regions −179 to −116 and −17 to +49 from the transcriptional start site. An IHF-binding site is located between the two OmpR binding sites at *flhDC*. A repression loop, similar to that predicted at *ompF* may also occur between the two binding sites at *flhDC* to inhibit transcription of this operon.

1.15.2 Curli Fimbriae Production

Many bacteria form biofilms on surfaces and life in a biofilm is a prevalent lifestyle for microbes in the environment. Biofilms are characterized as a community of microorganisms suspended in a polysaccharide matrix [111]. Bacteria in these environments are extremely resistant to most antimicrobial agents, and display different patterns of gene expression and morphology distinct from their free-swimming counterparts. A mutation in *ompR* (Leu-43 to Arg substitution; *ompR234* allele) causes a dramatic increase in bacterial surface binding and biofilm formation compared to the wild-type [98]. This phenotype is caused by an overproduction of curli, which are long, fimbrial structures composed of the products of the *csgBA* operon. In either wild-type or *ompR*$^-$ backgrounds, curli production is substantially reduced. Expression of *csgBA* is stimulated by the transcriptional activator CsgD, whose expression is dependent upon OmpR [97]. Although OmpR is proposed to act as a positive regulator at *csgD*, curli formation is repressed at high osmolality. *S. typhimurium* contains six OmpR binding sites upstream of the *csgD* transcriptional start site, one of which has an inhibitory effect on *csgD* expression [96].

1.15.3
Virulence

OmpR modulates the virulence properties of several pathogens and OmpR mutants have a reduced ability to infect hosts. In *Shigella flexneri*, OmpR alters expression of the *vir* genes necessary for invasion of epithelial cells [105]. Expression of the *vir* genes is reduced in an *envZ* null mutant, but can be restored by increasing medium osmolality. In an *ompR–envZ* double mutant, osmolality changes are uncoupled from *vir* expression and the mutants are attenuated in the invasion of HeLa cells. In *Yersinia enterocolitica*, an *ompR* null mutant is essentially avirulent and is unable to infect orally challenged BALB/c mice beyond 21 days post-inoculation [104]. Environmental stress experiments report that the *ompR* null strain exhibits decreased survival when exposed to high osmolality (1.5 M NaCl) or high temperature (55 °C) as compared to the wild-type. It remains to be determined whether the avirulent phenotype of *Y. enterocolitica ompR* null strains is caused by this increased susceptibility to environmental stress or, more likely, whether OmpR directly interacts with the regulatory regions of a subset of virulence genes necessary to sustain infection.

In *S. typhimurium*, *ompR* and *envZ* deletions are attenuated for virulence and are unable to kill phagosomes and spread to other macrophages. The avirulent phenotype is not a result of reduced expression of *ompF*, *ompC* or *tppB*; the virulence of the triple mutant is not as severely diminished as the *ompR* mutant [106]. A recent study reported that OmpR activated expression of SsrAB, a two-component regulatory system located on *Salmonella* pathogenicity island 2 [112]. SsrA/B subsequently regulates a type III secretion system that is required for *Salmonella* to replicate inside macrophages. This is one of only a few examples whereby a response regulator of one two-component system affects the transcription of another two-component system. Footprinting analysis identified several OmpR binding sites upstream of *ssrA* and both upstream and within the *ssrB* coding sequence [112, 113]. The expression of the kinase SsrA is uncoupled from the expression of the response regulator SsrB [114]. This is unusual compared to most two-component regulatory systems in which the ratio of kinase to response regulator is tightly controlled. It will be informative to understand the role of the various OmpR binding sites in the activation of transcription of *ssrA* and *ssrB*, given their novel location.

Acknowledgments

We acknowledge the following for financial support to L. J.K. (GM 58746 from the National Institutes of Health and MCB 0243085 from the National Science Foundation) and fellowship F32-GM68364 to D. W. L. J.K. is grateful to Tom Hope, UIC for help with the illustrations.

References

1 Robinson, V. L., Buckler, D. R., Stock, A. M. *Nature Struct Biol* **2000**, *7*, 626–633.

2 Stock, A. M., Robinson, V. L., Goudreau, P. N. *Annu Rev Biochem* **2000**, *69*, 183–215.

3 van Alphen, W., Lugtenberg, B. *J Bacteriol* **1977**, *131*, 623–630.

4 Nikaido, H., Vaara, M. *Microbiol Rev* **1985**, *49*, 1–32.

5 Dutta, R., Inouye, M. *Trends Biochem Sci* **2000**, *25*, 24–28.

6 Parkinson, J. S., Kofoid, E. C. *Ann Rev Genet* **1992**, *26*, 71–112.

7 Swanson, R. V., Alex, L. A., Simon, M. I. *Trends Biochem Sci* **1994**, *19*, 485–490.

8 Appleman, J. A., Chen, L. L., Stewart, V. *J Bacteriol* **2003**, *185*, 4872–4882.

9 Park, H., Saha, S. K., Inouye, M. *Proc Natl Acad Sci USA* **1998**, *95*, 6728–6732.

10 Ikegami, T., Okada, T., Ohki, I., Hirayama, J., Mizuno, T., Shirakawa, M. *Biochemistry* **2001**, *40*, 375–386.

11 Mourey, L., Da Re, S., Pedelacq, J. D., Tolstykh, T., Faurie, C., Guillet, V., Stock, J. B., Samama, J. P. *J Biol Chem* **2001**, *276*, 31074–31082.

12 Falke, J. J., Hazelbauer, G. L. *Trends Biochem Sci* **2001**, *26*, 257–265.

13 Tanaka, T., Saha, S. K., Tomomori, C., Ishima, R., Liu, D., Tong, K. I., Park, H., Dutta, R., Qin, L., Swindells, M. B., Yamazaki, T., Ono, A. M., Kainosho, M., Inouye, M., Ikura, M. *Nature* **1998**, *396*, 88–92.

14 Zhu, Y., Qin, L., Yoshida, T., Inouye, M. *Proc Natl Acad Sci USA* **2000**, *97*, 7808–7813.

15 Jung, K., Hamann, K., Revermann, A. *J Biol Chem* **2001**, *276*, 40896–40902.

16 Tokishita, S., Kojima, A., Aiba, H., Mizuno, T. *J Biol Chem* **1991**, *266*, 6780–6785.

17 Jin, T., Inouye, M. *J Mol Biol* **1993**, *232*, 484–492.

18 Utsumi, R., Brissette, R. E., Rampersaud, A., Forst, S. A., Oosawa, K., Inouye, M. *Science* **1989**, *245*, 1246–1249.

19 Biemann, H. P., Koshland, D. E., Jr. *Biochemistry* **1994**, *33*, 629–634.

20 Baumgartner, J. W., Kim, C., Brissette, R. E., Inouye, M., Park, C., Hazelbauer, G. L. *J Bacteriol* **1994**, *176*, 1157–1163.

21 Ward, S. M., Delgado, A., Gunsalus, R. P., Manson, M. D. *Mol Microbiol* **2002**, *44*, 709–719.

22 Head, C. G., Tardy, A., Kenney, L. J. *J Mol Biol* **1998**, *281*, 857–870.

23 Ames, S. K., Frankema, J., Kenney, L. J. *Proc Natl Acad Sci USA* **1999**, *96*, 11792–11797.

24 Qin, L., Yoshida, T., Inouye, M. *Proc Natl Acad Sci USA* **2001**, *98*, 908–913.

25 Mattison, K., Kenney, L. J. *J Biol Chem* **2002**, *277*, 11143–11148.

26 Li, J., Swanson, R. V., Simon, M. I., Weis, R. M. *Biochemistry* **1995**, *34*, 14626–14636.

27 Cai, S. J., Inouye, M. *J Biol Chem* **2002**, *277*, 24155–24161.

28 Batchelor, E., Goulian, M. *Proc Natl Acad Sci USA* **2003**, *100*, 691–696.

29 Tabatabai, N., Forst, S. *Mol Microbiol* **1995**, *17*, 643–652.

30 Leonardo, M. R., Forst, S. *Mol Microbiol* **1996**, *22*, 405–413.

31 Slauch, J., Silhavy, T. J. *J Mol Biol* **1989**, *210*, 291–292.

32 Forst, S., Delgado, J., Rampersaud, A., Inouye, M. *J Bacteriol* **1990**, *172*, 3473–3477.

33 Lukat, G. S., McCleary, W. R., Stock, A. M., Stock, J. B. *Proc Natl Acad Sci USA* **1992**, *89*, 718–722.

34 McCleary, W. R., Stock, J. B. *J Biol Chem* **1994**, *269*, 31567–31572.

35 Matsubara, M., Mizuno, T. *Biosci Biotechnol Biochem* **1999**, *63*, 408–414.

36 Mizuno, T. *DNA Res* **1997**, *4*, 161–168.

37 Kondo, H., Nakagawa, A., Nishihira, J., Nishimura, Y., Mizuno, T., Tanaka, I. *Nature Struc Biol* **1997**, *4*, 28–31.

38 Martinez-Hackert, E., Stock, A. M. *Structure* **1997**, *5*, 109–124.

39 Martinez-Hackert, E., Stock, A. M. *J Mol Biol* **1997**, *269*, 301–312.

40 Kim, S. K., Makino, K., Amemura, M., Nakata, A., Shinagawa, H. *Mol Gen Genet* **1995**, *248*, 1–8.

41 Makino, K., Amemura, M., Kim, S. K., Nakata, A., Shinagawa, H. *Genes Dev* **1993**, *7*, 149–160.

42 Pratt, L. A., Silhavy, T. J. *J Mol Biol* **1994**, *243*, 579–594.

43 Walthers, D., Go, A., Kenney, L. J. **2004**, in preparation.

44 Blanco, A. G., Sola, M., Gomis-Ruth, F. X., Coll, M. *Structure* **2003**, *10*, 701–713.

45 Go, A., Mattison, K., Kenney, L. J. **2004**, in preparation.

46 Buckler, D. R., Zhou, Y., Stock, A. M. *Structure* **2002**, *10*, 153–164.

47 Robinson, V. L., Wu, T., Stock, A. M. *J Bacteriol* **2003**, *185*, 4186–4194.

48 Mattison, K., Oropeza, R., Kenney, L. J. *J Biol Chem* **2002**, *277*, 32714–32721.

50 Walthers, D., Tran, V. K., Kenney, L. J. *J Bacteriol* **2003**, *185*, 317–324.

51 Maeda, S., Mizuno, T. *J Bacteriol* **1990**, *172*, 501–503.

52 Harlocker, S. L., Bergstrom, L., Inouye, M. *J Biol Chem* **1995**, *270*, 26849–26856.

53 Rampersaud, A., Harlocker, S. L., Inouye, M. **1994**, *J Biol Chem 269*, 12559–12566.

54 Huang, K. J., Igo, M. M. *J Mol Biol* **1996**, *262*, 615–628.

55 Huang, K. J., Schieberl, J. L., Igo, M. M. *J Bacteriol* **1994**, *176*, 1309–1315.

56 Ostrow, K. S., Silhavy, T. J., Garrett, S. *J Bacteriol* **1986**, *168*, 1165–1171.

57 Maeda, S., Takayanagi, K., Nishimura, Y., Maruyama, T., Sato, K., Mizuno, T. *J Biochem* **1991**, *110*, 324–327.

58 Mattison, K., Oropeza, R., Byers, N., Kenney, L. J. *J Mol Biol* **2002**, *315*, 497–511.

59 Harrison–McMonagle, P., Denissova, N., Martinez-Hackert, E., Ebright, R. H., Stock, A. M. *J Mol Biol* **1999**, *285*, 555–566.

60 Ozawa, Y., Mizuno, T., Mizushima, S. *J Bacteriol* **1987**, *169*, 1331–1334.

61 Igarashi, K., Hanamura, A., Makino, K., Aiba, H., Aiba, H., Mizuno, T., Nakata, A., Ishihama, A. *Proc Natl Acad Sci USA* **1991**, *88*, 8958–8962.

62 Slauch, J. M., Russo, F. D., Silhavy, T. J. *J Bacteriol* **1991**, *173*, 7501–7510.

63 Sharif, T. R., Igo, M. M. *J Bacteriol* **1993**, *175*, 5460–5468.

64 Gourse, R. L., Ross, W., Gaal, T. *Mol Microbiol* **2000**, *37*, 687–695.

65 Benoff, B., Yang, H., Lawson, C. L., Parkinson, G., Liu, J., Blatter, E., Ebright, Y. W., Berman, H. M., Ebright, R. H. *Science***2002**, *297*, 1562–1566.

66 Boucher, P. E., Maris, A. E., Yang, M. S., Stibitz, S. *Mol Cell* **2003**, *11*, 163–173.

67 Busby, S., Ebright, R. H. *J Mol Biol* **1999**, *293*, 199–213.

68 Chen, H., Tang, H., Ebright, R. H. *Mol Cell* **2003**, *11*, 1621–1633.

69 Czarniecki, D., Noel, R. J., Jr., Reznikoff, W. S. *J Bacteriol* **1997**, *179*, 423–429.

70 Flatow, U., Rajendrakumar, G. V., Garges, S. *J Bacteriol* **1996**, *178*, 2436–2439.

71 Ross, W., Schneider, D. A., Paul, B. J., Mertens, A., Gourse, R. L. *Genes Dev* **2003**, *17*, 1293–1307.

72 Kato, N., Aiba, H., Mizuno, T. *FEMS Microbiol Lett* **1996**, *139*, 175–180.

73 Bowrin, V., Brissette, R., Tsung, K., Inouye, M. *FEMS Microbiol Lett* **1994**, *115*, 1–6.

74 Aiba, H., Kato, N., Tsuzuki, M., Mizuno, T. **1994**, *FEBS Lett 351*, 303–307.

75 Kato, N., Tsuzuki, M., Aiba, H., Mizuno, T. *Mol Gen Genet* **1995**, *248*, 399–406.

76 Itou, H., Tanaka, I. *J Biochem (Tokyo)* **2001**, *129*, 343–350.

77 Russo, F. D., Silhavy, T. J. *J Mol Biol* **1991**, *222*, 567–580.

78 Nogami, T., Mizuno, T., Mizushima, S. *J Bacteriol* **1985**, *164*, 797–801.

79 Birck, C., Mourey, L., Gouet, P., Fabry, B., Schumacher, J., Rousseau, P., Kahn, D., Samama, J. P. *Structure Fold Des* **1999**, *7*, 1505–1515.

80 Lewis, R. J., Brannigan, J. A., Muchova, K., Barak, I., Wilkinson, A. J. *J Mol Biol* **1999**, *294*, 9–15.

81 Cho, H. S., Lee, S. Y., Yan, D., Pan, X., Parkinson, J. S., Kustu, S., Wemmer, D. E., Pelton, J. G. *J Mol Biol* **2000**, *297*, 543–551.

82 Halkides, C. J., McEvoy, M. M., Casper, E., Matsumura, P., Volz, K., Dahlquist, F. W. *Biochemistry* **2000**, *39*, 5280–5286.

83 Huang, L., Tsui, P., Freundlich, M. *J Bacteriol* **1990**, *172*, 5293–5298.

84 Sheridan, S. D., Benham, C. J., Hatfield, G. W. *J Biol Chem* **1998**, *273*, 21298–21308.

85 Pratt, L. A., Hsing, W., Gibson, K. E., Silhavy, T. J. *Mol Microbiol* **1996**, *20*, 911–917.

86 Andersen, J., Forst, S. A., Zhao, K., Inouye, M., Delihas, N. *J Biol Chem* **1989**, *264*, 17961–17970.

87 Andersen, J., Delihas, N., Ikenaka, K., Green, P. J., Pines, O., Ilercil, O., Inouye, M. *Nucleic Acids Res* **1987**, *15*, 2089–2101.

88 Ramani, N., Huang, L., Freundlich, M. *Mol Gen Genet* **1992**, *231*, 248–255.

89 Tsui, P., Helu, V., Freundlich, M. *J Bacteriol* **1988**, *170*, 4950–4953.

90 Tran, V. K., Oropeza, R., Kenney, L. J. *J Mol Biol* **2000**, *299*, 1257–1270.

91 Tsui, P., Huang, L., Freundlich, M. *J Bacteriol* **1991**, *173*, 5800–5807.

92 Oshima, T., Aiba, H., Masuda, Y., Kanaya, S., Sugiura, M., Wanner, B. L., Mori, H., Mizuno, T. *Mol Microbiol.* **2002**, *46*, 281–291.

93 Zhou, L., Lei, X. H., Bochner, B. R., Wanner, B. L. *J Bacteriol* **2003**, *185*, 4956–4972.

94 Hirakawa, H., Nishino, K., Hirata, T., Yamaguchi, A. *J Bacteriol* **2003**, *185*, 1851–1856.

95 Shin, S., Park, C. *J Bacteriol* **1995**, *177*, 4696–4702.

96 Gerstel, U., Park, C., Romling, U. *Mol Microbiol* **2003**, *49*, 639–654.

97 Prigent-Combaret, C., Brombacher, E., Vidal, O., Ambert, A., Lejeune, P., Landini, P., Dorel, C. *J Bacteriol* **2001**, *183*, 7213–7223.

98 Vidal, O., Longin, R., Prigent-Combaret, C., Dorel, C., Hooreman, M., Lejeune, P. *J Bacteriol* **1998**, *180*, 2442–2449.

99 Fernandez-Mora, M., Oropeza, R., Puente, J. L., Calva, E. *Gene***1995**, *158*, 67–72.

100 Oropeza, R., Sampieri, C. L., Puente, J. L., Calva, E. *Mol Microbiol* **1999**, *32*, 243–252.

101 Bang, I. S., Kim, B. H., Foster, J. W., Park, Y. K. *J Bacteriol* **2000**, *182*, 2245–2252.

102 Bang, I. S., Audia, J. P., Park, Y. K., Foster, J. W. *Mol Microbiol* **2002**, *44*, 1235–1250.

103 Dorman, C. J., Chatfield, S., Higgins, C. F., Hayward, C., Dougan, G. *Infect Immun* **1989**, *57*, 2136–2140.

104 Dorrell, N., Li, S. R., Everest, P. H., Dougan, G., Wren, B. W. *FEMS Microbiol Lett* **1998**, *165*, 145–151.

105 Bernardini, M. L., Fontaine, A., Sansonetti, P. J. *J Bacteriol* **1990**, *172*, 6274–6281.

106 Lindgren, S. W., Stojiljkovic, I., Heffron, F. *Proc Natl Acad Sci USA* **1996**, *93*, 4197–4201.

107 Armitage, J. P. *Adv Microb Physiol* **1999**, *41*, 229–289.

108 Wolfe, A. J., Chang, D. E., Walker, J. D., Seitz-Partridge, J. E., Vidaurri, M. D., Lange, C. F., Pruss, B. M., Henk, M. C., Larkin, J. C., Conway, T. *Mol Microbiol* **2003**, *48*, 977–988.

109 Pruss, B. M. *Arch Microbiol* **1998**, *170*, 141–146.

110 Pruss, B. M., Wolfe, A. J. *Mol Microbiol* **1994**, *12*, 973–984.

111 Stoodley, P., Sauer, K., Davies, D. G., Costerton, J. W. *Annu Rev Microbiol* **2002**, *56*, 187–209.

112 Lee, A. K., Detweiler, C. S., Falkow, S. *J Bacteriol* **2000**, *182*, 771–781.

113 Feng, X., Oropeza, R., Kenney, L. J. *Mol Microbiol* **2003**, *48*, 1131–1143.

114 Feng, X., Walthers, D., Oropeza, R., Kenney, L. J. *Mol Microbiol*, **2004**, in Press.

2
The Structures of General Porins

Georg E. Schulz

2.1
Bacterial Outer Membrane Proteins

A large group of bacteria possesses an outer membrane in addition to the common inner or cytoplasmic membrane. The outer membrane functions as a shield against an unfavorable environment and is permeable for small polar solutes. In contrast to the inner membrane, it contains only few proteins. As a general rule, the outer membrane proteins consist of β-barrels, whereas the inner membrane proteins have mostly α-helices. Due to the tight network of strong main-chain hydrogen bonds, the β-barrels are much more stable than the inner membrane α-helix bundles, which have only isolated strings of main-chain hydrogen bonds within each of the α-helices. The intrinsically strong structure of the β-barrels resists chemical and mechanical stress, and it corresponds to the relatively good crystallizability of these proteins resulting in numerous established structures.

According to their functions and structures, the outer membrane proteins can be subdivided into the five groups of Table 2.1. The general porins yielded the first X-ray-grade membrane protein crystals [1], indicating that their intrinsic stability approaches that of soluble proteins [2, 3]. They constitute the largest group [4–11]. The specific porins are structurally closely related to the general porins although their β-barrels have two more strands [12–14]. All of these porins are trimers (Figure 2.1). A quite different group is composed of pores where the β-barrel itself is a homo-oligomer, a dimer in gramicidin A [15], a heptamer in α-hemolysin [16] and a trimer in TolC [17]. In α-hemolysin and TolC the contact within the β-barrel is obviously of a secondary nature because the major oligomer interface lies between the globular domains of the monomers. In the monomer, the polypeptide segment which forms the barrel has a very different structure or none at all [18]. The group of outer membrane transport proteins has a much larger β-barrel and additional internal domains which may work as a stopper [19–23]. They do not form a pore, but open and close for the passage of relatively large molecules such as iron-containing siderophores or cobalamins through the outer membrane. Another group of outer membrane proteins contains small β-barrels that are not likely to

Bacterial and Eukaryotic Porins. Edited by Roland Benz

ISBN 3-527-30775-3

Table 2.1 Available structures of bacterial outer membrane porins all of which consist of β-barrels (according to Figure 2.2 the β-barrels are described by the number of strands n and by the shear number S).

Porin type	*Organism*	*Number of β-strands*	*Shear number*	*PDB code*	*Reference*
General porins					
Rc-porin	*Rhodobacter capsulatus*	16	20	2POR	5
OmpF	*Escherichia coli*	16	20	2OMF	6
PhoE	*E. coli*	16	20	1PHO	6
Rb-porin	*Rhodobacter blasticus*	16	20	1PRN	8
Pd-porin	*Paracoccus denitrificans*	16	20	–	9
OmpK36	*Klebsiella pneumoniae*	16	20	1OSM	10
Omp32	*Comamonas acidovorans*	16	20	1E54	11
Specific porins					
LamB	*E. coli*	18	20	1MAL	12
LamB	*Salmonella typhimurium*	18	20	1MPR	13
ScrY	*S. typhimurium*	18	20	1OH2	14
Composed pores					
Gramicidin A	*Bacillus brevis*	(2)	(6)	1MAG	15
α-hemolysin	*Staphylococcus aureus*	14	14	7AHL	16
TolC	*E. coli*	12	20	1EK9	17
Transport Omp					
FhuA	*E. coli*	22	24	1BY3	19
FhuA	*E. coli*	22	24	2FCP	20
FepA	*E. coli*	22	24	1FEP	21
FecA	*E. coli*	22	24	1KMO	22
BtuB	*E. coli*	22	24	1NQH	23
Poreless Omp					
OmpA	*E. coli*	8	10	1QJP	25
OmpX	*E. coli*	8	8	1QJ8	26
OmpLA	*E. coli*	12	12	1QD5	27
OmpT	*E. coli*	10	12	1I78	28
OpcA	*Neisseria meningitidis*	10	10	1K24	29
NspA	*N. meningitidis*	8	10	1P4T	30

form a pore in their native conformation. Rather, they function as membrane anchors, and fulfill numerous purposes using their extracellular and periplasmic surfaces [24–30]. Their small β-barrels have a non-polar exterior and a polar interior, and can therefore be considered inverse micelles.

The β-barrels of the outer membrane proteins show common construction principles which are sketched in Figure 2.2. The descriptive parameters are tilting angle α, shear number S [31], strand number n and barrel radius R, which are related to each other according to the given formulae. The β-barrels of outer membrane proteins obey the six rules stated below.

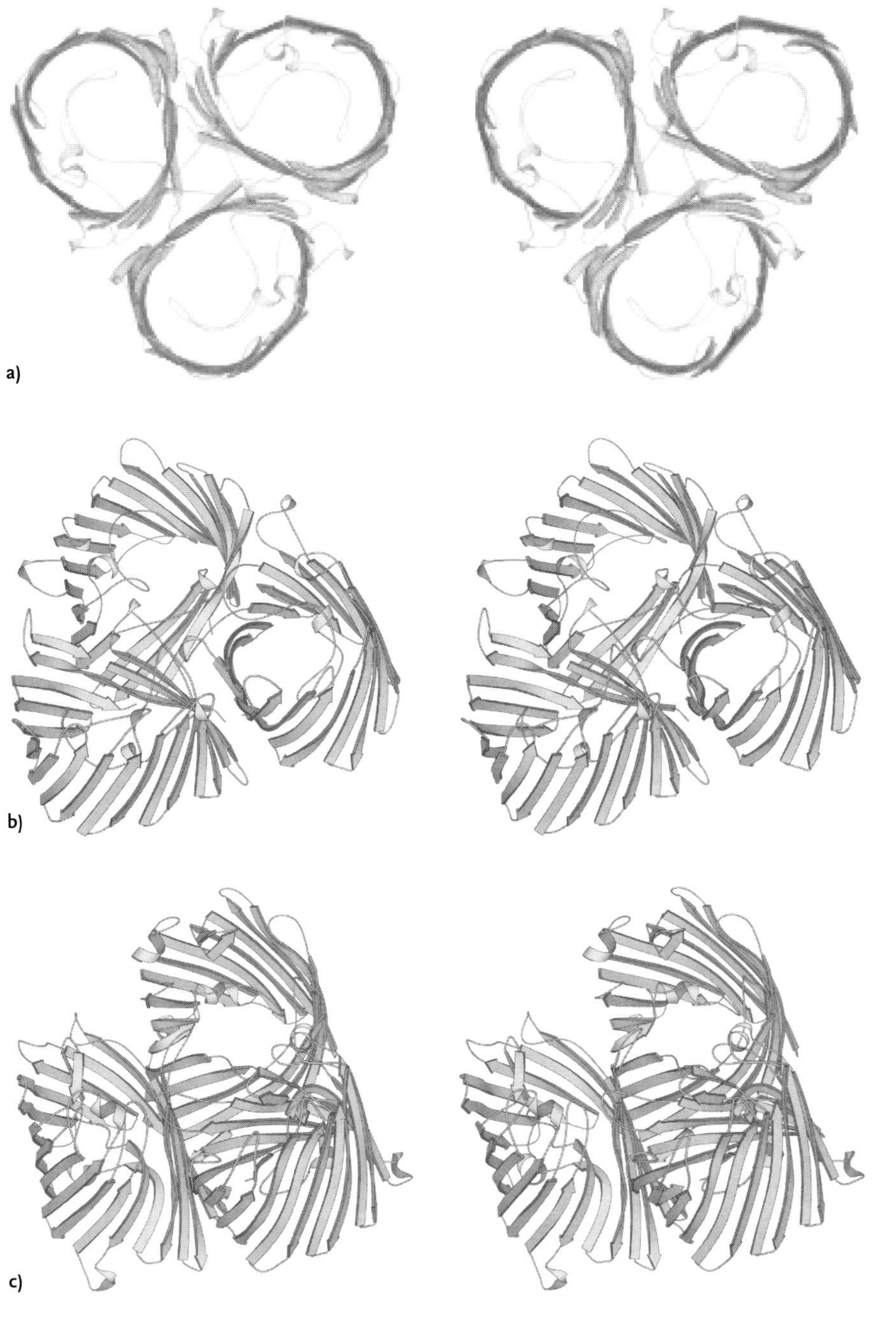

Figure 2.1 (Legend page 28)

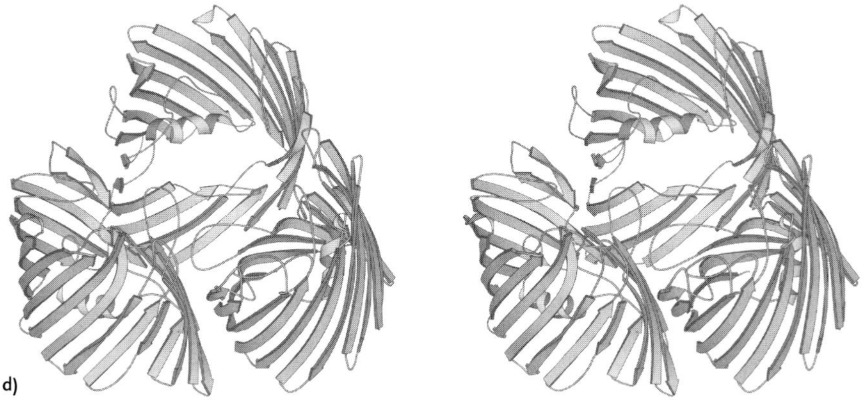

d)

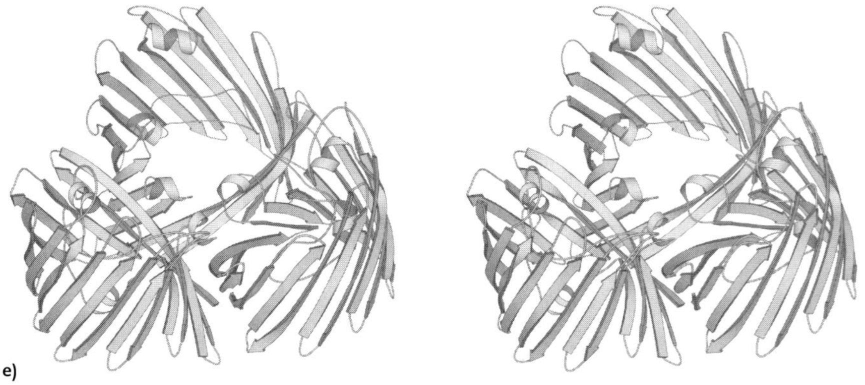

e)

Figure 2.1 Stereoviews of ribbon plots of the four distinct general porins. (a) Rb-porin as depicted in a projection along the membrane normal, showing that the membrane-facing part approximates closely to an upright cylinder. The central hub is not hollow, but filled by residue side-chains; the spikes contain the interfaces between barrels. Hub and spikes have a non-polar interior and a polar surface, like soluble proteins. The other pictures are views from the external side: (b) Rb-porin, (c) OmpF, (d) Omp32 and (e) Rc-porin.

(i) The N- and C-termini are at the periplasmic end, and the number of β-strands is even.

(ii) The tilting angle α varies between 30° and 60°. The tilt is always to the right when viewed from the outside, rendering the barrel construction chiral. The mirror image is disfavored by the steric hindrance at the C_α atoms as demonstrated by the asymmetry of the Ramachandran plot [32].

(iii) All β-strands are connected locally to their immediate neighbors along the chain, resulting in an antiparallel β-sheet pattern that maximizes the neighborhood correlation [32]. Viewed from the outside, the polypeptide runs from right to left.

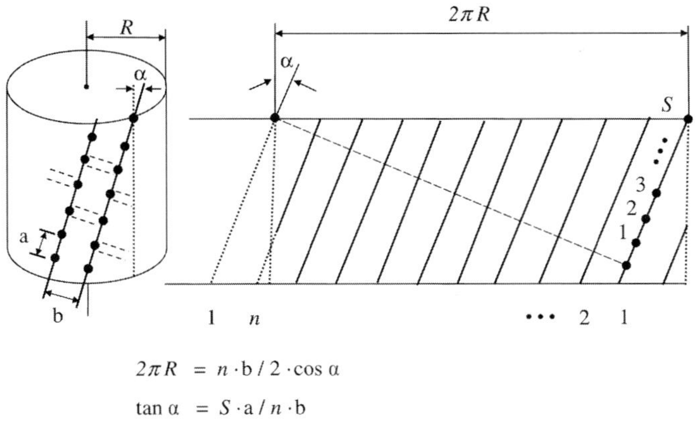

Figure 2.2 General construction principle of β-barrels. The tilt angle α is assumed to be constant, although it tends to vary around the observed barrels. The shear number *S* [31] is the displacement, counted as number of residues, when going around the barrel along the hydrogen bonds (dashed lines). The β-barrel radius *R* depends on α (or on *S*) and the number of β-strands *n*. The peptide parameters are $a = 3.3$ Å and $b = 4.4$ Å.

(iv) The strand connections at the periplasmic barrel end are short turns named T1, T2, etc., whereas those at the external barrel end are mostly long loops named L1, L2, etc.

(v) The external loops show a very high sequence variation and are often very mobile.

(vi) The outer barrel surface contacting the membrane has a 20-Å wide ribbon of aliphatic residues that is lined by an upper and by a lower girdle or aromatic residues.

Here, we are concerned with the general porins that are the most abundant species of outer membrane proteins. Seven structures have been published and, due to close sequence similarity, further structures are available with less accuracy. Among the six structures deposited in the Protein Data Bank, those of porins OmpF, PhoE and OmpK36 are closely related, showing more than 60 % sequence identity, and we use OmpF as a representative of the group in Figure 2.1.

2.2
Construction of General Porins

All structurally known general porins are trimers formed by monomers consisting of 16-stranded β-barrels. All barrel axes are approximately perpendicular to the membrane plane and all barrel interiors function as channels. OmpF and its homologs PhoE and OmpK36 are somewhat special as their barrel axes deviate by about 10° from the membrane normal. Figure 2.1 shows ribbon plots of the four distinct porins. The barrels have ellipsoidal cross-sections with axes ratios of 1.2:1, which is rather close to circular. In all cases the external loop L3 is of a particular length and folds into the barrel interior forming the pore eyelet. In some cases, especially in Omp32, further external loops assist L3 in constricting the available space inside the barrel.

The monomers are always closely associated, forming trimers which can be considered as wheels comprising a central globular hub, three spikes and a hoop. At the hub, the tilting angle α in the β-barrel is usually larger than at the hoop and an external loop folds over a neighboring subunit fortifying the oligomer. Hub and spikes contain a non-polar interior and a polar surface, part of which lines the three channels. All hubs start at the periplasmic barrel end and extend over less than the full barrel length so that the three channels merge at the external side of the eyelets to a common large channel that has been named the vestibule because the solutes gather there before entering the bacterium.

The Rc-porin was the second structurally established membrane protein [4]. Therefore, its interface to the membrane caught particular attention. The expected ribbon of non-polar residues was readily detected. This ribbon is also apparent in the statistics of the general porin surfaces shown in Figure 2.3. Interestingly, the ribbon splits into three parts, an aliphatic central ribbon surrounding the trimer and two girdles of aromatic residues lining the ribbon on both sides [33, 34]. The width of the non-polar ribbon facing the non-polar membrane layer is only 20–25 Å, which approximates the length of a single fatty acid chain. This is only half the thickness displayed in the textbooks which commonly show two opposing extended fatty acid chains forming the membrane. Accordingly, the porin surfaces show that the fatty acids are extensively intertwined, halving the generally assumed thickness of the non-polar layer. It was further recognized that the benzene ring polarity is intermediate between that of aliphatic and polar side-chains, and most suitable for contacting the membrane layer of ester bonds between the fatty acids and the polar heads of the lipids [33–35].

A statistical analysis of the six available general porin structures in Figure 2.3 shows a further peculiarity of the membrane-exposed porin surface. When subdividing the distribution of Tyr and Phe residues into one for their C_β atoms and one for their C_ζ atoms it became apparent that the aromatic ring orientations are not random. As recognized earlier [33, 34], the tyrosines are located closer to the membrane center and point outward, whereas the phenylalanines are closer to the membrane surfaces and point inwards. The tryptophans are also in the aromatic girdles, but their frequency is negligible in comparison with Tyr and Phe. This distribution confirms that the aromatic residues act as floating stabilizers in the membrane [33, 34]. In a β-barrel the C_α–C_β bonds point radially to the outside so that the aromatic side-chains can rotate freely around them. As demonstrated in Figure 2.4, this construction mitigates the consequences of membrane fluctuation because the aromatic rings can keep the contact with the ester bond layers between the non-polar and polar parts of the membrane so that the β-barrel will not be destroyed when its non-polar surface faces a polar environment or vice versa. Such fluctuations are likely to occur frequently due to mechanical stress on the bacterial surface.

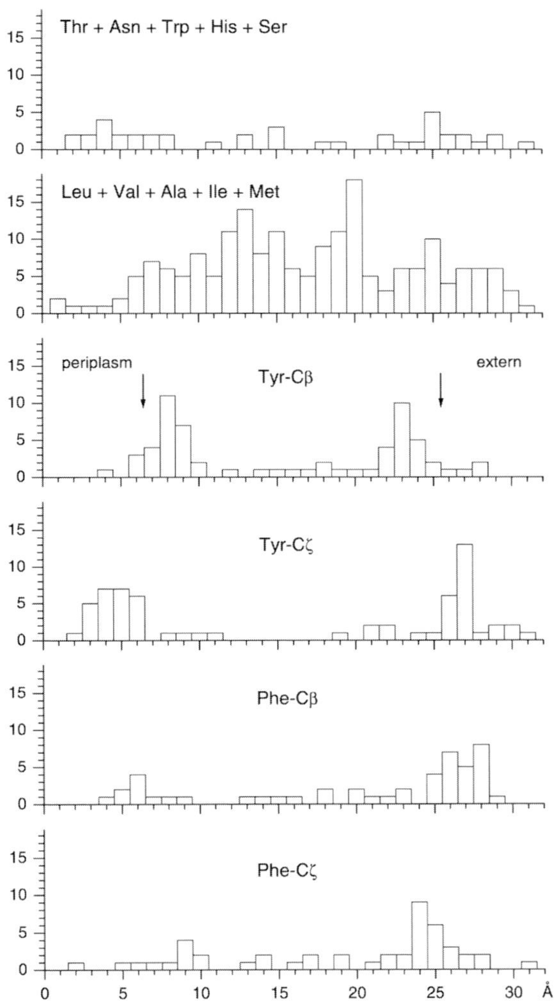

Figure 2.3 Average residue locations at the membrane-facing surfaces of the six available general porin structures listed in Table 2.1. The residues are plotted according to their position along the membrane normal. The two aromatic girdles were used as a reference, their positions are indicated by arrows. The girdle centers are 19 Å apart. On each side the statistics extend 6 Å beyond the girdle centers. Three prolines and 21 glycines are not shown (Table 2.2) because prolines disturb the barrel and glycines are neither polar nor non-polar. The Tyr and Phe statistics are given for the C_β and C_ζ atoms. The difference in the respective distribution indicates clearly the side-chain orientations: Tyr away from the membrane center and Phe towards the center.

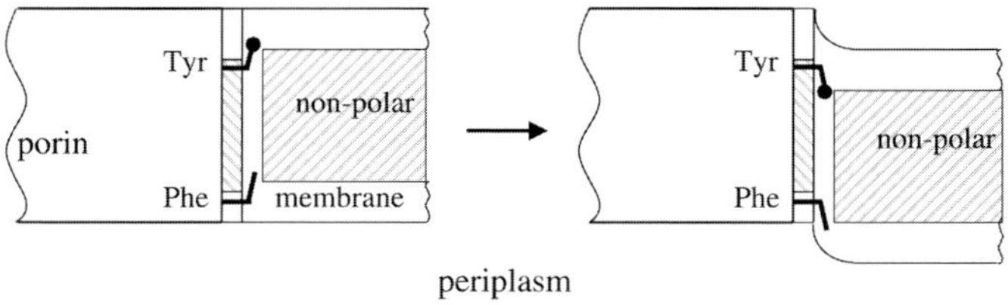

Figure 2.4 Sketch of the porin–membrane interface showing a usual (left) and a displaced (right) membrane position. The non-polar parts of the porin surface and of the membrane are striated. The aromatic girdles function as swim-stabilizers in the membrane. As the C_α–C_β bonds point out radially, the aromatic rings can rotate easily and thus mitigate the effect of membrane fluctuations by preventing a polar versus non-polar contact between the barrel and membrane, which would disrupt the β-sheet structure.

Table 2.2 Preferences for amino acid residues at the membrane-exposed surface as derived from the six trimeric general porins [the counting is for the C_γ atom positions (C_α for Gly, C_β for Ala) over a 31-Å stretch centered around the two aromatic girdles (Figure 2.3)].

Residue	*Local count*	*Local percentage*	*E. coli* overall percentage [36]	*Ratio*
Tyr	62	17.0	3.2	5.3
Leu	61	16.7	9.0	1.9
Val	52	14.2	6.4	2.2
Phe	46	12.6	3.9	3.2
Ala	43	11.8	7.8	1.5
Ile	29	7.9	5.2	1.5
Thr	23	6.3	5.8	1.1
Gly	21	5.8	7.4	0.8
Met	7	1.9	2.2	0.9
Asn	7	1.9	4.5	0.4
Trp	5	1.4	1.3	1.1
His	4	1.1	2.2	0.5
Pro	3	0.8	5.2	0.2
Ser	2	0.6	7.1	0.8
Total	365			

The residue counts of the distributions depicted in Figure 2.3 are given in Table 2.2. The preference for Tyr and Phe and the dominance of the aliphatic residues becomes most obvious in comparison with the overall residue abundance in *Escherichia coli* [36]. The depletion of the polar residues is also very clear. These preferences can be used as propensities for the prediction of membrane interfaces from protein sequences. They seem to be characteristic for all bacterial outer membrane proteins. The aromatic girdle is also present in cytoplasmic membranes where, however, it does not seem to be equally pronounced with respect to the Tyr and Phe orientations.

2.3
Trimer Association and Folding

The chain folds of the general porins depicted in Figure 2.1 show an almost circular high wall surrounding the trimer and, thus, facing the membrane. When describing the trimer as a wheel (see above), this wall is the hoop. The outer surface of the broad hoop consists of the non-polar ribbon lined by the two girdles of aromatic residues (Figure 2.3) that leave no further surface area on the periplasmic side (Figure 2.4). However, there is still plenty of surface on the external side where the porins face the lipopolysaccharides of the external leaflet of the bacterial outer membrane. This additional surface is covered by ionogenic residues, in particular by aspartates and glutamates. Presumably, these side-chains bind the porins tightly to the lipopolysaccharides, either through direct hydrogen bonds or through divalent ions such as Ca^{2+} which are abundant in the external medium. This anchoring should mitigate the mechanical stress from the environment so that the remaining fluctuations can be buffered more easily by the aromatic girdles (Figure 2.4).

The three spikes protruding from the central hub face the polar external medium on one side and the polar periplasm on the other (Figure 2.1). They are constructed like soluble proteins; their surfaces are covered with polar residues, whereas their interior is non-polar. Each spike contains about half a dozen non-polar residues and the hub has an additional dozen of non-polar residues. Most prominent are phenylalanines and leucines. The side-chains leave essentially no open space at the trimer axis running through the hub.

The construction of hub and spikes involves a non-polar trimer interface like those of most soluble oligomeric proteins. Such non-polar protein contacts are easily formed in a polar environment, but they are dispersed within a non-polar membrane. It was therefore suggested that the porin monomers associate in the periplasm and that the resulting trimer is inserted into the membrane [33, 34]. The association should start with the folding of hub and spikes in the manner of soluble proteins starting with the usual non-polar core covered by polar residues. For all porins such an intermediate consisting of hub and spike would contain the N- und C-termini (Figure 2.1). Consequently, the remaining polypeptide of around 150 residues per monomer would be fixed at both ends. Presumably this chain segment is mobile and unstructured in the periplasm and forms the second part of the β-barrel and, thus, the hoop only when inserted into the membrane.

The folding of a chain that is fixed at both ends is highly restrained because neither of the chain termini can be wrapped around any other part. This constraint does not affect the folding of the porins, however, because the chain segment which forms the hoop is an all-antiparallel and all-next-neighbor β-sheet structure (Figure 2.1) that can be easily formed by mere contraction and meandering. Actually, this exceptional β-barrel construction of all outer membrane proteins of Table 2.1, which is not observed in the β-barrels of soluble proteins [37], indicates that the barrels of outer membrane proteins are only formed after the chain ends have been fixed.

2.4 Pore Geometry

Within the established general porins we observe the three types of channel geometries described in Figure 2.5. Rc-porin and Rb-porin have a rather wide eyelet, OmpF, PhoE and OmpK36 show intermediate cross-sections, and Omp32 forms a particularly small pore. The general construction is always the same. The side-chains at the periplasmic end of the three channels (A in Figure 2.5) protrude by about 8 Å into the periplasm. The channels become narrower towards the external end until they are constricted by three eyelets after about 18 Å. Approximately 10 Å further up, the three channels join to form the vestibule (B in Figure 2.5), which after a further 8 Å opens up to the external medium (C in Figure 2.5).

The channel geometry should be related to the electric conductivity of these porins. When neglecting the radial components of the electric currents, the resistance of a pore should be proportional to the area under the reciprocal cross-section distribution of Figure 2.5, which is 0.39, 0.26 and 0.16 $Å^{-1}$ for Omp32, OmpF and Rc-porin, respectively. Accordingly, the conductivity should show the ratios of $0.39^{-1}{:}0.26^{-1}{:}0.16^{-1} = 0.66{:}1.00{:}1.68$. The actually measured conductivity for

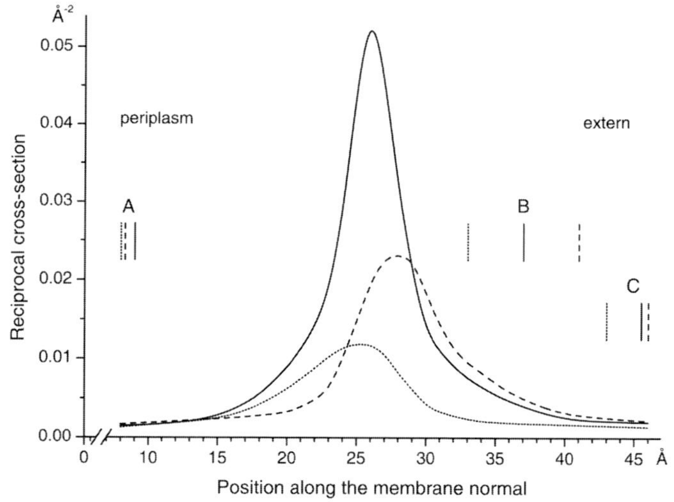

Figure 2.5 The geometry of three general porins as described by the reciprocal of the total channel cross-section plotted along the membrane normal. Each distribution along the normal is started from the first atom at the periplasmic end. The analysis shows three distinct examples Rc-porin (dotted line), OmpF (dashed line) and Omp32 (solid line). The two aromatic girdles level around the 13 and 32 Å values of the abscissa. The cross-sections correspond to the solvent-accessible space outside 2.5 Å radius spheres around the non-hydrogen atoms for one of the three channels. The three channels start around 8 Å above the first atoms (A), combine to form the vestibule around 29 Å further up (B), which opens up to the external space after a further 8 Å (C).

1 M KCl [38–40] is 0.5, 1.9 and 3.3 nS for Omp32, OmpF and Rc-porin, respectively, which converts to the ratios 0.27:1.00:1.74. These values fit each other only roughly. The fit improves, however, if one takes into account that Omp32 (in contrast to OmpF and Rc-porin) shows saturation effects at a high salt concentration [38] so that the first ratio increases appreciably at salt concentrations closer to the physiological level.

Since the geometry of the channel is known in absolute terms, its conductivity in a 1 M KCl solution can be deduced from the specific conductivity of the salt solution according to the formula:

$$(\text{pore conductivity}) = (\text{specific conductivity}) / \int Q(z)^{-1} dz \qquad (1)$$

where $Q(z)$ is the cross-section distribution along the membrane normal in the z-direction. The specific conductivity of a 1 M KCl solution is 0.113 S cm^{-1} = 1.13 nS $Å^{-1}$ and the integral is the area under the respective distribution in Figure 2.5 as stated above. This results in a calculated porin trimer conductivity of 9, 13 and 21 nS for Omp32, OmpF and Rc-porin, respectively. The OmpF and Rc-porin values exceed the experimental values by a factor of about 6 and the Omp32 value is too high by a factor of about 18. At this point it should be remembered that the $Q(z)^{-1}$ distribution of Figure 2.5 was deduced using a molecular surface 2.5 Å away from the non-hydrogen atoms, which is obviously not enough to ensure that the solutes are in an environment comparable to the bulk solvent. Enlarging this 2.5 Å distance to about 3.8 Å reduces the calculated conductivities to the observed values. Reversing the argument, the K^+ and Cl^- ions can be considered as mobile as in the bulk solvent if they are 3.8 Å away from the non-hydrogen atoms of the protein. This is a rather rough estimate. Much more detailed permeation analyses were performed with the specific porins [41, 42].

Pore conductivity is determined by applying a voltage across the membrane and measuring the resulting current. Assuming bulk solvent conditions, the electric potential follows the integral over the reciprocal cross-section distribution of Figure 2.5 and the voltage between any two points along the membrane normal equals the area under the distribution between these two points. Accordingly, the strongest electric field lies across the eyelet. For OmpF, the area under a 10-Å slice at the eyelet amounts to 64% of the total area so that a 100 mV voltage across the membrane translates to a 64 mV voltage across the eyelet. The displacement of a single charge over this distance corresponds to an energy of 1.6×10^{-19} A·s·64 $\times 10^{-3}$ V·6 $\times 10^{23}$ = 6 kJ mol^{-1} which approaches the energy of a hydrogen bond. A voltage in such a range can therefore disrupt weak hydrogen bonds and, thus, destroy the eyelet structure. These effects are known as voltage closures [43]. They are observed with all porins, although at different voltages depending on the strength of the eyelet construction. *In vivo*, any static voltage across a bacterial outer membrane would decay rapidly by electric currents through the porins. However, transient voltages in form of Donnan potentials in the 100 mV range may occur on rapid changes of the environment and thus may threaten the porin integrity.

2.5
Permeation

Porins are passive channels showing a transport rate proportional to the concentration gradient. The rate itself depends strongly on the construction of the particular eyelet and on the properties of the solute. General porins, for instance, have eyelet sizes that limit the size of permeating solutes to about 600 kDa. Clearly, the mass is a rather coarse and not a very characteristic property of a solute. Solutes are better described by their net charges, their dipoles and their shapes. The dependence of the permeability on the net charge can be rather easily determined for small ions. Ion selectivities of general porins are the rule. They correlate well with a predominance of the opposite charge at the eyelet.

As a further rule, porins discriminate against non-polar solutes even if the eyelet is much larger than the solute so that it cannot be expelled by unfavorable contacts with the protein. An explanation for this effect was found for the Rc-porin where a wide eyelet (Figure 2.5) is lined by six positive charges on one side and by six negative charges on the other [33, 34]. As sketched in Figure 2.6, this charge distribution gives rise to a transversal electric field across the eyelet. With its water filling, this is a capacitor at a low energy level because the oriented water molecules correspond to a dielectric constant as high as 80. As soon as a non-polar molecule with its low dielectric constant of about 3 intrudes this capacitor, the energy level is raised (similar to lifting a mass) and a force develops to expel this solute (similar to the gravitational force pulling a lifted mass down again). As a result such an electric field permits a wide eyelet without allowing the passage of non-polar solutes. On the other hand, the passage of polar solutes through this transversal electric field is facilitated because they stop tumbling around and pass the eyelet in a fixed orientation (Figure 2.6). The diffusion becomes one-dimensional and therefore faster. Such an electric field is also present in Rb-porin, OmpK36, PhoE and

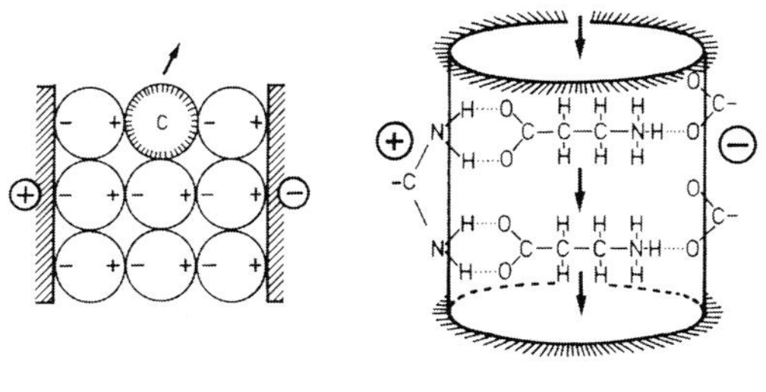

Figure 2.6 Sketch of the transversal electric field across the pore eyelet, which is most prominent in Rc-porin with a large pore (Figure 2.5) and least prominent in Omp32 with a small pore. As shown on the left-hand side, the positively and negatively charged walls of the eyelet form a capacitor that is usually filled with oriented water molecules giving rise to a low energy level. The energy level of the capacitor increases when a non-polar molecule (C, dielectric constant around 3) enters, thus developing an expelling force. The right-hand sketch indicates that the field orients polar solutes so that their permeation is facilitated by restricting the diffusion to only one dimension. Taken together, the electric field acts as a polarity filter.

OmpF where, however, it is based on fewer charges than in the Rc-porin. The small eyelet of Omp32 (Figure 2.5) lacks such a charge distribution. This seems reasonable because the electric field is only required for wide pores.

A survey of the residues of the eyelet wall reveals in all cases three arginines at characteristic positions of the β-barrel (Arg-42, Arg-82 and Arg-132 of OmpF). In several eyelets two guanidinium groups form a sandwich that stabilizes the side-chains mechanically. In addition, there are numerous lysines with their side-chains extending into the solvent. Many of these are laterally stabilized by other side-chains. Aspartates and glutamates are abundant giving rise to numerous negative charges. Several eyelets contain tyrosines pointing with their hydroxyl dipoles towards the solvent. In general, the eyelets are highly polar and stabilized by hydrogen bonding networks.

The Rc-porin shows two peculiarities. First, the negative side of the eyelet wall contains two Ca^{2+} ions. The charges of these Ca^{2+} are compensated by numerous acids giving rise to a negative net charge of 6. Obviously the eyelet size is regulated by the Ca^{2+} concentration in the external medium. As a second peculiarity, Rc-porin harbors a binding site for rather non-polar solutes close to the external end of the eyelet: a bound molecule of the size of adenosine would clog the eyelet. Such an arrangement gives rise to a Michaelis–Menten-type kinetics for the permeation for adenosine. The molecule binds at low concentrations to this site and then diffuses from there through the eyelet at a certain rate corresponding to k_{cat} of an enzyme. This results in a saturation effect for adenosine permeation which resembles the saturation effects of enzymes. Consequently, Rc-porin should be considered a general porin for all the polar solutes that do not bind to this particular site, showing a linear relationship between the permeating solutes and their concentration gradient. For solutes using the binding site, however, Rc-porin acts like a specific protein. Binding to this site increases the permeation at low concentrations but shows a saturation effect at high concentrations.

2.6 Conclusion

The structures of the general porins resemble each other closely and they are also related to all other bacterial outer membrane proteins with their all-next-neighbor all-antiparallel β-barrels with narrowly distributed tilt angles and barrel lengths. All of them are trimeric, showing the general structure of a wheel where the hub and the three spikes have polar surfaces and non-polar cores like soluble proteins. The hoop is broad and shows a characteristic surface structure consisting of an aliphatic ribbon lined by aromatic girdles at each side. The periplasmic rim of the hoop is very narrow. In contrast, the external rim harbors numerous ionogenic side-chains which bind tightly to the lipopolysaccharide cores. The mechanical stress on the outer membrane is buffered by the tight connection of the porins to the lipopolysaccharides in the outer leaflet and also by the two aromatic girdles that prevent unfavorable contacts at the polar and the non-polar outer surfaces of the β-barrels, which would destroy them. The porin channels have no uniform cross-section. Rather, they are wide at both ends and have a small eyelet extending over a length of about 10 Å approximately in the middle of the membrane. All general porins have a spacious vestibule at the external channel end. On the basis of the known structures, the porins are most suitable for channel engineering [44–49]. Such endeavors should help to correlate the permeation properties with the atomic structure of the pore. Moreover, they may give rise to porins of technical interest.

Acknowledgments

I thank Linda Böhm and Markus Krömer for their help with the preparation of the manuscript.

References

1 R. M. Garavito, J. P. Rosenbusch, Three-dimensional crystals of an integral membrane protein: an initial X-ray analysis. *J Cell Biol* **1980**, *86*, 327–329.
2 A. Kreusch, M. S. Weiss, W. Welte, J. Weckesser, G. E. Schulz, Crystals of an integral membrane protein diffracting to 1.8 Å resolution. *J Mol Biol* **1991**, *217*, 9–10.
3 A. Pautsch, J. Vogt, K. Model, C. Siebold, G. E. Schulz, Strategy for membrane protein crystallization exemplified with OmpA and OmpX. *Proteins: Struct Funct Genet* **1999**, *34*, 167–172.
4 M. S.Weiss, T. Wacker, J. Weckesser, W. Welte, G. E. Schulz, The three-dimensional structure of porin from *Rhodobacter capsulatus* at 3 Å resolution. *FEBS Lett* **1990**, *267*, 268–272.
5 M. S. Weiss, G. E. Schulz, Structure of porin refined at 1.8 Å resolution. *J Mol Biol* **1992**, *227*, 493–509.
6 S. W. Cowan, T. Schirmer, G. Rummel, M. Steiert, R. Ghosh, R. A. Pauptit, J. N. Jansonius, J. P. Rosenbusch, Crystal structures explain functional properties of two *E. coli* porins. *Nature* **1992**, *358*, 727–733.
7 A. Kreusch, A. Neubüser, E. Schiltz, J. Weckesser, G. E. Schulz, The structure of the membrane channel porin from *Rhodopseudomonas blastica* at 2.0 Å resolution. *Protein Sci* **1994**, *3*, 58–63.
8 A. Kreusch, G. E. Schulz, Refined structure of the porin from *Rhodopseudomonas blastica*: comparison with the porin from *Rhodobacter capsulatus*. *J Mol Biol.* **1994**, *243*, 891–905.
9 Hirsch, J. Breed, K. Saxena, O. M. H. Richter, B. Ludwig, K. Diederichs, W. Welte, The structure of porin from *Paracoccus denitrificans* at 3.1 Å resolution. *FEBS Lett* **1997**, *404*, 208–210.
10 R. Dutzler, G. Rummel, S. Alberti, S. Hernández-Allés, P. S. Phale, J. P. Rosenbusch, V. J. Benedi, T. Schirmer, Crystal structure and functional characterization of OmpK36, the osmoporin of *Klebsiella pneumoniae*. *Structure* **1999**, *7*, 425–434.
11 K. Zeth, K. Diederichs, W. Welte, H. Engelhardt, Crystal structures of Omp32, the anion-selective porin from *Comamonas acidovorans*, in complex with a periplasmic peptide at 2.1 Å resolution. *Structure* **2000**, *8*, 981–992.
12 T. Schirmer, T. A. Keller, Y. F. Wang, J. P. Rosenbusch, Structural basis for sugar translocation through maltoporin channels at 3.1 Å resolution. *Science* **1995**, *267*, 512–514.

13 J. E. W. Meyer, M. Hofnung, G. E. Schulz, Structure of maltoporin from *Salmonella typhimurium* ligated with a nitrophenyl-maltotrioside. *J Mol Biol* **1997**, *266*, 761–775.

14 D. Forst, W. Welte, T. Wacker, K. Diederichs, Structure of the sucrose-specific porin ScrY from *Salmonella typhimurium* and its complex with sucrose. *Nature Struct Biol* **1998**, *5*, 37–46.

15 R. R. Ketchem, W. Hu, T. A. Cross, High-resolution conformation of gramicidin A in a lipid bilayer by solid-state NMR. *Science* **1993**, *261*, 1457–1460.

16 L. Song, M. R. Hobaugh, C. Shustak, S. Cheley, H. Bayley, J. E. Gouaux, Structure of staphylococcal α-hemolysin, a heptameric transmembrane pore. *Science* **1996**, *274*, 1859–1865.

17 V. Koronakis, A. Scharff, E. Koronakis, B. Luisi, C. Hughes, Crystal structure of the bacterial membrane protein TolC central to multidrug efflux and protein export. *Nature* **2000**, *405*, 914–919.

18 R. Olson, H. Nariya, K. Yokota, Y. Kamio, E. Gouaux, Crystal structure of Staphylococcal LukF delineates conformational changes accompanying formation of a transmembrane channel. *Nature Struct Biol* **1999**, *6*, 134–140.

19 K. P. Locher, B. Rees, R. Koebnik, A. Mitschler, L. Moulinier, J. P. Rosenbusch, D. Moras, Transmembrane signaling across the ligand-gated FhuA receptor: crystal structures of free and ferrichrome-bound states reveal allosteric changes. *Cell* **1998**, *95*, 771–778.

20 A. D. Ferguson, E. Hofmann, J. W. Coulton, K. Diederichs, W. Welte, Siderophore-mediated iron transport: crystal structure of FhuA with bound lipopolysaccharide. *Science* **1998**, *282*, 2215–2220.

21 S. K. Buchanan, B. S. Smith, L. Venkatramani, D. Xia, L. Esser, M. Palnitkar, R. Chakraborty, D. van der Helm, J. Deisenhofer, Crystal structure of the outer membrane active transporter FepA from *Escherichia coli*. *Nature Struct Biol* **1999**, *6*, 56–63.

22 A. D. Ferguson, R. Chakraborty, B. S. Smith, L. Esser, D. van der Helm, J. Deisenhofer, Structural basis of gating by the outer membrane transporter FecA. *Science* **2002**, *295*, 1715–1719.

23 D. P. Chimento, A. K. Mohanty, R. J. Kadner, M. C. Wiener, Substrate-induced transmembrane signaling in the cobalamin transporter BtuB. *Nature Struct Biol* **2003**, *10*, 394–401.

24 A. Pautsch, G. E. Schulz, Structure of the outer membrane protein A transmembrane domain. *Nature Struct Biol.* **1998**, *5*, 1013–1017.

25 A. Pautsch, G. E. Schulz, High resolution structure of the OmpA membrane domain. *J Mol Biol* **2000**, *298*, 273–282.

26 J. Vogt, G. E. Schulz, The structure of the outer membrane protein OmpX from *Escherichia coli* reveals possible mechanisms of virulence. *Structure* **1999**, *7*, 1301–1309.

27 H. J. Snijder, I. Ubarretxena-Belandia, M. Blaauw, K. H. Kalk, H. M. Verhij, M. R. Egmond, N. Dekker, B. W. Dijkstra, Structural evidence for dimerization-regulated activation of an integral membrane phospholipase. *Nature* **1999**, *401*, 717–721.

28 L. Vandeputte-Rutten, R. A. Kramer, J. Kroon, N. Dekker, M. R. Egmond, P. Gros, Crystal structure of the outer membrane protease OmpT from *Escherichia coli* suggests a novel catalytic site. *EMBO J.* **2001**, *20*, 5033–5039.

29 S. M. Prince, M. Achtman, J. P. Derrick, Crystal structure of the OpcA integral membrane adhesin from *Neisseria meningitidis*. *Proc Natl Acad Sci USA* **2002**, *99*, 3417–3421.

30 L. Vandeputte-Rutten, M. P. Bos, J. Tommassen, P. Gros, Crystal structure of neisserial surface protein A (NspA), a conserved outer membrane protein with vaccine potential. *J Biol Chem* **2003**, *278*, 24825–24830.

31 W. M. Liu, Shear numbers of protein β-barrels: definition, refinements and statistics. *J Mol Biol* **1998**, *275*, 541–545.

32 G. E. Schulz, R. H. Schirmer, *Principles of Protein Structure*. Springer, New York, **1979**.

33 G. E. Schulz, Structure–function relationships in the membrane channel

porin as based on a 1.8 Å resolution crystal structure, In: A. Pullman, J. Jortner, B. Pullman (Eds), *Membrane Proteins: Structures, Interactions and Models*. Kluwer, Dordrecht, **1992**, pp. 403–412.

34 G. E. Schulz, Structure–function relationships in porins as derived from a 1.8 Å resolution crystal structure, In: J. M. Ghuysen, R. Hakenbeck (Eds), *New Comprehensive Biochemistry, Vol 27: Bacterial Cell Wall*. Elsevier, Amsterdam, **1994**, pp. 343–352.

35 W. C. Wimley, S. H. White, Experimentally determined hydrophobicity scale for proteins at membrane interfaces, *Nature Struct Biol* **1996**, *3*, 842–848.

36 A. B. Robinson, L. R. Robinson, Distribution of glutamine and asparagine residues and their neighbors in peptides and proteins. *Proc Natl Acad Sci USA* **1991**, *88*, 8880–8884.

37 C. Brändén, J. Tooze, *Introduction to Protein Structure*. Garland, New York, **1999**, pp. 67–88.

38 A. Mathes, H. Engelhardt, Nonlinear and asymmetric open channel characteristics of an ion-selective porin in planar lipid membranes. *Biophys J* **1998**, *75*, 1255–1262.

39 R. Benz, Porin from bacterial and mitochondrial outer membranes, In: *CRC Critical Reviews in Biochemistry*. CRC Press, Boca Raton, FL, **1986**, pp. 145–190.

40 R. Benz, K. Bauer, Permeation of hydrophilic molecules through outer membrane of gram-negative bacteria. *Eur J Biochem* **1988**, *176*, 1–19.

41 J. E. W. Meyer, G. E. Schulz, Energy profile of maltooligosaccharide permeation through maltoporin as derived from the structure and from a statistical analysis of saccharide–protein interactions. *Protein Sci* **1997**, *6*, 1084–1091.

42 T. Schirmer, P. S. Phale, Brownian dynamics simulation of ion flow through porin channels. *J Mol Biol* **1999**, *294*, 1159–1167.

43 D. J. Müller, A. Engel, Voltage and pH-induced channel closure of porin OmpF visualized by atomic force microscopy. *J Mol Biol* **1999**, *285*, 1347–1351.

44 B. Schmid, M. Krömer, G. E. Schulz, Expression of porin from *Rhodopseudomonas blastica* in *Escherichia coli* inclusion bodies and folding into exact native structure. *FEBS Lett* **1996**, *381*, 111–114.

45 P. VanGelder, N. Saint, R. van Boxtel, J. P. Rosenbusch, J. Tommassen, Pore functioning of outer membrane protein PhoE of *Escherichia coli*: mutagenesis of the constriction loop L3. *Protein Eng* **1997**, *10*, 699–706.

46 N. Liu, A. H. Delcour, The spontaneous gating activity of OmpC porin is affected by mutations of a putative hydrogen bond network or of a salt bridge between the L3 loop and the barrel. *Protein Eng* **1998**, *11*, 797–802.

47 B. Schmid, L. Maveyraud, M. Krömer, G. E. Schulz, Porin mutants with new channel properties. *Protein Sci* **1998**, *7*, 1603–1611.

48 L. Q. Gu, O. Braha, S. Conlan, S. Cheley, H. Bayley, Stochastic sensing of organic analytes by a pore-forming protein containing a molecular adapter. *Nature* **1999**, *398*, 686–690.

49 K. Saxena, V. Drosou, E. Maier, R. Benz, B. Ludwig, Ion selectivity reversal and induction of voltage-gating by site-directed mutations in the *Paracoccus denitrificans* porin. *Biochemistry* **1999**, *38*, 2206–2212.

3
Role of Bacterial Porins in Antibiotic Susceptibility of Gram-negative Bacteria

Jean-Marie Pagès

3.1
Introduction

The cell membrane forms an efficient adaptive barrier that protects the cytoplasmic machinery of prokaryotic and eukaryotic organisms against external harmful agents, including heavy metals atoms and acidic or other toxic substances.

At this hydrophobic boundary, the cells organize proteinous structures that form hydrophilic channels; these restricted entry checkpoints allow the selective uptake of required cell nutrients (see Chapters 1 and 5). The study of these special transmembrane inserts led to the identification and purification of the first bacterial pore-forming membrane proteins. The first identified porins were OmpC and OmpF of the Gram-negative bacterium *Escherichia coli*; the existence of porins was reported in Gram-positive and Gram-negative bacteria several years later. At present, both specific and non-specific pore-forming proteins have been found in prokaryotes and eukaryotes.

Independently of their sequence, their physicochemical properties, and their conformational and environmental characteristics, non-specific channel-forming bacterial porins display a trimeric organization (see Chapters 2, 9 and 13). Among the trimeric porins, OmpF of *E. coli* and the porin of *Rhodobacter capsulatus* were the first β-barrel structures solved [1, 2]. The three identical monomers making up the functional porin trimer are closely associated through numerous surface loop (subunit) interactions [1, 2] and only drastic conditions, e. g. harsh detergent such as sodium dodecyl sulfate at high temperature (up to 70 °C), can dissociate them. This tight conformation, inserted into the outer membrane, organizes a compact molecule particularly resistant to degradative attacks which exposes highly variable domains involved in specific activities at the cell surface (for reviews, see [3, 4]).

A striking feature is the high level of expression of this class of outer membrane protein. Depending on the bacterial species and the environmental conditions, porins can reach about 10^4–10^6 copies per cell [5]. The synthesis of porins may be up- or down-regulated by the presence or absence of special molecules or effectors in the growth medium (see Chapters 1 and 5). The presence of such a high number

Bacterial and Eukaryotic Porins. Edited by Roland Benz

ISBN 3-527-30775-3

of hydrophilic channels in the membrane may allow the entry of deleterious compounds and, consequently, bacteria must carefully regulate the balance of outer membrane permeability under special external conditions.

The aim of this chapter is not to list all Gram-negative bacteria showing a porin-dependant resistance mechanism, but to point out the documented relationships between porins and antibiotic molecules, the molecular basis of the channel–antibiotic "dialogue" and its impact on the resistance level.

Decreased access to periplasmic targets can enhance bacterial drug resistance and allow cell adaptation to antibiotic-containing environments [6]. Thus, decreasing the porin number or altering the pore efficiency may be an easy and fruitful response to antibiotic use, conferring the bacterium with a first line of resistance to be eventually followed by the acquisition or expression of additional resistance mechanisms. Three bacterial responses have been documented:

(i) *Controlling the porin level* by decreasing the expression of porins (e. g. OmpC in *E. coli*, OprD in *Pseudomonas aeruginosa*).
(ii) *Controlling the porin type* by selecting the expressed porin (e. g. OmpK35, OmpK36, OmpK37, OmpF, OmpC and other porin types in *Klebsiella*, *Salmonella*).
(iii) *Expressing a mutated porin* with restricted channel activities (e. g. *Enterobacter aerogenes* Omp36 3, *E. coli* OmpC and *Neisseria gonorrhoeae* PorIB).

These responses may involve various regulation pathways of porin expression (*envZ–ompR*, *ompX*, *marA*, *sox*, *ramA* or yet unknown systems in clinical isolates) and may also include external factors, e. g. salicylate, chloramphenicol or polyamines, in the control of porin production or function.

3.2
Role of Porins in Antibiotic Resistance

The number, nature and functional conformation of porins is likely to control the diffusion of hydrophilic antibiotic molecules in Gram-negative bacteria [6–10] as illustrated in Figure 3.1. Consequently, in many cases, the acquired resistance to β-lactams of clinical strains can be attributed to porin loss or porin functional modification in addition to enzymatic production [6, 8]. In *P. aeruginosa*, intrinsic β-lactam resistance is associated with reduced outer membrane protein permeability [11, 12]. The nature and number of porins produced by *P. aeruginosa* explain the reduced membrane permeability to charged antibiotic molecules [13–16]. Moreover, acquired resistance is also caused by porin loss as *P. aeruginosa* clinical isolates resistant to β-lactam show alterations of outer membrane profiles. Concerning carbapenem resistance, with or without production of carbapenem hydrolyzing enzyme, OprD deficiency or modification have been reported in various resistant strains isolated *in vitro* or *in vivo* [14]. Several mutations have been reported in the *oprD* gene which induce a reduced carbapenem susceptibility [13, 16–18]. Interestingly, the main cellular function of OprD is the passive uptake of basic amino acids [19] as well as small peptides [for a review, see [12]). An OprD model [20] postulated that the putative loop 7 located in the C-terminal position of this porin is involved in antibiotic susceptibility: deletions of several residues in this loop increase meropenem, cephalosporin and quinolone susceptibility without modifying imipenem CMI [20]. This loop, together with loops 5 and 8, probably contributes to the constriction of the OprD channel [20, 18]. Regarding loops 2 and 3, they seem to be involved in imipenem passage through OprD [17]. Various reported modifications of these putative loops participate in *P. aeruginosa* carbapenem resistance. In addition, overproduction of efflux pumps, conjointly with the decrease of OprD expression, is involved in high resistance levels to imipenem and carbapenem [11]. In other Gram-negative bacteria, such as *Citrobacter*, *Enterobacter*, *Escherichia*, *Klebsiella*, *Proteus*, *Salmonella*, *Serratia* and *Shigella*, imipenem resistance is closely related to the loss of non-specific porin [21–31].

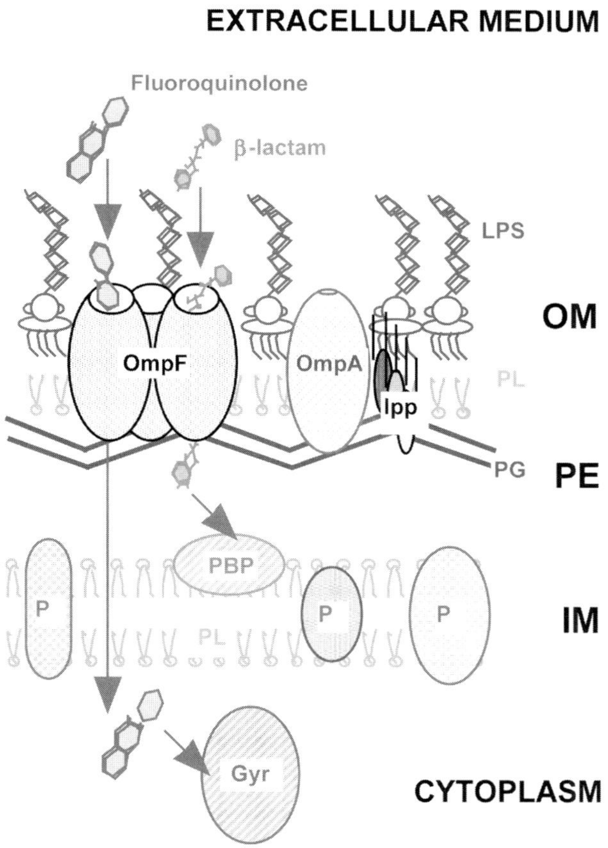

Figure 3.1 *E. coli* envelope. This illustration represents the drugs, β-lactam and fluoroquinolones passing through the OmpF channel. IM, inner membrane; OM, outer membrane; PE, periplasmic space; Gyr, gyrase (quinolone target); lpp, lipoprotein; LPS, lipopolysaccharide; P, protein; PBP, penicillin binding protein (β-lactam target); PG, peptidoglycan; PL, phospholipids.

Table 3.1 Gram-negative bacteria and porin alteration.

Bacteria	*Porin alteration (loss or mutation*)*	*β-Lactam susceptibility*
Citrobacter freundii	major porin	↘
Enterobacter cloacae	major porin	↘
Enterobacter aerogenes	Omp36, Omp 36*	↘
Escherichia coli	OmpC, OmpF, OmpC*	↘
Klebsiella pneumoniae	OmpK35, OmpK36	↘
Morganella morganii	major porin	↘
Neisseria gonorrhoeae	major porin	↘
Proteus rettgeri	porin	↘
Proteus mirabilis	porin	
Providencia spp.	major porin	↘
Pseudomonas aeruginosa	porins OprD, OprD*	↘
Salmonella typhimurium	major porins	↘
Salmonella typhi	major porins	
Salmonella enterica	major porins	
Serratia marcescens	major porin	↘
Shigella dysenteriae	major porins	↘

This table lists some Gram-negative pathogens for which the alteration of porin function, due to decreased porin synthesis or to the expression of a mutated porin (indicated by an asterisk), has been associated with a decrease (↘) of β-lactam susceptibility [13–31].

Intensive studies report the relevance of a lack of porin (Table 3.1) in strains expressing a characteristic pleiotopic β-lactam-resistance phenotype (for recent reviews, see [3, 10]). This porin phenotype is usually associated with enzymatic expression, e. g. β-lactamase and cephalosporinase [6]. It is interesting to note that one study reported that 6.4 % of highly β-lactam-resistant isolates of *E. aerogenes* (20 strains out of 311) collected over a 1-year period showed an altered porin expression profile [32]. Such a prevalence of porin synthesis alteration, which did not include possible channel alteration caused by mutation (see below), indicates the strong impact of porin expression, in conjunction with β-lactamase production, in the emergence of *E. aerogenes* β-lactam-resistant strains. It is worthy of note that, in addition to β-lactam molecule uptake, bacterial non-specific porins are also involved in fluoroquinolone entry as described in *E. coli, Enterobacter cloacae, P. aeruginosa* and *Serratia marcescens* [33–36]. Moreover, the diffusion rate of this antibiotic family seems to be related to the porin type synthesized by the strain [34]. The role of porins in antibiotic susceptibility, in combination with β-lactamase production, led to the analysis of the evolution of Gram-negative bacteria in response to antibiotic treatment.

3.2.1 Evolution of Clinical Isolates

Long-term antibiotic treatments offer the opportunity to analyze the evolution of individual bacterial strains *in vivo*. The acquisition of antibiotic resistance, a major challenge of antibiotherapy practice, represents a unique insight into bacterial evolution and for the development of new therapies.

Various hypotheses have been proposed to explain the spread of antibiotic resistance. A patient may acquire distinct, unrelated sequential strains during drug treatment that exhibit a narrow range of resistance behaviors. However, bacterial infections may also involve host colonization from a single strain which sequentially acquires or expresses numerous antibiotic resistance mechanisms. In this case, the original strain possesses or sequentially acquires the mutations and/or the genetic equipment necessary to build the resistance. In a protected environment, such as hospital wards (e. g. intensive care units), antiseptic and antibiotic concentrations may reach subinhibitory values, and consequently may favor the strong and continuous selection of resistant bacteria. The effect of the antibiotic pressure on the expression of bacterial porins has been poorly studied due to the ethical and clinical problems associated with this kind of investigation. Some data are available concerning *E. coli, E. aerogenes, K. pneumoniae, Haemophilus influenzae, P. aeruginosa* and *Salmonella typhimurium*. Among them, only three *a posteriori* studies reported the evolution of porin expression in a unique strain during antibiotherapy.

Regarding *E. coli*, several isolates were collected from a patient for 2 years [23]. Over this period, the antibiotic therapy comprised successively: ciprofloxacin + ceftazidime, ciprofloxacin + cefotaxime, trimethoprim + ceftazidime, gentamicin + cefotaxime + meropenem, gentamicin + imipenem + meropenem and meropenem + gentamicin (Table 3.2). Seven isolates from blood samples or liver abscesses were characterized. They exhibited a progressive increase in antibiotic resistance levels in correlation with the therapy used on the patient. It is important to mention that the last two isolates, from blood samples, showed a strong reduction of OmpC synthesis compared to the first isolates [23]. This altered porin phenotype followed the carbapenem treatment, meropenem and imipenem (Table 3.2). Interestingly, these two isolates, in addition to decreased OmpC levels, presented a R124H substitution which is located in the OmpC internal loop 3 [23]. The sequence comparison of *ompC* in the seven isolates indicated the preservation of a common sequence modification (100 base changes compared to *E. coli* K12 *ompC*). The molecular analyses carried out on the various isolates support their evolution from a common founder population [23] – a single infecting strain which exhibits a special *ompC* sequence.

In comparison to this work which analyses the effect of very long-term antibiotic therapy on *E. coli* porin evolution, it is interesting to observe the effect of imipenem therapy on *E. aerogenes* strains collected from different patients on a shorter timescale (2–9 weeks). *E. aerogenes* isolates were collected from different patients during the clinical courses of their respective therapy (Table 3.3). Molecular epidemio-

Table 3.2 Evolution of *E. coli* antibiotic susceptibility and porin expression (from [23]).

Isolate	***1***	***2***	***3***	***4***	***5***	***6***	***7***
Imi (μg/ml)	0.25	0.125	0.5	0.25	2.0	1.0	8.0
Caz (μg/ml)	0.125	0.25	16	32	4.0	32	128
Cip (μg/ml)	1.0	1.0	128	8.0	16	16	16
OmpC	+	+	+	+	+	–	–

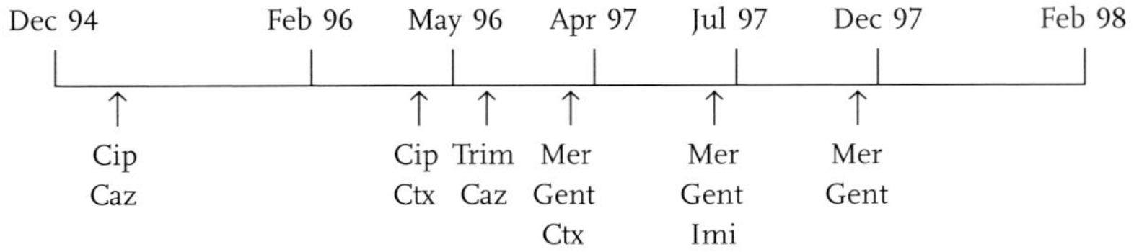

Antibiotic susceptibility testing was performed by broth microdilution in Mueller–Hinton broth according to NCCLS guidelines. The following antibiotics were studied: cefotaxime (Ctx), ceftazidime (Caz), ciprofloxacin (Cip), gentamicin (Gent), imipenem (Imi), meropenem (Mer) and trimethoprim (Trim). Shading is used to indicate whether the strain is sensitive, intermediate or resistant to the antibiotic. The isolate was considered resistant when the corresponding minimum inhibitory concentrations (MICs) (μg/ml) were ≥16 (cefotaxime), ≥32 (ceftazidime), ≥4 (ciprofloxacin), ≥16 (gentamicin, imipenem, meropenem) and ≥16 (trimethoprim). *E. coli* strain ATCC 25922, which has known MIC values, was used as a control and values obtained for this strain were within the established limits. The antibiotic therapy received by the patient is indicated on a time line below the MIC table with an arrow to indicate dates on which the antibiotic combination was changed. Antibiotic therapy was almost continuous, as the patient quickly developed septicemia if treatment was halted.

logical typing indicates the step-by-step *in vivo* emergence of resistant isolates from a susceptible *E. aerogenes* founder strain belonging to the prevalent type [25]. The emergence of such a resistant variant occurred under imipenem treatment and the restoration of imipenem susceptibility was rapidly recovered a few days after the treatment was stopped [25]. An important observation concerns the major Omp36 porin during the antibiotic treatment. A strong correlation was reported between Omp36 detection and β-lactam susceptibility of the various isolates: the absence of porin always correlated with a β-lactam-resistant phenotype. In contrast, the presence of porin was generally associated with imipenem susceptibility [25]. This data clearly showed that antibiotic treatment may select or favor the emergence of a resistant phenotype from a pre-existent susceptible pathogenic strain. Relative to imipenem, the prevalent clone of *E. aerogenes* in the clinical environment regulates outer membrane permeability more rapidly than other enterobacteria. In addition, other studies described the close relationship between porin and high β-lactam resistance in *E. aerogenes* isolated from various hospital wards [37–39].

Another escape mechanism used by Gram-negative bacteria in response to antibiotherapy is the modulation of porin type synthesis. *S. typhimurium* is an illustra-

Table 3.3 Evolution of *E. aerogenes* antibiotic susceptibility and porin expression (from [25]).

A

Isolate	*1*	*2*	*3*	*4*	*5*
Cefepime (μg/ml)	1	64	8	16	64
Imipenem (μg/ml)	1	16	2	2	16
Source	TA	TA	TA	U	TA
Date (1997)	Apr. 3	Apr. 9	Apr. 10	Apr. 10	Apr. 14
Treatment	–	Imi	Imi	Imi	Imi
Porin	+	–	+	+	–

B

Isolate	*1*	*2*	*3*	*4*	*5*	*6*	*7*	*8*
Cefepime (μg/ml)	4	4	4	128	32	32	2	2
Imipenem (μg/ml)	1	1	1	16	16	16	1	1
Source	U	U	TA	TA	TA	C	U	U
Date (1997)	Aug. 18	Aug. 25	Sep. 01	Sep. 08	Sep. 15	Sep. 25	Oct. 01	Oct. 19
Treatment	–	–	Imi	Imi	Imi /	–	–	–
Porin	ND	ND	ND	–	(+)	–	+	+

Two clinical courses are presented (A and B). TA, tracheal aspiration; U, urine; C, catheter; Imi, imipenem. The respective MICs (cefepime and imipenem) are indicated in μg/ml. Shading is used to indicate whether the strain is sensitive, intermediate or resistant to the antibiotic.

ND, not determined.

tion of this resistance strategy. A patient was infected by *S. typhimurium*, and all isolates collected before the antibiotherapy were cephalexin, cefazolin and cefoxitin susceptible [22]. A cephalexin treatment was consequently applied and a few days later a cephalosporin-resistant *S. typhimurium* strain was isolated from a drain. This post-therapy isolate was a descendent of the pre-therapy strain and exhibited complete resistance to all cephalosporins. No significant increase of β-lactamase activity was observed, whereas, in contrast, an alteration of porin expression was noted in the resistant isolate. The parental susceptible strain presented a normal osmoregulation of porin synthesis: only OmpC was expressed in high osmolarity medium, and both OmpC and OmpF were present in low osmolarity medium.

In contrast, the resistant isolate expressed only OmpF in low ionic strength conditions, whose synthesis was fully repressed at high osmolarity. The absence of OmpC observed under conditions which mimicked that of the human body explains the treatment failure [7, 22]. These observations pinpoint a major step in antibiotic uptake: the expression of the narrow-channel OmpC porin is generally favored in the human body, with the *in situ* conditions of temperature and osmolarity, and naturally restricts the entry of large charged molecules. The difference of cephalosporin diffusion through OmpC and OmpF has been reported [7, 22]. The rapid balance of porin expression is a noticeable advantage for the pathogen, conferring a key response to the antibiotherapy compared to the commensal flora and putative bacterial competitors which express an efficient porin pathway for β-lactam uptake [23, 25].

Today, these results highlight the importance of the adaptive response of the bacterial pathogens to antibiotherapy. They explain the sequential selection and dissemination of porin-less isolates from the original susceptible porin-producer strain and also show the rapid switch of porin synthesis in resistant strains.

Moreover, these studies also focused on the role of some antibiotics which select resistant strains during treatment. Some antibiotics, e. g. chloramphenicol or imipenem, may also induce specific bacterial responses via a regulation cascade [40–45]. It is generally assumed that antibiotic molecules possess biological activities other than inhibition of bacterial targets [46]. In some cases, antibiotics favor the expression of efflux pumps [47] and the repression of porin synthesis in some resistant variants [23, 25]. The data suggests that environments somewhat protected from antibiotics may allow a similar adaptation of Gram-negative pathogens and that the antibiotherapy used to target this ecological niche may orientate bacterial evolution. Some aspects of the regulatory network which govern this evolution remain quite obscure. To date, only a few reports have been published on the role of genetic elements, such as the *mar* and *sox* operons [44, 45] or *ramA* and *ompX* genes [43–46, 48, 49], in porin regulation of clinical strains. Concerning this point, it is of interest to mention that imipenem, one of the most powerful antibiotics used today, can select resistant variants exhibiting conjointly a porin deficiency and the over-expression of a drug efflux pump from susceptible strains [47]. As a consequence of widespread antibiotic use, and with the structural criteria now applied to drug design, e. g. stability toward enzymatic attack, the emergence of strains devoid of porins can be a fruitful escape mechanism for bacterial pathogens.

3.2.2
Expression of a Modified Porin

Only a few β-lactam-resistant Gram-negative isolates have been shown to synthesize a modified porin exhibiting a restricted channel. Modified porins have been identified in *E. coli, N. gonorrhoeae, P. aeruginosa, E. aerogenes, K. pneumoniae* and *H. influenzae* [16, 23, 50–55]. The substitutions in the *E. aerogenes* and *K. pneumoniae* porins are interesting in light of the known three-dimensional (3-D) structure of bacterial porins. In the *E. coli* clinical strains collected during a 2-year period [23], a R124H substitution in OmpC produced by the strain isolated after carbapenem treatment was identified. The 124 position corresponds to R132 in OmpF – a residue characterized as strategic for cephalosporin diffusion and β-lactam susceptibility (Table 3.4).

Table 3.4 Point mutations in the OmpF, OmpC and PhoE eyelet and antibiotic susceptibility.

Porin mutation	***β-Lactam susceptibility (CMI or inhibition diameter)***	***β-Lactam diffusion***
OmpF		
G119D	↘ β-lactam	↘ (+, −)
G119E	↘ β-lactam	↘ (+, −)
R132A	=	↗ (+, −)
R132D	=	↗ (+, −)
K16A		=
K16D		↘ (+, −)
R42S	↗ penicillin	
R82S	↗ penicillin	
R82H	↗ penicillin	
R82C	=	
D113G	↗↗ penicillin	
R132C	↗ penicillin	
R132S	↗ penicillin	
R132P	↗ penicillin	
OmpC		
R37H	↗ penicillin	
R37C	↗↗ penicillin	
R74S	↗ penicillin	
R74G	↗ penicillin	
R74C	↗ penicillin	
D105G	↗↗ penicillin	
PhoE		
E110C		↗ (+, −) ↗ (+, − −)
F111C		↘ (+, −) ↘ (+, − −)
G112A, G113A		↘ (+, − −); ↘ (+, −); ↗ (− −) depending on antibiotic
K18C		↗ (+, −) ↘ (+, − −) depending on antibiotic
R37C		↗
R75C		↗

This table summarizes the mutations located in the OmpC, OmpF and PhoE porins for which an effect on β-lactam susceptibilities or diffusion has been documented: =, no effect; ↘, decreased susceptibility; or ↗↗, increased susceptibility; the antibiotic charge is indicated [(+, −), (+, − −), (− −)].

An interesting porin mutation has been detected in a population of *E. aerogenes* collected from various hospital wards, harboring high cross-resistance against a number of standard β-lactam antibiotics. Among these isolates, one strain exhibited a peculiar porin phenotype – a normal level of Omp36 expression associated with a heat-labile trimer conformation [37]. The molecular and functional analyses indicated that several mutations had occurred in the porin gene compared to the porin gene of a reference strain [50, 56]. Mass spectrometry analyses identified a G to D substitution located inside the postulated internal loop 3 constituting the eyelet of the channel [50]. In fact, this substitution is similar to the G119D mutation previously obtained by mutagenesis of the *E. coli ompF* (see below); *E. coli* G119D OmpF exhibits strong structural and biological modifications [57, 58]. In the case of the *E. aerogenes* Omp36, G112D drastically impairs β-lactam diffusion through the channel and strongly participates in the *E. aerogenes* antibiotic resistance [50]. Interestingly, the kinetics of cefepime uptake, a recent cephalosporin molecule, is highly reduced in a strain producing such a modified porin. The initial rate of drug uptake is closely similar to that observed in porin-deficient strains [56]. The modifications of the electrophysiological properties of the G112D porin compared to wild-type *E. aerogenes* Omp36 are very similar to the results obtained with *E. coli* OmpF G119D (see below). At the time, this was the first example of a porin mutation described in a clinical strain for which the effect on antibiotic susceptibility has been previously hypothesized by mutagenesis screening. In a recent study carried out on *E. aerogenes* hospital strains, A. Thiolas *et al.* (*Biochem Biophys Res Comm*, 2004, in press) described other clinical isolates which exhibit a similar resistant phenotype associated with the Omp36 G112D substitution. With the general use of antibiotics especially designed to overcome enzymatic degradation and to efficiently use the porin channel, this kind of mutation will become more frequent.

Concerning *K. pneumoniae* porins, the description of OmpK37 illustrates an efficient bacterial response to toxic compounds [51]. Omp37 contains a Tyr residue inserted in the postulated loop 3 (in comparison to the OmpK36 and OmpF 3D structures) – this modification induces a reduction of sugar permeation, and a drastic reduction of cephalosporin susceptibility compared to OmpK36 and OmpK35 [51]. A Tyr insertion is also observed in this region of OmpS2, a putative porin of *S. typhi* [51]. However, no clinical isolate producing only OmpK37 has been fully characterized at this moment to demonstrate the role of OmpK37 in strain adaptation. It was recently reported that most isolates of *K. pneumoniae* lacking expression of extended spectrum β-lactamases expressed OmpK36 and OmpK35, whereas most isolates producing β-lactamases expressed only OmpK36; OmpK35 was either very low or absent, or neither OmpK36 nor OmpK35 were produced [27, 59, 60].

Several mutations in the porin gene of *N. gonorrhoeae* have been described to be involved in penicillin and tetracycline resistance. The *Por* gene sequence analyses indicated that three amino acid differences are located in the putative loop 3 in positions 101, 102 and 126 in various resistant strains [53, 54, 61]. Among them, mutations G101D and G102D may be responsible for reduced antibiotic susceptibility.

In addition, substitutions are reported in the same locations in a resistant group of clinical isolates, showing the importance of these residues in the penicillin-resistant phenotype [54]. The substitutions of these amino acid residues by targeted mutagenesis clearly showed their involvement in antibiotic diffusion [53, 62]. The analyses of the PorIB sequence demonstrate the protective effect of charged residues, substituted in these locations, against antibiotic entry [53].

3.3 *In Vitro* Mutagenesis Analyses of Porins and Modeling

Extensive data described the prominent role of porins in antibiotic susceptibility of Gram-negative bacteria. Several approaches have been used to identify the precise molecular basis of the porin–β-lactam interaction. With the resolution of the OmpF 3-D structure and the description of the internal L3 loop region [2], i. e. the eyelet, several mutated porins have been obtained by site-directed mutagenesis of this strategic region (Figure 3.2). In this constriction part of the channel, a strong transverse electrostatic field may govern the diffusion of charged molecules through the porin [63]. Consequently, to understand the role of residues and charges, numerous substitutions have been introduced in the cluster of acidic residues exhibited by L3 and in the facing ring of basic residues protruding from the barrel wall.

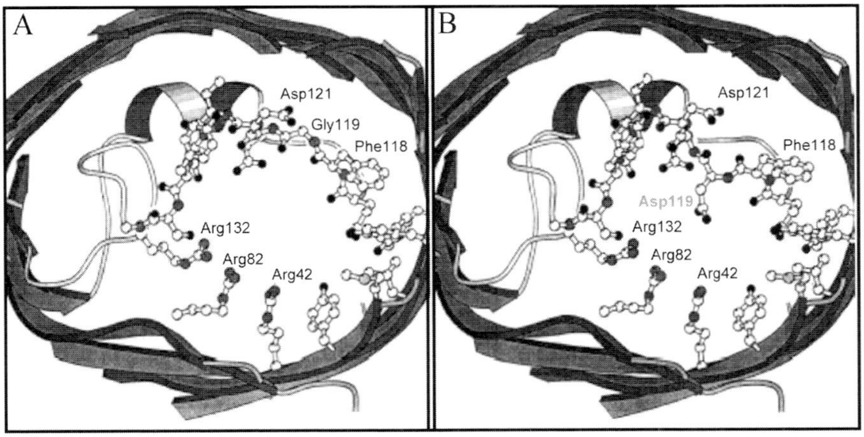

Figure 3.2 Structure of the OmpF eyelet (from [57]). Model of the G119D OmpF porin structure is presented in the channel constriction (A). The view is approximately along the pore axis. The slab shown is approximately 25 Å thick. Strands in the β-barrel (periphery) are presented by broad arrows; the short helical segment is represented by a ribbon. Residues lining the pore constriction are shown in full and other loop segments are indicated by double lines. For comparison, the model of wild-type porin [10] is shown (B).

3.3.1
Mutations in the Loop 3 Domain

From the study of mutated porins (OmpF, OmpC and PhoE), three types of modification can be summarized according to their impact on the β-lactam phenotype (Table 3.4): silent mutations which cause no significant modification of antibiotic resistance (=), positive mutations which increase β-lactam susceptibility (↗) and negative mutations which confer a significant resistance level (↘).

Regarding the substitutions that decreases β-lactam susceptibility, two kinds of effects may be responsible: (i) a steric hindrance due to side-chain protrusion into the channel lumen or/and the formation of a new transverse/lateral bond (hydrogen or else) decreasing the channel size, and (ii) the insertion of a bulky charge which perturbs the electrostatic field located in the eyelet. It is interesting to mention that the strongest effect is observed with the G119D or G119E substitutions; cefepime diffusion is drastically reduced to about 5% compared to wild-type (Figure 3.2). These changes cause a partition of the eyelet into two small subcompartments [57]. In addition, the charge distribution and the transverse electrostatic field that exist across the pore are further altered by the negative charge added in position 119 [57, 58]. The synergy of these two effects, i. e. steric hindrance and alteration of charge distribution, can explain the drastic reduction of cefepime uptake considering the zwitterionic and size properties of this cephalosporin [58].

In the case of PhoE, substitutions G112A or G113A cause an increase in the anionic selectivity of the channel, and conjointly decrease the uptake of charged β-lactam molecules (Table 3.4) such as cephaloridine (+, –, molecular weight 415) and cefsulodin (+, – –, molecular weight 550) [64]. In contrast, the same substitutions confer a better uptake of the double negatively charged CENTA (– –, molecular weight 535) probably associated to the anionic selectivity of the mutated pore, while only a small decrease is noted with ampicillin (–, molecular weight 371) [64]. The F111C substitution located at the tip of the PhoE loop 3 [65] has a drastic effect on the antibiotic entry into the cell: this mutation similarly decreases the uptake of zwitterionic cephaloridine (+, –, molecular weight 415) and negatively charged cefsulodin (+, – –, molecular weight 550) to the same extend (50% compared to wild-type), suggesting a reduction of the PhoE pore size [65].

In the same channel region, it is noteworthy that substitutions D113G in OmpF and D105G in OmpC, which correspond to similar positions in their respective porin structure [66, 67], induce an increase of susceptibility to β-lactam such as for penicillin (Table 3.4) [66, 67]. This observation suggests a possible interaction of the aspartate residue carboxyl charge with passing β-lactams.

The negative charge located on this eyelet side seems to play a key role in β-lactam diffusion and consequently in antibiotic susceptibility. In addition, substitutions resulting in the insertion of a negative charge in the L3 loop destabilize the functional structure of the channel [58].

3.3.2
Mutations in the Anti-loop 3 Domain

The available information about antibiotic susceptibility and this eyelet domain concerns mainly amino acid residues K16, R42, R82 and R132 (in the OmpF nomenclature [68]). It appears that the substitution of positively charged residues by neutral, acidic or basic amino acids generally induces an increase of β-lactam susceptibility (Table 3.4). For instance, the increased susceptibility to penicillin observed with R132C, R132S or R132P OmpF mutants highlights the role of R132 in antibiotic uptake [66]. In addition, the phenotype of substitutions R132A and R132D clearly indicates that the guanidium group of arginine plays an important role in cefepime diffusion since the removal of the Arg side-chain noticeably accelerates cefepime entry [58, 69]. Similarly, mutations removing the positive charge of any residue of the arginine triplet enhance β-lactam diffusion through OmpF, OmpC and PhoE (Table 3.4). The role of this arginine cluster was confirmed in recent molecular dynamics simulations (see below).

K16 is another critical residue, as the K16A transition does not affect cephalosporin diffusion, in contrast to K16D, which drastically perturbs cefepime uptake (Table 3.4) [69]. Moreover, the K18C substitution in PhoE suggests that this Lys residue, only present in the PhoE constriction [2, 63], controls the diffusion of zwitterionic cephalosporin through the phosphoporin eyelet [65].

3.3.3
Modeling of β-Lactam in the OmpF Eyelet

The notion of electrostatic interaction between the β-lactam molecule and the porin channel was established by the pioneering work of Nikaido [6, 7]. With the resolution of the OmpF 3D structure and the data obtained from mutagenesis studies, it became possible to decipher the interplay between the antibiotic and porin partners. Now, we have to take into account the amino acid residues and the electrostatic field located inside the channel in conjunction with the electrostatic profile of the β-lactam molecule. The electrostatic properties of the antibiotic, monoanionic, zwitterionic, dianionic or complex with one positive and two negative charges strongly participate to define the diffusion rate through the porin channel. The number and the nature of the charges on a β-lactam compound have been previously discussed [6, 7, 70]. Recently, Nestorovich and co-authors presented a molecular dynamic simulation of the ampicillin interaction inside the eyelet region of OmpF [71]. The ampicillin conformation was first determined by energy minimization of solvated zwitterionic forms of the drug and then the molecule was manually docked inside the channel. The initial location was selected from experimental data that suggested a transverse orientation of the molecule interacting with amino acid residues belonging to the constriction zone [71]. Following energy minimization and molecular dynamic refinements, the ampicillin acidic group was located close to the arginine cluster (positions 42, 82 and 132), whereas the other end came close to the negative charge (E117) of loop 3 [71]. Although this data is a milestone in the understanding of β-lactam diffusion through the channel, the molecular mechanism remains quite obscure. For instance, it is important to elucidate the timing and number of interactions generated by diffusing antibiotic with the residues located before the constriction region, with those protruding into the pore lumen, and analyze the effect of the various steric hindrances on the orientation of the β-lactam molecule. The affinity level between the various reactive groups, encountered successively along the channel, participates to the hydrodynamic flux and regulates the diffusion rate inside the pore. Moreover, the presence of water molecules or other charged solutes in the channel may modulate the antibiotic flux rate.

3.4 Conclusion

Antibiotic resistance mechanisms such as the mutation in the antibiotic target or the production of a drug-inactivating enzyme affect only a single class of antibacterial molecules. In contrast, mechanisms affecting membrane permeability (porin alteration or efflux pump expression) have a general consequence on the drug arsenal efficiency. The permeability of the bacterial outer membrane is a key factor in antibiotic activity. Drug uptake results from the relative natural intrinsic permeability of Gram-negative species and the porin type expressed [7, 72]. While the permeation barrier alone does not usually confer high antimicrobial resistance, in many cases porin alteration is associated with other resistance mechanisms [8, 73]. The loss of porin or some porin functional alteration severely decreases the uptake of several antibiotic families, e. g. β-lactam drugs and fluoroquinolones.

In the presence of some antibiotics, bacterial pathogens switch off or decrease the expression of the porins [22, 23, 25, 75]. Causality between the loss of OmpC and OmpF and antibiotic resistance has been demonstrated in several reports, especially for *E. coli* and *S. typhimurium*. The loss of porins in ESBL-producing *K. pneumoniae* and *E. aerogenes* strains has been shown to cause resistance to cefoxitin, to increase resistance to third-generation cephalosporins, monobactams and carbapenems, and to decrease susceptibility to fluoroquinolones. In other members of the *Enterobacteriaceae* family, resistance to carbapenems has been related to diminished outer membrane permeability and hydrolysis by overproduced chromosomal β-lactamase. More recently, the alteration of the functional structure of the pore has been demonstrated for β-lactam-resistant isolates.

In addition to the study of the channel function of porins via genetic approaches, several physiological and biochemical investigations reported a wide variability in the structure of the channel of Gram-negative resistant bacteria involved in nosocomial infections. Some porins produced by clinical strains such as *N. meningitidis*, *H. influenzae*, *E. coli* and *E. aerogenes* exhibit some substitutions in the residues located in the channel domain and more precisely in the eyelet area. The high resistance observed in clinical isolates is often associated with substitutions that introduce charged residues in place of neutral ones, or of amino acids with short side-chains. This additional charge is projected inside the lumen and consequently perturbs the diffusion of antibiotics.

In addition to the restriction of pore size, the modulation of porin expression plays a significant role in antibiotic resistance, favoring the onset of disease. Regulation variants exhibit an antibiotic resistance phenotype associated to a rapid switch of porin synthesis which explains the persistence of bacterial colonization despite the antibiotic therapy. Moreover, a recent report describing polyamine cell protection against acidic pH [74] suggests that polyamines could play a role in antibiotic resistance: cadaverine influx may be an antibiotic uptake competitor conferring resistance to excreting cells. This additional resistance mechanism may be important when bacterial pathogens colonize special niches allowing the induction of polyamine synthesis. All these data clearly indicate that Gram-negative bacteria are able to use various adaptive strategies, including regulation of porin synthesis or production of mutated porins, in response to environmental stimuli such as antibiotic molecules.

Acknowledgments

I thank E. Pradel for discussion and careful reading of this manuscript, and A. Favard for typing it. I apologize to those whose papers and studies were not cited due to the space limitations.

References

1 M. S. Weiss, A. Kreusch, E. Schiltz, U. Nestel, W. Welte, J. Weckesser, G. E. Schulz, The structure of porin from *Rhodobacter capsulatus* at 1.8 Å resolution. *FEBS Lett* **1991**, *280*, 379–382.

2 S. W. Cowan, T. Schirmer, G. Rummel, M. Steiert, R. Ghosh, R. A. Pauptit, J. N. Jansonius, J. P. Rosenbusch, Crystal structures explain functional properties of two *E. coli* porins. *Nature* **1992**, *358*, 727–733.

3 W. Achouak, T. Heulin, J.-M. Pagès, Multiple facets of bacterial porins. *FEMS Microbiol Lett* **2001**, *9887*, 1–7.

4 J. Lin, S. Huang, Q. Zhang, Outer membrane proteins: key players for bacterial adaptation in host niches. *Microbes Infect* **2002**, *4*, 325–331.

5 H. Nikaido, Outer membrane, In: Neidhardt, F. C. (Ed.) *Escherichia coli and Salmonella Cellular and Molecular Biology*. ASM Press, Washington, DC, **1996**, pp. 29–47.

6 H. Nikaido, S. Normark, Sensitivity of *Escherichia coli* to various β-lactams is determined by the interplay of outer membrane permeability and degradation by periplasmic β-lactamases: a quantitative predictive treatment. *Mol Microbiol* **1987**, *1*, 29–36.

7 F. Yoshimura, H. Nikaido, Diffusion of β-lactam antibiotics through the porin channels of *Escherichia coli* K-12. *Antimicrob Agents Chemother* **1985**, *27*, 84–92.

8 Nikaido H., Prevention of drug access to bacterial targets: permeability barriers and active efflux. *Science* **1994**, *264*, 382–388.

9 H. Nikaido, E. Y. Rosenberg, J. Foulds, Porin channels in *Escherichia coli*: studies with β-lactams in intact cells. *J Bacteriol* **1983**, *153*, 232–240.
10 K. Poole, Outer membranes and efflux: the path to multidrug resistance in Gram-negative bacteria. *Curr Pharm Biotechnol* **2002**, *3*, 77–98.
11 D. M. Livermore, Of *Pseudomonas*, porins, pumps and carbapenems. *J Antimicrob Chemother* **2001**, *47*, 247–250.
12 R. E. Hancock, F. S. Brinkman, Function of *Pseudomonas* porins in uptake and efflux. *Annu Rev Microbiol* **2002**, *56*, 17–38.
13 J. Trias, J. Dufresne, R. C. Levesque, H. Nikaido, Decreased outer membrane permeability in imipenem resistant mutants of *P aeruginosa*. *Antimicrob Agents Chemother* **1989**, *33*, 1201–1206.
14 J.-P. Pirnay, D. De Vos, D. Mossialos, A. Vanderkelen, P. Cornelis, M. Zizi, Analysis of the *Pseudomonas aeruginosa oprD* gene from clinical and environmental isolates. *Environ Microbiol* **2002**, *4*, 872–882.
15 K. Okamoto, N. Gotoh, T. Nishino, *Pseudomonas aeruginosa* reveals high intrinsic resistance to penem antibiotics: penem resistance mechanisms and their interplay. *Antimicrob Agents Chemother* **2001**, *45*, 1964–1971.
16 H. Pai, J. W. Kim, J. Kim, J. H. Lee, K. W. Choe, N. Gotoh, Carbapenem resistance in *Pseudomonas aeruginosa* clinical isolates. *Antimicrob Agents Chemother* **2001**, *45*, 480–484.
17 M. M. Ochs, M. Bains, R. E. Hancock, Role of putative loops 2 and 3 in imipenem passage through the specific porin OprD of *Pseudomonas aeruginosa*. *Antimicrob Agents Chemother* **2000**, *44*, 1983–1985.
18 S. F. Epp, T. Kolher, P. Plesiat, M. Michea-Hamzehpour, J. Frey, J. C. Pechere, C-terminal region of *Pseudomonas aeruginosa* outer membrane porin OprD modulates susceptibility to meropenem. *Antimicrob Agents Chemother* **2001**, *45*, 1780–1787.
19 J. Trias, H. Nikaido, Protein D2 channel of the *P aeruginosa* outer membrane has a binding site for basic amino acids and peptides. *J Biol Chem* **1990**, *265*, 15680–15684.
20 H. Huang, D. Jeanteur, F. Pattus, R. E. Hancock, Membrane topology and site-specific mutagenesis of *Pseudomonas aeruginosa* porin OprD. *Mol Microbiol* **1995**, *16*, 931–941.
21 J. L. Mainardi, P. Mugnier, A. Coutrot, A. Buu-Hoi, E. Collatz, L. Gutmann, Carbapenem resistance in a clinical isolate of *Citrobacter freundii*. *Antimicrob Agents Chemother* **1997**, *41*, 2352–2354.
22 A. A. Medeiros, T. F. O'Brien, E. Y. Rosenberg, H. Nikaido, Loss of OmpC porin in a strain of *Salmonella typhimurium* causes increased resistance to cephalosporins during therapy. *J Infect Dis* **1987**, *156*, 751–757.
23 A. S. Low, F. M. MacKenzie, I. M. Gould, I. R. Booth, Protected environments allow parallel evolution of a bacterial pathogen in a patient subjected to long-term antibiotic therapy. *Mol Microbiol* **2001**, *42*, 619–630.
24 E. H. Lee, M. H. Nicolas, M. D. Kitzis, G. Pialoux, E. Collatz, L. Gutmann, Association of two resistance mechanisms in a clinical isolate of *Enterobacter cloaccae* with high-level resistance to imipenem. *Antimicrob Agents Chemother* **1991**, *35*, 1093–1098.
25 C. Bornet, A. Davin-Regli, C. Bosi, J.-M. Pagès, C. Bollet, Imipenem resistance of *Enterobacter aerogenes* mediated by outer membrane impermeability. *J Clin. Microbiol* **2000**, *38*, 1048–1052.
26 C. Ardanuy, J. Linares, M. A. Dominguez, S. Hernandez-Alles, V. J. Benedi, L. Martinez-Martinez. Outer membrane profiles of clonally related *Klebsiella pneumoniae* isolates from clinical samples and activities of cephalosporins and carbapenems. *Antimicrob Agents Chemother* **1998**, *42*, 1636–1640.
27 L. Martinez-Martinez, A. Pascual, S. Hernandez-Alles, D. Alvarez-Diaz, A. I. Suarez, J. Tran, V. J. Benedi, G. A. Jacoby, Roles of β-lactamases and porins in activities of carbapenems and cephalosporins against *Klebsiella pneumoniae*. *Antimicrob Agents Chemother* **1999**, *43*, 1669–1673.

28 A. Raimondi, A. Traverso, H. Nikaido, Imipenem- and meropenem-resistant mutants of *Enterobacter cloacae* and *Proteus rettgeri* lack porins. *Antimicrob Agents Chemother* **1991**, *35*, 1174–1180.

29 L. Armand-Lefevre, V. Leflon-Guibout, J. Bredin, A. Amor, J.-M. Pagès, M.-H. Nicolas-Chanoine, Imipenem resistance in *Salmonella enterica* serovar when related to porin loss and CMY-4 β-lactamase production. *Antimicrob Agents Chemother* **2003**, *47*, 1165–1168.

30 A. S. Ghosh, A. K. Kar, M. Kundu, Impaired imipenem uptake associated with alterations in outer membrane proteins and lipopolysaccharides in imipenem-resistant *Shigella dysenteriae*. *J Antimicrob Chemother* **1999**, *43*, 195–201.

31 K. Marumo, T. Nagaki, Y. Nakamura, Evaluation of high-level carbapenem resistance in atypical *Serratia marcescens* by a comparison with its revertants. *J Antimicrob Chemother* **1996**, *38*, 47–58.

32 R. N. Charrel, J.-M. Pagès, P. De Micco, M. Malléa, Prevalence of outer membrane porin alteration in β-lactam-antibiotic-resistant *Enterobacter aerogenes*. *Antimicrob Agents Chemother.* **1996**, *40*, 2854–2858.

33 K. Hirai, H. Aoyama, S. Suzue, T. Irikura, S. Iyobe, S. Mitsuhashi, Isolation and characterization of norfloxacin-resistant mutants of *Escherichia coli* K-12. *Antimicrob Agents Chemother* **1986**, *30*, 248–253.

34 J. Chevalier, M. Malléa, J.-M. Pagès, Comparative aspects of the diffusion of norfloxacin, cefepime and spermine through the F porin channel of *Enterobacter cloacae*. *Biochem J* **2000**, *348*, 223–227.

35 M. Berlanga, N. Ruiz, J. Hernandez-Borrel, T. Montero, M. Vinas, Role of the outer membrane in the accumulation of quinolones by *Serratia marcescens*. *Can J Microbiol* **2000**, *46*, 716–722.

36 M. Michea-Hamzehpour, Y. X. Furet, J. C. Pechere, Role of protein D2 and lipopolysaccharide in diffusion of quinolones through the outer membrane of *Pseudomonas aeruginosa*. *Antimicrob Agents Chemother* **1991**, *35*, 2091–2097.

37 M. Malléa, J. Chevalier, C. Bornet, A. Eyraud, A. Davin-Régli, C. Bollet, J.-M. Pagès, Porin alteration and active efflux: two *in vivo* drug resistance strategies used by *Enterobacter aerogenes*. *Microbiology* **1998**, *144*, 3003–3009.

38 C. Bosi, A. Davin-Régli, C. Bornet, M. Malléa, J.-M. Pagès, C. Bollet, Most *Enterobacter aerogenes* strains in France belong to a prevalent clone. *J Clin Microbiol* **1999**, *37*, 2165–2169.

39 H. Yigit, G. J. Anderson, J. W. Biddle, C. D. Steward, J. K. Rasheed, L. L. Valera, J. E. McGowan Jr, F. C. Tenover, Carbapenem resistance in a clinical isolate of *Enterobacter aerogenes* is associated with decreased expression of OmpF and OmpC porin analogs. *Antimicrob Agents Chemother* **2002**, *46*, 3187–3822.

40 S. P. Cohen, L. M. McMurry, S. B. Levy, *marA* locus causes decreased expression of OmpF porin in multiple-antibiotic-resistant (Mar) mutants of *Escherichia coli*. *J. Bacteriol.* **1988**, *170*, 5416–5422.

41 S. P. Cohen, S. B. Levy, J. Foulds, J. L. Rosner, Salicylate induction of antibiotic resistance in *E coli*: activation of the *mar* operon and a *mar*-independent pathway. *J. Bacteriol.* **1993**, *175*, 7856–7862.

42 L. A. Pratt, W. Hising, K. E. Gibson, T. J. Silhavy, From acids to *osmZ*: multiple factor influence synthesis of the OmpF and OmpC porins in *Escherichia coli*. *Mol Microbiol* **1996**, *20*, 911–917.

43 A. M. George, R. M. Hall, H. W. Stokes, Multidrug resistance in *Klebsiella pneumoniae*: a novel gene, *ramA*, confers a multidrug resistance phenotype in *Escherichia coli*. *Microbiology* **1995**, *141*, 1909–1920.

44 M. N. Alekshun, S. B. Levy, The *mar* regulon: multiple resistance to antibiotics and other toxic chemicals. *Trends Microbiol* **1999**, *7*, 410–3.

45 P. F. Miller, M. C. Sulavik, Overlaps and parallels in the regulation of intrinsic multiple-antibiotic resistance in *Escherichia coli*. *Mol Microbiol* **1996**, *21*, 441–448.

46 E.-B. Goh, G. Yim, W. Tsui, J. A. McClure, M. G. Surette, J. Davies, Transcriptional modulation of bacterial gene expression by subinhibitory concentrations of antibiotics. *Proc Natl Acad Sci USA* **2002**, *99*, 17025–17030.

47 C. Bornet, R. Chollet, M. Malléa, J. Chevalier, A. Davin-Régli, J.-M. Pagès, C. Bollet, Imipenem and expression of multidrug efflux pump in *Enterobacter aerogenes*. *Biochem Biophys Res Commun* **2003**, *301*, 985–990.

48 J. Stoorvogel, J. A. W. M. van Bussel, J. A. M. van de Klundert, Biological characterization of *Enterobacter cloacae* outer membrane protein (OmpX). *J Bacteriol* **1990**, *173*, 161–167.

49 S. Gayet, R. Chollet, G. Molle, J.-M. Pagès, J. Chevalier, Modification of outer membrane protein profile and evidence suggesting an active drug pump in *Enterobacter aerogenes* clinical strains. *Antimicrob Agents Chemother* **2003**, *47*, 1555–1559.

50 E. Dé, A. Baslé, M. Jaquinod, N. Saint, M. Malléa, G. Molle, J.-M. Pagès, A new mechanism of antibiotic resistance in Enterobacteriaceae induced by a structural modification of the major porin. *Mol Microbiol* **2001**, *41*, 189–198.

51 A. Domenech-Sanchez, S. Hernandez-Alles, L. Martinez-Martinez, V. J. Benedi, S. Alberti, Identification and characterization of a new porin gene of *Klebsiella pneumoniae*: its role in β-lactam antibiotic resistance. *J Bacteriol* **1999**, *181*, 2726–2732.

52 M. B. Martinez, M. Flickinger, L. A. Higgins, T. Krick, G. L. Nelsestuen, Reduced outer membrane permeability of *Escherichia coli* O157:H7: suggested role of modified outer membrane porins and theoretical function in resistance to antimicrobial agents. *Biochemistry* **2001**, *40*, 11965–11974.

53 M. Olesky, M. Hobbs, R. A. Nicholas, Identification and analysis of amino acid mutations in porin IB that mediate intermediate-level resistance to penicillin and tetracycline in *Neisseria gonorrhoeae*. *Antimicrob Agents Chemother* **2002**, *46*, 2811–2820.

54 W. L. Veal, R. A. Nicholas, W. M. Shafer, Overexpression of the MtrC–MtrD–MtrE efflux pump due to an *mtrR* mutation is required for chromosomally mediated penicillin resistance in *Neisseria gonorrhoeae*. *J Bacteriol* **2002**, *184*, 5619–5624.

55 M. A. Arbing, J. W. Hanrahan, J. W. Coulton, Altered channel properties of porins from *Haemophilus influenzae*: isolates from cystic fibrosis patients. *J Membr Biol* **2002**, *189*, 131–141.

56 J. Chevalier, J.-M. Pagès, M. Malléa, *In vivo* modification of porin activity conferring antibiotic resistance to *Enterobacter aerogenes*. *Biochem Biophys Res Commun* **1999**, *266*, 248–251.

57 D. Jeanteur, T. Schirmer, D. Fourel, V. Simonet, G. Rummel, C. Widmer, J. P. Rosenbusch, F. Pattus, J.-M. Pagès, Structural and functional alterations of a colicin-resistant mutant of OmpF porin from *Escherichia coli*. *Proc Natl Acad Sci USA* **1994**, *91*, 10675–10679.

58 V. Simonet, M. Malléa, J.-M. Pagès, Substitutions in the eyelet region disrupt cefepime diffusion through the *Escherichia coli* OmpF channel. *Antimicrob Agents Chemother* **2000**, *44*, 311–315.

59 L. Martinez- Martinez, A. Pascual, M. del Carmen Conejo, I. Garcia, P. Joyanes, A. Domenech-Sanchez, V. Javier Benedi, Energy-dependant accumulation of norfloxacin and porin expression in clinical isolates of *Klebsiella pneumoniae* and relationship to extended-spectrum β-lactamase production. *Antimicrob Agents Chemother* **2002**, *46*, 3926–3932.

60 S. Hernandez-Alles, S. Alberti, D. Alvarez, A. Dominguez-Sanchez, L. Martinez-Martinez, J. Gil, J. M. Tomas, V. J. Benedi, Porin expression in clinical isolates of *Klebsiella pneumoniae*. *Microbiology* **1999**, *145*, 673–679.

61 M. J. Gill, S. Simjee, K. Al-Hattawi, B. D. Robertson, C. S. F. Easmon, C. A. Ison, Gonococcal resistance to β-lactams and tetracycline involves mutation in loop 3 of the porin encoded at the *penB* locus. *Antimicrob Agents Chemother* **1998**, *42*, 2799–2803.

62 N. Carbonetti, V. Simnad, C. Elkins, P. F. Sparling, Construction of isogenic gonococci with variable porin

structure: effects on susceptibility to human serum and antibiotics. *Mol Biol* **1990**, *4*, 1009–1018.
63 A. Karshikoff, V. Spassov, S. A Cowan, R. Ladenstein, T. Schirmer, Electrostatic properties of two porin channels from *Escherichia coli*. *J Mol Biol* **1994**, *240*, 372–377.
64 P. Van Gelder, N. Saint, R. van Boxtel, J. P. Rosenbusch, J. Tommassen, Pore functioning of outer membrane protein PhoE of *Escherichia coli*: mutagenesis of the constriction loop L3. *Protein Eng* **1997**, *10*, 699–706.
65 E. F. Eppens, N. Saint, P. Van Gelder, R. van Boxtel, J. Tommassen, Role of the constriction loop in the gating of outer membrane porin PhoE of *Escherichia coli*. *FEBS Lett* **1997**, *415*, 317–320.
66 S. A. Benson, J. L. L. Occi, B. A. Sampson, Mutations that alter the pore function of the OmpF porin of *Escherichia coli* K12. *J Mol Biol* **1988**, *203*, 961–970.
67 R. Misra, S. A. Benson, Genetic identification of the pore domain of the OmpC porin of *Escherichia coli* K-12. *J Bacteriol* **1988**, *170*, 3611–3617.
68 T. Mizuno, M. Y. Chou, M. Inouye, A comparative study on the genes for three porins of the *Escherichia coli* outer membrane. *J Biol Chem* **1983**, *11*, 6932–6940.
69 J. Bredin, N. Saint, M. Malléa, E. Dé, G. Molle, J.-M. Pagès, V. Simonet. Alteration of pore properties of *Escherichia coli* OmpF induced by mutation of key residues in anti-loop3 region. *Biochem J* **2002**, *363*, 521–528.
70 Y. Nitzan, E. Bokobza Deutsch, I. Pechatnikov, Diffusion of β-lactam antibiotics through oligomeric or monomeric porin channels of some Gram-negative bacteria. *Curr Microbiol* **2002**, *45*, 446–455.
71 E. K. Nestorovich, C. Danelon, M. Winterhalter, S. M. Bezrukov, Designed to penetrate: time-resolved interaction of single antibiotic molecules with bacterial pores. *Proc Natl Acad Sci USA* **2002**, *99*, 9789–9794.
72 P. Plesiat, H. Nikaido, Outer membranes of Gram-negative bacteria are permeable to steroid probes. *Mol Microbiol* **1992**, *6*, 1323–1333.
73 R. E. Hencock, The bacterial outer membrane as a drug barrier. *Trends Microbiol* **1997**, *5*, 37–42.
74 H. Samartzidou, M. Mehrazin, Z. Xu, M. J. Benedik, A. H. Delcour, Cadaverine inhibition of porin plays a role in cell survival at acidic pH. *J Bacteriol.* **2003**, *185*, 13–19.
75 D. Hocquet, X. Bertrand, T. Köhler, D. Talon, P. Plésiat, Genetic and phenotypic variations of a resistant *Pseudomonas aeruginosa* epidemic clone. *Antimicrob Agents Chemother* **2003**, *47*, 1887–1894.

4
Porins of the Outer Membrane of *Pseudomonas aeruginosa*

Robert E. W. Hancock and Sandeep Tamber

4.1
Introduction

Pseudomonas aeruginosa is an unusually versatile organism that has become a major opportunistic pathogen in our society [1]. It is the third most prevalent cause of nosocomial (hospital-acquired) infections, leading to more than 200000 infections in the USA annually, and is associated with serious and rapidly progressing lung infections. It is also the principal cause of lung infections in older patients with cystic fibrosis. There appear to be many potential reasons why this organism persists in the clinic, but there is no doubt that its ability to resist the action of antibiotics is a major factor. For the past two decades, researchers have been attempting to solve the mechanisms of intrinsic, inducible (adaptive) and mutational resistance in *P. aeruginosa* with some success [1–3]. In the past 3 years this has been assisted greatly by the publication of the complete DNA sequence of this organism [4]. Our knowledge of all of the genes in *P. aeruginosa* is permitting a more global approach to deciphering antibiotic resistance mechanisms in this organism. It is our contention that an understanding of the mechanisms of resistance will permit the development of novel strategies for overcoming *P. aeruginosa* infections.

A primary determinant explaining the high intrinsic resistance of this organism is its low outer membrane permeability [1, 5], which is itself determined by the properties of the porins of this organism. This property probably arose due to competition for an ecological niche, e. g. the soil, where many of the organisms produce antibiotic substances. Thus, *Pseudomonas,* in addition to producing its own antimicrobial chemicals, has developed and maintained the intrinsic ability to resist the chemicals secreted by other organisms. In this chapter we describe how the properties of the porins of *P. aeruginosa* are consistent with this interpretation and explain in part the high intrinsic antibiotic resistance of this organism, and thus its success as a human pathogen.

Bacterial and Eukaryotic Porins. Edited by Roland Benz

ISBN 3-527-30775-3

4.2
The Outer Membrane Permeability Defect in *P. aeruginosa*

Recently, we reviewed the current developments in antibiotic resistance in *P. aeruginosa* [1], and enumerated, classified and described the outer membrane proteins of this organism [6, 7]. Rather than reiterating this information, we will give just an overview here. There are basically three types of resistance in *P. aeruginosa*, i. e. intrinsic, adaptive/inducible and mutational resistance. Each of these shares elements in common. The major common element underlying all resistance in *P. aeruginosa* is low outer membrane permeability [5, 8]. Although the actual porins involved in determining general diffusion across the outer membrane were quite controversial [9, 10], it is now quite well accepted that this low outer membrane permeability involves a conglomerate of weakly effective porins, including the major porin OprF, the imipenem-selective porin OprD, the sugar-selective inducible porin OprB, etc.

Low outer membrane permeability does not *per se* equate with resistance. Since passage of most antibiotics occurs by simple diffusion and since the periplasm of cells is a rather small compartment, an antibiotic will equilibrate across the *P aeruginosa* membrane in about 1–200 s (depending on its rate of permeation). Even though this is 100-fold slower than the equilibration time across the outer membrane of *E. coli* [5, 10], it is still rapid relative to the doubling time of *P. aeruginosa* (approximately 3000 s). Thus, slow outer membrane penetration must work in synergy with a secondary resistance mechanism that operates more efficiently in the face of slower antibiotic uptake. The major secondary resistance mechanisms are efflux [11] and an inducible β-lactamase [5]. The importance of low outer membrane permeability in *Pseudomonas* was confirmed by the enhancement of susceptibility to antibiotics upon treatment with outer membrane permeabilizing polycations [12] and by introducing into *P. aeruginosa*, a porin with enhanced permeability (e. g. an OprD loop 5 deletion [13]). The importance of the MexAB–OprM efflux mechanism in intrinsic resistance to multiple classes of antibiotics has been demonstrated by mutations in this operon that make cells supersusceptible to the antibiotics [11]. Similarly, the MexXY efflux system is involved in intrinsic resistance to aminoglycosides [14, 15], while the AmpC β-lactamase is involved in intrinsic resistance to β-lactams [5].

High intrinsic resistance also impacts on other forms of resistance since it requires a relatively modest mutational change to increase the efficiency of a secondary resistance determinant and to make cells adaptively or mutationally resistant, thus explaining the reason why it has been so difficult to identify or develop novel antibiotics with lasting anti-pseudomonal activity. Adaptive resistance generally involves the transient (reversible) enhanced effectiveness of one of these secondary resistance mechanisms [1]. For example, aminoglycoside adaptive resistance is dependent on pre-exposure to aminoglycosides leading to decreased net uptake, possibly through enhanced efflux [15]. Other mechanisms include the induction of chromosomal AmpC β-lactamase by certain β-lactams [16] and the induction of the two-component regulator PmrAB by cationic peptides [17]. Similarly, mutational resistance often involves the mutation of one of the regulatory genes that influence these genes. Some progress is being made in defining certain elements of regulation of antibiotic resistance, e. g. regulation of AmpC β-lactamase [16, 18], the efflux systems MexAB–OprM [19], MexCD–OprJ [20], MexEF–OprN [21, 22] and MexXY [15], OprD alone [22, 23] or with conjoint MexEF–OprN regulation [22] and OprF [24].

Three other uptake routes are known for antibiotics in *P. aeruginosa* and two of these involve pore-forming proteins. The zwitterionic carbapenems imipenem and meropenem utilize the substrate specific porin OprD [25, 26]. A third uptake route is provided by a gated channel (for catechol-derivatized antibiotics); however, the specific channel involved has not been determined and there are a large number of candidate channels in the genome [4]. A fourth uptake route is the self-promoted uptake route [27] that involves the interaction of polycations with the polyanionic lipopolysaccharide (LPS) that occupies the outer monolayer of the outer membrane. Such polycations, including the polymyxins, aminoglycosides [28] and polycationic peptides [29] can competitively displace the divalent cations, Mg^{2+} or Ca^{2+}, that normally partially neutralize the polyanionic LPS. The bulky polycations disrupt the outer membrane creating lesions through which they can cross the outer membrane, thus "self-promoting" their own uptake. We have also extended this hypothesis to consider a mechanism involving largely hydrophobic interactions [30].

4.3
Porins Identified in the Genome Sequence

The publication of the *P. aeruginosa* genome sequence provided a tremendous boost to our understanding of the outer membrane of this organism [4, 6] (Table 4.1). For example, nearly 10% of all *P. aeruginosa* genes are regulatory genes (3 times the proportion and nearly 5 times the number in *E. coli*), indicating that this is a highly regulated organism, and our studies to date with microarrays have indicated that this holds true for antibiotic resistance genes.

Recently the putative porins of *P. aeruginosa* were described in some detail [6] and we refer the reader to this article. The most interesting observation was the existence of three large paralogous families of outer membrane proteins, the OprM, OprD and TonB families of 18, 19 and 35 genes, respectively. These are briefly discussed below.

Table 4.1 Functionally characterized porins of *P. aeruginosa*.

Name	*Porin class*	*PAO no.*	*Substrates*[a]
OprF	general	PA1777	various
OprM	efflux	PA0427	BL, CAM, FQ, NOVO, SULF, TC, TP, AC, AH, CV, EB, SDS
OprJ	efflux	PA4597	BL, CAM, FQ, ML, NOVO, TC, TP, AC, AH, CV, EB, SDS
OprN	efflux	PA2495	CAM, CARB, FQ, TC, TP, AH
OpmG	efflux	PA5158	aminoglycosides
OpmH	efflux	PA4974	multiple antibiotics; Tol C homolog
OpmI	efflux	PA3894	not known
AprF	efflux/secretion	PA1248	alkaline protease
OprB	specific	PA3186	glucose, carbohydrates
OprP	specific	PA3279	phosphate
OprO	specific	PA3280	pyrophosphate
OprD	specific	PA0958	basic amino acids, imipenem
FpvA	gated (Ton B)	PA2398	ferric-pyoverdin
FptA	gated (Ton B)	PA4221	Fe, Co, Mb-pyochelin
PfeA	gated (Ton B)	PA2688	ferric-enterobactin
PirA	gated (Ton B)	PA0931	ferric-enterobactin
HasR	gated (Ton B)	PA3408	heme
PhuR	gated (Ton B)	PA4710	heme
OprC	gated?	PA3790	copper

[a] BL, β-lactams; CAM, chloramphenicol; CARB, carbapenems; FQ, fluoroquinolones; ML, macrolides; NOVO, novobiocin; TC, triclosan; TP, trimethoprim; AC, acriflavin; AH, aromatic hydrocarbons; CV, crystal violet; EB, ethidium bromide; SDS, sodium dodecylsulfate.

4.4 The General Porins

The general or non-specific porins are those that do not have any particular substrate specificity. Unfortunately, we have been unable to identify any porins with significant homology to the archetypical β-barrel porin OmpF of *E. coli*, in contrast to the situation for other *Enterobacteriaceae* or *Neisseria* sp. [31]. This probably was responsible for early issues [9, 10, 32] regarding the nature of the major pore-forming proteins of *P. aeruginosa*. Retrospectively, we can reconstruct the following situation. The major outer membrane protein of *P. aeruginosa* is OprF. When reconstituted into available model systems (planar bilayers [33], liposome leakage [34] or liposome swelling, [32]), it forms large channels, but the amount of protein required to reconstitute such channels is much larger than for prototype porins like *E. coli* OmpF. The transport of smaller substrates, of the approximate size of an amino acid or a monosaccharide, appears to be handled primarily by other porins, including the OprD family and OprB. In this regard, it is important to note that although *P. aeruginosa* is one of the most nutritionally diverse bacteria known, it only grows on simple sugars like monosaccharides. For larger compounds like antibiotics, it seems probable that OprF is the predominant channel. This has not been conclusively proven, and given the low overall efficiency of OprF [10, 32], there may be no method for obtaining conclusive proof.

4.4.1 OprF

Despite some initial controversy, two papers conclusively confirmed a role for OprF in transport of larger molecules. The first involved a re-examination of the channel properties of OprF in model membranes [32]. The second provided convincing *in vivo* proof [35]. *P. aeruginosa*, which normally is unable to grow on any sugar larger than a monosaccharide was given the ability to grow on the disaccharide melibiose and trisaccharide raffinose by cloning in a cytoplasmic transport system and processing enzymes from an *E. coli* raffinose operon. Growth, however, depended on the ability of these sugars to diffuse across the outer membrane. As predicted for a bacterium with a larger exclusion limit, the growth of *P. aeruginosa* invested with this raffinose operon was less affected by an increase in sugar size than was growth of the equivalent *E. coli* construct. Importantly, an isogenic OprF mutant grew more slowly on these sugars than did wild-type (Table 4.2), a result inferring that OprF was the major (but not the only) channel for these larger sugars.

An alternative strategy for demonstrating the role of OprF employed measurements of the ability to plasmolyse *P. aeruginosa* strains. Plasmolysis occurs when osmolytes can cross the outer, but not the cytoplasmic, membrane, thus conferring on the periplasm (area between these membranes) a higher osmotic strength than in the bacterial cytoplasm. To equilibrate the osmotic gradient, water transits from the cytoplasm across the cytoplasmic membrane, causing the cytoplasm to shrink, a phenomenon that can be followed by measuring light scattering. In this system, the OprF mutant demonstrated a lower rate of plasmolysis with di-, tri and tetra-

Table 4.2 Role of OprF in uptake of larger substrates (data from [35]).

Carbon source/ osmolyte	*Growth K_m (mM)*		*Plasmolysis rate*	
	P. aeruginosa wild-type	*P. aeruginosa* OprF deficient	*P. aeruginosa* wild-type	*P. aeruginosa* OprF deficient
Gluconate/glucose[a]	0.15	0.16	15	14
Sucrose	–[b]	–[b]	9	7
Melibiose	13	35	9	7
Raffinose	98	>150	1.1	0.2
Stachyose	–[c]	–[c]	0.4	<0.1

[a]Gluconate was used for the growth K_m experiments and glucose for the plasmolysis experiments.
[b]Sucrose is not a substrate for *P. aeruginosa* even when the raffinose operon is present.
[c]Not tested.

saccharides as the osmolytes (Table 4.2), a result consistent with OprF being the major channel for these large (but non-physiological) substrates. In contrast, the physiological monosaccharide substrate, glucose demonstrated similar plasmolysis rates for both wild-type and OprF mutant strains.

In principle, therefore, OprF has been established as the major conduit across the outer membrane for larger molecules and anti-pseudomonal antibiotics generally have molecular weights ranging around that of a trisaccharide. However, we have not observed any outstanding changes in the minimal inhibitory concentrations of any antibiotic when OprF mutants are compared to their isogenic parent strains [26, 37]. There is more than one possible reason for this, e.g. after interposon mutants are constructed environmental pressures may cause phenotypic dilution (through acquisition of secondary mutations) leading to the loss of any anticipated antibiotic resistance phenotype and/or since many antibiotics have some hydrophobic character they may be able to transit across areas of the outer membrane denuded by the loss of OprF. However, one might also question whether a strong resistance phenotype would be anticipated. *P. aeruginosa* has very low outer membrane permeability and, thus, the wild-type itself behaves like a porin-deficient mutant of a more susceptible bacterium like *E. coli*. We know also that OprF is not the only channel for larger substrates, but merely the rate-limiting channel (see above). Thus "proof" that OprF is the major antibiotic channel is unlikely to come. Nevertheless, additional circumstantial evidence has been obtained through the identification of OprF-deficient mutants that arose in the face of antibiotic selective pressure [38, 39].

Model membrane studies are consistent with the suggestion that OprF is a heterogeneous porin that can form a majority of small, antibiotic-impermeable channels and a couple of hundred large channels per bacterial cell [33, 35, 40]. We confirmed through functional and model building studies that the N-terminus of OprF forms a small, antibiotic-impermeable eight-stranded β-barrel channel [40], consistent with its identification as a member of the OmpA family of outer membrane proteins (although this classification is based on C-terminal not N-terminal homology, and the N-termini of *E. co*li OmpA and OprF share only 9% identity). Furthermore, OprF expression appears to be regulated by an upstream ECF σ factor, SigX [24], and we have recently defined the regulon for SigX using both proteomic and microarray methodologies (Rommling, Brazas and Hancock, manuscript in preparation).

In contrast, the basis for the minor species of larger channels [32, 33] is currently unknown. We considered that there were two possibilities, the first of which was that another protein that contaminates OprF preparations is the actual large channel porin of *P. aeruginosa*. This possibility, consistent with our interpretation of published model membrane studies [33], does not agree with the above-described intact cell studies showing that OprF increases trisaccharide uptake [35]. Nevertheless, we have been testing this hypothesis, without success, through gene knockout studies of outer membrane proteins (particularly OprD homologs since OprD is a weak homolog of the major antibiotic channel OmpF of *E. coli*). A second possibility is that an alternative form of OprF, e.g. folding into a 16-stranded β-barrel [41] due to alternative pairing of the four OprF cysteine residues, results in a large channel.

4.4.2
Other General Porins

As discussed above, the lack of members of the general porin family in *P. aeruginosa* and inefficiency of OprF in forming large antibiotic-permeable channels contributes to the high intrinsic resistance that this bacterium displays towards many toxic compounds. The passage of smaller molecules across the outer membrane is probably accomplished by other porins. For example, Nakae and collaborators suggested, based on model membrane studies, that porins OprD, OprE and OprC were involved in the uptake of small molecules [9]. This was partly confirmed by demonstrating that *oprD* mutants had a defect in growth on gluconate [13], indicating that it facilitated uptake of this molecule, despite the apparent lack of a gluconate binding site. Thus, although OprD is a basic amino acid-specific porin, it appears to mediate uptake of anionic gluconate and may well be the major conduit for molecules of less than a molecular weight of about 200. Other data from Wylie and Worobec indicated that OprB appears to be the major conduit for saccharides [42]. The function of another putative porin, OprG, that is a member of an eight-stranded β-barrel family, is not clear [43].

4.5
Efflux

As indicated above, a key element in determining antibiotic resistance in *P. aeruginosa* is multidrug efflux, which works in synergy with slow uptake across the outer membrane. Known elements involved in efflux in *P. aeruginosa* are relatives of the resistance nodulation cell division (RND) efflux family of three-component efflux systems [44]. The three proteins that comprise these systems include a cytoplasmic membrane pump protein (MexB, MexD, MexF and MexY), a periplasmic linker protein (MexA, MexC, MexE and MexX, respectively) and an outer membrane efflux channel (OprM, OprJ, OprN and, as we recently demonstrated [45], OpmG). The publication of the crystal structures of the TolC efflux channel [46] and the AcrB pump [47] has dramatically enhanced our understanding of these RND efflux systems. TolC [46] is a trimer of subunits that together form a 12-stranded β-barrel in the outer membrane, with 12 long strands of coiled α-helices that stretch across the periplasm and form an iris diaphragm (see Chapter 8). It is envisaged that the AcrA linker protein, which also forms a trimer and is strongly associated with the cytoplasmic pump protein, transiently engages the TolC protein and with the assistance of energy, the pump protein, and possibly substrate, expels the substrate molecule across both membranes and the periplasm into the surrounding medium. The AcrB pump structure reveals that this protein is a trimer that contains a central chamber connected to a pore region by a (presumably) gated constriction. Entry to the chamber [48] is achieved from the membrane through vestibules that are located towards the outer monolayer of the cytoplasmic membrane.

P. aeruginosa OprM, despite only 21 % identity with TolC, appears to have a very similar structure ([49]; based on modeling using the process of threading, i. e. fixing in space the identical residues between TolC and OprM, and using the model building program Insight II to place the other residues by energy minimization). Insertion [50] and deletion [49] mutagenesis studies on OprM, combined with minimum inhibitory concentration assays and the functional reconstitution of OprM channels in planar bilayer membranes [50], provided experimental evidence for this model.

4.5.1 OprM

OprM is the outer membrane component of at least one (MexAB–OprM) and possibly other RND efflux systems. The MexAB–OprM efflux system is the major efflux system involved in intrinsic antibiotic resistance and its deletion renders *P. aeruginosa* far more susceptible to most antimicrobial agents [51]. The OprM structure model indicates that OprM, like TolC, comprises a trimer with a 12-stranded (four per monomer) open β-barrel crossing the outer membrane, atop a 12-stranded, coiled α-helical barrel crossing the periplasm [49]. Our data with deletion mutants indicated that the helical barrel of the protein is critical for both the function and the integrity of the protein, while a C-terminal domain localized around the equatorial plane of this helical barrel is dispensable. Extracellular loops appeared to play a lesser role in substrate specificity for this efflux protein compared to classical porins, and there appeared to be a correlation between the change in antimicrobial activity for OprM mutants and the pore size. Interestingly OprM was first shown to have a single channel conductance (at 50 mV applied voltage in 1 M KCl solution) of approximately 82 pS, very similar to that of TolC (80 pS) [49, 50]. However, in subsequent studies, using several fresh samples of OprM, we observed, under the same conditions, a 10-fold higher (850 pS) mean single-channel conductance. We proposed that this discrepancy reflected the age of the sample, since older samples gave rise to very noisy current tracings and the apparent channel size varied considerably. Interestingly, however, this could reflect the proposal, first made for TolC [46], that the periplasm-spanning coiled α-helical barrel acts as an iris diaphragm creating two states (open and restricted) equivalent to the large and small channel sizes respectively.

4.5.2 OprM Homologs

There are 17 known homologs of OprM [4, 6]. Two of these, OprJ and OprN, like OprM give rise to clinical resistance when overexpressed in clinical isolates [47]. Phylogenetic analysis of the 18 OprM family members [6] has suggested two clusters of outer membrane proteins, a relatively closely related OprM-like cluster of 11 pumps (which includes the known multidrug effluxers OprM, J and N), and a more divergent AprF-like cluster of seven pumps (which includes AprF involved in type I secretion of alkaline protease [52]). One member of this latter cluster, OpmN, appears to be a homolog of *Ralstonia eutropha* CzcC, which is involved in cation efflux, while OpmH from this cluster is the closest homolog of *E. coli* TolC, which is involved in efflux of hemolysin and multiple drugs. Knockouts in 17 of the 18 OprM homologs were obtained and it was demonstrated that OpmG (and to a lesser extent OpmH and I) appear to have roles in efflux of aminoglycoside antibiotics [45].

4.6 Specific Porins

Specific porins have stereo-specific binding sites for particular classes of substrates. The binding sites are saturable and uptake through these channels exhibits Michaelis–Menten kinetics, i. e. uptake is accelerated at low concentrations and plateaus at high external concentrations of solute. Faster rates of influx under nutrient limiting conditions enable the high-affinity transporters of the cytoplasmic membrane function at close to their maximal capacity and thus support the growth of the organism [53, 53]. The binding sites may also be involved in orienting larger substrates so they can pass through the channels unhindered as seen with the LamB maltodextrins-specific channel of *E. coli* [55].

Pseudomonas species have evolved a large number of specific porins (Table 4.1) to take up the diverse array of small molecules that they metabolize. These compounds include monosaccharides, amino acids, carboxylic acids, inorganic ions and aromatic compounds derived from or secreted by plants. Because of the specialized nature of the uptake process, specific porins tend to exclude large, hydrophobic compounds such as antibiotics, dyes and detergents [5]. However, they permit the non-specific diffusion of smaller compounds, thus supplementing the uptake activities of general porins.

Although the three-dimensional structures of the specific *Pseudomonas* porins have not yet been solved, it is believed that they will be similar to the solved crystal structures of porins from other bacteria [6, 56, 57]. Despite sharing very little sequence identity and taking up structurally diverse substrates (ranging from ions to maltodextrins), these porins all form trimeric molecules with the monomers each forming 16- or 18-stranded β-barrels. The β-strands have a general amphipathic character, and are interconnected by short periplasmic turns and the long extracellular loops. The extracellular loops stabilize the porin structure while a specific loop (usually loop 3) folds into the central channel to form a constriction that determines the molecular size exclusion and ion selectivity of the channel and, for the specific channels, forms part of the substrate binding site [56, 57].

4.6.1 OprB

OprB was first identified as a possible glucose uptake porin based on its induction in *P. aeruginosa* grown on glucose as a carbon source. It is co-regulated with a periplasmic glucose-binding protein of an ABC transporter. Based on its regulation and on mutational studies, it has been shown clearly that OprB is a component of the high-affinity glucose uptake system of *P. aeruginosa*. Subsequent work with interposon mutants [42] demonstrated the role of this porin in the uptake of other simple carbohydrates able to support the growth of *P. aeruginosa*, including mannitol, fructose and glycerol.

Circular dichroism spectroscopy of *P. aeruginosa* OprB indicates that the β-sheet content of this protein is 40%, similar to other porins [58]. Also, modeling of OprB, the closest homolog *P. aeruginosa* has to the *E. coli* maltodextrins-specific LamB channel, indicated that like LamB, OprB monomers fold into 18-stranded β-barrels. In addition, OprB has a group of aromatic residues that cluster into a motif resembling the greasy slide of LamB [6, 57].

OprB is positively regulated by its substrates glucose, fructose, glycerol, and mannitol [59]. Induction in response to glucose is via a two component regulatory system. The response regulator of this system, GltR, PA3192, [60] has been characterized, while the putative sensor kinase (PA3191) remains to be studied. Salicylate, a compound released by plants upon infection, and citric acid cycle intermediates such as succinate repress the expression of OprB. The effects of pH and temperature on OprB expression were also examined by Adewoye *et al.* [59]. They found that OprB is expressed optimally at pH 7 and that expression increases with temperature up to 42 °C.

P. aeruginosa has an OprB homolog, OpbA (PA2291) that is 96% identical to OprB at the amino acid level. The gene for this porin lies directly downstream of a glucose dehydrogenase gene, encoding an enzyme involved in the low-affinity glucose uptake pathway [61]. Thus it may be possible that *P. aeruginosa* uses two porins for glucose uptake– OprB at low external glucose concentrations and OpbA at higher glucose concentrations.

4.6.2 OprP and OprO

OprP is the anion-selective, phosphate uptake porin of *P. aeruginosa*. In contrast to OprB, which is induced by its substrate, OprP was first identified based on its induction in phosphate-limiting conditions [61]. This regulation is mediated by the PhoB regulator which also controls the expression of a periplasmic phosphate-binding protein [63]. Studies with a transposon mutant lacking *oprP* confirmed that in *P. aeruginosa*, OprP is involved in high-affinity phosphate uptake [64].

Studies of OprP in the planar lipid bilayer model membrane system showed that when phosphate is bound to the inside of the channel, the conductance of chloride ions through it is decreased [65]. In an attempt to identify the phosphate-binding site in OprP, the lysine residues in the N-terminal half of the protein were systematically changed to glutamates by site-directed mutagenesis. Three lysine residues were found to be important in phosphate binding. In the K121E mutant, phosphate binding, as assessed by inhibition of Cl^- conductance, was blocked. In the K74E and K126E mutants, phosphate binding was slightly reduced [64]. A topology model of OprP, generated by molecular modeling and insertion mutagenesis [66], predicted that both K121 and K126 are in loop 3, while K74 is found in the fourth β-strand. It is proposed that K121 forms the major phosphate-binding site of OprP, while K74 and K126 form secondary binding sites.

The *oprO* gene was identified immediately upstream of *oprP* [67]. The two proteins are homologous, sharing 76 % identity. OprO is also an anion-selective channel, like OprP, but preferentially binds pyrophosphate rather than phosphate. OprO expression requires that the cells be in stationary phase in addition to being under phosphate-limiting conditions.

4.6.3
OprD

Interest in OprD began when it was noticed that several clinical *P. aeruginosa* isolates resistant to the carbapenem antibiotic imipenem were missing this outer membrane protein [68]. Inspection of the chemical structure of imipenem revealed that it resembled a dipeptide containing a basic residue. Subsequent work with OprD in model membrane systems showed that this porin, indeed, was specific for the basic amino acids arginine and lysine, as well as dipeptides containing these residues [69].

The OprD substrate binding site was localized by first creating a topology map using molecular modeling followed by site-specific deletion mutagenesis [70]. This model predicted 16 β-strands per monomer and eight extracellular loops. Each of the OprD loop mutants was then tested for imipenem binding in a planar lipid bilayer apparatus. Mutants in loops 2 and 3 showed a decrease in imipenem binding [13, 71]. Mutants in loops 5, 7 and 8 were more susceptible to β-lactams, quinolones, chloramphenicol and tetracycline [13], suggesting that these loops fold over the mouth of the wild-type OprD channel and constrict its opening.

OprD is a highly regulated protein. In addition to being induced by its substrate arginine by the ArgR regulator, it is also induced by alanine and glutamate [23]. Like OprB, OprD is under the control of catabolite repression and is downregulated by salicylate. Imipenem also represses OprD expression via the MexT regulator, which concomitantly induces the expression of the *mexEF–oprN* efflux operon [22].

4.6.4
OprD Homologs

Prior to the release of the *P. aeruginosa* genome sequence, two OprD homologs, OprE [72] and OprQ [73], had been identified. Sixteen other homologs have since been discovered forming the 19 members of the OprD family (cmdr.ubc.ca/bobh/OprDfamily.html). Based on the amino acid sequences of its members, this family clusters into two subgroups. Preliminary studies on the role of this family suggest that they are specific porins with one subgroup responsible for the uptake of amino acids and dipeptides, and the other for a variety of substrates (such as small, organic acids) commonly encountered by *P. aeruginosa* (Tamber and Hancock, manuscript in preparation). The presence of this family probably gives *P. aeruginosa* a competitive edge in the environment where it can take up many different metabolites in a variety of environmental conditions. Indeed, it appears that the regulation of the OprD homologs is quite complex. The OprD proteins from *P. aeruginosa* cultures grown under different conditions appear to vary in their migration on SDS–PAGE gels, a result consistent with them almost certainly being OprD homologs.

4.7
TonB-dependent Receptors

TonB-dependent receptors, or gated porins, take up large macromolecular complexes. Examples of these channels' substrates include; iron-siderophore complexes, other metal–chelator complexes (Mb, Co, Cu), vitamin B_{12} and sulfate esters [6, 74]. To accommodate such large substrates, gated porins are much larger than their specific counterparts with molecular weights ranging from 75 to 92 kDa (*cf.* 42–53 kDa for specific porins). The crystal structures of two *E. coli* TonB-dependent receptors, FhuA and FecA [75, 76], have been solved. These proteins are monomers of 22-stranded β-barrels. Folded into the mouth of the barrel is a plug domain that provides a substrate-binding "gate" and thus prevents potentially harmful compounds from entering the cell.

TonB-dependent receptors are unique among outer membrane proteins because they require an energy source to translocate their substrates. The energy is provided by the TonB protein, which is found in the periplasm. *P. aeruginosa* has two TonB homologs, TonB2 (PA0197, 44 % similar to TonB) and PAO695 (39 % similarity to TonB); however, the role of these proteins in iron uptake is not well defined [77]. The exact mechanism of energy transduction is not clear but probably, involves a series of conformational changes in the TonB receptor upon docking of the iron–siderophore complex. These changes activate the TonB protein that then acts as an energy transducer, causing a further series of conformational changes in the receptor, culminating in the production of a translocation pathway and the release of the iron–siderophore complex from the binding pocket of the receptor [75–77].

TonB-dependent receptors have an exquisite specificity for their substrates. The general rule is one siderophore per receptor, even if it is produced by a different strain of bacteria [78]. This specificity may have arisen in an attempt to avoid the uptake of bacteriocins, various antibiotics [63] and other toxic compounds. However, since iron is an essential cofactor for many enzymes, *P. aeruginosa* has evolved an arsenal of 35 TonB-dependent receptors. Such a large number of iron receptor proteins gives these organisms a competitive edge in environments where iron is present in limiting quantities due to both the insolubility of the predominant ferric form and intense competition from other organisms. In addition to producing its own siderophores, (pyoverdin and pyochelin being the predominant ones), *P. aeruginosa* can also use heterologous siderophores (such as enterobactin, produced by *E. coli*). In addition, in the human host, this pathogen can acquire iron directly from heme. Below, we discuss some of the TonB receptors involved in the uptake of these siderophores.

4.7.1 FpvA

FpvA is the receptor for the type I pyoverdin produced by *P. aeruginosa. P. aeruginosa* produces two other pyoverdin types (type II and type III), and their receptors have been recently identified [79]; however, FpvA remains the most extensively characterized pyoverdin receptor. Molecular modeling of FpvA by comparison with the crystallized *E. coli* receptors FhuA and FecA indicates that it has 26 β-strands [80], and insertion mutagenesis has implicated two residues, Y350 and A402, from the sixth extracellular loop, in substrate binding and uptake [81].

The uptake of ferric-pyoverdin through FpvA differs slightly from the mechanism of other TonB-dependent receptors investigated to date [77]. In its resting state the channel is bound by free pyoverdin. Both iron-free pyoverdin and ferric-pyoverdin bind to FpvA with equivalent affinities [82]. Therefore, it has been proposed that TonB mediates the displacement of the iron-free pyoverdin allowing the ferric-pyoverdin to bind the receptor and enter the cell. Once in the cell, the ferric-pyoverdin complex dissociates and the siderophore is recycled back to the cell surface where it re-docks onto FpvA [77].

4.7.2 FptA

FptA is the receptor for *P. aeruginosa* siderophore pyochelin. Pyochelin is considered a secondary iron siderophore because its activity is masked by pyoverdin, which has a higher affinity for the ferric ion. However, pyochelin can coordinate other metals such as cobalt and molybdenum [83], and can obtain iron directly from transferrin [84].

4.7.3 PfeA and PirA

As mentioned above, *P. aeruginosa* has the ability to use heterologous siderophores produced by other *Pseudomonas* species, other bacteria and fungi. PfeA is the major TonB-dependent receptor involved in the uptake of the major *E. coli* siderophore enterobactin [77]. PfeA is 60 % similar at the amino acid level to the *E. coli* enterobactin uptake receptor FepA and can complement a FepA mutant. PirA shares

72 % amino acid similarity with PfeA. Although the role of PirA has yet to be investigated, it is presumed that it is involved in ferric-enterobactin uptake in the absence of PfeA due to either the loss or repression of the gene.

4.7.4 HasR, PhuR and OptI

Both HasR and PhuR are able to acquire iron directly from heme [85]. In addition, the *P. aeruginosa* genome encodes two other TonB-dependent receptors that may be involved in heme utilization. OptI is 52 % similar to that of HasR at the amino acid level and there is another protein, PA1302, which is 57 % similar to a protein from *Hemophilus influenzae* that is involved in heme acquisition.

4.7.5 Other TonB-dependent Receptors

In addition to the enterobacin receptors and heme utilization proteins, there are several other TonB-dependent receptors that share greater than 50 % similarity at the amino acid level. It is postulated that these alternative receptors take up their respective substrates either under different environmental conditions from their counterparts and/or serve as back-up systems in case the primary receptor is lost either due to mutation or inhibited by some other means. The presence of these redundant receptors, then, allows *P. aeruginosa* to optimize and maintain its uptake activities under a variety of different conditions, both environmental and physiological.

Of course, it is also expected that many as yet uncharacterized TonB-dependent receptors in *P. aeruginosa* that may be involved in the uptake of substrates other than iron–siderophore complexes. So far, in *P. aeruginosa*, this list includes four possible members, OprC, which takes up Cu^{2+} [6], PA1271, which is 46 % similar to the *E. coli* BtuB vitamin B_{12} uptake receptor, and two homologs to SftP, a gated porin in *P. putida* that presumably takes up sulfate esters [86], OptO, PA2335 (84 % similar) and PA0192 (75 % similarity).

4.8 Conclusions

P. aeruginosa appears to have many porins, which in turn help to explain the physiologically of this clinically important, nutritionally diverse organism. With the publication of the genome sequence we have identified many other putative porins and the coming years will permit us to fill in the details of how these porins contribute to the biology of this organism.

Acknowledgments

The authors' own porin work is supported by grants from the Canadian Cystic Fibrosis Foundation and the Canadian Institutes of Health Research. Genomic research is funded by Genome Canada through the Functional Pathogenomics of Mucosal Immunity project. The maintenance of the www.pseudomonas.com web page is funded by the US Cystic Fibrosis Foundation. R. E.W. H. has a Canada Research Chair and S. T. has a studentship from the Canadian Cystic Fibrosis Foundation.

References

1 R. E. W. Hancock, D. P. Speert. *Drug Resist Updates* **2000**, *3*, 247–255.

2 P. A. Lambert. *J Royal Med Soc* **2002**, *195* (Suppl 41), 22–26.

3 F. Bellido, R. E. W. Hancock. In: M. Campa, M. Bendinelli, H. Friedman (Eds), Pseudomonas aeruginosa *as an Opportunistic Pathogen*. Plenum Press, New York, **1994**, pp. 321–348.

4 K. C. Stover, X. Q. Pham, A. L. Erwin, S. D. Mizoguchi, P. Warrener, M. J. Hickey, F. S. L. Brinkman, W. O. Hufnagle, D. J. Kowalik, M. Lagrou, [Multi-author contribution]. *Nature* **2000**, *406*, 959–964.

5 R. E. W. Hancock. *Trends Microbiol* **1997**, *5*, 37–42.

6 R. E. W. Hancock, F. S. L. Brinkman. *Annu Rev Microbiol* **2002**, *56*, 17–38.

7 S. Tamber, R. E. W. Hancock. In: J. Ramos (Ed.), The *Pseudomonads. Vol. I. Genomics, Life Style and Molecular Architecture*. Academic Press/Plenum, New York, **2003**, in press.

8 T. I. Nicas, R. E. W. Hancock. *J Bacteriol.* **1983**, *153*, 281–285.

9 E. Yoshihara, T. Nakae. *J Biol Chem* **1989**, *264*, 6297–6301.

10 H. Nikaido, R. E. W. Hancock. In: J. Sokatch (Ed.), *The Bacteria: A Treatise on Structure and Function* X. Academic Press, London, **1986**, pp. 145–193.

11 X. Z. Li, H. Nikaido, K. Poole. *Antimicrob Agents Chemother* **1995**, *39*, 1948–53.

12 M. G. Scott, H. Yan, R. E. W. Hancock. *Infect Immun* **1999**, *67*, 2005–2009.

13 H. Huang, R. E. W. Hancock. *J Bacteriol* **1996**, *178*, 3085–3090.

14 J. R. Aires, T. Kohler, H. Nikaido, P. Plesiat. *Antimicrob Agents Chemother* **1999**, *43*, 2624–8.

15 S. Westbrock-Wadman, D. R. Sherman, M. J. Hickey, S. N. Coulter, Y. Q. Zhu, P. Warrener, L. Y. Nguyen, R. M. Shawar, K. R. Folger, C. K. Stover. *Antimicrob Agents Chemother* **1999**, *43*, 2975–83.

16 W. E. Sanders, C. C. Sanders. *Rev Infect Dis* **1988**, *10*, 830–838.

17 J. B. McPhee, S. Lewenza, R. E. W. Hancock. *Mol Microbiol* **2003**, *50*, 205–217.

18 N. D. Hanson, C. C. Sanders *Curr Pharmaceut Design* **1999**, *5*, 881–894.

19 K. Saito, H. Yoneyama, T. Nakae, *FEMS Microbiol Lett* **1999**, *179*, 67–72.

20 K. Poole, N. Gotoh, H. Tsujimoto, Q. Zhao, A. Wada, T. Yamasaki, S. Neshat, J. Yamagishi, X. Z. Li, T. Nishino. *Mol Microbiol* **1996**, *21*, 713–724.

21 T. Kohler, S. F. Epp, L. K. Curty, J. C. Pechere. *J Bacteriol* **1999**, *181*, 6300–6305.

22 M. M. Ochs, M. P. McCusker, M. Bains, R. E. W. Hancock. *Antimicrob Agents Chemother* **1999**, *43*, 1085–1090.

23 M. M. Ochs, C. D. Lu, R. E. W. Hancock, A. T. Abdelal. *J Bacteriol* **1999**, *181*, 5426–5432.

24 F. S. L. Brinkman, G. Schoofs, R. E. W. Hancock, R. DeMot. *J Bacteriol* **1999**, *181*, 4746–4754.
25 J. Trias, H. Nikaido. *J Biol Chem* **1990**, *265*, 15680–15684.
26 H. Huang, R. E. W. Hancock. *J Bacteriol.* **1993**, *175*, 7793–7800.
27 R. E. W. Hancock, A. Bell. *Eur J Clin Microbiol Infect Dis.* **1988**, *7*, 713–720.
28 R. E. W. Hancock, V. J. Raffle, T. I. Nicas. *Antimicrob Agents Chemother* **1981**, *19*, 777–785.
29 K. Piers, R. E. W. Hancock. *Mol Microbiol* **1994**, *12*, 951–958.
30 R. E. W. Hancock, D. S. Chapple. *Antimicrob Agents Chemother* **1999**, *43*, 1317–1323.
31 D. Jeanteur, J. H. Lakey, F. Pattus. *Mol Microbiol* **1991**, *5*, 2153–2164,
32 H. Nikaido, K. Nikaido, S. Harayama. *J Biol Chem* **1991**, *266*, 770–779.
33 R. Benz, R. E. W. Hancock. *Biochim Biophys Acta* **1981**, *646*, 298–308.
34 R. E. W. Hancock, G. M. Decad, H. Nikaido. *Biochim Biophys Acta* **1979**, *554*, 323–331.
35 F. Bellido, N. L. Martin, R. J. Siehnel, R. E. W. Hancock. *J Bacteriol* **1992**, *174*, 5196–5203.
36 W. A. Woodruff, R. E. W. Hancock. *J Bacteriol* **1988**, *170*, 2592–2598.
37 T. I. Nicas, R. E. W. Hancock. *J Bacteriol* **1983**, *153*, 281–285.
38 L. J. V. Piddock, M. C. Hall, F. Bellido, M. Bains, R. E. W. Hancock. *Antimicrob Agents Chemother* **1992**, *36*, 1057–1061.
39 S. Chamberland, F. Malouin, H. R. Rabin, T. Schollaardt, T. R. Parr, L. E. Bryan. *J Antimicrob Chemother* **1990**, *25*, 995–1010.
40 F. S. L. Brinkman, M. Bains, R. E. W. Hancock. *J Bacteriol.* **2000**, *182*, 5251–5255.
41 R. S. Y. Wong, H. Jost, R. E. W. Hancock. *Mol Microbiol* **1993**, *10*, 283–292.
42 J. L. Wylie, E. A. Worobec. *J Bacteriol* **1995**, *177*, 3021–3026.
43 K. Gensberg, A. W. Smith, F. S. L. Brinkman, R. E. W. Hancock. *J Antimicrob Chemother* **1999**, *43*, 607–608.
44 H. Nikaido. *Antimicrob Agents Chemother* **1989**, *33*, 1831–1836.
45 J. T. H. Jo, F. S. L. Brinkman, R. E. W. Hancock. *Antimicrob Agents Chemother* **2003**, *47*, 1101–1111.
46 V. Koronakis, A. Sharff, E. Koronakis, B. Luisi, C. Hughes. *Nature* **2000**, *405*, 914–919.
47 S. Murakami, R. Nakashima, E. Yamashita, A. Yamaguchi. *Nature* **2002**, *419*, 587–593.
48 E. W. Yu, G. McDermott, H. I. Zgurskaya, H. Nikaido, D. E. Koshland. *Science* **2003**, *300*, 976–980.
49 K. K. Y. Wong, F. S. L. Brinkman, R. Benz, R. E. W. Hancock. *J Bacteriol* **2001**, *183*, 367–374.
50 K. K. Y. Wong, R. E. W. Hancock. *J Bacteriol* **2000**, *182*, 2402–2410.
51 K. Poole, *J Mol Microbiol Biotechol* **2001**, *3*, 255–264.
52 J. Tommassen, A. Filloux, M. Bally, M. Murgier, A. Lazdunski. *FEMS Microbiol Rev* **1992**, *9*, 73–90.
53 R. Hengge, W. Boos. *Biochim Biophys Acta* **1983**, *737*, 443–478.
54 H. Nikaido, M. Vaara. *Microbiol Rev* **1985**, *49*, 1–32.
55 P. Van Gelder, F. Dumas, I. Bartoldus, N. Saint, A. Prilipov, M. Winterhalter, Y. Wang, A. Philippsen, J. P. Rosenbusch, T. Schirmer. *J Bacteriol* **2002**, *184*, 2994–2999.
56 R. Koebnik, K. P. Locher, P. Van Gelder. *Mol Microbiol* **2000**, *37*, 239–253.
57 T. Schirmer, T. A. Keller, Y. F. Wang, J. P. Rosenbusch. *Science* **1995**, *267*, 512–514.
58 J. L. Wylie, C. Bernegger-Egli, J. D. O'Neil, E. A. Worobec. *J Bioenerg Biomembr* **1993**, *25*, 547–556.
59 L. O. Adewoye, E. A. Worobec. *Can J Microbiol* **1999**, *45*, 1033–1042.
60 A. E. Sage, W. D. Proctor, P. V. Phibbs, Jr. *J Bacteriol* **1996**, *178*, 6064–6066.
61 L. F. Guymon, R. G. Eagon. *J Bacteriol* **1974**, *117*, 1261–1269.
62 R. E. W. Hancock, K. Poole, R. Benz. *J Bacteriol* **1982**, *150*, 730–8.
63 R. E. W. Hancock, E. A. Worobec. In: T. C. Montie (Ed.), *Biotechnology Handbooks 10: Pseudomonas.* Plenum Press, New York, **1998**, pp. 139–167.
64 A. Sukhan, R. E. W. Hancock. *J Biol Chem* **1996**, *271*, 21239–21242.

65 R. Benz, C. Egli, R. E. W. Hancock. *Biochim Biophys Acta* **1993**, 1149, 224–230.

66 A. Sukhan, R. E. W. Hancock. *J Bacteriol* **1995**, *177*, 4914–4920.

67 R. J. Siehnel, C. Egli, R. E. W. Hancock. *Mol Microbiol* **1992**, *6*, 2319–2326.

68 J. P. Quinn, E. J. Dudek, C. A. DiVincenzo, D. A. Lucks, S. A. Lerner. *J Infect Dis* **1986**, *154*, 289–294.

69 J. Trias, H. Nikaido. *J Biol Chem* **1990**, *265*, 15680–15684.

70 H. Huang, D. Jeanteur, F. Pattus, R. E. W. Hancock. *Mol Microbiol* **1995**, *16*, 931–941.

71 M. M. Ochs, M. Bains, R. E. W. Hancock. *Antimicrob Agents Chemother* **2000**, *44*, 1983–1985.

72 Y. Yamano, T. Nishikawa, Y. Komatsu. *Mol Microbiol* **1993**, *8*, 993–1004.

73 K. Okamoto, N. Gotoh, H. Tsujimoto, H. Yamada, E. Yoshihara, T. Nakae, T. Nishino. *Microbiol Immunol* **1999**, *43*, 297–301.

74 A. Gudmundsdottir, C. Bradbeer, R. J. Kadner. *J Biol Chem* **1988**, *263*, 14224–14230.

75 A. D. Ferguson, R. Chakraborty, B. S. Smith, L. Esser, D. van der Helm, J. Deisenhofer. *Science* **2002**, *295*, 1715–1719.

76 A. D. Ferguson, E. Hofmann, J. W. Coulton, K. Diederichs, W. Welte. *Science* **1998**, *282*, 2215–2220.

77 I. J. Schalk, M. A. Abdallah, F. Pattus. *Biochem Soc Trans* **2002**, 30:702–705.

78 J. M. Meyer. In: T. C. Montie (Ed.), *Biotechnology Handbooks 10: Pseudomonas*. Plenum Press, New York, **1998**, pp. 201–244.

79 M. de Chial, B. Ghysels, S. A. Beatson, V. Geoffroy, J. M. Meyer, T. Pattery, C. Baysse, P. Chablain, Y. N. Parsons, C. Winstanley, S. J. Cordwell, P. Cornelis. *Microbiology* **2003**, *149*, 821–831.

80 N. Folschweiller, I. J. Schalk, H. Celia, B. Kieffer, M. A. Abdallah, F. Pattus. *Mol Membr Biol* **2000**, *17*, 123–133.

81 L. Kilburn, K. Poole, J. M. Meyer, S. Neshat. *J Bacteriol* **1998**, *180*, 6753–6756.

82 N. Folschweiller, J. Gallay, M. Vincent, M. A. Abdallah, F. Pattus, I. J. Schalk. *Biochemistry* **2002**, *41*, 14591–14601.

83 P. Visca, G. Colotti, L. Serino, D. Verzili, N. Orsi, E. Chiancone. 1992. *Appl Environ Microbiol* **1992**, *58*, 2886–2893.

84 J. Wang, A. Mushegian, S. Lory, S. Jin. *Proc Natl Acad Sci USA* **1996**, 93:10434–10439.

85 U. A. Ochsner, Z. Johnson, M. L. Vasil. *Microbiology* **2000**, *146*, 85–98.

86 A. Kahnert, P. Mirleau, R. Wait, M. A. Kertesz. *Environ Microbiol* **2002**, *4*, 225–237.

5
Regulation of Bacterial Porin Function

Arnaud Baslé and Anne H. Delcour

5.1
Introduction

Over a decade ago, the first crystal structures of general diffusion porins were solved at high resolution [1, 2]. The pictures showed the beautiful trimeric arrangement of monomeric β-barrels, each highlighted by its own open pore. The large pore diameter, even at its narrowest point, and the precise constellation of charges at the constriction zone defined the structural and electrostatic features that permit the fast flow of hydrophilic or charged solutes, while providing molecular exclusion capability [3–5]. This structure immediately suited the model derived from microbiological and biochemical studies of transport through the bacterial cell envelope, that porins confer to the outer membrane its molecular sieving properties while providing permanently open conduits for the rapid flux of small nutrient and waste molecules [6]. Hence, general diffusion porins are essentially responsible for the selectivity and permeability of the outer membrane. However, the sophisticated regulation of porin gene expression according to environmental conditions emphasizes that the cell needs to control outer membrane permeability. Thus, shutting down porin expression or switching dominant porin type has been used by cells as a means of acquiring antibiotic resistance or adapting to a new milieu [7–10].

This chapter focuses on a different mechanism to control outer membrane permeability, i. e. the rapid modulation of porin function. Most of the work described here stems from biophysical studies of the channel proteins themselves and essentially highlights the dynamic nature of these pores. Although the physiological impact of these forms of regulation is only starting to be investigated, biophysicists have provided much information on the effect of many parameters, such as membrane potential, polycations, pH, bile, membrane tension, ionic strength and lipopolysaccharide (LPS), on the open probability of porins. These will be described here, along with the mechanistic insights obtained from mutant studies and modeling, and with the few cases so far where *in vivo* correlations have been established.

Bacterial and Eukaryotic Porins. Edited by Roland Benz

ISBN 3-527-30775-3

5.2
Voltage Dependence

First described by Schindler and Rosenbusch 25 years ago [11], the voltage dependence of general diffusion porins has been extensively studied and reproducibly observed by numerous laboratories. It is defined as the change in open probability that occurs in response to varying the transmembrane potential. A common way of studying voltage dependence of porins reconstituted in planar lipid bilayers is by applying continuous voltage ramps. Figure 5.1(A) shows five cycles of such a voltage protocol and the resulting current response. As the voltage increases, the current increases due to the progressively larger driving force for ion movement. This ohmic behavior, however, ceases once the voltage reaches a certain threshold, called the critical voltage (V_c) (Figure 5.1B). This deviation is due to the voltage-induced closure of a number of channels. Even as the voltage is ramped down, channels remain in this inactivated state and re-open only once the voltage is low enough again. The same pattern is duplicated somewhat symmetrically in the voltage regime of opposite sign. Asymmetries in voltage-dependent gating have been observed in some cases (see below), but would not be expected in this experiment since the purified porin was added to both sides of the bilayer. As seen in Figure 5.1(A),

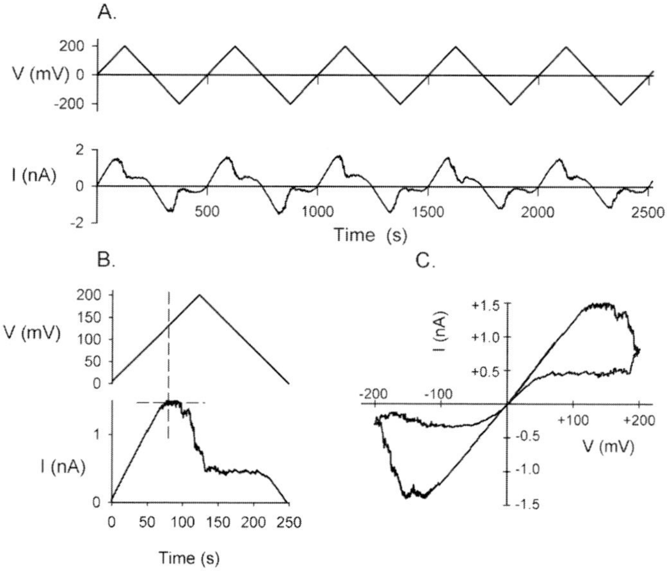

Figure 5.1 Voltage dependence of porins seen with voltage ramps. Purified OmpF (1 μg) was added to both sides of a planar lipid bilayer in 1 M KCl, 0.1 mM K-EDTA, 10 μM $CaCl_2$, 5 mM HEPES, pH 7.2. (A) Once channel insertion was complete, triangular ramps were applied between ±200 mV at a rate of 1.6 mV s^{-1} and the resulting current response was recorded. (B) The detail of the first 250 s is shown. V_c is defined as the voltage at which channel closure first occurs (intersection of the vertical dashed line with the voltage curve). (C) The current response of the first 500 s is re-plotted against voltage to show the hysteresis loops.

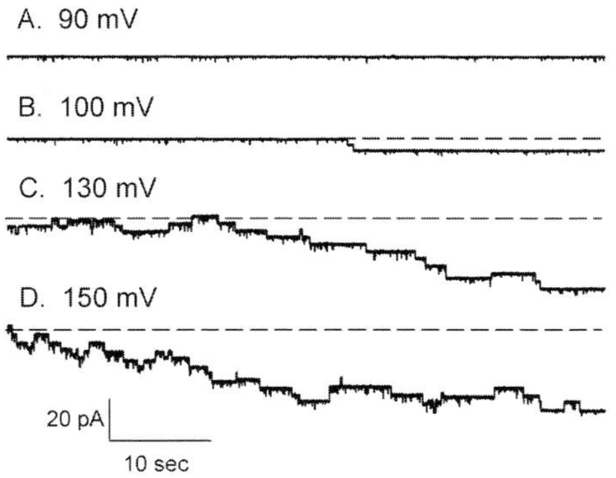

Figure 5.2 Voltage dependence of porins at constant voltage. Purified OmpF (1 μg) was added to both sides of a planar lipid bilayer in 1 M KCl, 0.1 mM K-EDTA, 10 μM $CaCl_2$, 5 mM HEPES, pH 7.2. Representative current traces are shown at the given voltages. The dashed line indicates the maximum current. Stepwise downward deflections represent closures.

this behavior is quite reproducible. Most publications do not present this raw data (current versus time in Figure 5.1A and B), but rather re-plot one such cycle on a current versus voltage plot. The resulting graph shows the typical figure-of-eight with hysteresis loops that has become a hallmark of porin voltage-dependent behavior (Figure 5.1C).

Another way of documenting voltage dependence is by following the progressive and stepwise decrease in current at a maintained potential. The traces shown in Figure 5.2 illustrate this behavior and were obtained in a bilayer experiment. Similar traces can also be observed with the patch-clamp technique, although fewer channels (only one to three trimers maximum) are typically present in a patch. Current jumps to prolonged states are believed to represent successive closures of single monomers, because their conductance is about a third of the average conductance obtained from insertion events. Voltage dependence here is manifested by the more rapid onset and time course of successive closures at higher voltages. Note that although a closing trend is obvious, the channels do oscillate between closed and open states, and thus display a fair amount of dynamic behavior. Attempts to fit these types of relaxation have been reported and suggest multiple channel states, some of which possibly act cooperatively [12, 13].

The critical voltage is often used as a characteristic fingerprint for a given channel. For example, the V_c of OmpF is around 130 mV in either the positive or negative range [14]. This value is quite high, relative to eukaryotic channels that display voltage dependence in a voltage range much below 100 mV. Some porins, however, have critical voltages in more "physiological" ranges, such as the *Vibrio cholerae* OmpT porin with a V_c of around 90 mV [15] or the *Neisseria* porin (PorB), which starts to close between 40 and 50 mV [16].

What is a "physiological" range of potential for the bacterial outer membrane? The reported weak selectivity of OmpF [17] does not allow the maintenance of any gradient of permeant ions, as the flow of cations is essentially balanced by the flux of counter-ions. Only the presence of fixed and impermeant charges on one side of a membrane would allow a gradient of permeant ions to be sustained in order to preserve bulk neutrality [18]. This sustained gradient creates a diffusion potential called Donnan potential. When cells grow in low osmolarity medium, they synthesize membrane-derived oligosaccharides (MDOs) – large, impermeant, negatively charged metabolites that accumulate in the periplasm [19]. In 1988, Nikaido and colleagues were able to control the magnitude of Donnan potentials by inducing MDO synthesis and changing the ionic strength of the external medium [20]. Potentials ranging from –30 to –100 mV (negative on the periplasmic side) were generated and thus can be taken as "physiological", in these specific conditions. If OmpF were a highly cation-selective channel, it may behave as a K^+ ionophore, like valinomycin, known to generate electrical potentials in a membrane that separates solutions of different K^+ concentrations. In this case, ionic gradients are maintained because the system reaches the Nernst equilibrium and the flux of K^+ driven by the concentration gradient is exactly compensated by the opposite flux of K^+ driven by the diffusion potential. Interestingly, recent molecular dynamics calculations of K^+ and Cl^- flux suggest that OmpF becomes highly cation-selective for K^+ at low ionic concentrations [5], a habitat that *Escherichia coli* might encounter naturally outside the host. These calculations, however, still await experimental validation.

Voltage dependence of general porins remains a highly debated subject, since its physiological relevance is not established. Nikaido and colleagues demonstrated that the Donnan potentials generated as described above did not reduce the porin-mediated influx of β-lactam antibiotics and, thus, concluded that porins are not affected by Donnan potentials [20]. Conditions leading to positive-inside membrane potentials, that might be more conducive to porin closure since the voltage-dependence of OmpF is asymmetric in patch-clamp experiments [21], have not yet been found. However, it has been proposed that voltage gating is a protective mechanism for the bacterium to close porins wrongfully inserted in the inner membrane [22].

Regardless of its physiological relevance, voltage dependence of porins has been – and is likely to continue being – a dominant subject of investigation, in particular with respect to the still unknown mechanism of voltage-induced closing. The paragraphs below summarize some of current information regarding the mechanistic aspects of porin voltage dependence.

5.2.1
L3 and the Constriction Zone

Two types of approaches have been used to obtain information on the molecular basis of voltage dependence: modeling and site-directed mutagenesis. The inwardly folded L3 loop became a prime target for these investigations because of its unique location and its participation in the intrinsic electrostatic field generated at the constriction zone [3]. The most obvious hypothesis to be tested was whether a large motion of L3 across the eyelet might underlie voltage-dependent pore closure.

In 1995, Soares et al. reported a molecular dynamics simulation of the L3 loop of the porin of *R. capsulatus* [23]. They suggested that an applied electrical field might redistribute ions and water molecules inside the pore, and lead to a conformational change of L3 in response to the change in electrostatic screening. They described a first closed state involving a 6-Å displacement of residues Ile-102 to Ala-113 from their crystal positions towards the center of the pore. A second closed state involved the additional movement of residues Thr-92 to Asp-96 toward the center of the pore. In this latest configuration the pore is completely closed. The validity of these results was later questioned, because the simulations were performed on a single monomer, and in the absence of membrane modelization and solvent. Two years later, Watanabe et al. [24] reproduced the L3 loop movement using an OmpF trimer in a vacuum, but the simulation failed to show such a displacement when crystal water molecules were added to the system. Finally, an extensive molecular dynamics simulation was performed on trimeric OmpF embedded in a palmitoyloleoylphosphatidylcholine bilayer as an approximation of the *E. coli* membrane, and in the presence of water and sodium ions [25]. These studies also failed to show a large motion of the L3 loop.

The hypothesized gross movement of L3 during voltage-induced gating was experimentally tested by engineering disulfide bonds in double-cysteine mutants. The rationale was to tether loop L3 to the surrounding β-barrel with a covalent bond and hence prevent drastic conformational changes of L3. Rosenbusch et al. tethered the tip of L3 to the adjacent barrel in the E117C/A333C OmpF mutant and showed that the crystal structure was essentially identical to the wild-type structure despite the removal of the hydrogen bonds stabilizing L3 [26]. A mild 30% increase in V_c was observed in this mutant. Non-significant difference in V_c was obtained between wild-type PhoE and mutants containing disulfide bridges between the pairs K18C/E110C, E110C/D302C and F111C/D302C [22]. In another study of OmpF, Lakey's group engineered disulfide bridges in various positions, such as at the tip of L3 (E117C/D312C), within the L3 loop (Y124C/D127C and V105C/F129C), and between L3 and the adjacent barrel wall (D107C/S177C). None of these mutants demonstrated a loss in voltage-gating capability [27]. Interestingly, the only mutant believed to fail in forming a disulfide bridge (Y111C/G217C) displayed a rather strong decrease in voltage gating sensitivity ($V_c > 300$ mV), which was also seen with the single residue mutant Y111C. Unfortunately, this lead was not pursued any further.

Early work of Benson et al. in 1988 demonstrated that L3 deletion led to functionally altered OmpF porins. OmpF mutants deleted for 6–15 residues in L3 displayed hyperpermeable phenotypes, but expression was not altered, thus ruling out a structural role of L3 [28]. Several years later, crystallographic studies of OmpF mutants deleted in the L3 residues 109–114 and 116–120 confirmed that the structure is not affected (although these are much shorter deletions than some of those reported by Benson et al.). In addition, these deletions did not alter significantly V_c [26, 29]. Altogether, these and the loop-tethering results argue against L3 motion as the conformational change underlying voltage-induced closure of the channel.

Even the participation of eyelet charges in voltage gating remains elusive. The influence of charged residues on voltage gating was first discovered by Lakey et al. in 1991 with the *ompC*(DEX) mutants of Misra and Benson [30, 31]. These mutants were obtained by selecting for growth on maltodextrins as the sole carbon source. These nutrients are normally too large to permeate through OmpC and thus mutants that displayed a larger pore size were obtained [30]. A similar selection was performed on OmpF as well [28]. All OmpF and OmpC mutants had mutations in a region found to correspond to the constriction zone when the OmpF crystal structure was published a few years later. Five of these mutations involved the loss of a positive charge and resulted in an increased voltage sensitivity compared to wild-type OmpC [31]. Many other studies point to an apparent correlation between the loss of positive charge by mutation on specific residues and an increased voltage sensitivity. In particular, mutations of single basic residues in the barrel wall facing L3 of OmpF lead to decreased V_c (increased sensitivity): K16A [32], R42C and R82C [29]. On the other hand, the loss of negative charge of L3 residues D113 and E117 resulted in a mild decrease in voltage sensitivity (increased V_c) [29, 33]. Similar results were found with OmpC, where removal of basic residues (R37C and R74C, equivalent to R42 and R82 in OmpF, respectively) decreased V_c [34].

Interestingly, anion-selective porins show opposite trends. For example, the removal of positive charges in PhoE (R37C, R75C and K18C) increased V_c and, reciprocally, the removal of a negatively charged residue of the L3 loop (E110C) resulted in a decreased of V_c [35]. The *Paracoccus denitrificans* porin, which shows a voltage-independent behavior up to around 230 mV, could be converted into a voltage-dependent channel when negatively charged residues of the L3 loop were mutated to neutral ones [36]. If the comparison of this porin with PhoE holds, a decreased sensitivity should be expected in mutants of the basic cluster facing L3. Of course, this was not detected in the voltage range investigated, since the wild-type *P. denitrificans* channel is already insensitive to voltage in this range. This opposite effect of similar mutations on OmpF and PhoE voltage dependence might correlate with their opposite asymmetric voltage dependence, which is readily observable in patch-clamp experiments [21]. However, the exact nature of the correlation between voltage sensitivity and ion selectivity has not yet been defined.

More recent studies emphasize the need to be careful when attempting to strictly correlate voltage sensitivity and the nature and location of charges of the constriction zone. For example, although the single OmpF mutants D113N and E117Q have decreased voltage sensitivity relative to wild-type, the double mutant D113N/E117Q shows an opposite trend [33]. Removal of all the major charges (both positive and negative) at the constriction zone (mutant R42A/R82A/D113N/E117Q/R132A) resulted in a slightly more voltage-sensitive channel than wild-type (V_c is 78 % of wild-type). This mutant had an identical V_c to an OmpF mutant where two positive charges had been introduced (V18K and G13K). The G119D mutant, which was isolated from a selection for colicin N-resistant strains, is less voltage sensitive that wild-type [37]. This is remarkable, because this mutation, which introduces a negative charge in the eyelet, affects the voltage dependence in the same way as the D113G and E117Q mutations, which remove negative charges [29, 33]. The same mutation was identified in the major porin of a multidrug-resistant clinical isolate of the nosocomial pathogen *Enterobacter aero-*

genes [38], which displays no voltage dependence up to 230 mV [39]. Finally, even mutations at uncharged residues affect voltage dependence. For example, two conserved mutations in OmpF that changed tyrosine to phenylalanine (Y102F and Y106F), without any structural or electrostatic modifications, drastically increased V_c [33]. In the *P. denitrificans* porin as well, the Y94G mutation affected V_c, but in the opposite way [36]. Interestingly, the introduction of bulky molecules, such as the fluorescent probes NBD-Cl and IAEDANS, by labeling the eyelet residues R37 and R74 of OmpC, also increased the voltage sensitivity [34]. Hence, it appears that charges at the eyelet are not the sole determinants in the channel voltage sensitivity, and that no clear correlations can be established between the magnitude of V_c and the exact charge state and constellation at the constriction zone.

In conclusion, the role of L3 in voltage dependence remains undefined. It is the general consensus that the L3 loop does not undergo large movement during voltage gating. Although it may play a role in voltage sensing, the exact contribution of its resident amino acids still remains to be defined within the context of other regions surrounding the constriction zone.

5.2.2 Extracellular Loops

Attention has now shifted to other regions of the porin protein, in particular to the extracellular loops. Loops L1 and L4–L8 form a somewhat compact structure that projects outwards about 13 Å away from the plane of the membrane. An atomic force microscopy study of two-dimensional OmpF-phospholipid crystals suggested that this extracellular domain is more flexible than originally observed by X-ray crystallography [40]. The atomic force microscopy (AFM) observation that the protein might exist under two distinct conformations in this region prompted Engel and his group to investigate whether conformational changes could be detected by AFM under specific conditions [41]. Their study demonstrated that the loops experience a voltage-dependent conformational change, where the 13 Å protrusions collapsed into 6-Å high doughnut-like structures. The high voltage used (500 mV) may be questionable, but the authors were able to reproduce these results by decreasing the ionic strength or the pH of the external solution. They envisioned that this conformational change might represent a bending of the extracellular domain about a hinge at the rim of the barrel and, thus, lead to channel closure.

Coulton's laboratory has also provided evidence in favor of a role of extracellular loops in voltage gating. Based on the observation that different subtypes of Hib porins displayed different V_c and diverge in sequences in the extracellular L4 [42], the authors hypothesized that charged residues on extracellular loops are involved in voltage gating. They used chemical modification of surface-accessible lysine residues by succinic anhydride to replace the positive charge of these residues by a negative one and showed that the modification correlated with a decrease in V_c (+40 mV compared to +75 mV of wild-type Hib) [43]. Their study was further completed using site-directed mutagenesis on lysine residues predicted to be located on the extracellular loops L4 and L6 [44]. Five out of six of the single lysine mutants showed alterations in voltage sensitivity. Thus, it appears that the extracellular loops have a role in voltage-dependent gating, possibly by undergoing conformational changes that bring about pore closure. In this model, the roles played by the constriction zone and the extracellular loops are not mutually exclusive, as amino acids from these two regions may be brought in closer contact and interact during gating [44, 45]. This intriguing and highly speculative scenario was never reported in computational studies, and still needs to be put to the experimental test.

5.2.3 Modulation of Voltage Gating

The outer leaflet of the outer membrane is comprised of a typical prokaryotic lipid, LPS. The LPS inner core is more hydrophobic that phospholipids, but the charge density of LPS is also about 2.5 greater than phospholipids [46]. It is estimated that a difference in surface potential of around 75 mV exists between a monolayer made of exclusively phospholipid or exclusively LPS at an ionic strength of 65 mM [46]. This asymmetry leads to profoundly different intrinsic membrane potential profiles, which are themselves influenced by the ionic strength of the solution bathing each leaflet [46]. LPS is required for porin biogenesis [47] and is known to remain strongly associated with porins, even during their purification [48, 49]. Thus, the impact of this molecule naturally found in the vicinity of porins should not be underestimated. Seydel's group has been successful in creating asymmetric planar lipid bilayers for electrophysiological measurements in a lipid system resembling the natural outer membrane [46, 50]. The LPS-free porin of *P. denitrificans* was used as a source of biological material. They demonstrated that, although symmetric bilayers made of phospholipids still allow the voltage-dependent gating to be observed, the critical voltage and shape of the hysteresis loops are influenced by the nature of lipids. When the membrane potential is more negative on the LPS side, the critical voltage is smaller and the extent of channel closure is much more pronounced, leading to an increased area of the hysteresis loop. These trends appear to be correlated with the difference in surface potential of the two leaflets [46, 50]. Interestingly it was found that the voltage gating of porins is also influenced by other physicochemical characters of the bilayer, such as membrane rigidity [50] and length of the sugar moiety of different LPS types [51]. LPS is also reported to affect the rate of channel incorporation [46] and the conductance [52].

Asymmetries in voltage-dependent gating have also been reported for some porins, even in symmetrical bilayers. Uncertainty about the unidirectional insertion of general diffusion porins using the bilayer technique renders the interpretation of this data difficult. In some cases, the channels that are added on one side only of the bilayer and appear to insert unidirectionally into the artificial membrane do display strong asymmetries [13, 53, 54]. In other cases, addition of porins to the *cis* side only produces a symmetric [44] or a mildly asymmetric behavior [32]. When outer membrane fractions are reconstituted into liposomes, where randomization is unlikely, an asymmetric and opposite voltage dependence of is found for OmpF and PhoE [21].

Several studies have also documented the effect of pH on voltage dependence. These will be discussed in the following section, where the effects of pH on various functional features of porins are presented as well.

5.3
Effect of pH

There are numerous reports that acidic pH affects porins, but, except for a few studies, the information is often anecdotal and some discrepancies between publications have been noted. A molecular mechanism for the pH modulation of various porin function is also completely lacking. The effect of acidic pH on conductance was reported in a one of the very first electrophysiological studies on OmpF [17]. From pH 3 to 9, the specific conductance increased by a factor of 1.25 and 1.60, in 1 M KCl and 1 M NaCl, respectively. Similar results were reported by some groups [55], but others did not find any effect of pH on conductance in the investigated range of pH 4.3 to 9.4 [14, 56]. However, it appears that new populations of channels with distinct conductances are revealed by decreasing the pH. This is seen as new or additional peaks in amplitude histograms taken from insertion events [55] or monomeric closing events [52]. The appearance of these new populations may correlate with the greatly enhanced gating to substates that we observed in the patch-clamp technique with OmpF upon decreasing pH (Baslé and Delcour, unpublished). McGroarty et al. also report that the frequency of closing events of both OmpF and OmpC is higher at pH 5.4 that at pH 9.2 [55]. Surprisingly, the opposite trend was observed with the strongly anion-selective Omp34 porin of *Acidovorax delafieldii*, where the conductance increases with decreasing pH [54]. Effects on selectivity have been reported as well: at very acidic pH (pH 2–3), the permeability of K^+ (P_K) is identical to that of Cl^- (P_{Cl}), but the permeability ratio P_K/P_{Cl} steadily increases with pH, until it reaches a plateau of about 2.5–3.0 around pH 5.0 [17].

An interesting aspect of porin modulation is the interplay between pH- and voltage-induced regulations. The enhancement of voltage sensitivity at acidic pH was first reported by McGroarty's group [57]. The effect was documented mainly by looking at the frequency of closing events and the shift in size distribution. A quantitative analysis of the critical voltage was performed later and indicated that OmpF becomes less voltage sensitive at higher pH, as V_c increases from around 120 to 200 mV between pH 5.0 and 9.0 [29]. Faster time constants for the voltage-induced current decay of *E. coli* 0111:B4 porin were also found at pH 5.5 than pH 7.2 [12]. Again, the omp34 porin of *A. delafieldii* is reported to behave differently, since decreasing or increasing pH from 8.3 has the same effect, i. e. a loss of voltage sensitivity [54].

The molecular basis for these effects of pH remains unknown. The switch in porin channel size was inhibited by chemical modification of the single histidine residue present in OmpC and OmpF (His-21) by diethylpyrocarbonate [58]. The involvement of His-21 was, however, refuted in another study using site-directed mutagenesis of this residue [14]. The participation of the constriction zone is suggested by some observations, such as effects on selectivity [17] and modifications in the V_c titration curves in eyelet mutants of OmpF [29]. However, pH-induced conformational changes at homologous residues in OmpC were not observed [34]. Muller and Engel reported that a conformational change in the extracellular domain occurs upon acidification of the external milieu [41]. It is hoped that the residues and protein regions involved in this mode of regulation will soon been found, which will allow mutants to be constructed. Indeed, the inhibitory effect of acidic pH on porins is strong enough to lead to a decrease in overall outer membrane permeability [59]. The availability of mutants would allow the physiological relevance of this mode of porin regulation to be tested.

5.4
Polyamine Modulation

Polyamines are polycationic molecules ubiquitously distributed among all organisms (Figure 5.3A). Our work on polyamine modulation was stimulated by the publication of Koski and Vaara [60] documenting the presence of polyamines associated with the outer membrane of *Salmonella typhimurium* and *E. coli*. Treatment of intact cells with 1 N NaCl released 8–13 % of the total polyamine content (putrescine, cadaverine and spermidine), without concomitant cell lysis. Since polyamines are known to bind LPS [61], the authors proposed that LPS is the major binding site for polyamines in the envelope of Gram-negative bacteria. It was particularly striking that the amount of cadaverine extracted from LPS preparations was much greater than the amount extracted from isolated outer membrane, suggesting that this polyamine is lost upon membrane purification. The association of spermidine with the outer membrane is surprising as this polyamine is not excreted in the medium [62, 63]. In *E. coli*, the synthesis of spermidine is catalyzed by the cytoplasmically located spermidine synthase, which transfers the aminopropyl moiety of the decarboxylated adenosylmethionine to putrescine. Intracellular concentrations of spermidine reach around 6 mM [62, 64], with 90 % bound to RNA.

Of particular interest with respect to porin function is the environmental regulation of synthesis of putrescine and cadaverine in *E. coli*, and the excretion of these compounds in specific conditions. Putrescine is synthesized either by ornithine decarboxylase or by the sequential action of arginine decarboxylase and agmatine ureohydrolysase [62, 65]. Both ornithine decarboxylase and arginine decarboxylase exist as constitutive and induced enzymes, the latter being induced under semi-aerobic conditions at low pH. Thus, intracellular concentrations of putrescine, which have been estimated at around 20 mM under normal conditions [64], can be greatly increased at acidic pH. Since about 40 % exist as a free form (the rest being essentially bound to nucleic acids), increases in intracellular concentrations can easily lead to excretion of the polyamine. This might be particularly true as the

A.

Spermine

$NH_2\text{-}(CH)_3\text{-}NH\text{-}(CH)_4\text{-}NH\text{-}(CH)_3\text{-}NH_2$

Spermidine

$NH_2\text{-}(CH)_3\text{-}NH\text{-}(CH)_4\text{-}NH_2$

Cadaverine

$NH_2\text{-}(CH)_5\text{-}NH_2$

Putrescine

$NH_2\text{-}(CH)_4\text{-}NH_2$

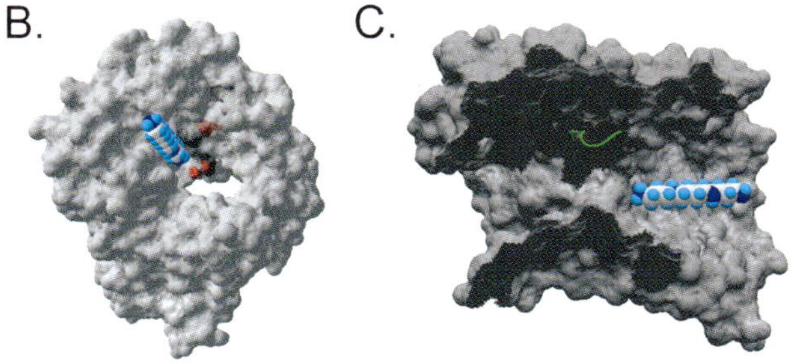

Figure 5.3 Polyamines and porin. (A) Formulae of the major natural polyamines (spermine is not found endogenously in *E. coli*). Amine groups are show in blue and bear a positive charge when protonated in solution. (B) Molecular surface of OmpF obtained form the crystal structure coordinates of 1OPF and viewed from the extracellular side. The atoms colored in red represent the oxygen of the side chains of Y294, D121 and D113, three residues facing the pore and believed to participate in the interactions between the porin and the polyamine molecule. A single spermine molecule is shown entering the pore in a head-on direction (white atoms are carbon, light blue atoms are hydrogen and dark blue atoms are nitrogen). (C) Cross-section of the OmpF molecular surface showing the constriction zone, highlighted by the green L3 loop. A single spermine molecule is shown entering the pore from the periplasmic side. The oxygen atoms of the three residues mentioned in (B) are in a pore-exposed crevice on the extracellular side of the L3 loop and are not seen in this view (this orientation was chosen to give the best view of the constricted region at the eyelet). The figures were made with the PDB Viewer program.

putrescine transporter PotE, used for either uptake or excretion, is also acid inducible [65]. Excretion of putrescine has been reported when cells are growing in minimal medium [63] and stimulation of putrescine release has been demonstrated upon osmotic up-shock [63, 66]. Thus, putrescine excretion across the cell envelope and exposure to porins from both periplasmic and extracellular sides could be significant in conditions of low pH and/or osmotic up-shock. Similarly, cadaverine is excreted when cells are grown at acidic pH [67]. The decarboxylation of lysine is catalyzed by the cytoplasmic enzyme CadA. The resulting cadaverine is excreted through the membrane-bound lysine-cadaverine antiporter CadB. The acid induction of the *cadBA* operon is initiated by the activation of the environmental sensor CadC, which binds to the promoter region. Increase in *cadA* mRNA is quite rapid (within 5 min) upon a pH drop [68]. As a result, cadaverine accumulates in the external milieu.

Finally, significant amounts of polyamines are found in the human gut [69, 70]. In particular, putrescine and cadaverine are present in the lumen of the proximal small intestine, at concentrations in the low millimolar range [69]. These are presumed to originate from the metabolism of the microflora [69], but the diet also supplies measurable amounts [71].

In conclusion, polyamines are found in the vicinity of porins, either because of the natural presence in the surrounding medium or because of the environmentally regulated excretion through the cell envelope and into the extracellular space. As discussed below, an important avenue of research is to establish the role of the modulation of porin function by these endogenous compounds within the physiological contexts that dictate the timing and location of their synthesis.

We initially discovered that cadaverine inhibits porin-mediated ion fluxes in patch-clamp experiments [72]. Since then, we have used a variety of assays to study the effects of polyamines on porin function, including antibiotic flux assays, chemotaxis assays, patch-clamp and planar lipid bilayer electrophysiology. We included different polyamines in our survey: the natural polyamines putrescine, cadaverine, spermidine and spermine [72, 73], synthetic polyamines [74], and polyamine toxins [75]. Different potencies and types of effects were found with these compounds. In addition, analogs that have the same length and bear the same number of charges, but contain quaternary rather than tertiary amines, are not effective [74]. We also found that simply titrating the charge on cadaverine by performing experiments at pH 9.5 was sufficient to annihilate the effect [72]. Finally, we reported that the anion-selective PhoE porin was insensitive to polyamine modulation [76]. The molecular weights of the natural polyamines are well below the size cut-off of 600 Da for porin permeability and, thus, it is unlikely that the molecules get stuck within the pore. Taken together, these findings lend strong support to the notion that the compounds specifically bind to porins and do not merely exert their effects by physically obstructing the pores.

The kinetic pattern of polyamine-mediated modulation is also suggestive of a distinct molecular mechanism from open-channel block. Examples of patch-clamp traces recorded on liposomes containing a single purified OmpF trimer are shown in Figure 5.4. The current level marked by a solid line is the total current flowing through all three monomers. The dashed line represents the current level attained when one monomer has closed (upward deflections represent closures). In control conditions, closing transitions are quite rare. In the presence of 100 μM spermine, the frequency of the closing transitions is greatly increased. Interestingly, the majority of these transitions have a size that is below that of the single monomer. This is in contrast to the pattern found when single antibiotic molecules penetrate trough OmpF [77]. In this latter case, a clear disruption of the whole monomer conductance is found, as would be expected for an open-channel blocker. With any of the natural polyamines, we find transitions of multiple sizes. For example, the events marked by the numbers 1, 2 and 3 have conductances (in 150 mM KCl) of 200, 120 and 160 pS, respectively, all smaller that the monomeric conductance of 350 pS (calculated from the known trimeric conductance). Our original publications refer to these events as representing the cooperative gating of multiple monomers of around 30–60 pS conductance. Since then we have been able to obtain the trimeric conductance of purified OmpF from insertions events into planar lipid bilayer using the very same solutions as those of patch-clamp experiments. We have been able to assign around 350 pS for the monomeric conductance in 150 mM KCl, a value that is in agreement with that of others in slightly different salt conditions [32]. Thus, we have revised our prior working hypothesis that the smallest transition of 30–60 pS observed in patch-clamp represents the monomer, and come to the conclusion that a large amount of subconductance gating is observed in modulated traces such as the one shown in Figure 5.4(B). This type of gating is remarkable, as different conductance levels can be attained in what appears as single smooth transitions (i. e. without a staircase aspect). In addition, the channel can dwell at any of these levels for prolonged periods of time. A similar type of gating pattern is found when the channels are studied in planar lipid bilayers, as shown in Figure 5.4(D). The traces are somewhat less crisp because of the increased noise level with this technique. Despite increased subconductance gating being a hallmark of the modulated mechanism, at high enough concentrations total closure of all the pores can be observed, emphasizing again the allosteric, rather than obstructive, nature of the inhibition.

We found that the polyamines spermine, spermidine and cadaverine all gave rise to this type of kinetic signature [73]. Differences were found in the potencies of the compounds [73]: spermine was the most effective polyamine, with detectable effects

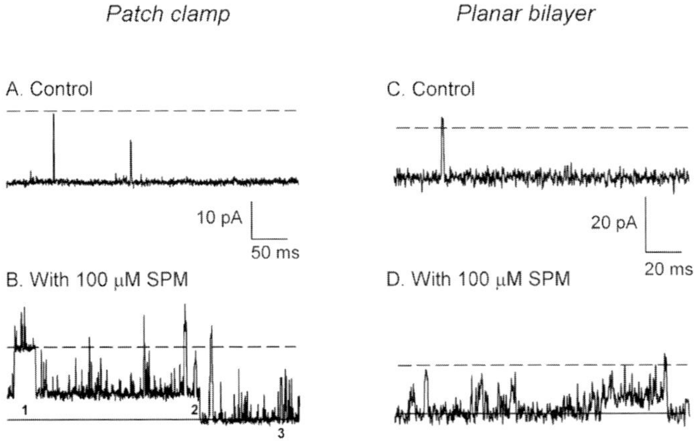

Figure 5.4 Modulation of OmpF by spermine. Representative current traces obtained with the patch-clamp technique (A and B) or the planar lipid bilayer technique (C and D) of a single trimer of purified OmpF in the absence ("control") and the presence of 100 μM spermine (SPM) applied to the same patch or bilayer (in 150 mM KCl, 0.1 mM K-EDTA, 10 μM $CaCl_2$, 5 mM HEPES, pH 7.2). The dashed line represents the current level obtained upon closure of a single monomer from the maximum current (solid line). For all traces, upward deflections are closures and the voltage is −60 mV. In both spermine traces, most deflections represent gating to substates: events marked 1, 2 and 3 have a conductance of 200, 120 and 160 pS (for events 1 and 2, the size is measured from the current level of the prolonged closed state that starts the trace). Note the different scale bars for the patch-clamp and planar bilayer experiments.

in the low nanomolar range, while cadaverine required millimolar concentrations. Spermidine was intermediate, and putrescine was ineffective up to 100 mM. The inhibition of OmpF and OmpC by polyamines is voltage dependent [73]. Membrane potentials that are negative on the opposite side to the side of polyamine application enhance the inhibitory effect, because the positively charged polyamine is electrophoresed to its site of action, which we believe is inside the pore. It is clear, however, that polyamines do not play the role of gating particles to confer voltage dependence to porins, since the voltage-dependence is observed with purified proteins.

Our study of natural and synthetic polyamines allowed us to identify the molecular properties required for compounds to act as inhibitors of OmpF in the nanomolar range [74]. Efficient inhibition requires a 12-carbon chain triamine with terminal primary amines and replacement of the eighth methylene by a secondary amine. The need for this type of molecular architecture suggests that inhibition is governed by interactions between specific amine groups and protein residues. By analogy with polyamine binding proteins of the *E. coli* Pot system [65], we have made the assumption that the charged groups of the polyamine molecules interact with negatively charged residues in the pore. Together with previous observations from site-directed mutagenesis studies [78] and inspection of the crystal structure of OmpF, these results allowed us to propose three residues (D113, D121 and Y294) as putative sites of interaction between the channel and spermine. Alanine substitution at each of these three residues resulted in a total loss of inhibition by spermine, while mutations of only D113 and D121 affected inhibition by spermidine [74]. These observations support a model whereby spermine would enter the pore in a head-on conformation and effectively bridge the periplasmic and extracellular branches of the L3 loop and connect them to the adjacent barrel wall (Figure 5.3B and C). Spermidine, being shorter, would only interact with the L3 loop branches. This saddling over the L3 loop would bring about the allosteric modulation leading to enhanced subconductance gating, stabilization of closed states and porin inhibition. A computational analysis of spermine docking also suggested possible interactions with residues across the pore from the L3 loop [79], but this has not yet been confirmed experimentally.

The inhibition by polyamines initially revealed in electrophysiology has physiological consequences. The addition of any of the four natural polyamines to cells leads to a measurable decrease in overall outer membrane permeability in a concentration-dependent manner. This was evidenced by a reduction in chemotactic ability of motile *E. coli* [80], due to the decreased availability of chemoeffectors that require porins for entry into the periplasm [81]. Importantly, polyamines affect the ability of antibiotics to penetrate inside the cell, as demonstrated for the β-lactam antibiotic cephaloridine in *E. coli* [80], and cefepime and norfloxacin in *Enterobacter cloacae* [82]. The next logical step was to assess whether endogenous polyamines were able to modulate outer membrane permeability *in vivo*. For this, we turned to the well-characterized *cadBA* operon [67]. By using inducible plasmids with the *cadA* and *cadB* genes, we were able to show that increased levels of excreted cadaverine correlate with a decreased outer membrane permeability to cephaloridine, without any change in porin expression [83]. Cadaverine appears to promote a sustained inhibition of porins, since the effect remains even after removal of the exogenously added or excreted polyamine.

These results supported the notion that polyamines can act as endogenous modulators of outer membrane permeability, possibly as part of an adaptive response to acidic conditions. To test this hypothesis, we isolated an OmpC mutant (G195D) that showed resistance to spermine in growth, and had lost sensitivity to polyamine inhibition as measured by antibiotic flux assays and electrophysiology [84]. If the modulation of porins by cadaverine plays a role in survival at acidic pH, the cadaverine-resistant porin mutant should be more sensitive to low pH than wild-type cells. Thus, we performed competition experiments between strains that solely express the wild-type *ompC* gene and those that solely express the G195D mutant *ompC* gene [84]. Wild-type and mutant cells were mixed together in a single flask and grown until mid-log phase, at either pH 7 or 5. At the end of this growth, an aliquot of this culture was used to start a fresh culture and the cycle was repeated 6 times. Thus, relative survival of wild-type and mutant cells is compared as the cultures go through several generations. We found that the wild-type cells out-competed the mutant cells only at pH 5, and not at pH 7 [84]. We confirmed that cadaverine was secreted only at pH 5. Since the lack of cadaverine sensitivity is the only known phenotypic difference between the wild-type and the G195D OmpC porin, these results substantiate the model whereby inhibition of porin by secreted cadaverine at low pH is a component of the strategy for cell survival at acidic pH.

These types of studies illustrate that modulation of porin function can provide rapid means of controlling outer membrane permeability and that this phenomenon is important to the cell's adaptive responses. It is anticipated that similar work on putrescine in *E. coli*, for example, or on polyamines in other microorganisms will unravel additional important situations where porin modulation plays a role in the adaptation of bacteria to external environments.

5.5 Others

Voltage, pH and polyamine are the three modulators that have received the most attention. However, over the years, other forms of modulation have also been documented, and will be briefly mentioned in this section. The investigations conducted in our own laboratory started with the observation that MDOs induced a great amount of gating of porin channel in patch-clamp experiments [85] and appeared to behave as open channel blockers. These traces are similar to those recently recorded in the presence of β-lactam antibiotics [77]. In this latter case, it has been interpreted that the permeation of the ampicillin molecule causes transient interruptions in the ionic current flowing through single monomers and, thus, the trace displays a large amount of flickers whose size correspond to that of a single monomer. Whether this effect, which is basically a competitive inhibition of transport, should be called "modulation" can be debated. At this point in time, we believe that the inhibition exerted by polyamines is mostly of a different nature because the induced closing events are subconductance states and because the channels displayed prolonged dwell times in closed states. However, it is likely that some competitive inhibition by polyamines occurs as well, as the kinetic patterns and concentration dependencies are rather complex [73]. On the other hand, the change in kinetic pattern of PhoE that is induced by the addition of ATP is highly reminiscent of an increase in channel noise due to open channel block and competitive inhibition of transport [76]. Similarly, the inhibition of *Vibrio cholerae* porins by bile might be simply due to temporary block of the pore [86]. This effect has been documented only with antibiotic flux assays and more precision on the mechanism of bile inhibition is still awaited, since electrophysiological experiments testing the effect of this detergent-like compound are rather difficult.

The modulation of the pathogenic *Neisseria* species PorB by purine nucleoside triphosphate (i. e. ATP, GTP) involves both a decrease in channel conductance (possibly due to open channel block) and modulation of V_c [16]. Similarly to what we have observed with polyamines, the effect on voltage dependence was always in conditions where the sign of the voltage was negative on the side of addition of nucleotides (for polyamines the effect occurs when the voltage is negative on the side opposite to the side of addition). These results, together with the reported effect of the nucleotides on selectivity, suggest that the nucleotides interact with pore-exposed residues. Interestingly, this modulation is not observed in the nonpathogenic species nor is the spontaneous transfer of the porins molecules from the outer membrane of whole cells to the target cell membrane [16]. Thus the authors propose that endogenous nucleotides found in the cytoplasm of the target eukaryotic cell would regulate the open probability of PorB, such that the channel remains closed. The results would provide an explanation for the lack of toxic effect of PorB insertion in the eukaryotic cytoplasmic membrane, despite the fact that PorB produces such large channel.

Changes in ionic strength have been reported to affect porins. Decreasing ionic strength in planar bilayer experiments with OmpF or OmpC leads to the appearance of additional conductance levels that are likely to represent substates [52]. Similar findings have been made in patch-clamp experiments as well, where the OmpC porin and some of its mutants show very distinct gating patterns at 10, 150 and 1000 mM KCl [87]. These effects emphasize that permeating ions do interact with the pore, not only because of the expected thermodynamics and electrostatics of binding [5, 33], but in complicated ways that may lead to conformational or dynamic changes within the pore itself. Such ion flow might lead to perturbations in the intrinsic electrostatic field that is created by the charge constellation of L3 and the opposite barrel wall [3]. These perturbations might lead to conformational changes or disruption in ion transport that lead to kinetic changes in electrophysiological traces [23, 88]. This model was proposed as a possible explanation for the unidirectional effect of the pressure differential (around 1 kPa) across the membrane, which increased the gating of OmpC (but not PhoE) to closed states and acted in synergy with the voltage-dependent modulation [89]. Here the authors proposed that the applied hydraulic force might act directly on counter-ions occupying the pore and lead to disturbance in their distribution. The resulting perturbation in the electrostatic interactions between various regions of the eyelet might result in the observed increased gating activity towards closed states. On the contrary, high hydrostatic pressure (around 90 kPa), acting uniformly in all directions, led to openings of OmpC [90]. Investigating the impact of this form of regulation in *E. coli* and in barophiles might be interesting in light of the known effect of hydrostatic pressure on expression of outer membrane proteins [91].

5.6 Concluding Remarks

Our own study of porin modulation started out with the observation that high resistance seals could be made with the patch-clamp technique on giant cells or spheroplasts of *E. coli* that appeared intact by electron microscopy [92]. Such electrically tight seals would not have been possible on the outer membrane unless porins were closed. Thus, along with many other researchers, we set out to identify and characterize the natural parameters that might cause porins to be closed. In retrospect, it is not clear whether our early studies investigated intact outer membrane, since we often found activities from porins and the now well-established inner membrane channel MscL [93] in the same patches. Some membrane randomization might have occurred during patch formation. Thus, it is likely that in live cells, most porins are in the open state, at least in the conditions in which transport assays have been performed [94]. There is no question, however, that porins are dynamic, and oscillate between open and closed states, an activity which we have referred to as "gating". We have described here how some parameters can modulate this gating, as well as other features of porin function. Because of these forms of modulation, the outer membrane permeability will not be constant *in vivo*.

There is still much to be learned about porin modulation. What is the molecular basis for the forms of regulation described so far? Are there conformational changes? What is the nature of these molecular motions? How can porin modulation help bacteria adapt to a variable environment? What type of porin regulatory mechanisms exists in pathogenic species? Do they contribute to the pathogenesis, adaptation or antibiotic resistance of these strains?

Answering these questions will keep researchers active in this field. Biophysicists and biochemists will continue to provide the molecular and mechanistic framework for this aspect of porin function. However, these questions also beckon microbiologists to return to this field and to re-consider the physiology of the outer membrane under a new light, with the goal of defining the biological circumstances under which porin regulation takes place naturally, and the physiological impact of this biochemical process. A thorough biophysical study of the effect of pH on OmpF conductance, selectivity and kinetics has recently been published by Bezrukov and colleagues (*Biophys. J.* **85**, 3718, (2003)), where it was shown that acidic pH promotes an increase in closing probability and gating to substates, and that extremely acidic pH converts OmpF from a slightly cation-selective channel to an anion-selective channel.

Acknowledgements

Our own work on porins is supported by NIH grant AI34905.

References

1 M. S. Weiss, U. Abele, J. Weckesser, W. Welte, E. Schiltz, G. E. Schulz, *Science* **1991**, *254*, 1627–1630.
2 S. W. Cowan, T. Schirmer, G. Rummel, M. Steiert, R. Ghosh, R. A. Pauptit, J. N. Jansonius, J. P. Rosenbusch, *Nature* **1992**, *358*, 727–733.
3 A. Karshikoff, V. Spassov, S. W. Cowan, R. Ladenstein, T. Schirmer, *J Mol Biol* **1994**, *240*, 372–384.
4 W. Im, B. Roux, *J Mol Biol* **2002**, *319*, 1177–1197.
5 W. Im, B. Roux, *J Mol Biol* **2002**, *322*, 851–869.
6 H. Nikaido, In: Escherichia coli *and* Salmonella *Cellular and Molecular Biology*, Neidhardt, F. C. (Ed. in Chief). ASM Press, Washington, DC, **1996**.
7 H. Yigit, G. J. Anderson, J. W. Biddle, C. D. Steward, J. K. Rahseed, L. L. Valera, J. E. McGowan Jr, F. C. Tenover, *Antimicrob Agents Chemother* **2002**, *46*, 3817–3822.
8 A. Oliver, L. M. Weigel, J. K. Rasheed, J. E. McGowan Jr, P. Raney, F. C. Tenover, *Antimicrob Agents Chemother* **2002**, *46*, 3829–3836.
9 L. Armand-Lefèvre, V. Leflon-Guibout, J. Bredin, F. Barguellil, A. Amor, J.-M. Pagès, M. H. Nicolas-Chanoine, *Antimicrob Agents Chemother* **2003**, *47*, 1165–1168.
10 L. A. Pratt, T. J. Silhavy, in *Two-component Signal Transduction*, Hoch, J. A., Silhavy, T. J. (Eds). ASM Press, Washington, DC, **1995**.
11 H. Schindler, J. P. Rosenbusch, *Proc Natl Acad Sci USA* **1978**, *75*, 3751–3755.
12 C. M. Jones, D. M. Taylor, *Thin Solid Films* **1996**, *284*, 748–751.
13 A. Mathes, H. Engelhardt, *J Membr Biol* **1998**, *165*, 11–18.
14 N. Saint, A. Prilipov, A. Hardmeyer, K. L. Lou, T. Schirmer, J. P. Rosenbusch, *Biochem Biophys Res Commun* **1996**, *223*, 118–122.
15 V. C. Simonet, A. Baslé, K. E. Klose, A. H. Delcour, *J Biol Chem* **2003**, *278*, 17539–17545.
16 T. Rudel, A. Schmid, R. Benz, H. A. Kolb, F. Lang, T. F. Meyer, *Cell* **1996**, *85*, 391–402.
17 R. Benz, K. Janko, P. Läuger, *Biochim Biophys Acta* **1979**, *551*, 238–247.
18 J. B. Stock, B. Rauch, S. Roseman, *J Biol Chem* **1977**, *252*, 7850–7861.
19 E. P. Kennedy, In: Escherichia coli *and* Salmonella *Cellular and Molecular Biology*, Neidhardt, F. C. (Ed. in Chief). ASM Press, Washington, DC, **1996**.
20 K. Sen, J. Hellman, H. Nikaido, *J Biol Chem* **1988**, *263*, 1182–1187.
21 H. Samartzidou, A. H. Delcour, *EMBO J* **1998**, *17*, 93–100.
22 E. F. Eppens, N. Saint, P. Van Gelder, R. van Boxtel, J. Tommassen, *FEBS Lett* **1997**, *415*, 317–320.
23 C. M. Soares, J. Björkstén, O. Tapia, *Protein Eng* **1995**, *8*, 5–12.
24 M. Watanabe, J. Rosenbusch, T. Schirmer, M. Karplus, *Biophys J* **1997**, *72*, 2094–2102.
25 D. P. Tieleman, H. J. C. Berendsen, *Biophys J* **1998**, *74*, 2786–2801.
26 P. S. Phale, T. Schirmer, A. Prilipov, K. L. Lou, A. Hardmeyer, J. P. Rosenbusch, *Proc Natl Acad Sci USA* **1997**, *94*, 6741–6745.
27 G. Bainbridge, H. Mobasheri, G. A. Armstrong, E. J. A. Lea, J. H. Lakey, *J Mol Biol* **1998**, *275*, 171–176.
28 S. A. Benson, J. L. L. Occi, B. A. Sampson, *J Mol Biol* **1988**, *203*, 961–970.

29 N. Saint, K. L. Lou, C. Widmer, M. Luckey, T. Schirmer, J. P. Rosenbusch, *J Biol Chem* **1996**, *271*, 20676–20680.

30 R. Misra, S. A. Benson, *J Bacteriol* **1988**, *170*, 522–533.

31 J. H. Lakey, E. J. A. Lea, F. Pattus, *FEBS Lett* **1991**, *278*, 31–34.

32 J. Bredin, N. Saint, M. Malléa, E. Dé, G. Molle, J.-M. Pagès, V. Simonet, *Biochem J* **2002**, *363*, 521–528.

33 P. S. Phale, A. Philippsen, C. Widmer, V. P. Phale, J. P. Rosenbusch, T. Schirmer, *Biochemistry* **2001**, *40*, 6319–6325.

34 H. Mobasheri, E. J. A. Lea, *Eur Biophys J* **2002**, *31*, 389–399.

35 P. Van Gelder, N. Saint, P. Phale, E. F. Eppens, A. Prilipov, R. van Boxtel, J. P. Rosenbusch, J. Tommassen, *J Mol Biol* **1997**, *269*, 468–472.

36 K. Saxena, V. Drosou, E. Maier, R. Benz, B. Ludwig, *Biochemistry* **1999**, *38*, 2206–2212.

37 D. Jeanteur, T. Schirmer, D. Fourel, V. Simonet, G. Rummel, C. Widmer, J. P. Rosenbusch, F. Pattus, J.-M. Pagès, *Proc Natl Acad Sci USA* **1994**, *91*, 10675–10679.

38 M. Malléa, J. Chevalier, C. Bornet, A. Eyraud, A. Davin-Regli, C. Bollet, J.-M. Pagès, *Microbiology* **1998**, *144*, 3003–3009.

39 D. Dé, A. Baslé, M. Jaquinod, N. Saint, M. Malléa, G. Molle, J.-M. Pagès, *Mol Microbiol* **2001**, *41*, 189–198.

40 F. A. Schabert, C. Henn, A. Engel, *Science* **1995**, *268*, 92–94.

41 D. J. Müller, A. Engel, *J Mol Biol* **1999**, *285*, 1347–1351.

42 D. Dahan, V. Vachon, R. Laprade, J. W. Coulton, *Biochim Biophys Acta* **1994**, *1189*, 204–211.

43 M. A. Arbing, D. Dahan, D. Boismenu, O. A. Mamer, J. W. Hanrahan, J. W. Coulton, *J Membr Biol* **2000**, *178*, 185–193.

44 M. A. Arbing, J. W. Hanrahan, J. W. Coulton, *Biochemistry* **2001**, *40*, 14621–14628.

45 A. H. Delcour, *Front Biosci* **2003**, *8*, d1055–1071.

46 Wiese, J. O. Reiners, K. Brandenburg, K. Kawahara, U. Zähringer, U. Seydel, *Biophys J* **1996**, *70*, 321–329.

47 H. de Cock, J. Tommassen, *EMBO J* **1996**, *15*, 5567–5573.

48 W. J. Rocque, R. T. Coughlin, E. J. McGroarty, *J Bacteriol* **1987**, *169*, 4003–4010.

49 D. Fourel, A. Bernadac, J.-M. Pagès, *Eur J Biochem* **1994**, *222*, 625–630.

50 A. Wiese, G. Schröder, K. Brandenburg, A. Hirsch, W. Welte, U. Seydel, *Biochim Biophys Acta* **1994**, *1190*, 231–242.

51 S. O. Hagge, H. de Cock, T. Gutsmann, F. Beckers, U. Seydel, A. Wiese, *J Biol Chem* **2002**, *277*, 34247–34253.

52 L. K. Buehler, S. Kusumoto, H. Zhang, J. P. Rosenbusch, *J Biol Chem* **1991**, *266*, 24446–24450.

53 H. Morgan, J. T. Lonsdale, G. Alder, *Biochim Biophys Acta* **1990**, *1021*, 175–181.

54 M. Brunen, H. Engelhardt, *Eur J Biochem* **1993**, *212*, 129–135.

55 J. C. Todt, W. J. Rocque, E. J. McGroarty, *Biochemistry* **1992**, *31*, 10471–10478.

56 N. Liu, A. H. Delcour, *FEBS Lett* **1998**, *434*, 160–164.

57 D. Xu, B. Shi, E. J. McGroarty, H. T. Tien, *Biochim Biophys Acta* **1986**, *862*, 57–64.

58 J. C. Todt, E. J. McGroarty, *Biochemistry* **1992**, *31*, 10479–10482.

59 J. C. Todt, E. J. McGroarty, *Biochem Biophys Res Commun* **1992**, *189*, 1498–1502.

60 P. Koski, M. Vaara, *J Bacteriol* **1991**, *173*, 3695–3699.

61 A. Peterson, R. W. Hancock, E. J. McGroarty, *J Bacteriol* **1985**, *164*, 1256–1261.

62 C. W. Tabor, H. Tabor, *Microbiol Rev* **1985**, *49*, 81–99.

63 D. Schiller, D. Kruse, H. Kneifel, R. Krämer, A. Burkovski, *J Bacteriol* **2000**, *182*, 6247–6249.

64 S. Miyamoto, K. Kashiwagi, K. Ito, S. Watanabe, K. Igarashi, *Arch Biochem Biophys* **1993**, *300*, 63–68.

65 K. Igarashi, K. Kashiwagi, *Biochem J* **1999**, *344*, 633–642.

66 F. Munro, K. Hercules, J. Morgan, W. Sauerbier, *J Biol Chem* **1972**, *247*, 1272–1280.
67 E. Olson, *Mol Microbiol* **1993**, *8*, 5–14.
68 M. N. Neely, E. R. Olson, *J Bacteriol* **1996**, *178*, 5522–5528.
69 D. L. Osborne, E. R. Seidel, *Am J Physiol* **1990**, *258*, G576–G584.
70 R. Benamouzig, S. Mahé, C. Luengo, J. Rautureau, D. Tomé, *Am J Clin Nutr* **1997**, *65*, 766–770.
71 S. Bardocz, *Eur J Clin Nutr* **1993**, *47*, 683–690.
72 A. L. delaVega, A. H. Delcour, *EMBO J* **1995**, *14*, 6058–6065.
73 R. Iyer, A. H. Delcour, *J Biol Chem* **1997**, *272*, 18595–18601.
74 R. Iyer, Z. Wu, P. M. Woster, A. H. Delcour, *J Mol Biol* **2000**, *297*, 933–945.
75 A. Baslé, A. H. Delcour, *Biophys Biochem Res Commun* **2001**, *285*, 550–554.
76 H. Samartzidou, A. H. Delcour, *FEBS Lett* **1999**, *444*, 65–70.
77 E. M. Nestorovich, C. Danelon, M. Winterhalter, S. M. Bezrukov, *Proc Natl Acad Sci USA* **2002**, *99*, 9789–9794.
78 N. Liu, M. J. Benedik, A. H. Delcour, *Biochim Biophys Acta* **1997**, *1326*, 201–212.
79 S. Vidal, P. Brouant, J. Chevalier, M. Malléa, J. Barbe, J.-M. Pagès, *In vivo* **2002**, *16*, 111–116.
80 A. L. delaVega, A. H. Delcour, *J Bacteriol* **1996**, *178*, 3715–3721.
81 C. Ingham, M. Buechner, J. Adler, *J Bacteriol* **1990**, *172*, 3577–3583.
82 D. Chevalier, M. Malléa, J.-M. Pagès. *Biochem J* **2000**, *348*, 223–227.
83 H. Samartzidou, A. H. Delcour, *J Bacteriol* **1999**, *181*, 791–798.
84 H. Samartzidou, M. Mehrazin, Z. Xu, M. J. Benedik, A. H. Delcour, *J Bacteriol* **2003**, *185*, 13–19.
85 A. H. Delcour, C. Kung, J. Adler, B. Martinac, *FEBS Lett* **1992**, *304*, 216–220.
86 J. A. Wibbenmeyer, D. Provenzano, C. Landry, K. Klose, A. H. Delcour, *Infect Immun* **2002**, *70*, 121–126.
87 N. Liu, H. Samartzidou, K.-W. Lee, J. Briggs, A. H. Delcour, *Protein Eng* **2000**, *13*, 491–500.
88 D. M. Robertson, D. P. Tieleman, *Biochem Cell Biol* **2002**, *80*, 517–523.
89 A. C. Le Dain, C. C Häse, J. Tommassen, B. Martinac, *EMBO J* **1996**, *15*, 3524–3528.
90 G. Macdonald, B. Martinac, *FEMS Microbiol Lett* **1999**, 327–334.
91 D. H. Bartlett, C. Kato, K. Horikoshi, *Res Microbiol* **1995**, *146*, 697–706.
92 M. Buechner, A. H. Delcour, B. Martinac, J. Adler, C. Kung, *Biochim Biophys Acta* **1990**, *1024*, 111–121.
93 P. Blount, S. I. Sukharev, P. C. Moe, M. J. Schroeder, H. R. Guy, C. Kung, *EMBO J* **1996**, *15*, 4798–4805.
94 H. Nikaido, E. Y. Rosenberg, *J Gen Physiol* **1981**, *77*, 121–135.

6
Reconstitution of General Diffusion Pores from Bacterial Outer Membranes

Christophe Danelon and Mathias Winterhalter

6.1
Introduction

In this chapter, we briefly introduce our technique to characterize channel-forming proteins and discuss the results obtained with reconstituted porins from *Escherichia coli*. We conclude this chapter with possible application of porins in material science.

Gram-negative bacteria consist of an asymmetrical bilayer with an inner leaflet composed of phospholipids and an outer leaflet containing lipopolysaccharides (LPS). The LPS headgroups are cross-linked by divalent cations, providing an impermeable network for hydrophilic solutes, and protecting the cell from damaging agents such as bile salts, lipases and proteases. Uptake of nutrients or secretion through this barrier is accomplished by several pore-forming proteins [1]. These proteins can be divided into three different classes: (i) general diffusion pores, (ii) specific channels containing a binding site for a certain solute and (ii) actively transporting, energy-consuming channels. The general diffusion pores allow the unspecific diffusion of hydrophilic molecules and may show slight selectivity towards ions. Examples are OmpF, OmpC and PhoE from *E. coli*. In our group, we have mainly characterized OmpF (outer membrane protein F). This channel is permeable for molecules up to a molecular weight of approximately 600 Da. The functional unit contains three water-filled channels [2]. In each monomer, 16 β-strands span the outer membrane, and form a barrel with short turns at the periplasmic side and large loops at the outside of the cell. Unlike the other loops, the third loop, L3, is not exposed at the cell surface, but folds into the barrel, forming a constriction zone at half the height of the channel. Therefore, this loop determines mainly the permeability properties of the channel such as size exclusion or selectivity [3]. These channels play an important role in the permeation of antibiotics across the outer membrane. Therefore, studying their transport properties is of great interest for a better understanding of the efficiency of antibacterial agents.

Bacterial and Eukaryotic Porins. Edited by Roland Benz

ISBN 3-527-30775-3

6.2
Planar Lipid Bilayer Technique

Conductance recording across a planar lipid bilayer [also called a black lipid bilayer (BLM)] is the technique of choice for a first structural and functional study of membrane channels. Different techniques have been described to form planar lipid bilayers [4]. In the experiments presented here, membranes are prepared according to the technique of Montal and Mueller [5] with some slight modifications [6]. With the help of a discharge spark (10–50 kV at 500 kHz) using a "BD-10 Vacuumtester" (Electrotechnic Products, North Ravensburg, PA) we burn small holes, typically less than 100 μm diameter, into a 25-μm thick Teflon film (Goodfellow, Cambridge, UK). This septa is used to separate two black Delrine half cells. The quality of the hole, here about 60 μm diameter, is visualized under a microscope and shown in Figure 6.1. The cleaned septum is sandwiched between two Delrine chambers, each containing 1–2 ml volume. Prior to membrane formation, the aperture was "pre-painted" with approximately 1 μl of a 2 % solution *n*-hexadecane in *n*-hexane. After drying both chambers are filled with buffer, typically 1 M KCl, 1 mM $CaCl_2$ and 10 mM Tris buffered to pH 7.4. On top of both solutions a lipid multilayer is prepared from 4 mg ml^{-1} solution of 1,2-diphytanoyl-*sn*-glycero-3-phosphatidylcholine in a solvent mixture of *n*-hexane:chloroform (9:1, v/v); the amount of spread lipids corresponds to about 10 times that necessary for a dense surface monolayer. Bilayer membranes are formed after allowing 30 min

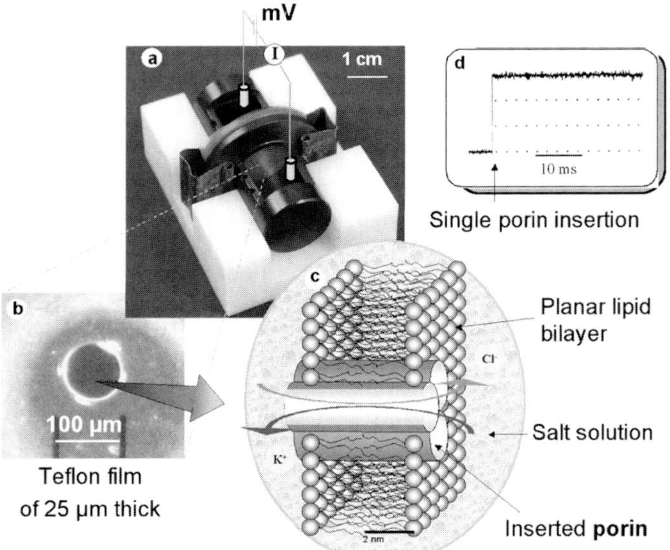

Figure 6.1 Set-up of the planar lipid bilayer. A planar bilayer is formed across an aperture separating two aqueous solutions. A single porin spontaneously inserts, allowing a constant ionic current. (a) Measurement cell, (b) view of the hole under the microscope, (c) scheme of the bilayer with an incorporated porin and (d) typical current recording of a single porin insertion.

for solvent evaporation. A small magnetic agitator is placed in each compartment to quickly homogenize the solution after adding a new compound. Easy access to both sides facilitates direct electrical measurement and the manipulation of the composition of the bathing solutions.

The electrode on the *cis* side of the measuring cell is grounded, whereas the other one (*trans*) is connected to the headstage of an Axopatch 200B amplifier (Axon Instruments, Foster City, CA) in voltage clamp mode. The applied transmembrane voltage refers to the potential on the *cis* side relative to the *trans* side. The output signal is filtered with the low-pass Bessel filter of the amplifier and then monitored with a Lecroy LT342 digital storage oscilloscope. Relevant data are recorded on a digital tape recorder. The data is later transferred to a personal computer via a GPIB card (National Instruments) and LabVIEW 4.01.

Purified, detergent-solubilized channel-forming proteins are added to the chamber and spontaneously insert into the artificial membrane. In the case of OmpF reconstitution, a small volume (about 1 µl) of porins from a 0.1–10 µg/ml buffer solution with 1 % octyl-polyoxyethylene (octyl-POE) detergent (Alexis, Lauchringen, Switzerland) is injected into the *cis*-side compartment. We have optimized the protocol to observe single porin insertion in less than 10 min, but rarely a second event, by choosing a very diluted protein concentration. A single porin molecule in the bilayer could be kept stable for several hours without any significant change of its physical properties. With improved resolution it became obvious that each OmpF trimer has slightly different properties with respect to conductance and voltage gating. In order to elucidate effects due to buffer exchange, this requires us to carry out the experiments on the same protein in order to avoid possible divergences between different individual single protein molecules [7]. All measurements are performed at room temperature.

6.3 Intrinsic Properties of General Diffusion Channels

6.3.1 Single-channel Analysis of OmpF Gating

Ion conductance through OmpF channels has been investigated by several groups. This type of measurements allowed conclusions on structural properties like pore size as well as functional properties as ion selectivity. Conductance steps under applied voltages have been observed in reconstituted solvent-free planar lipid bilayers [8–11] or with the patch-clamp technique [12, 13]. The channel gating is influenced by pH [14–16], hydrostatic pressure [17], polyamines [18–21] and the presence of LPSs in the membrane [22–26], and depends on the bilayer formation technique [27]. OmpF trimers spontaneously insert into planar lipid membranes visualized by a single jump in conductance. Figure 6.2 shows typical recordings of the ion current at various external potentials in 1 M KCl aqueous solution. Working with the solvent-free membranes, OmpF shows an interesting feature: at sufficiently high transmembrane voltages sudden channel closure is observed. As seen in Figure 6.2(A–C), increasing the applied voltage increases the rate of OmpF closure. At +190 mV transmembrane voltage, all three channels are blocked after a few seconds (Figure 6.2C). In contrast, at 110 mV, only one monomer or two monomers are closed in the same time interval (Figure 6.2A). It is interesting to note that the lag-time to close the channels shows a large scatter even for repeated closure at the same molecule. Figure 6.2 reveals that the closures are not complete, leading to a residual conductance. For identical conducting channels one would expect four states of conductance in the amplitude histogram of current recordings: fully open, single-closed, double-closed and triple-closed states. However, statistical analysis of the amplitude of the current transitions on the same OmpF porin does not show well-defined states of conductance, but rather a distribution as seen in Figure 6.3.

A second parameter influencing the gating process is the electrolyte concentration. For example, application of 150 mV transmembrane voltage in 1 M KCl closes all three OmpF channels typically after 20 s (Figure 6.2B). In contrast, at 100 mM KCl (Figure 6.4) the same voltage did not cause a trimer closure within minutes. At 10 mM KCl simultaneous closure of two channels occurred fairly seldom. Furthermore, we modified also the electrolyte composition and, surprisingly, we observed pronounced effects on the time lag for closure, e. g. for an applied voltage of

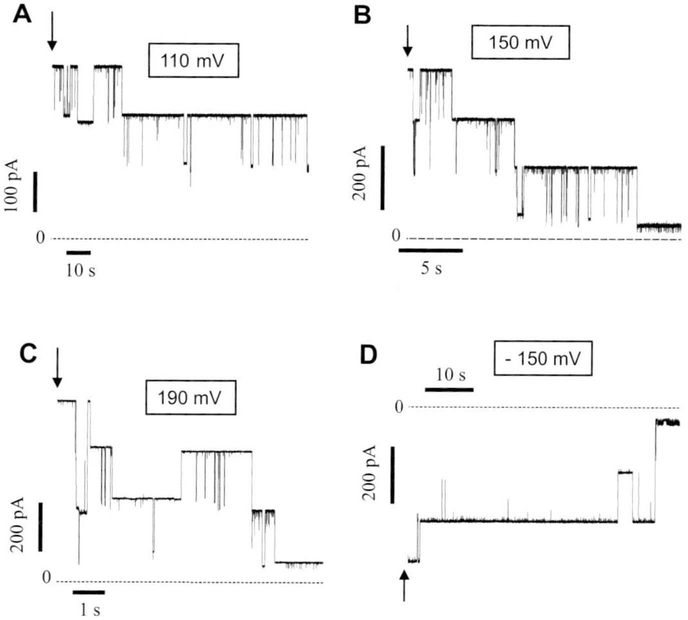

Figure 6.2 Typical recordings of the gating process at various applied voltages. The vertical arrows indicate constant voltage application. The OmpF proteins were added to the *cis* side of the membrane. The voltage refers to the potential on the *trans* side minus the potential on the *cis* side. The aqueous solution contained symmetrical 1 M KCl and was unbuffered. The pH was about 6.

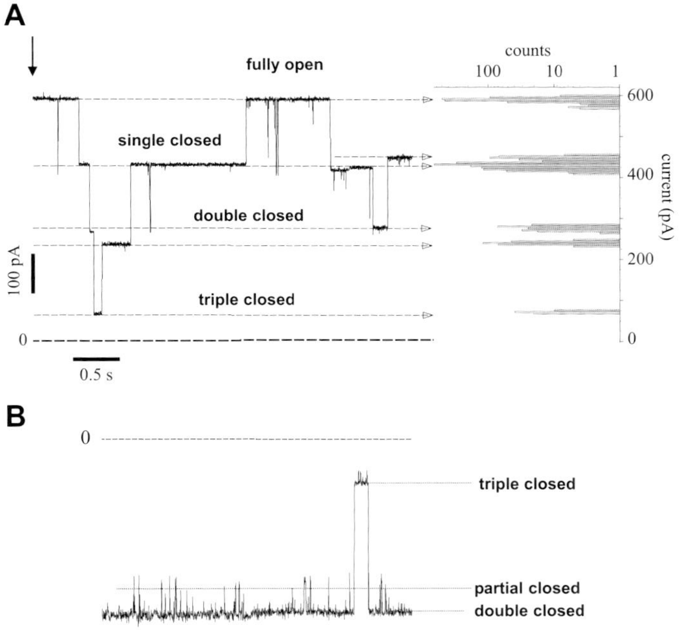

Figure 6.3 Details on the opening–closing transitions. Various conductance steps can be observed demonstrating that a given channel can exhibit different states of conductance. Small current interruptions occur preferentially at negative applied voltages. (A) The applied voltage was 190 mV. (B) The applied voltage was –190 mV. The voltage refers to the potential on the *trans* side minus the potential on the *cis* side. The aqueous solution contained symmetrical 1 M KCl and was unbuffered with pH of about 6. Time resolution was 0.5 ms.

190 mV it was about 8 s for CsCl and 25 s for LiCl. This suggests a faster kinetics of closure with increasing size of the cations.

6.3.2
Molecular Origin of Voltage Gating

The availability of the high-resolution X-ray structure for OmpF or PhoE porins [2] simplified the design of site-directed mutants. To understand the function of the channel, mutants were designed to enlarge pores [15, 28], to modify hydrogen bond formation [29, 30], to constrain parts [29, 31, 32] or to change the charge constellation at the channel interior [31, 33–36]. Replacement of the positively charged arginine residues in the constriction zone of OmpF [15] and OmpC [28] resulted in an increased voltage sensitivity. However, substitution of positively charged residues in PhoE [33] and chemical modification of arginine and lysine residues in the anion-selective porin of *Acidovorax delafieldii* decreased its voltage sensitivity [37]. Replacement of negatively charged residues in PhoE had a reverse effect, i. e. increased voltage sensitivity, whereas mutations of negatively charged residues

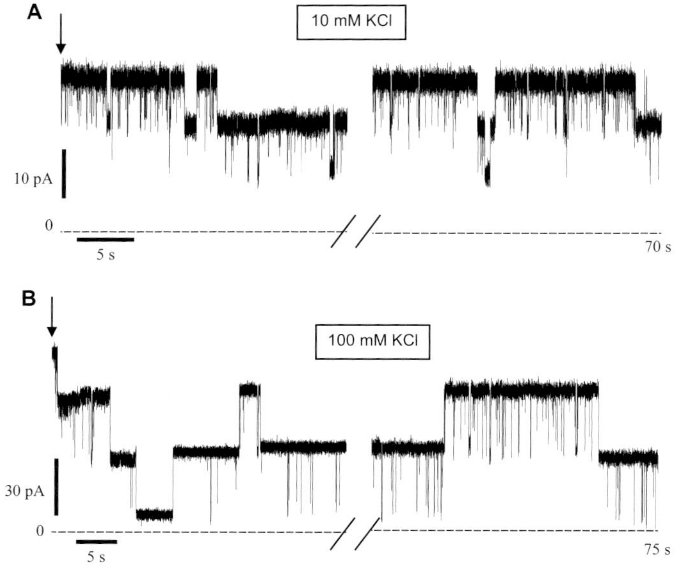

Figure 6.4 Typical voltage gating of a single OmpF trimer in 10 (A) and 100 (B) mM KCl solutions, pH 6. The applied voltage was 150 mV. The vertical arrows indicate constant voltage application. Time resolution was 0.5 ms.

in OmpF resulted in an increased threshold potential. Indeed, implication of charged residues at the constriction zone as voltage sensors has been demonstrated suggesting subtle rearrangements of charges in the lumen of the channel [29, 36]. Mutagenesis of porin from *Haemophilus influenzae* type b identified charged residues on extracellular loops that influence voltage-controlled activity [38, 39]. However, no drastic modulatory effects on the opening/closing characteristics have been observed and the molecular mechanism of voltage gating is still speculative.

Most of the current hypotheses on the molecular origin of porin gating are consistent with an electromechanical coupling between the electric field in the pore generated by the externally applied voltage and some charged residues. Recent resolution of the crystal structure of a voltage-dependent K^+ channel [40, 41] permitted the identification of gating charges which sense the membrane voltage. The hourglass shape of the OmpF channel [2] with its restriction in the middle causes the transmembrane voltage to drop principally at the narrowest part of the pore. Therefore, charged amino acids at the eyelet region became prime candidates for site-directed mutagenesis (for review, see [42]). Substitution of specific residues modified the critical voltage, but the gating activity was not abolished. The responsibility of large-scale motion of the L3 loop for gating [43] has been ruled out in a series of experiments, where L3 was tethered to the barrel wall through disulfide bridges [29, 31, 32]. Recently, atomic force microscopy was able to visualize the conformational change from the open to the closed form [44]. In this study, the large extracellular loops were suspected to move into the channel lumen.

The Debye length λ_D, which modulates the potential generated by charges in solution, is given for monovalent ions by:

$$\lambda_D = (\epsilon\epsilon_0 kT/2ez^2C)^{1/2} \quad (1)$$

where ϵ is the relative dielectric constant of the aqueous phase, ϵ_0 is the dielectric constant of the vacuum, kT is the Boltzmann factor, z is the valency, e is the elementary charge and C is the ion concentration. Assuming that structural changes that gate the channel are triggered by the force acting on fixed protein charges in the electric field, one would expect that (i) the rate of channel closure is higher at low ionic strength where the charges screening effect is low and (ii) is independent on the nature of the monovalent cations. However, we have shown that OmpF gating increases with salt concentration and with the size of the alkali metal ion. This finding suggests that we have to consider the specific molecular properties of the electrolyte ions, not only the charge, but also hydration energy, hydrated radius and polarization forces.

Cation permeation may proceed by specific binding to affinity sites within the channel [45]. On the basis of the experimental results presented here, an alternative mechanism for voltage gating involving the electrolyte may be the stacking of ions at the constriction zone of the porin (I. Yamato, personal communication). Free energy change calculations for alkali metal ions binding to residue Asp-113 [46] showed that larger cations bind with higher affinity. Therefore, the influence of the cation on the gating kinetics may be correlated to specific affinity to residues in the channel lumen. The dependence of the rate of channel inactivation on the alkali metal ion can be attributed to a diffusion-dependent association step to penetrate inside the pore or to the dehydration step of the binding process [46]. This hypothesis is in agreement with the proposition that gating is an intrinsic property of β-barrel channels [47].

It would be interesting to apply conformational diffusion models developed for ion-channel gating [48–52] to study the complex gating dynamics exhibited by single OmpF channels. Diffusion models extend the standard Markovian nature of the gating kinetics proposing that ion-channel proteins have a very large number of closed substates all of similar energy, which is in agreement with our observation that a given OmpF channel closes with different conductance steps. Experimental validation of this open/close mechanism is difficult for two reasons: cooperativity effects between subunits within one OmpF trimer may complicate the model and the system is not in a stationary state.

6.3.3 Effect of Membrane Composition and OmpF–LPS Interactions

In addition to non-specific and specific electrostatic effects, the lipid environment also has a very important effect. The voltage-dependent activity of OmpF porin from *E. coli* has not been observed previously using solvent-containing membranes [53, 54]. One reason may originate from purification and reconstitution protocols.

The Mueller–Rudin technique forms membranes out of a decane droplet and gives rise to very soft membranes. As the hydrocarbon may help to adopt the hydrophobic part, this may result in less-stressed proteins in comparison to channels in Montal–Mueller-type bilayers. In addition, if one maintains a constant voltage, as a result of bilayer thickness differences the transmembrane electric field is smaller in Mueller–Rudin-type membranes. To clarify the situation, we have reconstituted single OmpF porins from the same preparation of purified proteins also in Mueller–Rudin films. The voltage gating occurred surprisingly in both preparation in agreement with an earlier investigation [27]. One may speculate if gating is due to purification of the sample.

The outer membrane of Gram-negative bacteria consists of a phospholipid inner leaflet and a LPS outer leaflet. In reconstitution experiments, the incorporation rate and the gating behavior have been shown to be influenced by the presence of LPS molecules in planar bilayers [22–26]. LPS contains negative charges on saccharide moieties. Asymmetrical repartition of LPS charged headgroups in bilayers may cause a membrane surface potential U according to the Gouy equation [55]. This intrinsic membrane potential is likely to contribute to voltage-controlled properties of channels, particularly to the gating asymmetry with the polarity of the external applied voltage [25]. In patch-clamp studies performed on outer membrane fractions, an asymmetrical voltage dependence of the OmpF has been observed [13]. The implication of LPS as a gating modulator can be proposed. In the present study, we provide evidence of gating asymmetry for purified OmpF from *E. coli* reconstituted into pure phospholipid bilayers (Figure 6.3). The method used to purify OmpF was as described by Garavito and Rosenbusch [56]. Porins were extracted and solubilized with octyl-POE. Acrylamide gel (10 %) analysis revealed that varying amounts of LPS (at least their lipid moieties) remain tightly associated with OmpF protein. The presence of negative charges at the extracellular surface of proteins may locally influence the electrical properties of the channel.

The relevance of voltage gating in the case of OmpF is still an open discussion. Moreover, the potentials needed to close the pores are larger than the naturally occurring Donnan potential across the outer membrane [57]. However, porin Omp34 of *A. delafieldii* was shown to close at potentials as low as 10 mV [58]. Variations of the transmembrane voltage have a minor influence on the uptake of substrates *in vivo* [59]. Therefore, the voltage dependence of the porins activity has been proposed to be an *in vitro* reconstitution effect; however, the underlying mechanism remains an exciting question.

6.3.4
Open-channel Conductance

The open-channel conductance of OmpF porin was studied in symmetrical solutions of the series LiCl–NaCl–KCl–RbCl–CsCl. The relationship between the single OmpF conductance and the specific conductance of the NaCl salt solution is given in Table 6.1. The channel conductance is not a linear function of the bulk conductivity, exhibiting particular high conductance at low concentration. Interestingly, at

Table 6.1 Open-channel conductance and voltage-induced asymmetry.

Salt	Concentration (M)	G[a] (pS)	%[b]
LiCl	0.01	85	35
	0.1	260	22
	1	1590	<1
NaCl	0.01	175	27
	0.1	535	22
	1	2820	1
KCl	0.01	240	39
	0.1	810	27
	1	3145	2
RbCl	0.01	220	26
	0.1	850	15
	1	3650	<1
CsCl	0.01	225	36
	0.1	855	16
	1	3510	2

[a] *G* is the conductance of the fully open trimer. The applied voltage was –150 mV referring to the potential of the *cis* side minus the potential of the *trans* side. It should be noted that the conductance values differ from other preparations [54].
[b] $\% = [(G^- - G^+)/G^-] \times 100$ represents the conductance asymmetry with the sign of the applied voltage where G^- and G^+ stand for the conductance at –150 and +150 mV, respectively. The data represent the mean of at least four independent experiments. The standard deviation was always less than 15 % of the mean value.

high concentration no apparent saturation effect was visible. Table 6.1 shows that the channel conductance values for the series of alkali metal ions were qualitatively parallel to the solution conductivity values, LiCl < NaCl < KCl < RbCl ~ CsCl. However, the conductance differences were quantitatively more pronounced at low concentrations and were larger than expected from the consideration of cation mobility. The channel conductance was 2.8-fold higher from 0.01 M LiCl to 0.01 M KCl, whereas the bulk conductivity of LiCl differs from KCl only by a factor of 1.3. The channel conductance was not a linear function of the bulk conductivity, demonstrating that the OmpF channel plays a specific role in the ion transport (compared to free diffusion of ions inside an inert water-filled pore). Under the assumption of free diffusion, the channel conductance can be expressed as:

$$G_{\text{diff.}} = \sigma[(f_c \pi R_{\text{eff}}^2/L)^{-1} + (2R_{\text{eff}})^{-1}]^{-1} \quad (2)$$

where R_{eff} is the effective radius of an OmpF pore directly determined from the crystal structure, L is the channel length, σ is the bulk conductivity, and f_C is the Faxen correction factor which takes into account frictional interactions between the permeant solvated ions and the channel walls [60]; the first term is the resistance of the cylindrical pore and the second term is the sum of the access resistance from each side of the channel [61]. For $R_{eff} = 0.7$ nm [62], $L = 4$ nm and $f_C = 0.5$ (corresponding to a ratio of solvated ions to channel effective radius of 0.25), we obtain $G_{diff.} = 644$ pS in 100 mM KCl as an estimation of the OmpF trimer conductance for passive ionic transport which is close to the experimentally observed one. However, in 0.01 M NaCl solution, the calculated value is $G_{diff.} = 59$ pS and about three times less than the experimentally observed one indicating that the OmpF channel enhances the conducting properties, whereas at 1 M NaCl $G_{exp} \sim 0.6\ G_{diff.}$, suggesting that free diffusion would be more efficient at high salt concentration, which corresponds to the observation by MD simulation [63, 64]. The values were compared with the theoretical predictions and showed a nice agreement but not a definite proof.

6.3.5
Voltage Effect and Channel Orientation

We found that the open-channel conductance was always higher when the potential at the *cis* side (the side of protein addition) was lower than the potential at the *trans* side. Table 6.1 shows that the ion transport asymmetry was more pronounced at high voltage and low ionic strength. Between +190 and −190 mV the conductance difference dropped from 43 % at 0.01 M to 4 % at 1 M KCl solutions. This voltage-induced asymmetry was taken as a quick probe for the directional insertion of OmpF into the membrane. Importantly, no correlation between the sign of the applied voltage and the orientation of the insertion was observed.

The computational observations of conductance asymmetry with the direction of ion transport through OmpF channels occured in absence of LPS molecules [64]. This suggests that the origin of the open-channel conductance asymmetry is inside the channel, the LPS having only minor influence. The unidirectional orientation of the OmpF porin probably originates from its structural asymmetry. Similar directional insertion has been reported for the specific maltoporin channels [65–68] suggesting an identical incorporation process. Recent computational studies [64] revealed that the OmpF conductance was higher when the potential at the extracellular side was lower than the potential at the periplasmic side. We conclude that the OmpF channel is oriented in our experimental procedure with its extracellular surface exposed to the *cis* side of the membrane, the side of protein addition, in analogy to the orientation found for incorporated maltoporin channels [68].

6.3.6
Ion Selectivity

An important physiological parameter is the selectivity towards ions. For selectivity measurements, salt gradients were generated by adding small amounts of a concentrated solution to one side of the bilayer, while the other side was maintained at a constant lower concentration. The aqueous phase was stirred for a few seconds and measurements were carried out after membrane potential stabilization at zero-current conditions (5–10 min). The whole titration was performed on the same OmpF porin to avoid variations due to different proteins. The ion selectivity was characterized by the ratio P_c/P_a (i. e. the ratio of the permeability for cations and the permeability for anions) calculated according to the Goldman–Hodgkin–Katz equation [69]. Zero-current membrane potentials, V_m, refer to the potential of the diluted side minus the potential of the concentrated side. The vectorial insertion of single OmpF porin into the bilayer permitted the control of the direction of the concentration gradient generated across the channels (from *cis* to *trans* side or the opposite). Two series of concentration gradients, first at low, then at high salt concentrations, were performed on individual OmpF porins. The derived selectivity coefficients showed no significant difference with the direction of the concentration gradient. The mean values of the permeability ratios of the alkali metal ions over the chloride ion are reported in Table 6.2. The results corroborate the cation selectivity of the OmpF porin and show that channel selectivity decreases along the sequence of the electrolytes LiCl–NaCl–KCl–RbCl–CsCl. At low concentration, the OmpF porin was highly cation selective with $P_{Li^+}/P_{Cl^-} = 37.0$ versus only 4.6 at high salt concentration.

For conductance and selectivity measurements, the electrolyte solutions were used unbuffered and had a pH of approximately 6. This avoids buffer effects on the selectivity, but requires special caution as ion-conduction properties of the OmpF channel are pH dependent [53, 54]. Using unbuffered solutions leads to significant variations of pH values during the experiment. We observed no clear difference in the selectivity coefficient with the direction of the concentration gradient, i. e. the reversal potential $V_m(cis \rightarrow trans) \approx V_m(trans \rightarrow cis)$, within the standard deviation of our measurements (around 20 %). However, experiments performed at fixed pH values (solution buffered with HEPES or MES) revealed clear asymmetry of the reversal potential when investigating the salt concentration gradient [70].

Table 6.2 Zero-current membrane potentials V_m and derived permeability ratios of the alkaline metal ions and chloride anion at low and high salt concentrations.

Salt	*Pauling cation radius (Å)*	*Low concentrations*		*High concentrations*	
		V_m (mV)[a]	P_c/P_a[c]	V_m (mV)[b]	P_c/P_a
LiCl	0.68	55 ± 6	37.0 ± 5.0	30 ± 3	4.5 ± 1.0
NaCl	0.97	48 ± 2	19.0 ± 2.5	30 ± 3	4.5 ± 1.0
KCl	1.33	46 ± 3	15.5 ± 2.5	25 ± 4	3.5 ± 0.5
RbCl	1.47	42 ± 3	10.5 ± 2.0	22 ± 3	3.0 ± 0.5
CsCl	1.67	40 ± 4	8.5 ± 1.5	19 ± 2	3.0 ± 0.5

[a] Here, the V_m values were measured at the 0.01 M/0.1 M condition.
[b] Here, the V_m values were measured at the 0.1 M/1 M condition.
[c] The selectivity values P_c/P_a were obtained by fitting the experimental V_ms determined for five increasing concentration gradients with the theoretical Goldman–Hodgkin–Katz equation. The zero-current membrane potentials V_m refer to the less-concentrated side potential minus the more-concentrated side potential. Each value is the mean of at least four independent experiments.

Based on pure electrostatic considerations, the decrease of the ionic selectivity coefficient, P_c/P_a, at high concentration (Table 6.2) may be qualitatively explained by the classical Debye screening effect. Note that the charge selectivity and the zero-current potential (V_m) determined at high NaCl concentrations (Table 6.2) gave similar results as previous data [15, 71] obtained under identical experimental conditions, but performed on multi-channels inserted into the bilayer. However, the results slightly differ from those previously obtained and one reason for the discrepancy may originate from using solvent-containing membranes, which results in less-stressed proteins [54, 72].

The physiological relevance of the weak ionic selectivity exhibited by most of the bacterial porins is unclear. However, the distinct properties of cation versus anion conduction of the *Vibrio cholerae* porins OmpU and OmpT have been proposed to play an integrated role with environmental regulation of their expression [73]. Strains expressing the cation-selective OmpU porin show higher bile salts (negatively charged) resistance compared to strains presenting solely the non-selective OmpT.

6.3.7
The Permeating Cations Interact with Specific Elements along the Ionic Pathway

At low concentration the OmpF channel has been found to be strongly cation selective. According to the Gouy–Chapman theory, the large conductance value may reflect an increase of the local cation concentration inside the channel due to negative surface charges distributed on the pore walls. However, the Gouy–Chapmann theory does not include ion specificity over monovalent cations. This suggests that we have to consider specific physicochemical properties of the cations, not only the charge but also hydration energy, hydrated radius and so on. Furthermore, Table 6.1 showed that the conductance changes along the alkali metal ion series are higher than expected looking at the bulk solution conductivity differences. The mobility of ions inside the channel is not the only parameter controlling the channel conductance, confirming the contribution of other cation-specific factors. Together with the above electrostatic considerations, we argue that the permeating cations specifically interact with lateral charges pointing toward the channel lumen.

Recent MD simulation with multiple free ions inside the pore showed that ions can pass freely through the constriction zone, probably because most charges are screened by counter-ions [63]. Moreover, counter-ions placed near charged residues arranged along the pore reduce the effective radius of the channel by steric hindrance ($R_{eff} < 0.7$ nm), which is in agreement with the lower experimental conductance at high concentrations compared with a passive channel of $R_{eff} = 0.7$ nm.

6.3.8
Single-channel Recordings versus Free Energy Calculation

It is noteworthy that the main ion transport properties, i. e. the conductance–concentration relationship and the magnitude of ionic screening, could be reproduced by Brownian dynamics and Poisson–Nernst–Planck electrodiffusion theory [64]. Their results were in excellent agreement with our experimental results. For a better understanding of the molecular mechanism involved during the cation transport through OmpF channels, we combined the conductance in the low-concentration regime and the free energy changes for the various alkali metal ions. The ions passing through the OmpF channels may encounter several potential barriers and we found by MD simulation that Asp-113 is the strongest binding site for a permeant cation at low salt concentration [45]. Free energy calculations showed that the affinity for Cs^+, Rb^+ or K^+ to Asp-113 is higher than for Na^+ and Li^+ [46], which is consistent with physicochemical data of the hydration energy for the alkali metal ions. It is seen that cation mobility alone cannot explain the conductance differences for the various salt solutions (Table 6.1). Combining the experimental data, stronger binding of a larger cation correlated with increasing channel conductance. This result supports the theory that a deep energy well increases the translocation rate by increasing the local cation concentration inside the channel: Cs^+ is bound with higher affinity, increasing the Cs^+ local concentration at the constriction zone, which enhances the conductance specifically more than for the case of Li^+. This participation of bound cations in the conductance may be enabled by the "push-out" mechanism as was observed in MD simulation [45].

Our computational study also suggests an explanation for the cation selectivity reduction along the series LiCl–NaCl–KCl–RbCl–CsCl. Stronger bound cations, on the one hand, exert the steric hindrance effect working as inert chemical species for permeation and, on the other hand, give the higher screening probability of the negative charge at Asp-113 modulating the energy profile inside the channel by decreasing the electrostatic repulsion for permeating anions. As a consequence the cation selectivity is lower for tightly bound alkali metal ions. Similar results were reported for the reversible binding of halide anions to the roflamycoin channel [74].

In the future, combinations of experiments and MD simulation can be applied to gain insight in the voltage-induced charge rearrangements along the ion pathway and also in the molecular mechanism of channel gating.

6.4
OmpF as a Specific Channel: Antibiotic Translocation

As shown above, the ion current reflects the interaction of the electrolyte with the channel inside. The ion current can also be influenced by the presence of uncharged molecules protruding into the channel. If the channel has a similar range of size as the diffusing particle, the ion flow will be hindered and cause fluctuations in the ion current, reflecting the molecular interactions with the channel wall. This observation was used about 10 years ago to reveal substrate translocation across maltoporin [75]. Since than, noise analysis has been widely used and the technique has been significantly improved. Just to name a few, e. g. channels have been used as a molecular Coulter counter to see the passage of single-stranded DNA molecules or to see the permeation of ATP on a single-molecule level. Recently, single-molecule chemistry initiated by light has been monitored by measuring the ionic current flowing through the pore [76–80].

Numerous molecules such as polyamines, colicin A and N, penicillins, and cephalosporins use OmpF to penetrate the cell. The efficiency of their diffusion rate depends on the interactions with specific residues of OmpF [81, 82]. These interactions induce fluctuations of porin-mediated ion fluxes in patch-clamp or BLM experiments. Below, we develop our recent work on antibiotic interaction with OmpF channels.

It is well known that membrane permeability barriers are among the main reasons for bacteria antibiotic resistance [83]. Ampicillin and other β-lactam antibiotics act on the peptidoglycan layer that is located between the outer and inner bacterial membranes, and protects bacteria cells mechanically. These antibiotics are structural analogues of the terminal D-Ala–D-Ala unit that participates in peptidoglycan synthesis. They inhibit the peptidoglycan biosynthesis by binding to transpeptidase and eventually lead to bacteria death. The general diffusion porin OmpF is the main pathway for β-lactam antibiotics to cross the bacterial outer membrane into the periplasmic space [83].

The questions we ask here are:

- Is the interaction between effectively penetrating drugs and the outer membrane channel strong enough to render the passage events resolvable?
- Is there correlation between the antibiotic efficacy and the strength of this interaction?

These two structures illustrate the common chemical structure of β-lactam antibiotics:

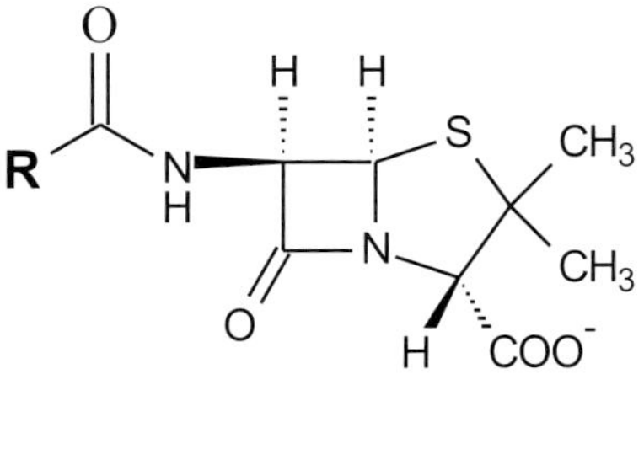

Penicillins

Structure 6.1

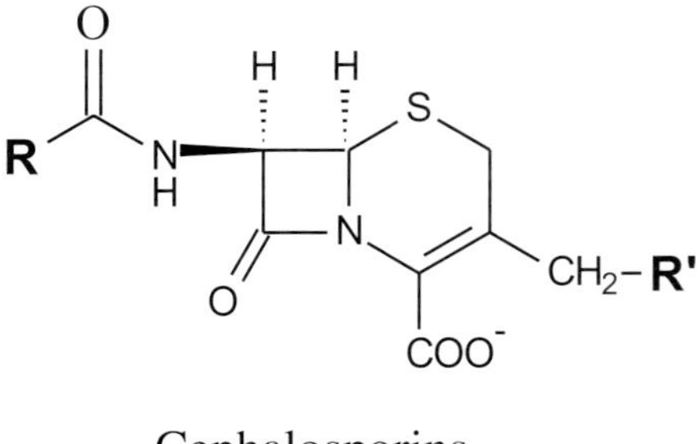

Cephalosporins

Structure 6.2

where R and R′ designate variable chains of different size and charge.

From a single OmpF protein reconstituted into a planar lipid membrane, we demonstrated that addition of β-lactam antibiotics to the membrane-bathing solution introduces transient interruptions in the small ion current through the channel [76]. Time-resolved measurements reveal that one antibiotic molecule blocks one of the monomers in the OmpF trimer for characteristic times in the range of tens to hundreds of microseconds. Figure 6.5 shows results of such measurements for a penicillin drug, ampicillin.

Noise analysis enables us to perform measurements on a wide range of parameters. Depending on voltage and pH, the characteristic residence time varied in the range of microseconds and hundreds of microseconds. For cephradine the residence time is 8.8 μs, for ampicillin about 110 μs. These times should be compared with several nanoseconds expected in the case of purely diffusional relaxation for molecules of this size.

Thus, we demonstrate that in analogy to substrate-specific channels that evolved to bind certain metabolite molecules, antibiotics have "evolved" to be channel-spe-

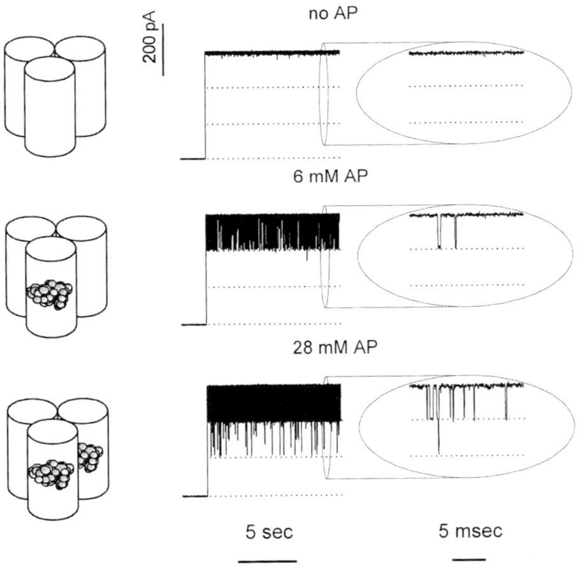

Figure 6.5 Penetrating ampicillin molecules modulate ion current through a single OmpF channel. Membrane bathing solution was 1 M KCl, pH 6.0 and the applied voltage was –100 mV. (Top) In the absence of antibiotics, the ion movement is mainly determined by the geometry and the surface properties of the channel pore. The ion current is stable; no interruptions are seen even at the high resolution recording shown on the right. (Middle) In the presence of ampicillin in small concentrations, one of the three OmpF pores gets spontaneously blocked by a translocating drug molecule. At the time resolution of 0.15 ms (left) blockage events look like downward spikes. At the higher resolution of 0.015 ms (right), they are seen as well-defined steps to two-thirds of the open channel current and back. Time between blockage events and their width report on kinetic parameters of antibiotic–pore interactions. (Bottom) At higher ampicillin concentrations, channel blockades are more frequent. Sometimes they statistically overlap in time, leading to transient reduction of the ion current to one-third of its initial value (from reference [76], with permission).

cific. To understand which structural features of β-lactam antibiotics determine their permeation properties through OmpF porin, we have simulated the ampicillin transport [76]. The charge distribution of an efficient antibiotic complements the charge distribution at the narrowest part of the bacterial porin. Interaction of these charges creates a zone of attraction inside the channel and compensates the penetrating molecule's entropy loss and dehydration energy. The phenyl ampicillin group (at the R end of the molecule) also gives favorable energetic contribution during translocation by interacting with hydrophobic environment. This facilitates antibiotic translocation through the narrowest part of the channel and accounts for higher antibiotic permeability rates. This analysis provides new tools to interpret results obtained with other antibiotics. We expect that these kinds of studies will have applications for directed drug design in the near future.

6.5 Application: Nanoreactor

One of the important research goals in current materials science is to devise new types of responsive materials on a nanometer scale. Porins may be very attractive as their natural function is to regulate the permeability of small water-soluble molecules of molecular weights up to 600. The basic requirements for applications of membrane proteins in materials science are their stability and the possibility to produce them in large quantities. OmpF is an attractive candidate and, in addition, genetic engineering allows modification of its properties. We have shown that encapsulation of enzymes into liposomes or polymer nanocapsules protects them against proteases and denaturation [84]. Encapsulation of enzymes was performed by adding the protein into the buffer prior to liposome formation. We tested several techniques to achieve high encapsulation yields [84–86]. We found that the most efficient way is to mix preformed liposomes with the enzyme and perform several rounds of freeze–thaw cycles. During each cycle the liposome wall ruptures and a few enzyme molecules diffuse into the inside. It turned out that this approach denatured the enzyme less [86]. It could be that during liposome formation lipid monomers are present which might act as a detergent to the enzyme. Higher encapsulation yields are achieved if the enzyme has an affinity to the liposome surface, e. g. electrostatic attraction or His-tag. Non-encapsulated enzyme molecules were removed chromatographically either by gel filtration on Sephadex G-150 (Pharmacia) or by inverse affinity. In this latter case, the solution containing liposomes is loaded on an affinity column – free enzyme molecules bind to the column ligand while the encapsulated enzyme passes through the column without any reaction and is recovered. A different possibility is to enzymatically digest the accessible enzyme by adding a sufficiently high concentration of pronase.

Functional reconstitution of porins in the capsule shell allows the control and selection of the substrate permeation, and thus the control of the enzyme reaction kinetics [84]. Depending on the number of porins, the diffusion through the channel can be the rate-limiting step for the enzyme activity. We suggest that this technique can be useful in the area of biosensors, providing enzymatic stability while keeping the functionality or even enhancing the sensitivity by substrate pre-selection. Another application of this kind of stabilization is in the field of single-enzyme activity recording or a new type of delivery device for applications in pharmacology and diagnostics. For the applications of this kind, it is especially important that such systems provide a constant release of substances over an extended period of time.

Acknowledgments

We would like to thank Didier Fournier and Marcus Lindemann for valuable discussion. Financial support through the Research Training Network "Nanocapsules les with functionalized surfaces and walls CT 2000 00159" is gratefully acknowledged.

References

1 H. Nikaido, M. Vaara, *Microbiol Rev* **1985**, 49,1–32.
2 S. W. Cowan, T. Schirmer, G. Rummel, M. Steiert, R. Ghosh, P. A. Pauptit, J. N. Jansonius, J. P. Rosenbusch, *Nature* **1992**, *358*, 727–733.
3 D. Jeanteur, T. Schirmer, D. Fourel, V. Simonet, G. Rummel, C. Widmer, J. P. Rosenbusch, F. Pattus, J.-M. Pagès, *Proc Natl Acad Sci USA* **1994**, *91*, 10675–10679.
4 W. Hanke, W.-R. Schlue, *Planar Lipid Bilayers: Methods and Applications.* Academic Press, San Diego, CA, **1993**.
5 M. Montal, P. Mueller, *Proc Natl Acad Sci USA* **1972**, *69*, 3561–3566.
6 M. Winterhalter, *Colloids and Surfaces* **1999**, *A149*, 547–551.
7 L. Kullmann, M. Winterhalter, S. M. Bezrukov, *SPIE Fluctuation and Noise* **2003**, *5110*, 50–56.
8 H. Schindler, J. P. Rosenbusch, *Proc Natl Acad Sci USA* **1978**, *75*, 3751–3755.
9 H. Schindler, J. P. Rosenbusch, *Proc Natl Acad Sci USA* **1981**, *78*, 2302–2306.
10 M. Buechner, A. H. Delcour, B. Martinac, J. Adler, C. Kung, *Biochim Biophys Acta* **1990**, *1024*, 111–121.
11 B. Dargent, W. Hofmann, F. Pattus, J. P. Rosenbusch, *EMBO J* **1986**, *5*, 773–778.
12 A. Delcour, H., B. Martinac, J. Adler, C. Kung, *J Membr Biol* **1989**, *112*, 267–275.
13 H. Samartzidou, A. H. Delcour, *EMBO J* **1998**, *17*, 93–100.
14 J. C. Todt, W. J. Rocque, E. J. McGroarty, *Biochemistry* **1992**, 31,10471–10478.
15 N. Saint, K.-L. Lou, C. Widmer, M. Luckey, T. Schirmer, J. P. Rosenbusch, *J Biol Chem* **1996**, *271*, 20676–20680.
16 N. Liu, A. H. Delcour, *FEBS Lett* **1998**, *434*, 160–164.
17 A. C. Le Dain, C. C. Häse, J. Tommassen, B. Martinac, *EMBO J* **1996**, *15*, 3524–3528.
18 R. Iyer, A. H. Delcour. *J Biol Chem* **1997**, *272*, 18595–18601.
19 H. Samartzidou, A. H. Delcour, *FEBS Lett* **1999**, 444, 65–70.
20 R. Iyer, Z. Wu, P. M. Woster, A. H. Delcour, *J Mol Biol* **2000**, *297*, 933–945.

21 A. Baslé, A. H. Delcour, *Biochem Biophys Res Commun* **2001**, *285*, 550–554.
22 U. Seydel, G. Schröder, K. Brandenburg, *J Membr Biol* **1998**, *109*, 95–103.
23 L. K. Buehler, S. Kusumoto, H. Zhang, J. P. Rosenbusch, *J Biol Chem* **1991**, *266*, 2446–24450.
24 A. Wiese, G. Schröder, K. Brandenburg, A. Hirsch, W. Welte, U. Seydel, *Biochim Biophys Acta* **1994**, *1190*, 231–242.
25 Wiese, J. O. Reiners, K. Brandenburg, K. Kawahara, U. Zähringer, U. Seydel, *Biophys J* **1996**, *70*, 321–329.
26 S. O. Hagge, H. de Cock, T. Gutsmann, F. Beckers, U. Seydel, A. Wiese, *J Biol Chem* **2002**, *277*, 34247–34253.
27 J. H. Lakey, F. Pattus, *Eur J Biochem* **1989**, *186*, 303–308.
28 J. H. Lakey, E. J. A. Lea, F. Pattus, *FEBS Lett* **1991**, *278*, 31–34.
29 P. S. Phale, T. Schirmer, A. Prilipov, K.-L. Lou, A. Hardmeyer, J. P. Rosenbusch, *Proc Natl Acad Sci USA* **1997**, *94*, 6741–6745.
30 N. Liu, A. H. Delcour, *Protein Eng* **1998**, *11*, 797–802.
31 E. F. Eppens, N. Saint, P. Van Gelder, R. van Boxtel, J. Tommassen, *FEBS Lett* **1997**, *415*, 317–320.
32 G. Bainbridge, H. Mobasheri, G. A. Armstrong, E. J. Lea, J. H. Lakey, *J Mol Biol* **1998**, *275*, 171–176.
33 P. Van Gelder, N. Saint, P. Phale, E. F. Eppens, A. Prilipov, R. van Boxtel, J. P. Rosenbusch, J. Tommassen, *J Mol Biol* **1997**, *269*, 468–472.
34 N. Liu, M. J. Benedik, A. H. Delcour, *Biochim Biophys Acta* **1997**, *1326*, 201–212.
35 N. Liu, H. Samartzidou, K. W. Lee, J. M. Briggs, A. H. Delcour, *Protein Eng* **2000**, *13*, 491–500.
36 H. Mobasheri, E. J. A. Lea, *Eur Biophys J* **2002**, *31*, 389–399.
37 M. Brunen, H. Engelhardt, *FEMS Microbiol Lett.* **1995**, 126127–126132.
38 M. A. Arbing, D. Dahan, D. Boismenu, O. A. Mamer, J. W. Hanrahan, J. W. Coulton, *J Membr Biol* **2000**, *178*, 185–193.
39 M. A. Arbing, J. W. Hanrahan, J. W. Coulton, *Biochemistry* **2001**, *40*, 14621–14628.
40 Y. Jiang, V. Ruta, J. Chen, A. Lee, R. MacKinnon, *Nature* **2003**, *423*, 33–41.
41 Y. Jiang, V. Ruta, J. Chen, A. Lee, R. MacKinnon, *Nature* **2003**, *423*, 42–48.
42 K. M. Robertson, D. P. Tieleman, *Biochem Cell Biol* **2002**, *80*, 517–523.
43 M. Watanabe, J. P. Rosenbusch, T. Schirmer, M. Karplus, *Biophys J* **1997**, *72*, 2094–2102.
44 D. J. Muller, A. Engel, *J Mol Biol* **1999**, *285*, 1347–1351.
45 A. Suenaga, Y. Komeiji, M. Uebayasi, T. Meguro, M. Saito, I. Yamato, *Biosci Rep* **1998**, *18*, 39–47.
46 C. Danelon, A. Suenaga, M. Winterhalter, I. Yamato, *Biophys Chem* **2003**, *104*, 591–603.
47 G. Bainbridge, I. Gokce, J. H. Lackey, *FEBS Lett* **1998**, *431*, 305–308.
48 P. Läuger, *Biophys J* **1988**, *53*, 877–884.
49 G. L. Millhauser, E. E. Salpeter, R. E. Oswald, *Proc Natl Acad Sci USA* **1998**, *85*, 1503–1507.
50 M. S. P. Sansom, F. G. Ball, C. J. Kerry, R. McGee, R. L. Ramsey, P. N. R. Usherwood, *Biophys J* **1989**, *56*, 1229–1243.
51 D. Sigg, H. Qian, F. Bezanilla, *Biophys J* **1999**, *76*, 782–803.
52 Goychuk, P. Hänggi, *Proc Natl Acad Sci USA* **2002**, *99*, 3552–3556.
53 R. Benz, K. Janko, W. Boos, P. Läuger, *Biochim Biophys Acta* **1978**, *511*, 305–319.
54 R. Benz, A. Schmid, R. E. W. Hancock, *J Bacteriol* **1985**, *162*, 722–727.
55 S. McLaughlin, *Annu Rev Biophys Biophys Chem* **1989**, *18*, 113–136.
56 R. M. Garavito, J. P. Rosenbusch, *Methods Enzymol* **1986**, *125*, 309–328.
57 J. B. Stock, B. Rauch, S. Roseman, *J Biol Chem* **1977**, *252*, 7850–7861
58 M. Brunen, H. Engelhardt, *Eur J Biochem* **1993**, *212*, 129–135.
59 K. Sen, J. Hellman, H. Nikaido, *J Biol Chem* **1988**, *263*, 1182–1187.
60 S. M. Bezrukov, *J Membr Biol* **2000**, *174*, 1–13.
61 B. Hille, *Ionic Channels of Excitable Membranes*, 2nd edn. Sinauer Associates, Sunderland, MA, **1992**.
62 T. K. Rostovtseva, E. M. Nestorovich, S. M. Bezrukov, *Biophys J* **2002**, *82*, 160–169.

63 W. Im, B. Roux, *J Mol Biol* **2002**, *319*, 1177–1197.

64 W. Im, B. Roux, *J Mol Biol* **2002**, *322*, 851–869.

65 S. M. Bezrukov, L. Kullman, M. Winterhalter, *FEBS Lett* **2000**, *476*, 224–228.

66 L. Kullman, M. Winterhalter, S. M. Bezrukov, *Biophys J* **2002**, *82*, 803–812.

67 G. Schwarz, C. Danelon, M. Winterhalter, *Biophys J* **2003**, *84*, 2990–2998.

68 Danelon, T. Brando, M. Winterhalter, *J Biol Chem* **2003**, *278*, 35542–35551.

69 A. L. Hodgkin, B. Katz, *J Physiol (Lond)* **1949**, *108*, 37–77.

70 A. Alcaraz, E. M. Nestorovich, V. M. Aguilella, S. M. Bezrukov, *Biophys J* **2003**, *84*, 533a.

71 P. S. Phale, A. Philippsen, C. Widmer, V. P. Phale, J. P. Rosenbusch, T. Schirmer, *Biochemistry* **2001**, *40*, 6319–6325.

72 R. Benz, K. Janko, P. Läuger, *Biochim Biophys Acta* **1979**, 238–247.

73 V. C. Simonet, A. Baslé, K. E. Klose, A. H. Delcour, *J Biol Chem* **2003**, *278*, 17539–17545.

74 P. A. Grigorjev, S. M. Bezrukov, *Biophys J* **1994**, *67*, 2265–2271.

75 S. Nekolla, C. Andersen, R. Benz, *Biophys J* **1994**, *66*, 1388–1397.

76 E. M. Nestorovich, C. Danelon, M. Winterhalter, S. M. Bezrukov, *Proc Natl Acad Sci USA* **2002**, *99*, 9789–9794.

77 J. J. Kasianowicz, E. Brandin, D. Branton, D. W. Deamer, *Proc Natl Acad Sci USA* **1996**, *93*, 13770–13773.

78 T. K. Rostovtseva, S. M. Bezrukov, *Biophys J* **1998**, *74*, 2365–2373.

79 T. Luchian, S. H. Shin, H. Bayley, *Angew Chem Int Ed* **2003**, *42*, 1926–1929.

80 S. M. Bezrukov, I. Vodyanoy, V. A. Parsegian, *Nature* **1994**, *370*, 279–281.

81 J. Bredin, V. Simonet, R. Iyer, A. H. Delcour, J.-M. Pagès, *Biochem J* **2003**, *376*, 245–252.

82 A. H. Delcour, *Front Biosci* **2003**, *8*, d1055–1071.

83 H. Nikaido, *Antimicrob Agents Chemother* **1989**, *33*, 1831–1836

84 M. Nasseau, Y. Boublik, W. Meier, M. Winterhalter, D. Fournier, *Biotech Bioeng* **2001**, *75*, 615–618.

85 J. P. Colletier, B. Chaize, M. Winterhalter, D. Fournier, *BMC Biotechnology* **2002**, *2*, 9–15.

86 B. Chaize, M. Winterhalter, D. Fournier, *Biotechniques* **2003**, *34*, 1158–1162.

7
OmpA/OprF: Slow Porins or Channels Produced by Alternative Folding of Outer Membrane Proteins

Etsuko Sugawara and Hiroshi Nikaido

7.1
Introduction

The "classical" porins, exemplified by the OmpF/OmpC/PhoE porins of *Escherichia coli*, have been studied intensively during the last 25 years or so since their discovery [1]. Such porins, typically composed of a tight trimer of 16-stranded β-barrels, produce relatively high permeability for small solutes and their properties were the central topic in the reviews of bacterial outer membrane channels [2–7]. In contrast, the major porin OprF of *Pseudomonas aeruginosa* did not seem to fit this pattern. It was identified as the major porin of the *P. aeruginosa* outer membrane by the liposome reconstitution approach back in 1979 [8] by Hancock et al. The difference between this porin and the trimeric, classical porins include the following. (i) There is no strong evidence that OprF exists as a stable oligomer. (ii) Upon reconstitution into proteoliposomes, it allows a much slower diffusion of small solutes such as monosaccharides (in one experiment the diffusion of arabinose was 50 times slower than in *E. coli* OmpF porin [9]; Figure 7.1), yet it allows diffusion of much larger solutes that cannot penetrate through the OmpF channel [8] (Figure 7.2). The slow permeation rates in reconstituted vesicles are in agreement with the very low permeability of intact *P. aeruginosa* outer membrane, measured by using hydrophilic solutes such as cephaloridine [11, 12]. (iii) In planar lipid bilayer experiments, Benz and Hancock [13] reported apparently large channels with single channel conductivity of 5 nS. With these large conductivity channels, the calculation of channel diameter from the conductivity is more justifiable and such calculation led to the predicted diameter of about 2 nm. With both the liposome and planar bilayer assays, however, the apparent contradiction between the large pore size and the slower penetration rates was unexpected, and some simplistic "solutions" were proposed (see below). (iv) Finally, when the *oprF* gene was sequenced [14], it was found to be a homolog of *E. coli* OmpA protein, which at that time was not known to have a channel-forming function.

Bacterial and Eukaryotic Porins. Edited by Roland Benz

ISBN 3-527-30775-3

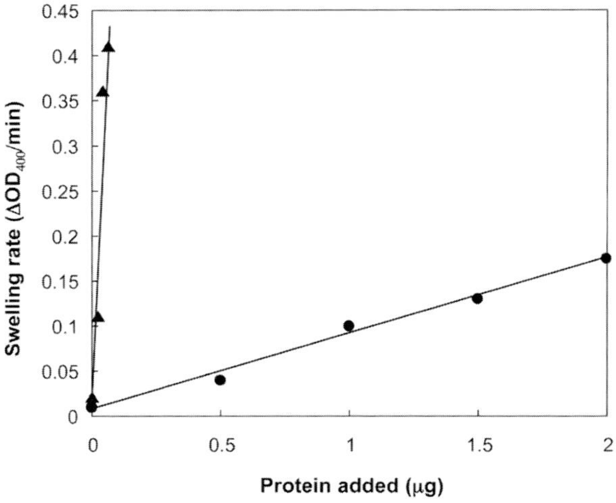

Figure 7.1 Diffusion rates of L-arabinose through the channels of *P. aeruginosa* OprF and *E. coli* OmpF porins. The diffusion rates were measured by using the osmotic swelling of proteoliposomes containing indicated amounts of either *E. coli* OmpF (▲) or *P. aeruginosa* OprF (●) per 2.25 μmol of phosphatidylcholine and 0.1 μmol of dicetylphosphate. The OprF preparation used was purified in lithium dodecylsulfate. This is a re-plotting of data published by Yoshimura et al. [9].

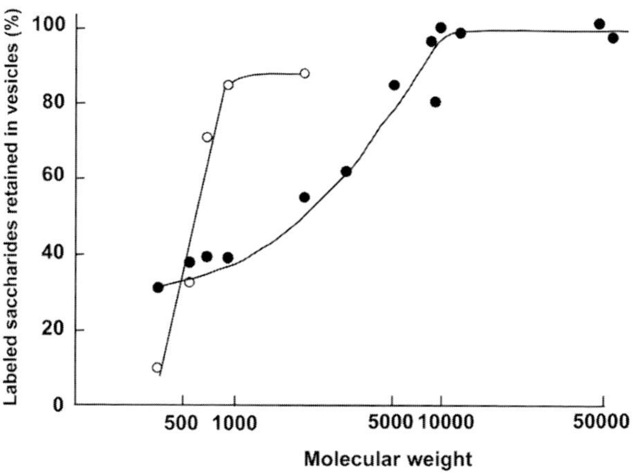

Figure 7.2 Diffusion of polysaccharides of different sizes through the OprF porin channel. Proteoliposomes were reconstituted from 15 μg of partially purified OprF, 0.5 μmol of *P. aeruginosa* phospholipids and 0.08 μmol of *P. aeruginosa* LPS in the presence of radioactively labeled oligosaccharides, inulins and dextrans of different sizes, and their retention in the intravesicular space was examined after gel filtration (●). From [8], with permission from Elsevier. For comparison, the results with proteoliposomes reconstituted with the fragments of *Salmonella typhimurium* outer membrane, containing classical, trimeric porins (○) [10] are also shown.

7.2
Controversies on OprF as the Major Porin of *P. aeruginosa*

The unusual and sometimes difficult-to-explain properties of OprF porin, mentioned above, led to widespread skepticism about the pore properties of OprF. Many papers were published, often in reputable journals, claiming to refute the idea that OprF is the major porin of *P. aeruginosa* with a wide channel. Because these papers are still cited occasionally and contribute to the perception that there may be something that is not quite right about the notion that OprF is the major porin in *P. aeruginosa,* we will examine these papers in some detail.

7.2.1
Is the OprF Channel Wider than those of *E. coli* Porins?

Caulcott et al. [15] found that the spontaneous efflux of various radiolabeled sugars from the periplasm of *P. aeruginosa* and from that of *E. coli* occurred at apparently similar rates, and concluded that the porin diameters must be similar in these two organisms. However, one can calculate the half-equilibration times for sugars across the outer membrane [16] from their known permeability coefficients [17]. These predicted half-equilibration times are of the order of 0.1 s for hexoses in *E. coli* and would be about 10 s for *P. aeruginosa* when we assume that its outer membrane permeability is only 1 % of that of *E. coli*. Thus, the observation of Caulcott *et al.* [15], made at the time scale of 15–30 min, is not likely to have anything to do with the permeability of the outer membrane.

The Caulcott *et al.* study [15] was followed by a study from Nakae et al. [18], who also claimed that the OprF channel was very narrow. This hypothesis, if true, seemed to solve the apparent dilemma of a large-channel porin producing low permeability and, apparently because of this consideration, it was received with enthusiasm by some workers. This study [18] examined, by using intact cells, the outer membrane permeability of sugars ranging in molecular weight from 150 to 666. They found first that, when added at 0.6 M concentration, pentoses produced plasmolysis, whereas sugars larger than 342 Da produced shrinkage of entire cells. However, these authors apparently were not aware of the fact that in Gram-negative bacteria, the periplasm is in osmotic equilibrium with the external medium [19]. Thus, any solute, when present at high concentrations, can produce shrinkage of whole cells regardless of its ability to permeate through the outer membrane. Only when the solute penetrates into the periplasm *rapidly* does it produce plasmolysis. Smaller sugars obviously penetrate through any porin channel more rapidly than large sugars and, thus, the data cannot be interpreted as evidence that the outer "membrane does not allow the penetration of saccharides larger than disacccharides". The next study by Yoneyama and Nakae [20] extended the preceding work, and measured the total amounts, in the centrifuged pellet of plasmolyzed bacterial cells, of various sugars and of the marker of extracellular space, the enzyme glucose 6-phosphate dehydrogenase. They obtained the normalized concentration ratio of 2:1 with pentose and methylhexosides, and concluded that these small sugars, but not the larger sugars, penetrate into the enlarged periplasmic space. This interpretation, however, does not appear convincing, because Decad and Nikaido [21] showed earlier, in the series of experiments that pioneered this experimental approach, that almost one-half of the volume of such cell pellet consists of intercellular space. Thus, the ratio of 2:1 means the distribution of the small sugars into all of the pellet volume, including the cytoplasm. This is an assumption not so unreasonable because pentoses are known to diffuse across phospholipid bilayers with significant rates [17]. A conspicuous deficiency of the study of Yoneyama and Nakae [20] is that there are no measurements of the total pellet volume as the control.

Nakae's laboratory then published a paper [22] in which sugar permeation was assayed by the osmotic swelling of proteoliposomes reconstituted from phospholipids and fragments of *P. aeruginosa*. In this technique, multilamellar proteoliposomes containing porin-impermeable molecules in the intravesicular space are diluted into iso-osmotic solutions of test solutes. If the latter can penetrate through the porin channel, their influx into the vesicles is followed by the influx of water, which causes osmotic swelling of the vesicles that can be followed by the changes in the turbidity of the vesicle suspension [17]. There were two major conclusions in the *P. aeruginosa* study [22]. First, the "exclusion limit", obtained by plotting the proteoliposome swelling rates versus the logarithms of solute molecular weight, was 220 Da. Second, comparisons with the swelling data with those obtained with proteoliposomes containing *E. coli* outer membrane showed that the slope was steeper with *P. aeruginosa*, i. e. the diffusion through the *P. aeruginosa* porin(s) was influenced more strongly by the changes in the size of diffusing solutes. This seemed to indicate that *P. aeruginosa* porin(s) contained narrower channels in comparison with *E. coli* porins.

These results are in striking contrast to the data from proteoliposome swelling assays conducted in our laboratory [9, 23]. We found that the slopes of swelling rate versus solute molecular weight plots to be much less steeper in vesicles containing either fragments of *P. aeruginosa* outer membrane or the purified OprF porin in comparison with vesicles containing *E. coli* OmpF porin (Figure 7.3). This is consistent with the results of spontaneous efflux assay using labeled saccharides, shown in Figure 7.2. It is also consistent with the observation that changes in the structures of β-lactams, which have profound influences on the diffusion rates through the narrower *E. coli* porin channels [16, 26], showed little effect on the influx through larger channels of *P. aeruginosa* outer membrane in intact cells [11]. Finally, a very convincing experiment was performed by Bellido et al. [27] in 1992. By expressing genes for the metabolism of raffinose (molecular weight 505) in *P. aeruginosa*, they showed that *oprF*$^+$ cells, but not an *oprF* mutant, can grow readily by utilizing this sugar. This puts to rest all the "controversy" about the "exclusion limits" of the OprF porin channel.

We can only speculate on how Yoshihara et al. [22] were led to the wrong conclusions. At the experimental level, they used a buffer solution of exceptionally low concentration (1 mM) for the proteoliposome-swelling assay. This very low ionic strength likely exaggerated artifacts caused by the Donnan potential and flux of

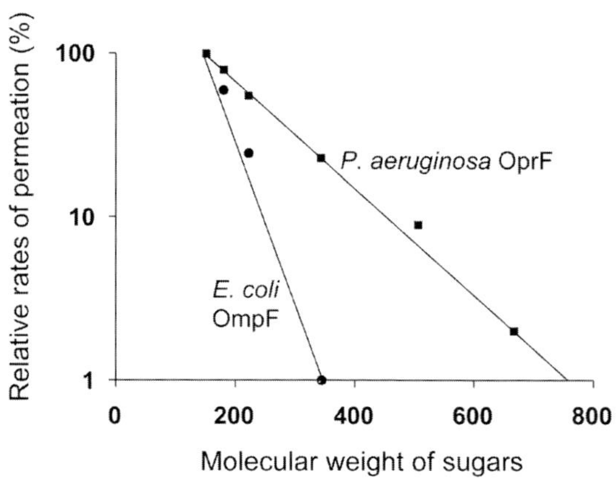

Figure 7.3 Relative osmotic swelling rates of OprF- and OmpF-containing proteoliposomes in sugars of different sizes. For both types of proteoliposomes, the swelling rate in L-arabinose solution was arbitrarily assigned the value of 100 and the rates in other sugar solutions were normalized to the rate in arabinose. This is a re-plotting of some of the data reported in Yoshimura et al. [9], with addition of data points for OmpF-containing vesicles from Nikaido and Rosenberg [25].

ions (see the Experimental Procedure section of [23]). In the comparison between *E. coli* and *P. aeruginosa* porins, Yoshihara et al. included stachyose (666 Da) in the intravesicular space as the "porin-impermeable" solute. Since stachyose can actually diffuse through the OprF porin channel [23], this must have led to the underestimation of the diffusion rates of larger solutes, as the swelling caused by the influx of sugars was counterbalanced by the shrinkage caused by the efflux of stachyose. At the level of interpretation, they plotted the diffusion rates in relation to the *logarithms* of molecular weight of solutes and obtained "exclusion limits" from the *x*-intercepts by extrapolation. There is no justification for the use of logarithmic scale on the *x*-axis or for the extrapolation procedure. Indeed, this procedure gave an exclusion limit of only 320 for *E. coli* porin [22], a prediction that is totally contradictory to the observation that *E. coli* can grow rapidly on lactose (342 Da) by using the classical porins for its uptake and that proteoliposomes containing *E. coli* porins allowed the transmembrane diffusion of raffinose (505 Da), a result published by Nakae himself [24]. Clearly the conclusions of the work of Yoneyama et al. [22] cannot be correct. There is now an overwhelming body of convincing data showing that the OprF channel is larger than those of the *E. coli* porins [23, 25]. The results from Nakae's laboratory and especially their interpretations have been criticized many times (most thoroughly in [23]), yet as far as we are aware Nakae and his associates have not published any rebuttal or re-examination of their previous data/conclusions. However, we note that a later publication from Nakae's laboratory [28] described the rapid diffusion of cephalosporin antibiotics, with molecular weights as high as 644, across the outer membrane of *P. aeruginosa*. Although this was ascribed to the activity of another outer membrane protein,

"C", it is in contradiction to the previous claim from Nakae's group that nothing larger than 220 Da can cross the *P. aeruginosa* outer membrane [22]. We can only assume that this group has now come to the conclusion that large solutes can cross the outer membrane of *P. aeruginosa*.

7.2.2 Is OprF a Porin?

Yoshihara et al. [22] examined the relative osmotic swelling rates of proteoliposomes reconstituted with the wild-type outer membrane and with the outer membrane of an OprF-deficient mutant strain. They saw little difference in the slope of the plots of swelling rates versus the logarithm of the solute molecular weight and this led them to the conclusion that OprF may play no role in solute permeation. In order to strengthen their claim, they purified OprF from the wild-type strain and found that the purified OprF had no detectable pore-forming activity [28]. In retrospect, this was a surprising claim because earlier work from our laboratory [9] showed that purified OprF had pore-forming activity, and because proteins could become inactivated during the long and complex purification steps and, therefore, lack of activity in purified proteins usually is not the automatic proof of the absence of presumed functions for proteins. In any case, when we repeated the purification procedures reported by Yoshihara and Nakae [28], we found that OprF had an activity that produced large channels [23] precisely as reported earlier by us [9]. Furthermore, when the same purification procedure was applied to the outer membrane of an OprF-deficient mutant, there was little pore-forming activity eluted at the corresponding positions from the ion-exchange column, a result showing that the pore-forming activity of the purified OprF was not due to contaminants. It is clear that OprF must have been inadvertently inactivated in the study of Yoshihara and Nakae [28].

7.2.3 Is OprF the Major Porin in *P. aeruginosa*?

Some scientists expressed doubts about whether OprF was the major non-specific porin, primarily because OprF-deficient mutants show no or only minor increase in resistance to various antibiotics [29, 30]. In the study of Woodruff and Hancock [29], the minimum inhibitory concentrations (MICs) of cefpirome, cefotaxime and cefepime showed about a 3-fold increase in the OprF-deficient mutant, with about 2-fold increases in the MICs of aztreonam, carbenicillin and ceftazidime. However, these small changes are not unexpected because the absence of OprF impairs the growth of the mutant in low-ionic strength media [30, 31] and makes the outer membrane "leaky" as shown by the leakage of periplasmic enzymes into the medium [30]. Thus, the decreased influx of antibiotics through OprF is presumably compensated by an increased influx through the leaky gaps in the outer membrane and it is not surprising that the MIC values do not show a strong increase. In fact, Woodruff and Hancock [29] showed a strong increase in the influx of a lipophilic fluorescent marker, *N*-phenylnaphthylamine, in the OprF-deficient mutant. In the study by Gotoh et al. [30], assay of antibiotic permeability was performed by osmotic swelling of proteoliposomes using the conditions devised for *E. coli* porins [32]. Unfortunately, the swelling assay under these conditions depends on the fact that both stachyose and NAD^+, which are incorporated into the intravesicular space, are essentially impermeable through *E. coli* porin channels, and the results obtained with *P. aeruginosa* are not valid, especially when strains containing OprF, which allows the diffusion of these solutes [23], were used.

Nakae's laboratory claimed that they identified the true, non-OprF porins in *P. aeruginosa* outer membrane [28]. By purifying various outer membrane proteins, they concluded that protein "E" had the highest specific activity in allowing the diffusion of L-arabinose, followed by protein "D" and protein "C". However, because the swelling rates were reported in arbitrary units, it is difficult to assess the magnitude of their channel-forming activity; it appears that all of them were very inefficient as they had to use 5 μg of purified proteins for reconstitution (for an efficient porin such as *E. coli* OmpF, about 0.1 μg is enough). We have shown that in terms of the total pore-forming activity OprF was most important, followed by protein E2 (presumably corresponding to OprE), with OprD and OprB (also called protein D1) playing very minor roles – at least an order of magnitude lower than OprF [23]. OprD is a specific channel that catalyzes the transport of basic amino acids and peptides, and also the antibiotic imipenem as their analog [33]. Its sequence seems to have a marginal homology with OmpF and the homology becomes somewhat more convincing within the predicted transmembrane segments [34]. However, this observation does not mean that OprD functions as a classical non-specific porin. It allows only a very slow non-specific permeation of sugars, at rates only one-third of already very inefficient OprF [23]; clearly it is a specific channel. OprE is strongly similar in its sequence to OprD and it is also likely to be a specific channel. Its specific activity, measured by the diffusion of small sugars, is similar to OprF, but it is less abundant than OprF in the outer membrane [23]. OprB is a specific channel for glucose [35, 36]. Finally, OprC was shown by Yoneyama and Nakae [37] to belong to the family of TonB-dependent-gated channels, apparently involved in the uptake of Cu^{2+}. Thus, all of the proteins postulated by Yoshihara and Nakae [28] to be the "real" non-specific porins in *P. aeruginosa* are actually specific or gated channels and none has the property of the efficient, classical porin. It should be emphasized again that none of these proteins had a specific activity (of forming nonspecific channels) higher than OprF. We showed that at least two-thirds of the total non-specific channel activity in the *P. aeruginosa* outer membrane comes from OprF [23] and OprF indeed appears to be the major porin in this organism. (The properties of various outer membrane proteins of *P. aeruginosa* have been reviewed by Hancock and Worobec [38] and Hancock and Brinkman [39].)

It appears that OprF homologs are the major porins in other species of fluorescent pseudomonads, which do not appear to contain high-efficiency trimeric porins similar to the *E. coli* porins. The other species have not been studied as intensively as *P. aeruginosa*, but *P. fluorescence* OprF and *P. syringae* OprF have been studied by planar bilayer methods [40–44]. OprF from both species showed channels of moderate conductivity, 250–270 and 280 pS (in 1 M NaCl and 1 M KCl) in *P. fluorescens* and *P. syringae*, respectively. These values are comparable to the conductivity of *E. coli* OmpF channel, 2 nS for a trimer or about 660 pS for a single channel in 1 M KCl [45]. Very interestingly, inactivation of the *oprF* gene in one strain of *P. fluorescens* produced a persuasive piece of genetic evidence that the OprF protein functions as the major porin [46]. In this particular strain, the isolation of *oprF*-inactivated mutant required much longer incubation in comparison with other strains and the isolated mutant strain contained suppressor mutations that resulted in the over-expression of other outer membrane channels of OprD family. This suggests that OprF was indeed the major porin in the parent organism and that the increased expression of other channel proteins was needed for the survival of the mutant deficient in this major porin.

OmpA/OprF homologs may be the major porins in other bacteria. A protein of apparently size 51 kDa was shown to be one of the three proteins with significant (but low) porin activity in the outer membrane of *Bacteroides fragilis* [47]. Subsequently, the protein was purified and characterized. It is a homolog of OmpA/OprF, with a size of 41 kDa [48]. The earlier assignment of the larger size was likely due to the "heat modifiability" phenomenon, which is characteristic of this class of proteins and produces slower mobility in SDS gels unless the proteins are thoroughly denatured by heating in SDS sample buffers. This 41-kDa protein was shown to act as a low-efficiency porin in the proteoliposome swelling assay. The outer membrane of the intact cells of *Bacteroides fragilis* was shown to have cephalosporin permeability at least an order of magnitude lower than that of *E. coli* [49]. If the 41-kDa OmpA/OprF homolog is the major porin in this species, this will explain the low permeability of the outer membrane of this species and, perhaps, the generally high levels of intrinsic resistance of this species to various antibiotics. Interestingly, disruption of the gene coding for this protein results in a much slower growth of the mutant [48]; this is consistent with the notion that the 41-kDa protein plays the role of a major porin, although other interpretations cannot be excluded at present.

A possibly similar situation was found in *Haemophilus ducreyi* [50]. The outer membrane of this organism contains small amounts of a 45-kDa protein, which was described as a homolog of trimeric porins of *Pasteurella multocida*. It also contains large amounts of two OmpA homologs. The disruption of the *momp* gene, coding for the more abundant homolog, resulted in the overproduction of the 45-kDa protein, suggesting that at least this OmpA homolog may play a major role in the uptake of nutrients in the wild-type organism.

7.3
Similarity between OmpA and OprF

7.3.1
OmpA is also an Inefficient Porin

We mentioned above that the homology between OprF and OmpA caused doubt about the pore-forming activity of OprF, because OmpA was not known to produce channels at that time. However, when we purified OmpA from a mutant of *E. coli* lacking both OmpF and OmpC, and reconstituted it into proteoliposomes, the OmpA protein showed a low pore-forming activity [51] (Figure 7.4). Importantly, this activity was not seen if the identical purification steps were carried out on a mutant defective in *ompA*; thus it is clear that the activity derived from OmpA, not from any contaminating proteins. In order to produce osmotic swelling rates similar to those seen with the classical porins OmpF/OmpC, it was necessary to use much more OmpA protein for reconstitution, indicating that the specific activity of OmpA is much lower than in the classical porins [51]. This is of course very similar to the property of OprF. One difference between OmpA and OprF in terms of function was that the channel size of OmpA is similar to that of the classical porins, whereas OprF, as described above, produces much larger channels.

The pore-forming activity of OmpA was also confirmed in planar lipid bilayer assays. Saint et al. [52] found some heterogeneity in single-channel conductance values, but with one preparation 600 pS in 1 M NaCl was the most frequently observed value. This is quite close to the conductance of a protomer of OmpF, as mentioned above. In the study of Arora et al. [53], OmpA produced channels

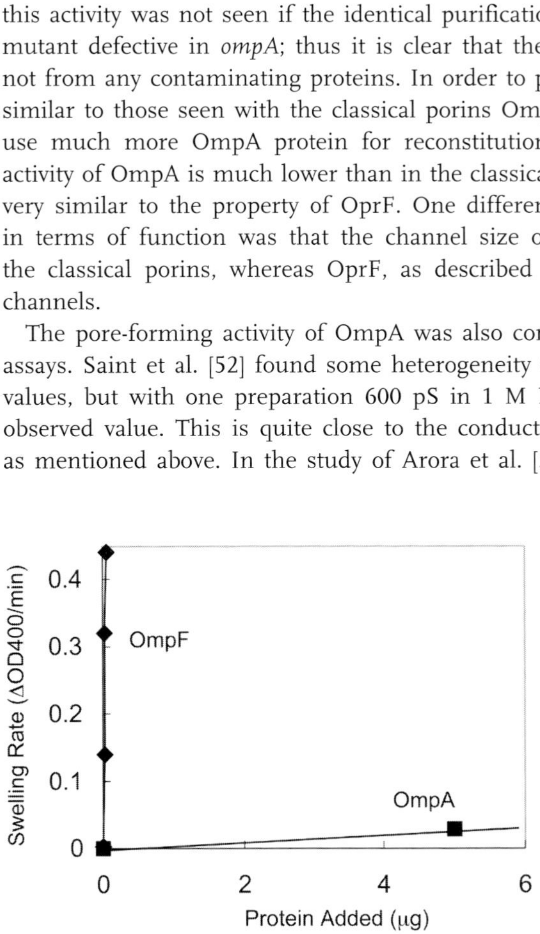

Figure 7.4 OmpA shows a low pore-forming activity in proteoliposome swelling assays. Purified OmpA and OmpF from *E. coli* were reconstituted into proteoliposomes and the osmotic swelling rate in a solution of L-arabinose was measured. In this particular experiment (Sugawara and Nikaido, unpublished), the difference in specific pore-forming activity between the two proteins was about 500. In some other experiments, the difference was as small as 100.

of 260–320 pS conductance in 1 M KCl, again in a range similar to the conductance of OmpF monomer. (In this study, smaller conductance steps of 60–80 pS were also seen; these will be discussed below.)

7.3.2
The Majority of OmpA and OprF Folds into Two-domain Conformers

OmpA has always been hypothesized to fold as a two-domain protein [54], with its N-terminal half spanning the outer membrane as an eight-stranded β-barrel and its C-terminal segment located in the periplasm and interacting with the peptidoglycan layer [55, 56] (Figure 7.5). This folding model is supported by a large amount of convincing data, beginning with the location of phage-resistance mutations in the predicted external loops [57] and culminating in the determination of crystal structure of the N-terminal domain of OmpA as a β-barrel [58, 59]. (One paper suggests an unconventional single-domain model containing 16 transmembrane strands [60]. As a model for the bulk population, this model is clearly incorrect, although we now believe that the minority, pore-forming conformer assumes this conformation; see below.). In contrast, earlier models for OprF assumed that it folds essentially as a continuous β-barrel, containing up to 16 transmembrane β-strands as in the classical porins [61, 62] (Figure 7.5). The surface exposure of residues in the C-terminal region was indeed confirmed, e. g. by inserting malarial antigen epitopes at various places in the OprF sequence [63]. The discovery that a very strong antigenic epitope exists within the C-terminal domain of (unmodified) OprF and that antibody directed against this epitope can react with whole cells of *P. aeruginosa* [64, 65] further appeared to support this folding model, because the C-terminal domain of OprF will not be exposed on the cell surface if all OprF molecules followed the OmpA-like folding pattern.

A strong piece of evidence showing that OprF shares the same folding pattern as OmpA came from the circular dichroism study of OprF protein isolated without the use of denaturing detergent [66]. OmpA contained a substantial fraction of helical structures (presumably coming from the periplasmic, globular domain of the two-domain conformation) in addition to β-structure (which obviously must come

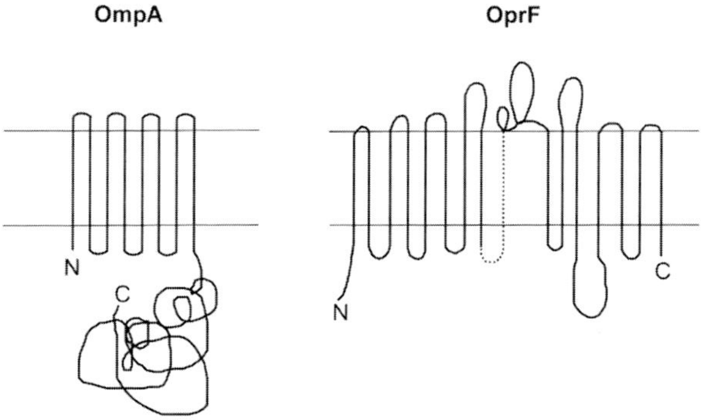

Figure 7.5 Published folding models of OmpA (left) and OprF (right). The OmpA folding model is basically the one proposed by [54]. The OprF folding model was drawn based on the scheme proposed in [61, 62].

mostly from the N-terminal β-barrel domain. Importantly, OprF also contained about the same fraction of α-helical structure – a result suggesting the two-domain conformation (just like OmpA), in contradiction to the essentially all β-structure conformation predicted from the earlier model. There is now little controversy on the folding of (the majority conformers of, see below) OprF, as Hancock, who advocated the 16-stranded β-barrel model as late as 1998 [38], now agrees with the OmpA-like, two-domain model [39]. We shall discuss below the reason why C-terminal epitopes of OprF can be recognized by antibodies in intact cells.

The eight-stranded β-barrel of OmpA, according to its crystal structure, is unlikely to function as a channel even for water molecules [58, 59]. Interestingly, however, planar bilayer studies showed that channels of small conductance values (about 70 pS) can be found in the preparations of the N-terminal domain of OmpA [42, 53]. Recent molecular dynamics simulation study [67] indeed suggests that movement of a charged residue inside the β-barrel would generate channels that would allow the permeation of water and small ions. Similarly, the N-terminal domain of OprF was shown to produce channels of somewhat larger conductance, centered around 360 pS [68, 69].

7.4 The Minority, Open-channel Conformers of OmpA/OprF

It was difficult to explain why solutes penetrated through the OprF or OmpA channels at rates about two orders of magnitude slower than through the classical porins such as OmpF or OmpC. The answer to this question came through the function-based separation of the two OmpA conformers [70]. In this study, unilamellar proteoliposomes each containing only a few molecules of OmpA were reconstituted by the octyl glucoside dilution method in a buffer containing 0.3 M urea. These vesicles were then added on top of a continuous, iso-osmolar gradient that contained 0.3 M urea at the top and 0.15 M urea–0.15 M sucrose at the bottom. During centrifugation, if any of the OmpA channels were open, sucrose entered into the vesicles, which then became heavier and sedimented toward the bottom. However, if all of the OmpA channels were closed in given vesicles, they retained their lower density because urea is lighter than sucrose and they remained on top of the gradient. This is a technique called "transport-specific isolation", developed for the isolation of mitochondrial porin or voltage-dependent anion-selective channel (VDAC) by Harris et al. [71]. Control vesicles containing no protein indeed all remained at the top and those containing OmpF porin mostly went to the bottom, as expected [70]. OmpA-containing vesicles were separated into two fractions, one close to the top and one close to the bottom (Figure 7.6). Both of these fractions contained roughly the same levels of OmpA per lipid. By using Poisson statistics, we could calculate that only a few percent of the OmpA molecules had the ability to form open channels [70]. When protein was isolated from both the sedimented and the lighter fractions, both fractions were shown to contain only OmpA. When these OmpA preparations were reconstituted into multilamellar proteoliposomes for osmotic swelling assay, the protein from the bottom fraction was shown to be about ten times more active in producing channels for a small sugar, L-arabinose. However, this activity was not due to a covalent modification of OmpA, because when the active OmpA preparation, isolated from the sedimented vesicles, was denatured in guanidium hydrochloride, renatured by dialysis and reconstituted again into unilamellar vesicles, most of the vesicles stayed on top during the gradient centrifugation. These experiments proved that OmpA existed in two different conformations, one "closed" and the other open, that the open conformers comprised only a minority of OmpA preparation and that this paucity of the open conformer completely explained the low permeability of vesicles containing unfractionated OmpA population. A similar experiment using OprF also showed that the OprF population was heterogeneous, only a minority containing the open channel (Sugawara et al., in preparation).

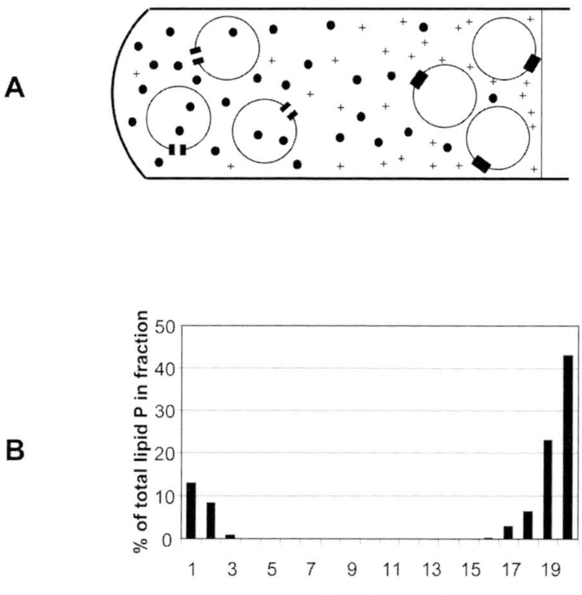

Figure 7.6 Transport-based separation of two OmpA conformers. (A) Principle of separation. The isoosmotic gradient contains higher concentrations of urea (+) closer to the top (right in the figure) and higher concentrations of sucrose (●) closer to the bottom (left). When vesicles each containing only a few molecules of OmpA are centrifuged through this gradient, those vesicles with no open channels remain on top, as they do not allow the influx of the heavier sucrose molecules. However, those vesicles containing open conformers of OmpA are sedimented to the bottom, as sucrose enters into the vesicles. (B) The results of the actual experiment, from Sugawara and Nikaido [70]. Fractions were collected from the bottom of the tube and were analyzed for vesicles by determining the amount of phospholipids. Analysis for the amount of OmpA produced a similar distribution. Controls gave expected results. Thus, liposomes containing no proteins stayed completely on top and those containing OmpF porin largely went to the bottom [70].

The finding of two stable conformations of OmpA encountered some skepticism in the field when our paper [70] was published nearly a decade ago. However, there has been an increased realization that alternative folding occurs with many proteins, sometimes causing diseases now called "protein-folding diseases" [72], including Alzheimer's disease and prion diseases. Still, in these examples the protein has one "correct" conformation and straying from the correct path of folding causes "misfolding" that often causes diseases. In the example of OmpA/OprF, both conformations were apparently selected as physiological ones during the course of evolution. Thus, with both proteins, the function of the more frequent conformer is apparently the stabilization of the cell envelope through the interaction of the C-terminal domain with the peptidoglycan [55, 56], as is suggested by the observation of spherical morphology of mutants defective in OprF [30, 31]. *E. coli,* however, contains an additional component, murein lipoprotein (which is absent in *P. aeruginosa*), that connects the outer membrane with the peptidoglycan. Thus, pronounced morphological alterations in *E. coli* requires the absence of both OmpA and murein lipoprotein [73]. The alternative folding pathway that generates the low porin activity is useful in fluorescent pseudomonads, which must often survive in soil and on plants, an environment full of toxic compounds. Enteric bacteria including *E. coli* have developed a high-efficiency, trimeric porin, which gives them an advantage in taking up nutrients in a highly competitive environment of the intestinal tract and here the low-level porin activity of OmpA, possibly a relic of evolution, presumably plays no major role in the uptake of nutrients.

It is not clear what determines the ratio between the open-channel conformer and the closed-channel conformer in OmpA/OprF, except in *P. fluorescens.* In this psychrophilic organism, cultivation at low temperature (8 °C) resulted in the marked increase of OprF producing smaller channels of 80–90 pS in contrast to the 250–270 pS conductance of OprF from cells grown at 28 °C [41]. It has been suggested that this effect on the folding pathway is caused by the modification of lipopolysaccharide (LPS) structure, with which OprF presumably interacts during the folding process [74].

7.5 The Nature of the Open Conformer

The results presented above explain the slow permeation produced by OmpA/OprF homologs, but leave open the question of the nature of the difference in conformation between the "open" and "closed" proteins. Since the eight-stranded β-barrel has little room inside for the passage of even small nutrient molecules, this conformer cannot serve as a large, open channel.

The open conformers are, thus, likely to contain a large β-barrel with more β-strands, as found in the classical porins, as shown in Figure 7.7. Such a conformation is actually the conformation predicted earlier for OprF (see above) and the recent change in the model is that it is now thought to be valid only for the minority conformer of this protein. The introduction of the two-conformer hypothesis now explains the low permeability of OprF and most other available data on this protein. One finding that was thought, earlier, to contradict the notion that the majority conformer of OprF folds as an OmpA-like two-domain protein was that epitopes in the C-terminal half of OprF are exposed on the surface of intact cells [64, 65]. However, this is now understandable because, as OprF is the major protein in the outer membrane that exists in probably 10^5 copies per cell, even its minority conformer must exist at the order of about 10^4 copies per cell, certainly enough to act as a strong immunogen and to bind antibodies on the surface of intact cells. In fact, some earlier data are consistent with this interpretation. Thus, in the foreign epitope insertion experiment [63], the reactivity in intact cells of the epitopes located in the C-terminal domain of OprF, which will be exposed on the cell surface only in the minority conformers, was indeed lower than that of the same epitopes inserted into the external loops of the N-terminal half, where they are exposed on cell surface in both the majority and minority conformers. With OmpA, an opposite situation has prevailed. Because OmpA was believed to fold as a two-domain protein, and because the N-terminal half was expressed as a truncated protein and was crystallized as a structure with no visible channel [58, 59], many workers expressed skepticism on the pore-forming activity of this protein, claiming that this function is controversial or unsubstantiated (e. g. [5]). However, a recent immunological study showed clearly that monoclonal antibodies directed toward the epitope in C-terminal domain do react with intact cells of *Salmonella typhimurium*, although the reactivity is increased strongly by making the outer membrane permeable to macromolecules, as expected [76]. Thus, the OmpA conformer whose C-terminal domain is exposed on cell surface does exist in intact cells, just like with OprF in *P. aeruginosa*.

It would be ideal if we could isolate the open-channel conformer in a pure state and compare its properties with the pure preparation of closed-channel conformer. The transport-specific isolation methodology [71], however, allows the isolation of only less than 1 μg of open conformers, which is contaminated with a much larger

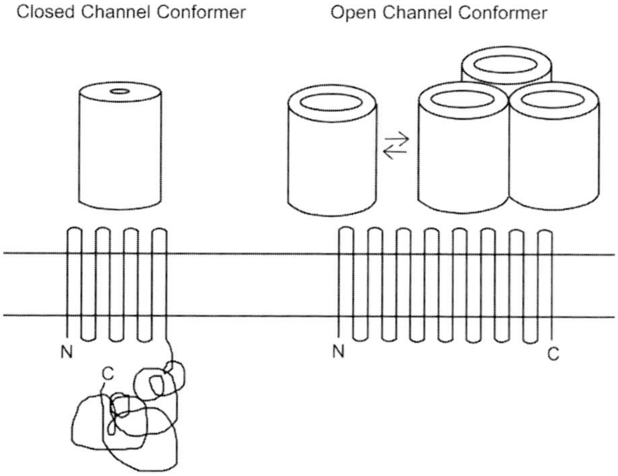

Figure 7.7 Current models of the two conformers of OmpA/OprF. Modified from [75], with the permission from Elsevier.

amount of closed conformers [70]. Interestingly, the open conformers of OmpA (and OprF) showed a tendency to exist apparently in an oligomeric form [77] and were thus eluted at the leading edge of the protein peak from gel-filtration columns. We have been able to utilize this property and enrich the OmpA and OprF preparation for their open conformers (Sugawara et al., in preparation).

The current model of OmpA/OprF received strong support from recent studies using planar bilayer reconstitution. The model predicts that the N-terminal fragments (alone) of these proteins will not produce large channels because they consist only of eight-stranded β-barrels and that the entire protein sequences are needed to produce approximately 16-stranded β-barrels that would allow the permeation of large molecules. Indeed, Arora et al. [53] found only very small channels of 60–80 pS conductance with the truncated N-terminal domain of OmpA, whereas large channels with approximately 300 pS single-channel conductance were found with our OmpA preparation that was enriched for open conformers by repeated gel filtration. Subsequently, Brinkman et al. [68, 69] and El Hamel et al. [43] obtained similar results with OprF from *P. aeruginosa* and *P. fluorescens*, respectively. These results are completely consistent with the prediction of the model (Figure 7.7). With the intact OmpA/OprF, the smaller channel (presumably coming from the insertion of the N-terminal eight-stranded β-barrel) is far more frequent than the large channel (which presumably comes from larger β-barrel of less frequent, open-channel conformers).

The two-conformer model of OmpA/OprF is, thus, quite useful and is likely to be correct. We have been able to produce further evidence that supports this model. If residues in the C-terminal domain of OprF are exposed on the cell surface only in the minority conformer containing open channels, one should be able to isolate (or at least enrich for) this conformer by inserting an additional cysteine residue in one of the predicted external loops of the C-terminal domain, by labeling, in intact cells, the residue with a biotinylation reagent that does not traverse the outer membrane rapidly, and by capturing this conformer (but not the majority conformer in which the cysteine residue is hidden in the periplasm) via its binding to avidin. We labeled the cysteine residue of the mutant OprF, in which the alanine residue at position 312 (the 15th residue from the C-terminus) was replaced with cysteine, with a short (5 min) treatment of intact cells with biotin-maleimide, a large (452 Da) and rather lipophilic reagent that is expected to diffuse through the porin channels of *E. coli* (an organism in which the labeling experiment was carried out). This experiment showed first that the vast majority of OprF protein was in a conformation in which the C-terminal domain was hidden in the periplasm and, therefore, was not labeled. However, in a minority population of OprF, the C-terminal domain was labeled, and we have indeed been able to enrich for this biotin-labeled species and to show that this species had about 10-fold higher specific activity for pore formation (Sugawara et al., in preparation).

There is, thus, no controversy any longer as to the porin functions of OprF/OmpA. It is remarkable that OprF/OmpA family proteins are found in almost every Gram-negative bacterial genome sequenced and possibly they serve as the major porin also in organisms other than the pseudomonads.

One remaining question, though, is whether the two folding pathways of these proteins are regulated. In this connection, we mentioned above the remarkable finding that *P. fluorescens* was found to produce OprF of smaller single-channel conductance when grown at low temperature [41]. Although small values of ionic conductance cannot be immediately equated with smaller channel size, the low-temperature conformer is more susceptible to protease hydrolysis [43], as expected for the two-domain conformer, and it seems likely that the growth temperature affects the folding of OprF at least in this organism.

7.6 Regulation of Expression of *ompA* and *oprF* Genes

The upstream sequence of *ompA* gene contains a cAMP receptor protein (CRP)-binding site [78] and the binding of the CRP–cAMP complex to the upstream sequence was demonstrated *in vitro* [79]. In intact cells, the synthesis of OmpA was shown to be down-regulated by cAMP in a CRP-dependent manner [80]. Nevertheless, it is not clear what potential advantages this regulation would bring to the cell. *E. coli* cells can possibly save energy by down-regulating the synthesis of one of the major proteins, which is needed less for the stabilization of the envelope in a slowly growing cell. On the other hand, another group reported the up-regulation, rather than down-regulation, of *E. coli* OmpA by the cAMP–CRP complex [81]. Since these authors do not cite the earlier publication on the down-regulation, it is unclear why these totally different results were obtained. However, their study relied on the expression of an *ompA′–lacZ* translational fusion and such a construct may have produced artifacts as the leader sequence in *ompA* would try to secrete large, secretion-incompetent LacZ protein, likely causing the jamming of the secretory apparatus. In fact the reported expression levels of the *ompA′–lacZ* fusion construct are unreasonably low.

The sequencing study [78] also uncovered two additional translation initiation sites in front of the ATG codon coding for the N-terminal end of the pro-OmpA protein. These sites were proposed to increase the translation rate of pro-OmpA by initiating the synthesis of two short peptides upstream. One outstanding feature of OmpA mRNA is the secondary structure in the upstream, untranslated region, which gives OmpA mRNA an exceptionally long half-life (see, e. g. [82]).

A proteomics study of *E. coli* grown at different values of pH [83] found that the intensity of OmpA spot was strongly (about 6-fold) increased in cells grown at pH 9, in comparison with cells grown at pH 7. However, in cells grown at neutral pH, the OmpA spot is barely visible, a finding unexpected for one of the major proteins in the outer membrane. It seems likely that the changes seen reflect more the altered extractability of OmpA from the outer membrane by the urea–SDS agent utilized in this study and the study of transcription levels is needed for comparison. At the other end of the pH spectrum, OmpA was reported to affect the acid tolerance of *E. coli*. Thus, Rowbury et al. [84] found that the sensitization of *E. coli* to a complex medium of pH 3.0, caused by the pretreatment of the cells with low concentrations (15–50 μg/ml) of leucine, did not occur in OmpA-deficient mutant cells. Unfortunately, the mechanism of this sensitization reaction is not understood nor is it known whether the sensitization involves the altered expression of OmpA. Leucine usually affects *E. coli* cells through the Lrp regulatory system [85], but mutational inactivation of the *lrp* locus has almost no effect in this response. In contrast, the regulator of cysteine biosynthesis, CysB, appears essential. The authors propose that OmpA functions as the major channel for proton influx across the outer membrane. However, this seems most unlikely as other classical porins, with much higher permeability in general, do exist in these cells.

Insertion of many proteins into the outer membrane appears to require the presence of LPS of the correct structure. This was shown by the observation that mutants producing LPS defective in the inner core region ("deep rough" LPS) have difficulty in incorporating major proteins into the outer membrane [86]. The defect appears most severe with the classical trimeric porins [86], and it was shown later that LPS of normal structure, but not deep rough LPS, is capable of facilitating the folding, insertion and trimerization of these porins in a cell-free system [87]. Although this LPS dependence is somewhat less with OmpA than with classical porins, a detailed later study [88] confirmed the earlier conclusions. A recent study showed further that a periplasmic chaperone Skp plays an important role in assisting the folding of OmpA [89]. It will be a most interesting future project to study if the two different folding pathways are affected differently by these "helper" molecules.

The regulation of transcription of *P. aeruginosa oprF* gene is apparently more complex. Just upstream of the *oprF* gene, a gene called *sigX* is found. SigX belongs to the family of ECF (extracytoplasmic function) σ factors and is involved at least partly in the transcription of *oprF* [90]. Interestingly, another member of the ECF σ factor family, AlgU, was also shown to enhance the transcription of *oprF* [84]. Much less is known on the biochemistry of insertion of OprF into the outer membrane and we do not yet know how the alternative folding pathways of OprF are regulated, although there exist the important data that in *P. fluorescence* this is dependent on temperature [41, 43] as mentioned already.

References

1 T. Nakae. *J Biol Chem* **1976**, *251*, 2176–2178.

2 R. Benz. *Annu Rev Microbiol* **1988**, *42*, 359–393.

3 H. Nikaido. *J Biol Chem*.**1994**, *269*, 3905–3908.

4 G. E. Schulz. *Curr Opin Cell Biol* **1993**, *5*, 701–707.

5 R. Koebnik, K. P. Locher, P. Van Gelder. *Mol Microbiol* **2000**, *37*, 239–253.

6 G. E. Schulz. *Curr Opin Struct Biol* **2000**, *10*, 443–447.

7 G. E. Schulz. *Biochim Biophys Acta* **2002**, *1565*, 308–317.

8 R. E. W. Hancock, G. M. Decad, H. Nikaido. *Biochim Biophys Acta* **1979**, *554*, 323–331.

9 F. Yoshimura, L. S. Zalman, H. Nikaido. *J Biol Chem* **1983**, *258*, 2308–2314.

10 R. E. W. Hancock, H. Nikaido. *J Bacteriol* **1978**, *136*, 381–390.

11 F. Yoshimura, H. Nikaido. *J Bacteriol* **1982**, *152*, 636–642.

12 B. L. Angus, A. M. Carey, D. A. Caron, A. M. Kropinski, R. E. W. Hancock. *Antimicrob Agents Chemother* **1982**, *21*, 299–309.

13 R. Benz, R. E. W. Hancock. *Biochim Biophys Acta* **1981**, *646*, 298–308.

14 M. Duchene, A. Schweizer, F. Lottspeich, G. Krauss, M. Marget, K. Vogel, B. U. von Specht, H. Domdey. *J Bacteriol* **1988**, *170*, 155–162.

15 C. A. Caulcott, M. R. W. Brown, I. Gonda. *FEMS Microbiol Lett* **1984**, *21*, 119–123.

16 H. Nikaido, E. Y. Rosenberg, J. Foulds. *J Bacteriol* **1983**, *153*, 232–240.

17 H. Nikaido, E. Y. Rosenberg. *J Gen Physiol* **1981**, *77*, 121–135.

18 H. Yoneyama, A. Akatsuka, T. Nakae. *Biochem Biophys Res Commun* **1986**, *134*, 106–112.

19 J. B. Stock, B. Rauch, S. Roseman. *J Biol Chem* **1977**, *252*, 7850–7861.

20 H. Yoneyama, T. Nakae. *Eur J Biochem* **1986**, *157*, 33–38.

21 G. M. Decad, H. Nikaido. *J Bacteriol* **1976**, *128*, 325–336.

22 E. Yoshihara, N. Gotoh, T. Nakae. *Biochem Biophys Res Commun* **1988**, *156*, 470–476.

23 H. Nikaido, K. Nikaido, S. Harayama. *J Biol Chem* **1991**, *226*, 770–779.

24 T. Nakae. *Biochem Biophys Res Commun* **1976**, *71*, 877–884.

25 H. Nikaido, E. Y. Rosenberg. *J Bacteriol* **1983**, *153*, 241–252.

26 F. Yoshimura, H. Nikaido. *Antimicrob Agents Chemother* **1985**, *27*, 84–92.

27 F. Bellido, N. L. Martin, R. J. Siehnel, R. E. W. Hancock. *J Bacteriol* **1992**, *174*, 5196–5203.

28 E. Yoshihara, T. Nakae. *J Biol Chem* **1989**, *264*, 6297–6301.

29 W. A. Woodruff, R. E. W. Hancock. *J Bacteriol* **1988**, *170*, 2592–2598.

30 N. Gotoh, H. Wakebe, E. Yoshihara, T. Nakae. *J Bacteriol* **1989**, *171*, 983–990.

31 W. A. Woodruff, R. E. W. Hancock. *J Bacteriol* **1989**, *171*, 3304–3309.

32 H. Nikaido, E. Y. Rosenberg. *J Bacteriol* **1983**, *153*, 241–252.

33 J. Trias, H. Nikaido. *J Biol Chem* **1990**, *265*, 15680–15684.

34 H. Huang, D. Jeanteur, F. Pattus, R. E. W. Hancock. *Mol Microbiol* **1995**, *16*, 931–941.

35 R. E. W. Hancock, A. M. Carey. *J Bacteriol* **1980**, *140*, 901–910.

36 J. Trias, E. Y. Rosenberg, H. Nikaido. *Biochim Biophys Acta* **1990**, *938*, 493–496.

37 H. Yoneyama, T. Nakae. *Microbiology* **1996**, *142*, 2137–2144.

38 R. E. W. Hancock, E. A. Worobec. In: T. C. Montie (Ed.), *Pseudomonas*. Plenum Press, New York, **1998**, pp. 139–167.

39 R. E. W. Hancock, F. S. L. Brinkman. *Annu Rev Microbiol* **2002**, *56*, 17–38.

40 E. Dé, R. De Mot, N. Orange, N. Saint, G. Molle. *FEMS Microbiol Lett* **1995**, *127*, 267–72.

41 E. De, N. Orange, N. Saint, J. Guerillon, R. De Mot, G. Molle. *Microbiology* **1997**, *143*, 1029–35.

42 N. Saint, C. El Hamel, E. Dé, G. Molle. *FEMS Microbiol Lett* **2000**, *190*, 261–5.
43 C. El Hamel, M. A. Freulet, M. Jaquinod, E. Dé, G. Molle, N. Orange. *Biochim Biophys Acta* **2000**, *1509*, 237–44.
44 C. A. Ullstrom, R. Siehnel, W. Woodruff, S. Steinbach, R. E. W. Hancock. *J Bacteriol* **1991**, *173*, 768–75.
45 R. Benz, K. Janko, P. Läuger. *Biochim Biophys Acta* **1979**, *551*, 238–247.
46 S. Chevalier, J. F. Burini, M. A. Freulet–Marriere, C. Regeard, G. Schoofs, J. Guespin-Michel, R. De Mot, N. Orange. *Res Microbiol* **2000**, *151*, 619–27.
47 K. Kanazawa, Y. Kobayashi, M. Nakano, M. Sakurai, N. Gotoh, T. Nishino. *FEMS Microbiol Lett* **1995**, *127*, 181–186.
48 H. M. Wexler, E. K. Read, T. J. Tomzynski. *Anaerobe* **2002**, *8*, 180–191.
49 A. Yotsuji, J. Mitsuyama, R. Hori, T. Yasuda, I. Saikawa, M. Inoue, S. Mitsuhashi. *Antimicrob Agents Chemother* **1988**, *32*, 1097–9.
50 R. E. Throm, J. A. Al-Tawfiq, K. R. Fortney, B. P. Katz, A. F. Hood, C. A. Slaughter, E. J. Hansen, S. M. Spinola. *Infect Immun* **2000**, *68*, 2602–2607.
51 E. Sugawara, H. Nikaido. *J Biol Chem* **1992**, *267*, 2507–2511.
52 N. Saint, E. Dé, S. Julien, N. Orange, G. Molle. *Biochim Biophys Acta* **1993**, *1145*, 119–23.
53 A. Arora, D. Rinehart, G. Szabo, L. K. Tamm. *J Biol Chem* **2000**, *275*, 1594–2000.
54 G. Ried, R. Koebnik, I. Hindennach, B. Mutschler, U. Henning. *Mol Gen Genet* **1994**, *243*, 127–135.
55 R. De Mot, J. Vanderleyden. *Mol Microbiol* **1994**, *12*, 333–334.
56 R. Koebnik. *Mol Microbiol* **1995**, *16*, 1269–1270.
57 R. Morona, M. Klose, U. Henning. *J Bacteriol* **1984**, *159*, 570–578.
58 A. Pautsch, G. E. Schulz. *Nat Struct Biol* **1998**, *5*, 1013–1017.
59 A. Pautsch, G. E. Schulz. *J Mol Biol* **2000**, *298*, 273–282.
60 C. Stathopoulos. *Protein Sci* **1996**, *5*, 170–173.
61 E. G. Rawling, N. L. Martin, R. E. W. Hancock. *Infect Immun* **1995**, *63*, 38–42.
62 R. S. Wong, H. Jost, R. E. W. Hancock. *Mol Microbiol* **1993**, *10*, 283–292.
63 R. S. Wong, R. A. Wirtz, R. E. W. Hancock. *Gene* **1995**, *158*, 55–60.
64 H. E. Gilleland, Jr, E. E. Hughes, L. B. Gilleland, J. M. Matthews-Greer, J. Staczek. *Curr Microbiol* **1995**, *31*, 279–286.
65 L. B. Gilleland, H. E. Gilleland, Jr. *Infect Immun* **1995**, *63*, 2347–2351.
66 E. Sugawara, M. Steiert, S. Rouhani, H. Nikaido. *J Bacteriol* **1996**, *178*, 6067–6069.
67 P. J. Bond, J. D. Faraldo-Gómez, M. S. P. Sansom. *Biophys J* **2002**, *83*, 763–775.
68 F. S. Brinkman, M. Bains, R. E. W. Hancock. *J Bacteriol* **2000**, *182*, 5251–5255.
69 F. S. Brinkman, M. Bains, R. E. W. Hancock. *J Bacteriol* **2000**, *182*, 6863.
70 E. Sugawara, H. Nikaido. *J Biol Chem* **1994**, *269*, 17981–17987.
71 A. L. Harris, A. Walter, J. Zimmerberg. *J Membr Biol* **1989**, *109*, 243–250.
72 C. M. Dobson. *Phil Trans R Soc Lond B Biol Sci* **2001**, *356*, 133–45.
73 I. Sonntag, H. Schwartz, Y. Hirota, U. Henning. *J Bacteriol* **1978**, *136*, 280–285.
74 M. A. Freulet–Marriere, C. El Hamel, S. Chevalier, E. Dé, G. Molle, N. Orange. *Res Microbiol* **2000**, *151*, 873–6.
75 H. Nikaido. *Semin Cell Dev Biol* **2001**, *12*, 215–223.
76 S. P. Singh, Y. U. Williams, S. Miller, H. Nikaido. *Infect Immun* **2003**, *71*, 3937–3946.
77 E. Sugawara, H. Nikaido. *Biophys J* **1997**, *72*, A138.
78 N. R. Movva, K. Nakamura, M. Inouye. *Proc Nat Acad Sci USA* **1980**, *77*, 3845–3849.
79 R. N. Movva, P. Green, K. Nakamura, M. Inouye. *FEBS Lett* **1981**, *128*, 186–190.
80 U. Mallick, P. Herrlich. *Proc Nat Acad Sci USA* **1979**, *76*, 5520–5523.
81 I. Gibert, J. Barbé. *FEMS Microbiol Lett* **1990**, *68*, 307–312.
82 M. J. Hansen, L.-H. Chen, M. L. S. Fejzo, J. G. Belasco. *Mol Microbiol* 1994, 12, 707–716.

83 L. M. Stancik, D. M. Stancik, B. Schmidt, D. M. Barnhart, Y. N. Yonchera, J. L. Slonczewski. *J Bacteriol* **2002**, *184*, 4246–4258.

84 R. J. Rowbury, Z. Lazim, M. Goodson. *Lett Appl Microbiol* **1996**, *23*, 426–430.

85 J. M. Calvo, R. G. Matthews. *Microbiol Rev* **1994**, *58*, 466–490.

86 G. F.-L. Ames, E. N. Spudich, H. Nikaido. *J Bacteriol* **1974**, *117*, 406–416.

87 K. Sen, H. Nikaido. *J Bacteriol* **1991**, *173*, 926–928.

88 M. G. Beher, C. A. Schnaitman. *J Bacteriol* **1981**, *147*, 972–985.

89 P. V. Bulieris, S. Behrens, O. Holst, J. H. Kleinschmidt. *J Biol Chem* **2003**, *278*, 9092–9099.

90 F. S. L. Brinkman, G. Schoofs, R. E. W. Hancock, R. de Mot. *J Bacteriol* **1999**, *181*, 4746–4754.

8
Drug Efflux and Protein Export through Channel-tunnels

Christian Andersen

8.1
Introduction

The outer membrane is an essential component of the cell envelope of Gram-negative bacteria. It forms an additional permeability barrier and provides the cells with an increased resistance to antibiotics, digestive enzymes, detergents, and host-defense proteins [1–3]. The advantage of this additional layer is purchased by more complex strategies for exporting substances into the surrounding environment. In addition to the cytoplasmic or inner membrane, exported substances like proteins or drugs have to pass the periplasmic space and have to overcome the second permeability barrier, the outer membrane. The outer membrane is not energized apart from a small Donnan potential [4]. Thus, for the latter export step there is no direct energy in the form of membrane potential, ATP or proton motif force as used for transporting substances across the inner membrane.

In the case of protein export out of the cell, one can distinguish roughly between sec-dependent and sec-independent export pathways. In the sec-dependent export pathways, proteins with a cleavable N-terminal secretion signal are directed to the sec-system in the inner membrane, which transports them into the periplasmic space [5, 6]. This route is also known as the general secretion pathway (GSP). After or during the translocation the secretion signal is cleaved resulting in processed periplasmic intermediates. For further export across the outer membrane one differentiates between six different groups. The chaperon/usher pathway, the single accessory pathway, the autotransporter, the type II secretion system (secreton), the type IV pilus biogenesis systems and the type IV secretion system [7–12].

There are two distinct sec-independent secretion machineries. Both systems form assemblies which span the whole cell envelope and export proteins without a periplasmic intermediate. The type III secretion system is known from the injection needle of pathogens and from the flagellar export apparatus. These are complex assemblies formed by up to 20 proteins [13, 14]. The other sec-independent export apparatus is the type I secretion system described in detail in the following section.

Bacterial and Eukaryotic Porins. Edited by Roland Benz

ISBN 3-527-30775-3

The export of drugs and cations is mediated by several distinct transporters in the inner membrane. They can be divided into five families [15, 16]. One group belongs to the ancient superfamily of ATP-binding cassette (ABC) transporters using ATP as an energy source. The other four families are energized by proton motif force and act as proton antiporters. They belong to the major facilitator superfamily (MFS), the small multidrug resistance (SMR) family, the resistance nodulation cell division (RND) family or the multidrug and toxin extrusion (MATE) family [17–20]. Without any accessory proteins these inner membrane transporters would not be able to export their substrates across the outer membrane. The exported drugs or cations would remain in the periplasmic space and could enter again immediately without any further hindrance. For three of the four transporter families it is known that members interact with two other proteins to expel drugs or cations out of the cell. These so-called efflux pumps are described in more detail in the following section.

8.2
Channel-tunnel-dependent Export Systems

Type I secretion systems and efflux pumps have in common that all systems engage an outer membrane protein of the channel-tunnel family to export substrates out of the cell (Figure 8.1). In addition to this outer membrane protein, the export machineries are composed of an inner membrane transporter as well as a third protein called an accessory protein also known as membrane fusion protein (MFP).

8.2.1
The Type I Secretion System

The type I protein secretion system is a sec-independent protein secretion system, which exports proteins without any periplasmic intermediate directly into the surrounding medium. Two subtypes of the type I secretion system can be classified depending on the nature of the exported substrate: the classical type for the export of substrates with a C-terminal secretion signal and a closely related system for substrates with an N-terminal secretion signal.

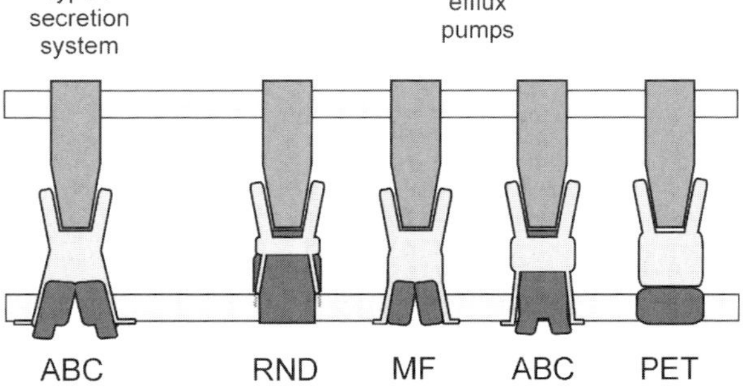

Figure 8.1 Schematic presentation of channel-tunnel dependent export systems. The tripartite export apparatus is composed of an inner membrane transporter of the ABC, RND, MF or PET family (dark gray), a periplasmic accessory protein (light gray) and an outer membrane channel-tunnel (gray).

8.2.1.1 Substrates

Substrates with a C-terminal secretion signal

One subtype of type I secretion systems exports proteins with a C-terminal secretion signal. This domain is located within the last 60 residues of the C-terminus and is not split off during or after translocation [21]. Hybrid constructs with the secretion signal fused to the C-terminus of various proteins were successfully secreted by the corresponding export systems [22–25]. This shows that the C-terminal secretion signal is sufficient to direct proteins to the secretion apparatus and no further domains within the exported substrates are needed.

There are various studies investigating the secretion signal [21, 26–28]. Comparison of the primary sequence of diverse C-terminal secretion signals revealed low similarity. However, three conserved motifs are found: an extreme C-terminal eight residues-long hydrophobic region, proximal to this a 13 residues-long uncharged sequence and proximal to both these features a 22-residues amphipathic α-helical domain [21]. Mutations in any of these regions affected the secretion efficiency. When subjected to circular dichroism and nuclear magnetic resonance (NMR) measurements the signal peptides are shown to be highly flexible and unstructured in aqueous solution and in mimetic environments [29–32].

The size of the secreted proteins varies between several hundred and more than 4000 residues. Examples for this group of substrates are RTX (Repeats in Toxin) toxins like the *Escherichia coli* hemolysin or *Bordetella pertussis* adenylate cyclase, S-layer proteins or extracellular enzymes like proteases, lipases or exopolysaccharide processing enzymes [33–41]. Recently, a secreted protein was characterized which is required for biofilm formation of *Pseudomonas fluorescens* [42].

The possibility to secrete hybrid proteins has been applied to immunological and vaccine research. The *E. coli* hemolysin secretion system could be used to export diverse antigens fused to the hemolysin secretion signal [43–49]. Because the hemolysin signal sequence represents a very weak antigen for B cells and T cells, this system is excellently suited for the presentation of antigens in attenuated Gram-negative bacterial live vaccines [23].

Substrates with an N-terminal secretion signal

The other subtype of type I secretion system has a secretion signal located at the N-terminus of the substrate. In contrast to the C-terminal secretion signal, it is cleaved during or after secretion [50, 51]. This makes it similar to the leader peptides recognized by the sec-system of the GSP. However, the N-terminal secretion signal is not hydrophobic and comprises a characteristic double-glycine motif between position 15 and 20 marking the cleaving site. Prediction of the secondary structure of the so-called double-glycine leader peptide revealed that most of the leader peptide might be folded as a α-helix followed by a β-turn directly upstream of the double-glycine motif [52].

The size of the substrates is also not comparable with that of substrates with a C-terminal secretion signal. The unprocessed proteins vary in size between 58 and 105 residues [53, 54]. Substrates belong to the family of microcins, which are defined as post-translationally modified peptides with antibiotic activity exported by Gram-negative bacteria. The best-characterized microcins derive from *E. coli* strains and *Klebsiella pneumoniae*, e. g. colicin V, microcin H47 or microcin E492 [51, 55, 56]. A databank search revealed potential microcins in other bacteria like *P. aeruginosa*, *Xylella fastidiosa* or *Neisseria meningitidis* [52].

Interestingly, there exist operons coding for similar substrates and export systems in Gram-positive bacteria. The substrates are also small peptides possessing a double-glycine leader peptide, which is cleaved off concomitant with export [57, 58]. They act either as antibacterial peptides called lantibiotics or bacteriocins (also termed non-lantibiotics) or as competence-stimulating peptides – so-called competence pheromones [59–61]. The parallels suggest that microcins secreted by Gram-negative bacteria might also serve as pheromones to sense the cell density of bacterial populations [52].

8.2.1.2 The Inner Membrane Transporters of Type I Secretion Systems

The translocation of substrates across the inner membrane is mediated by ABC transporters, which according to the transporter classification (TC) system of M. Saier belong to the groups TC 3.A.1.109 and TC 3.A.1.110 [62]. Microcin secretion systems employ exclusively ABC transporters of the latter group, while substrates with a C-terminal secretion signal are transported by both types of ABC transporter. ABC transporters which are part of type I secretion systems consist of two domains, an N-terminal transmembrane domain (TMD) embedded in the inner membrane and a C-terminal ATP-binding or nucleotide binding domain (NBD). It is thought that these transporters act as homodimers. There is no structural information of a complete ABC transporter involved in type I protein secretion, but the structures of two other members of the large ABC transporter superfamily are solved – the lipid A exporter MsbA and the vitamin B_{12} importer BtuBC [63, 64]. The structure of MsbA reveals that the membrane-spanning part of the functional unit consists of 12 transmembrane helices (six per monomer). Two ATP-binding domains (one per monomer) are linked to the cytoplasmic side of the membrane domain providing energy for the translocation process. Structural predictions show that this architecture can be transferred to that of ABC transporters involved in type I secretion. Indeed, the structure of the NBD of HlyB, an ABC transporter involved in hemolysin secretion by *E. coli*, was crystallized recently and shown to exhibit a similar overall structure to other known NBDs [65, 66].

The size of ABC transporters varies between 550 and 750 residues. Interestingly, it could be shown for transporters of lantibiotics of Gram-positive bacteria that the proteolytic function, which is responsible for cleaving off the double-glycine leader peptide, resides in the 150 N-terminal residues of the transporters [67]. Thus, the transporter has a dual function: in addition to the the translocation of the substrate across the inner membrane, it is also involved in processing the substrate. This extra domain upstream of the TMD is also found in ABC transporters for microcin secretion in Gram-negative bacteria, but does not exist in most of the transporters of the C-terminal signal dependent type I systems. One exception is HlyB, the transporter involved in hemolysin secretion by *E. coli*. This could be a hint that HlyB might also be proteolytically active. Another hint is that the hemolysin secretion system is capable of secreting colicin V, a microcin with a double-glycine leader peptide [68]. It is also shown that the export signal in colicin V recognized by the hemolysin secretion system is localized within the N-terminus, which means that the hemolysin secretion system is not only able to secrete substrates with a C-terminal secretion signal, but also substrates with an N-terminal secretion signal.

Recently, it could be shown that the NBD of the hemolysin transporter HlyB interacts specifically with the secretion signal of the substrate hemolysin HlyA [69]. This binding might be one of the first steps in the secretion process initiating the export of the substrate.

8.2.1.3 The Accessory Protein of the Type I Secretion System

The accessory proteins involved in the type I protein secretion also known as MFP are classified as TC 8.A.1.3.1 [62]. The length varies between 350 and 500 residues. The best-studied member of this family is HlyD, the accessory protein of the hemolysin secretion system of *E. coli*. In contrast to the putative dimeric inner membrane transporter HlyB, HlyD assembles as a homotrimer. Cross-linking experiments have shown that HlyB and HlyD form a stable complex in the inner membrane, which means that there may be a break in symmetry within the complex [70]. HlyD can be separated in three domains. The 59 N-terminal residues form a cytoplasmic domain, which is linked by a 21-residue transmembrane domain to the large periplasmic domain (residues 81–478). It could be shown by cross-linking that the cytoplasmic domain binds the substrate hemolysin independently of the presence of HlyB and that this interaction is crucial for secretion [70]. Structure prediction for the cytoplasmic domain revealed an amphipathic helix, which is most likely embedded in the inner membrane followed by a cluster of charged residues [71]. Deletion mutants lacking the cytoplasmic domain were still able to trimerize and to complex with HlyB, showing that this domain is not necessary for assembly of a proper HlyBD complex. The major part of the protein is located in the periplasmic space. Topology prediction of the transmembrane domain of the transporter HlyB reveals that the periplasmic loops connecting the transmembrane helices are rather small [72], which means that the transporter has no excessive periplasmic extension. One can assume that the interface between the trimeric periplasmic domain of the accessory protein and the transporter is almost inplane with the periplasmic border of the inner membrane. A close contact with a large surface between these two proteins is supported by the observation that mutations in the hemolysin transporter HlyB can be suppressed by a C-terminal mutation of the accessory protein HlyD [73]. This means that not only the N-terminal region but also the C-terminal region of the periplasmic domain must be in close proximity to the inner membrane transporter.

Secondary structure prediction for the periplasmic domain shows that following the transmembrane domain there is a region with a high probability for coiled-coils. The potential coiled-coil region is flanked by motifs found in domains of enzymes that transfer covalently attached lipoyl or biotinyl moieties between enzymatic components [74]. In these proteins two approximately 30 residues-long motifs form four β-strands, which assemble into a so-called lipoyl domain as shown in several structures solved by NMR or X-ray crystallography [75–77]. The two motifs are connected by a three residues long linker. In accessory proteins of the type I secretion system the two motifs are separated by 140–200 residues. The C-terminal end of the accessory protein is predicted to contain β-strands. The importance of this part is underlined by the observations that mutations in this region very often lead to instable or non-functional proteins [78, 79]. A rough model of the overall structure of accessory proteins proposes that the periplasmic domain adopts a ring-like structure [80]. Assuming that the inner membrane transporter releases the substrate at the periplasmic surface, one can expect that the substrate has to pass the accessory protein to reach the outer membrane component. A ring-like structure would fulfill the condition. Assuming that the coiled-coil domain is not involved in trimerization, it is the C-terminal part of the accessory protein which mediates trimerization into a ring-like structure.

8.2.2 Efflux Pumps

Efflux pumps mediate resistance against diverse noxious substances like antibiotics, dyes, detergents, bile salts or heavy metals. They may, for example, raise antibiotic resistance by several orders of magnitude, rendering antibiotics clinically useless [81]. The involvement of an outer membrane channel-tunnel makes it possible to export the substance out of the cell into the medium. In order to enter the cell again the expelled molecules must cross the outer membrane. Thus, the synergistic interplay between the efflux pump and the low permeability of outer membrane is highly advantageous for Gram-negative bacteria. Like type I secretion systems, the efflux pumps are tripartite export systems consisting of an inner membrane transporter, a periplasmic accessory protein and an outer membrane channel-tunnel.

8.2.2.1 The Inner Membrane Transporters of Channel-tunnel-dependent Efflux Pumps

Members of three distinct transporter families can be employed by channel-tunnel-dependent efflux pumps. Inner membrane transporters belong either to the RND superfamily, to the MF superfamily or to the superfamily of ABC transporters. Transporters of the RND and MF superfamilies are driven by proton motif force, while ABC transporters hydrolyze ATP as an energy source for transport. Genomic analysis revealed that there might exist a fourth transporter family, the putative extrusion transporters (PET), which might also act in cooperation with an accessory and a channel-tunnel protein.

Transporters of the RND superfamily

The RND transporter superfamily is a ubiquitous family with representation in all major kingdoms. It is grouped in seven families; two of them are involved in channel-tunnel-dependent efflux [19]. Dependent on the exported substrate one distinguishes between the heavy metal efflux transporters (HME, TC 2.A.6.1), and the hydrophobic and amphiphilic compounds efflux transporters (HAE; TC 2.A.6.2).

Recently, the structure of the transporter AcrB from *E. coli* was solved at 3.5 Å resolution [82]. It is the first structure of a transporter which is driven by proton motive force transducing free energy stored as an electrochemical proton gradient into a substrate concentration gradient. AcrB belongs to the HAE transporters and is part of the well-studied AcrAB/TolC multidrug efflux pump of *E. coli*. The crystal structure revealed that it functions as a homotrimer, which was somehow surprising because it was long been believed to function as monomer [83]. The 1049 residues-long peptide chain can be divided into two halves with homologous sequences implying that RND transporters arose as result of a tandem, intragenic duplication event. The duplication is also reflected in the tertiary structure of the transporter, which is composed of two halves with similar architecture. For a homolog RND transporter of *P. aeruginosa* it could be shown that the two halves could be expressed independently without loosing functionality [84].

As a whole, the trimer looks like a jellyfish (Figure 8.2). The 50 Å thick membrane embedded part has a diameter of 80 Å. It is composed of 36 transmembrane α-helices – 12 from each protomer. The three transmembrane domains of the protomers are arranged in a ring-like manner with a central hole, which spans the transmembrane domain. The hole has a diameter of about 30 Å and is most likely filled with phospholipids. Two periplasmic loops between transmembrane helices 1 and 2 as well as 7 and 8 of each protomer form the headpiece protruding 70 Å into the periplasm. The headpiece is divided into two stacked parts. The upper and lower parts are 30 and 40 Å thick, respectively. The bottom part is called the pore domain. Its outer diameter is 100 Å. In its center is a pore formed by three helices – one by each protomer. Seen from the top the pore is almost closed. At its proximal end the pore connects to a central cavity. Assuming that the hole in the transmembrane domain is filled with phospholipids, the height of the cavity measured from the membrane plane to the pore entrance is 15 Å. The total volume of the cavity is around 5000 $Å^3$. There are three openings, called vestibules, at the junctions of the protomer headpieces just outside the plane of the membrane. These vestibules connect the central cavity with the periplasm and might serve as conduits for substrate entry (see below). The side view of the upper part of the headpiece, called the TolC docking domain, has a trapezoidal appearance with a diameter of 70 Å at the bottom and 40 Å at the top. It forms a funnel with a maximal internal diameter of 30 Å. The bottom of the funnel is connected to the pore, which thus is the link between the funnel of the upper part and the cavity of the lower part of the periplasmic protrusion.

The crystal structure of AcrB greatly advanced the understanding of multidrug efflux pumps. The subsequent crystallization of AcrB complexed with four differ-

Figure 8.2 The crystal structure of the RND transporter AcrB from *E. coli*. The front protomer of the trimeric protein is omitted for clarity. The remaining two protomers are shaded differently. The jellyfish-shaped structure is divided into a transmembrane domain, a pore domain and a TolC-docking domain. The shaded background represents the surface representation of the transporter. The vestibule, cavity, pore, and funnel are shaded darker. The membrane plane is indicated by two bars. The figure was generated by using WEBLABVIEWER (Accelrys, Cambridge UK). The crystal data are taken from [82].

ent ligands [85] gave further insight into the mechanism of multidrug resistance by efflux pumps (see below).

Transporters of the MF superfamily

The MF superfamily is also a ubiquitous family of transporters which are driven by electrochemical gradients to import or export substances across membranes. One out of 17 families is involved in channel-tunnel-dependent export of drugs. The drug:H^+ antiporter (14 spanner) drug-efflux family (DHA14) is categorized as TC 2.A.1.3. The size of the transporters ranges between 500 and 550 residues, comprising 14 transmembrane helices. Like the RND transporters, MF transporters are most likely evolved by gene duplication. They consist of two halves with homologous sequences. The prototype of this family is EmrB of *E. coli*, which together with the accessory protein EmrA and the outer membrane channel-tunnel forms an efflux pump for substances like carbonyl cyanide *m*-chlorophenylhydrazone (CCCP) or methylviologen [86].

Recently, structures of two other MF transporters were solved [87, 88]. They belong to other families of the MF superfamily. Glycerol-3-phosphate transporter (GlpT) of *E. coli* is a member of the organophosphate:inorganic phosphate antiporter (OPA) family (TC 2.A.1.4) and lactose permease of *E. coli* (LacY) belongs to the oligosaccharide:H^+ symporter (OHS) family (TC 2.A.1.5). Despite the fact that these transporters belong to different MF families and have only 12 transmembrane helices one gets an idea what the overall structure of MF transporter looks like. The transporters are functional as monomers. The major part of the peptide chain resides in the membrane forming a compact structure. Between the N- and C-terminal domain there exists an internal hydrophilic cavity open to the cytoplasmic side. Similar to ABC transporters, the periplasmic loops between the transmembrane helices of MF transporter are small, meaning that the transporters have no extensive periplasmic protrusions. This is a clear difference to transporters of the RND superfamily.

Transporters of the ABC superfamily

It has long been known that efflux pumps involve transporters of the RND and MF superfamilies. The fact that ABC transporters also function in cooperation with an accessory protein and an outer membrane channel-tunnel for the export of drugs was first reported in 2001 [89]. MacB from *E. coli* is characterized as an inner membrane protein with an NBD, which acts together with MacA, a member of the accessory protein family coded by the same operon, and the channel-tunnel TolC as an efflux pump for macrolide antibiotics. MacB is the first characterized transporter of this family numbered TC 3.A.1.122. The size of transporters of this family is not different to that of ABC transporters involved in the type I secretion system (around 650 residues). However, a recent study has underlined structure predictions showing that the folding of both families is different [90]. The main difference is that the NBD is not C-terminal, but located at the N-terminus, and that the membrane-embedded part is composed of only four transmembrane helices and not six as in the case of ABC transporters involved in protein secretion. Another interesting feature of this family is that the topology study revealed that the transporter has a 230 residues-long periplasmic loop, which also distinguishes it from other ABC transporters. Nothing is known about the oligomeric state of the MacB transporter. It is speculated that it might act as hexamer [90]. This would mean that the periplasmic extension would consist of six periplasmic loops analogous to the six periplasmic loops of the AcrB trimer.

PET family

The search for members of the accessory protein family revealed that genes of this family could also be linked to genes of an uncharacterized transporter family, PETs (TC 9.B.4.1). There are also homologs found in yeast and plants, showing that it is a ubiquitous transporter family [91]. It seems reasonable that these operons in bacteria also code for channel-tunnel-dependent export machineries. This assumption that the proteins of the PET family function together with an accessory protein and a channel-tunnel as export machinery remains to be verified. It would mean that in addition to the three known transporter families there is a fourth family which is able to assemble with an accessory protein and an outer membrane channel-tunnel forming an efflux pump-like export apparatus. The size of these transporters is in the range of ABC transporters (around 650–700 residues), but there are no motifs for a NBD showing that they are not energized by ATP hydrolysis. The primary structure analysis exhibits a repeat sequence due to an internal gene duplication event. There are six putative transmembrane domains at the N-terminal end of each repeat followed by a large hydrophilic domain. Assuming that the N-terminus is located in the cytoplasm, this means that PET transporters have no periplasmic extension, but large cytoplasmic domains.

8.2.2.2 Accessory Proteins of Multidrug Efflux Pumps

The accessory proteins involved in efflux pumps are grouped in several families [62]. The classification correlates with the different transporter families they are interacting with. One distinguishes between the TC 8.A.1.1.1, TC 8.A.1.2.1 and TC 8.A.1.6.1 families of accessory proteins which act together with transporters of the MF, HAE RND, and the HME RND transporter family, respectively. Accessory proteins interacting with ABC or PET transporters are not yet categorized. There are some features which all accessory proteins have in common.

A molecular fingerprint of accessory proteins is the so-called lipoyl domain already described above. These two approximately 30 residues-long motifs, each forming β-strands, are found in all members of different accessory proteins [92, 93]. The structure of this domain is known from enzymes that transfer covalently attached lipoyl or biotinyl moieties between enzymatic components [74–77]. The two motifs are connected by a three residues-long linker in these proteins. In the case of accessory proteins, this linker is much longer. The size of the linker is 41–65 (accessory proteins of RND transporters), 64 (PET family), 79 (ABC transporters) and 111–121 residues (MF transporters), respectively. The secondary prediction for this linker region reveals for all different accessory protein families that there are two domains with a high coiled-coil probability separated by a gap of five to ten residues. It is predicted that the coiled-coil domains form an α-helical hairpin. This is confirmed by the fact that the two regions are of approximately equal length (four to five heptad repeats) and that among different accessory proteins the length of the regions frequently differ by integers of seven.

Accessory proteins possess a membrane anchor at the N-terminal end. This is either a transmembrane helix or a covalently bound fatty acid serving as a lipid anchor. Accessory proteins interacting with MF transporters are anchored in the membrane by a transmembrane helix. Accessory proteins interacting with RND transporters have a lipid anchor. There are no experimental data available for accessory proteins interacting with ABC transporters or with members of the PET family. There is no cysteine residue following the hydrophobic region at the N-terminus. Therefore, it needs to be experimentally shown if this hydrophobic N-terminus serves as a leader sequence or as a membrane anchor. It should be mentioned that the lipid anchor is not necessary for the function of accessory proteins interacting with RND transporters. Deletion of the acylation site had no effect on the functionality of the efflux pump [94, 95].

For the RND transporter AcrB and the corresponding accessory protein AcrA, it could be shown by cross-linking that both proteins form a complex. This is not surprising considering that the anchor keeps the accessory protein attached to the membrane. The specificity of interaction between the accessory protein and the transporter lies in periplasmic loops as shown by chimeric RND transporters [96]. For efflux pumps involving transporters of the other families, it seems reasonable to assume that these transporters also form complexes with the corresponding accessory proteins.

The oligomeric state of accessory proteins is only experimentally proven for AcrA, the accessory protein of the *E. coli* multidrug efflux pump AcrAB/TolC. It assembles into trimers as does the accessory protein HlyD of the type I protein secretion system [70, 97]. One can therefore assume that all accessory proteins assemble into trimeric structures. In the case of the AcrAB complex, with both partners forming trimers, there is no break in symmetry as is predicted for the inner membrane complex HlyBD from the hemolysin secretion system. Analytical ultracentrifugation and dynamic light scattering have shown that AcrA exists in solution as a highly asymmetric molecule with an axial ratio of eight and a predicted length of 170 Å [98]. This result has to be taken with care because the protein used in this study was not in its natural configuration as a trimer, but monomeric. Thus, it might be that the elongated structure does not represent the natural folding. Another study of the same protein using lipid-layer crystallization revealed a preliminary three-dimensional structure [99]. The protein was organized as a ring-like structure with a central opening of 30 Å. The oligomeric state of AcrA in this study is dimeric, which again leaves doubt if the structure corresponds with the native structure. Therefore, it is one of the forthcoming tasks to solve the structure of the members of the accessory protein families.

8.2.3
Comparison of Channel-tunnel-dependent Export Systems in *E. coli* and *P. aeruginosa*

Analysis of the genome of diverse bacteria reveals that the genes of the three families forming a functional export apparatus are often genetically linked. Table 8.1 lists operons of known and putative type I secretion systems and drug efflux pumps found in the genomes of *E. coli* and *P. aeruginosa*. These bacteria were chosen because they are the best-studied organisms with regard to these export mechanisms.

Comparison of the two organisms shows that in the case of *E. coli* only three of 17 operons include a gene coding for a channel-tunnel, while in *P. aeruginosa* 17 out of 26 operons contain genes for this outer membrane protein. For seven operons of *E. coli*, which do not comprise a gene for a channel-tunnel protein, it has been shown that functional export depends on the presence of TolC, a channel-tunnel protein coded by a gene of the stress induced *mar-sox* operon [33, 56, 89, 101–104]. For *P. aeruginosa* one might assume that the channel-tunnel OpmH, the gene which is not linked to any transporter gene, plays a similar role as TolC in *E. coli*. However, recent studies have shown that OpmH could only partially

Table 8.1 Operons of known and putative type I secretions systems and efflux pumps found in *E. coli* and *P. aeruginosa.*

Family	*Transporter*	*Accessory protein*	*Channel tunnel*	*Substrate*	*Operon*[c]
E. coli					
ABC	HlyB	HlyD	TolC[a]	hemolysin	TA
	CvaB	CvaA	TolC[a]	colicin V	AT
	MchF	MchE	TolC[a]	microcin	AT
	MtfB	MtfA	?	microcin	AT
	MacB	MacA	TolC[a]	drugs	AT
	YhiH, YhhJ	YhiI	?	?	ATT
	YbhS, YbhR, YbhF	YbhG	?	?	ATTT
RND	CusA	CusB	CusC	cations	CAT[b]
	YegN, YegO	YegM	YegB	drugs	ATTC
	AcrB	AcrA	TolC[a]	drugs	AT
	AcrF	AcrE	TolC[a]	drugs	AT
	YhiV	YhiU	?	?	AT
MFS	EmrB	EmrA	TolC[a]	drugs	AT
	EmrY	EmrK	?	drugs	AT
PET	YjcQ	YjcR	YjcP	?	ATC
	B1645	FusE	?	?	AT
	YhcP	YhcQ	?	?	AT
P. aeruginosa					
ABC	AprD	AprD	AprF	alkaline protease, AprX	TAC
	HasD	HasE	HasF	hem acquisition protein	TAC
	PA1876	PA1877	OpmL	hypothetical RTX protein	CTA
	PA4143	PA4142	OpmK	hypothetical colicin	ATC
	PA2390	PA2389	OpmQ	?	ATC
	PA4593–4595	PA4591	OpmF	?	TTTCA
RND	CzcA	CzcB	OpmN	cations	CAT
	MexB	MexA	OprM	drugs	ATC
	MexC	MexD	OprJ	drugs	CTA
	MexF	MexE	OprN	drugs	ATC
	MexY	MexX	OprMa	drugs	AT
	PA3522	PA3523	OpmE	?	ATC
	PA2526–2527	PA2528	OpmB	?	ATTC
	PA1436	PA1435	?	?	AT
	PA0158	PA0156, PA0157	?	?	AAT
	MexK	MexJ	OprM[a]	drugs	AT
	PA4375	PA4374	?	?	AT
	MexI	MexH	OpmD	cations	ATC

Table 8.1 Continued.

Family	*Transporter*	*Accessory protein*	*Channel tunnel*	*Substrate*	*Operon*[c]
MFS	PmrB	PmrA	OpmG	?	CAT
	PA2835	PA2836	OpmA	?	TAC
	PA3137	PA3136	?	?	AT
	PA1236	PA1237	OpmJ	?	CAT
PET	PA3893	PA3892	OpmI	drugs	CTA
	PA3305	PA3304	?	?	TA
	PA1232	PA1231	?	?	AT
	PA2431	PA2430	?	?	TA

[a]The channel-tunnel protein is essential for a functional export apparatus but is not part of the same operon.
[b]In the *cusABC* operon of *E. coli* there is a fourth gene *cusF* coding for a periplasmic 10-kDa protein, which has been shown to be involved in copper export [161].
[c]The operon structure is given in transcriptional order (T = transporter, A = accessory protein, C = channel tunnel).

serve as an outer membrane component for channel-tunnel-dependent export systems [105]. There are more hints that either OpmG, which is part of the *pmrAB/opmG* operon, or OprM, which belongs to the *mexAB/oprM* operon, are outer membrane proteins serving as the third component to assemble a functional export apparatus if a gene for a channel-tunnel is missing in the operon [105–107]. Generally, in *P. aeruginosa,* all three genes, which are necessary to express a functional efflux pump or type I secretion system, are normally coded in a single operon and are transcribed together. In *E. coli,* most operons lack genes for the outer membrane component. One can assume that the channel-tunnel TolC, which serves as an outer membrane component for several systems, is permanently present in the outer membrane so that the existence of a gene coding for an "own" channel-tunnel in operons coding for type I secretion systems or multidrug efflux pumps is not necessary. Assuming that the natural environment stresses the *E. coli* cells perpetually, it is understandable that stress-induced TolC expression provides enough molecules to saturate all systems which engage this outer membrane component. However, the engagement of TolC by several distinct transport systems means that TolC has to be compatible with all of them. As described in the next section, the channel-tunnel protein has direct contact with the accessory protein, which means that TolC must be able to interact with at least seven different proteins.

8.3 Channel-Tunnels

According to the classification (TC) system of Saier [62], the channel-tunnel family is named the outer membrane factor (OMF) categorized as TC 1.B.17. Channel-tunnels are found in almost all Gram-negative bacteria [108]. This confirms the importance of this outer membrane protein for the cells. A phylogenetic tree based on a sequence alignment of 36 TolC homologs shows that the channel-tunnel family splits into three subfamilies [109]. This division based on sequence alignments corresponds well with export functions the members of the three subfamilies are involved in. These are protein secretion, drug efflux and cation efflux. Prototypes of the three families are TolC of *E. coli*, OprM of *P. aeruginosa* and CzcC of *Ralstonia eutrophus*, respectively [33, 110, 111]. As mentioned above TolC is not only involved in protein secretion, but is also part of the diverse efflux pumps. This makes it an interesting protein to study because it must be able to interact with several distinct inner membrane complexes.

8.3.1
The Structure of TolC

In 2000, Koronakis et al. solved the structure of TolC by X-ray crystallography with a resolution of 2.1 Å [112]. TolC forms homotrimers, which are cannon-shaped with a long axis measuring 140 Å (Figure 8.3A). One end of the cylinder is open with an inner diameter of about 30 Å. The diameter is uniform for a length of 100 Å. Towards the other end, it becomes smaller, leading to almost closure of the cylinder. The interior of the cylinder is mostly solvent filled. The cavity has a volume of roughly 43000 $Å^3$ making it one of the largest known in protein structure.

The structure can be divided into three domains. The 40 Å long β or channel domain, the 100 Å long α-helical or tunnel domain and the equatorial domain, a mixed α/β domain, which forms a ‚strap' around the mid section of the tunnel

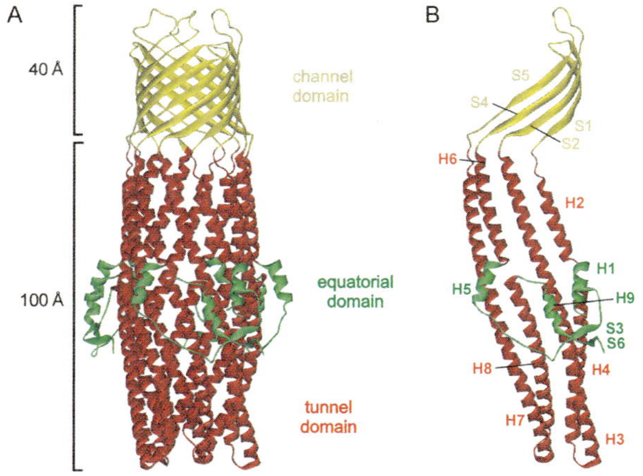

Figure 8.3 The crystal structure of the TolC protein from *E. coli*. The protein consists of three domains: a channel domain (yellow), a tunnel domain (red) and an equatorial domain (green). (A) Three TolC protomers are assembled into a cannon-shaped structure termed the channel-tunnel. (B) Structure of a TolC protomer. The secondary structural elements are labeled according Koronakis et al. [112].

domain. The channel domain and the tunnel domain form a continuous tube, which is the origin of the name of this protein family: channel-tunnels. The folding of a TolC monomer is shown in Figure 8.3B. Each monomer adds four β-strands (S1, S2, S4 and S5) to the channel domain, the tunnel domain is formed by two long (H3 and H7) and four shorter α-helices (H2, H4, H6 and H8), and the equatorial domain comprising the N- and C-terminus consists of small β-strand and α-helical structures (S3 and S6, and H1, H5 and H9, respectively).

It is evident that TolC comprises a structural repeat. The two halves (H1–H4 and H6–H9, respectively) are linked by parts of the equatorial domain. The structural repeat corresponds to a repeat in the primary structure, which was already recognized before the structure was solved [93, 113]. It is most likely that the family of channel-tunnels evolved by gene duplication from a common ancestor.

Over 80 members of the channel-tunnel family are listed in the databank so far. The overall length varies between 414 and 541 residues. This is due to variable extensions at the N- and C-termini. Gaps or insertions exist only in the extracellular loops or the equatorial domain. The length of the tunnel-forming helices is constant. It is clearly determined because residues that are important at the transitions between the different parts of the structure are highly conserved. These are glycine residues between the coiled-coils of the lower tunnel domain, which are necessary for the turn between the helices H3 and H4 and H7 and H8, respectively (Figure 8.3B). The transitions between the α-helices of the tunnel domain and the β-strands of the channel domain are characterized by highly conserved proline and glycine residues. The long helices (H3 and H7) consist of 67 residues, the four shorter helices consist of 23 (H2 and H6) and 34 (H4 and H8) residues, respectively. Insertions or deletions within this region are poorly tolerated [114, 115]. The tapering of the channel-tunnel is most likely also conserved within the family. Small residues like alanines and serines at the coiled-coil regions allowing dense packing of the coiled-coils are found in all homologs.

8.3.1.1 The Channel Domain

The channel domain anchors TolC in the outer membrane. It consists of 12 β-strands which are arranged in an antiparallel way to form a barrel. Apolar residues are located at the outside of the barrel facing the lipids, whereas polar residues are inside, resulting in typical amphipathic β-strands. At the base of the channel domain are aromatic side-chains, particularly tyrosine and phenylalanine, forming a ring facing outwards. Aromatic side-chains are known to localize at the membrane–water interface [116]. Thus, the aromatic ring anchors the barrel domain in the outer membrane delimiting the inner edge of the lipid bilayer.

The β-barrel is a common feature of all outer membrane proteins crystallized so far [117, 118]. The number of β-strands per barrel varies between eight and 22 [119, 120]. Nevertheless, the architecture of the β-barrel of channel-tunnels is different; β-barrels of all other known protein structures are formed by a single peptide chain. In the case of channel-tunnels, three protomers are necessary to form a single barrel each contributing four β-strands. Like proteins of the porin family, TolC forms water-filled barrels allowing diffusion of hydrophilic substrates across the outer membrane [121–123]. Despite a similar function, the interior of the TolC β-barrel is different from the interior of porins. The cross-sectional area of the TolC channel domain is 960 $Å^2$, making it 15-fold larger than that of the general porin OmpF. That is because TolC lacks the structural element of porins – an inward folded loop that constricts the internal diameter of the porin β-barrel. Residues located at this loop determine the channel properties, like the diameter, ion-selectivity or substrate affinity [124–126]. In addition to this important role, the inward folded loop is also important to stabilize the barrel. In the TolC channel-tunnel there are two extracellular loops per monomer enclosed by strands S1 and S2 and S4 and S5, respectively. The high crystallographic thermal disorder factors and the weaker electron density of the eight and 21 residues-long loops correspond with a high flexibility. It is unlikely that they fold into the barrel structure and form a stable barrier, restricting the channel entrance and stabilizing the barrel. The stabilization of the barrel in channel-tunnels is achieved by the elongation of the cylinder into the periplasm.

8.3.1.2 The Tunnel Domain

More than 80 % of the TolC peptide chain is located in the periplasmic space forming the equatorial and tunnel domain. This 100-Å periplasmic extension is the most striking difference to all other outer membrane proteins known so far [127]. The tunnel domain consists entirely of α-helices. Above the equatorial domain the helices assemble into an α-barrel. This is a structure which is not found in any other crystallized protein. Twelve helices (H2, H3, H6 and H7 of each monomer) arrange antiparallel to form an almost uniform cylinder with an inner diameter of around 20 Å [128]. The local packing of the helices is similar to a conventional coiled-coil, i. e. knobs-into-holes interaction [129]. However, in the α-barrel the helices do not wrap around each other, but form a cylinder. In this assembly one helix has two interfaces resulting in two sequence patterns that are phased to match the two contact sites. The formation of an α-barrel needs a special distribution of residues allowing bending and untwisting of the helices [128]. The α-barrel structure finishes with the equatorial domain, where the small helices H2 and H6 end. As mentioned above, the periplasmic extension is responsible for stabilization of the channel domain. The α-barrel is left twisted, different to the right-twisted β-barrel of the channel domain. The opposite twist might mutually stabilize the barrels so that the structures do not collapse.

Below the equatorial domain the continuous helices H3 and H7 pair with helices H4 and H8, respectively. H3 is straight and helix H4 coils around it forming the outer coiled-coil. The other pair, H7/H8, forms a conventional coiled-coil bending inwards. This inner coiled-coil is responsible for the tapering of the periplasmic end of TolC. Seen from the periplasmic side, the coiled-coils arrange like an iris to almost seal the periplasmic entrance (Figure 8.4A). The diameter of the periplasmic entrance is around 4 Å – too small to allow the passage of exported substrates.

8.3.2
Electrophysiological Characterization of TolC

Channel-tunnels provide an export duct across the periplasmic space and the outer membrane for substrates fed in by the inner membrane complexes. In 1993, the pore-forming ability of TolC was first described by Benz et al. [130]. TolC formed stable pores in the planar lipid bilayer with a single channel conductance of 85 pS in 1 M KCl. This is about 20-fold smaller than that of general diffusion pores like OmpF of *E. coli* [125]. This is somehow surprising considering that the cross-sectional area of the membrane-embedded β-barrel of TolC is 15-fold bigger than that of OmpF. The low single-channel conductance of TolC can be explained by the fact that the restriction zone is not in the membrane-embedded channel domain like in porins, but 100 Å apart at the periplasmic opening. Here, at the end of the tunnel domain, is the narrowest part of the conduit with a diameter of around 4 Å. Apart from the small diameter one has to consider that due to the distance to the membrane, the electric field across the tunnel entrance might be weaker than across a porin channel. This can also contribute to the lower conductance.

8.3.2.1 The Role of Aspartate Residues at the Periplasmic Entrance

It is already known from porins that residues lining the narrowest part of the channel have a strong influence on the electrophysiological properties [124–126]. As mentioned above, the narrowest part of the channel-tunnel is the periplasmic entrance. It is lined by six aspartate residues (two per monomer), establishing a highly electronegative area (Figure 8.4B). By comparison with a mutant having the aspartate residues exchanged for alanine residues ($TolC^{DADA}$) we demonstrated that the aspartate residues determine ion selectivity, pH dependence, and interaction with di- and trivalent cations [131]. Wild-type TolC is highly cation selective. Quantitative measurements show that it has a 16.5-fold preference of potassium ions over chloride ions. The exchange of the aspartate for alanine residues inverts the selectivity. $TolC^{DADA}$ is anion selective with a 10-fold preference of chloride ions over potassium ions. Another property of the TolC channel-tunnel also originates from the aspartate residues. We could show that the single-channel conductance of wild-type TolC becomes smaller when the pH decreased. At pH 3, the single-channel conductance was 20 % of the initial conductance at neutral pH. However, the substitution mutant did not show any sensitivity for low pH. The third property of the TolC channel-tunnel which can be attributed to the aspartate residues is the binding of di- and trivalent cations. The negatively charged residues at the periplasmic entrance form a binding site for di- and trivalent cations, which could be monitored by the blockage of the potassium ion flux through the channel-tunnel after addition of these cations.

Sequence comparison of all channel-tunnel proteins revealed that in most proteins at least one of the two negatively charged residues is conserved. This underlines the importance of these residues. One can predict that most of the TolC

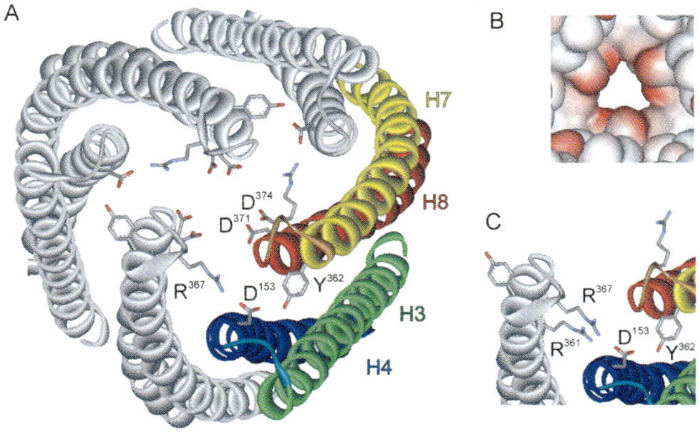

Figure 8.4 (A) The periplasmic entrance seen from the periplasmic side. One monomer is colored. Helices forming the outer coiled-coil are shown in green (H3) and blue (H4); helices forming the inner coiled-coil are yellow (H7) and red (H8). Residues establishing the circular network (D153, Y362 and R367) which keeps the tunnel in a closed confirmation are shown in detail. Aspartates (D371 and D374) lining the tunnel entrance are also presented. (B) Surface representation of the periplasmic entrance seen from the tunnel inside. The electronegative surface (red) of the aspartate residues is clearly visible. (C) Section of the circular network showing modeled arginine residues (R361) which might take over the role of R367 in TolC homologs lacking this residue.

homologs have similar electrophysiological properties. Cation selectivity is common for most of the pore-forming proteins of the outer membrane of *E. coli* [133]. It might be a general feature to protect cells from uptake of negatively charged, harmful substances like bile salts found in the environment of these bacteria. Furthermore, the pH-induced closure is known from other porins of *E. coli* [134, 135]. It is speculated that this is an immediate response to protect the cell against sudden acidification of the environment. The cation-binding site formed by the aspartates at the periplasmic entrance could be a possible target for new drugs aimed at the inhibition of efflux pumps or type I secretion systems. The conservation of these negative charges within the channel-tunnel family suggests a possibility that those drugs could act on a wide range of Gram-negative bacteria.

8.3.2.2 **Opening of the Periplasmic Entrance**

The diameter of the periplasmic entrance is 4 Å, which is too narrow for exported proteins or drugs to pass through. Therefore, the tunnel entrance has to be opened. Electrophysiological measurements have shown that the closed configuration is very stable. Neither treatment with high salt concentrations, high potentials nor urea opened the channel-tunnel [132]. In the case of the TolC homolog OprM from *P. aeruginosa*, protease treatment leads to a higher conductance, which underlined the assumption that channel-tunnels are gated and can adopt a conformation with higher conductance tantamount to a wider periplasmic entrance [136]. An open configuration of the TolC channel-tunnel was already proposed by Koronakis [112]. An iris-like rearrangement of the inner coiled-coil is predicted to open the periplasmic entrance. Evidence for the predicted outwards movement of the inner coiled-coils was locking of TolC in the closed state by introducing cysteine residues at the tunnel entrance, which were cross-linked by cysteine-specific bifunctional agents. Keeping TolC that way in the closed state completely abolished hemolysin secretion [137]. We have identified residues at the tunnel entrance which connect the coiled-coils of the tunnel domain and stabilize the closed confirmation [138]. An aspartate residue at the outer coiled-coil (D153) forms an intramolecular hydrogen bond with a tyrosine residue (Y362) located at the inner coiled-coil of the same monomer. Additionally, the same aspartate residue forms an intermolecular salt bridge with an arginine residue of the adjacent monomer. All together, these connections form a circular network which keeps the tunnel in the closed confirmation (Figure 8.4A). Disruption of the connections by substituting tyrosine against phenylalanine (YF) and arginine against serine (RS) widens the tunnel entrance. The single mutations $TolC^{YF}$ and $TolC^{RS}$ lead to an increase of the single-channel conductance by a factor of 1.3 and 2.5, respectively. The combination of both single mutants in $TolC^{YFRS}$ had a synergistic effect and increased the conductance 10-fold. This indicates most likely that the tunnel entrance of TolC is opened and it is estimated that the single-channel conductance corresponds with a diameter of around 16 Å, wide enough to allow export.

The circular network which keeps the TolC channel-tunnel in a closed conformation is conserved in almost all channel-tunnels. The aspartate and tyrosine residues involved in the intramolecular hydrogen bond are highly conserved. The arginine residue forming the intermolecular salt bridge is found in a minority of the homologs. However, it is most likely that in all other channel-tunnels its function is taken over by a lysine or arginine residue positioned previous to Y362 facing towards the aspartate (Figure 8.4C). Thus, the closed tunnel entrance is a general feature of channel-tunnels. This means that all export systems involving channel-tunnels need to open the periplasmic end in order to feed in the exported substrate.

8.4 Model for TolC-dependent Export

Channel-tunnel-dependent export systems are composed of three components. The inner membrane transporter, the accessory protein and the outer membrane channel-tunnel. The channel-tunnel structure is most likely conserved in all members of the family. The structure of the RND transporter AcrB, which is directly involved in channel-tunnel-dependent export, was solved recently. For transporters of the ABC and MF superfamily structural data exist for the homolog transporter, which can be transferred to transporters acting as part of type I secretion systems or efflux pumps. The structure of accessory proteins is still unknown. However, the known structure of the other components helps us to understand the function the accessory protein has to fulfill.

8.4.1
The Role of the Accessory Protein

The structure of the AcrB transporter revealed that the headpiece protrudes 70 Å into the periplasmic space. The funnel in the TolC docking domain at the top has the same diameter as the bottom of TolC and it is suggested that there is direct contact between AcrB and TolC. The sum of the periplasmic length of TolC and the headpiece of AcrB is 170 Å. This is the most accurate dimension of the periplasmic space. Previous studies with different invasive techniques have determined dimensions between 70 and 500 Å [139–142]. Thus, regarding AcrB and TolC, there is no gap between the inner membrane transporter and the outer membrane channel-tunnel. However, in the case of ABC transporters which are part of type I secretion systems, MF transporters or transporters of the PET family, which all have no periplasmic domain, the distance of around 70 Å between the transporter in the inner membrane and the channel-tunnel has to be bridged. This is one of the key roles of the accessory protein.

Bridging the gap implies interaction between the adaptor protein and the channel-tunnel. It is know that the tripartite export systems are no permanent assemblies. The transporter and the accessory protein form a complex, but interaction between this complex and the channel-tunnel is dynamic [70, 97]. Another role of the accessory protein is, therefore, to recruit and interact with the channel-tunnel. This function is most likely mediated by the coiled-coil domains of the accessory proteins. It is predicted that these domains form α-helical hairpins. The bottom of the tunnel domain could be enfolded by the three hairpin structures of the trimeric accessory protein. The interaction could be stabilized by close contact between the helices of the tunnel domain and the hairpins.

As mentioned above, a circular network at the bottom of the tunnel domain keeps the tunnel in a closed conformation. The contact between the channel-tunnel and the inner membrane complex could function as trigger for opening of the tunnel entrance. The central residue of the network, Asp-153, is accessible from the periplasmic side and might be contacted by parts of the accessory protein. This could lead to destabilization of the network and, finally, to opening of the tunnel entrance. Electrophysiological experiments have shown that the open state of the channel-tunnel is unstable [138]. It is likely that the interaction with the hairpins stabilizes the open configuration making an unhindered export possible.

8.4.2
The Mechanism of Protein Secretion

The prototype of the type I secretion system is the hemolysin export apparatus of *E. coli* consisting of the ABC transporter HlyB, the accessory protein HlyD and the channel-tunnel TolC. Hemolysin is a 110-kDa protein which is post-translationally activated by fatty acylation at two lysine residues [143]. It is a substrate with a C-terminal secretion signal. HlyB as well as the cytoplasmic domain of HlyD binds independently HlyA [69–71]. Interestingly, the binding of HlyA to the cytoplasmic domain of HlyD, which is essential for successful secretion, induces a conformational change in the periplasmic domain of the accessory protein [70, 71]. This means that there is a signal transduction from the cytosol into the periplasmic space possibly mediated by the transmembrane helix. It should be mentioned that the necessity of substrate binding to the cytoplasmic part of the accessory protein could only be shown for the hemolysin secretion system and might not be valid for all type I secretion systems. Experiments with hybrid secretion systems might suppose that there exist transporters which recognize a broader range of secretion signals and transporters that are more specific. For example, it was shown that the hemolysin secretion system of *E. coli* is also able to secrete colicin V and metalloprotease from *E. chrysanthemi* [30, 68]. In contrast, the natural export systems for colicin V and for the *E. chrysanthemi* metalloprotease were not able to secrete *E. coli* hemolysin [68]. What actually determines the specificity is not understood yet.

One can assume that the proteins are not secreted as a long peptide chain, but are at least partially folded. As mentioned above, there are substrates like the *E. coli* hemolysin or *B. pertussis* adenylate cyclase which are acylated post-translationally with fatty acid chains leading possibly to a certain folding of the substrate [143, 144]. It could also be shown that substrates with internal disulfide bonds formed in the cytoplasm are exported via this secretion system [145]. This means that secondary structure domains are tolerated when exported via this system. Nevertheless, there exists a size restriction for the exported substrates. The type I secretion system can be used to secrete hybrid proteins containing a C-terminal secretion signal fused to diverse other proteins. Some of those hybrid proteins are exported even when they contain disulfide bonds, but there are other examples where these internal bonds are not tolerated and no secretion takes place [25, 145].

HlyB belongs to the ABC transporter superfamily, which means that transport is energized by ATP hydrolysis. It is still not clear how ATP hydrolysis is coupled to the secretion process. It is known from a HlyB mutant which binds but not hydrolyzes ATP, that hemolysin was not exported but instead accumulated in the assembled export complex. This shows that release of hemolysin into the recruited TolC channel-tunnel is dependent upon ATP hydrolysis [70]. Proteins exported by the type I secretion system could be over 4000 residues long. It is questionable if the inner membrane transporter works steadily to push the long peptide chain into the extracellular space. Energy for secretion could also come from folding of the substrate at the cell surface. The majority of all long substrates secreted via the type I secretion system have hepta- or nonapeptide repeats that bind calcium to fold into a compact structure [37, 40, 41, 146, 147]. This calcium binding at the surface concomitant with immediate folding of the protein could contribute to the translocation by pulling the remaining peptide chain out of the export apparatus.

This assumption is confirmed by experiments showing that in some cases the repeats appear to be required for efficient secretion through the transport apparatus, especially for large, heterologous proteins [27, 148, 149].

8.4.3 The Mechanism of Efflux Pumps

It is most likely that there exist no general mechanisms by which substrates are exported via the different efflux pumps. There are four different inner membrane transporters involved, which have different substrate specificity, driving force and topology. RND and ABC transporters have a periplasmic domain; the two other transporters, MF and PET, have no periplasmic extensions. ABC transporters are energized by ATP hydrolysis; the other transporters are driven by proton motive force. The substrate specificity is much broader for RND transporters compared to transporters of the MF or ABC superfamily. The best-studied system is *E. coli*. Here, the AcrAB/TolC efflux pump is the most effective efflux pump, which is responsible for the efflux of a wide range of drugs. Substrates can be positively or negatively charged; neutral compounds are exported as well as zwitterionic substances. Efflux pumps involving transporters of the MF or ABC superfamily are much more specific [86, 101].

8.4.3.1 Substrate Binding of Different Transporters

The molecular basis of substrate binding to RND transporters could be elucidated by the crystallization of AcrB complexed with different substrates [85]. Each substance uses a slightly different subset of AcrB residues for binding. The binding site is located in the large central cavity in the periplasmic extension. There are three substrate molecules bound simultaneously, often interacting with each other, which stabilizes the binding. In contrast to other known drug binding proteins, where electrostatic interactions play an important role for binding [150], the drug binding to AcrB involves mostly hydrophobic interactions. Comparison of the substrate bound with the substrate free AcrB structure gave a first hint that the substrate binding might trigger the efflux process. The comparison revealed that binding induces a 1° rigid bottom rotation of each monomer approximately in parallel to the plane of the bilayer. This movement results in an increased diameter of the periplasmic domain of about 2.5 Å. It is possible that substrate binding also induces a conformational change in the accessory protein AcrA. This might further trigger drug transport by recruiting and opening the channel-tunnel TolC. How the drug molecules get from the AcrB cavity into the outer membrane channel-tunnel remains open. The most likely way is through the pore domain, which connects the central cavity with the funnel-domain at the top of the AcrB trimer. The pore, which is built by three helices (one per protomer), is too narrow to allow export and therefore one has to demand that it has to be opened if drugs are extruded through the pore. In both crystal structures, with and without bound substrate, the pore is closed. This means that another much larger conformational change has to occur. A movement which opens the pore domain in AcrB could be induced by the proton influx. Hereby, the protonation of membrane-embedded residues forming ion pairs between transmembrane helices could lead to conformational changes which are transduced towards the periplasmic domain.

For the other two transporter families there are no such details about substrate binding available. As mentioned above, transporters of the MF and ABC superfamily are more substrate specific than RND transporters. This implies that the binding site is smaller and involves fewer residues compared to the AcrB transporter. There are good hints that the binding site of MF transporters is located in the membrane domain. The *E. coli* multidrug transporter MdfA belonging to the MF superfamily exports lipophilic cationic drugs. It possesses a single acidic residue in its putative first transmembrane helix, which provides drug selectivity [151]. The substitution for a basic residue abolishes transport of positively charged ethidium bromide, but not of uncharged chloramphenicol. Furthermore, many multidrug transporters belonging to the MF superfamily possess acidic residues at a similar position in the first transmembrane helix, suggesting that they utilize a similar mechanism of drug selectivity [152].

Nothing is known about drug-transporters of the ABC superfamily involved in efflux pumps. However, studies with the ABC drug transporter from the Gram-positive *Lactococcus lactis* LmrA have revealed that the homodimeric LmrA possesses two drug-binding sites – a transport-competent site on the cytosolic surface and a drug-release site on the extracellular surface of the membrane. Drug transport is mediated by an alternating two-site transport (two-cylinder engine) mechanism driven by ATP hydrolysis [153]. One can assume that this mechanism may also be relevant for ABC transporters, which are part of multidrug efflux pumps.

8.4.3.2 Export of Substances by RND Transporters Exclusively from the Periplasm?

The solved crystal structure of the RND transporter AcrB raised the question if the transporter is able to transport substances out of the cytosol as proposed for all efflux pumps. There are more and more hints that efflux pumps involving RND transporters act exclusively as efflux machineries, which recruit their substrate from the periplasm or the outer leaflet of the inner membrane. The binding site for the different substrates is located within the cavity of the periplasmic domain [85]. There are no binding sites within the transmembrane domain or the cytoplasmic part of the transporter, which could support the idea that the transporter is able to translocate substrates across the inner membrane. Also, the specificity of the transporter lies within the periplasmic domain. This could be shown by hybrids of the RND transporters MexB and MexY of *P. aeruginosa*. Exchange of the transmembrane domain of the MexB protein by the corresponding parts of MexY did not change the preferential export of β-lactams. Thereby, when the two large periplasmatic loops of MexY were replaced by the ones of MexB, the specificity of the transporter changed from aminoglycoside antibiotics, characteristic for the MexY transporter, to β-lactam [154]. These results clearly demonstrate that RND transporters do not select substrates with the transmembrane or cytoplasmic part. They expel substrates directly from the periplasmic space before potential hazards penetrate into the cytoplasmic membrane. In AcrB, a groove in the membrane-exposed peripheral surface of the transmembrane domain of each protomer is suggested to serve as a pathway for translocation from the cytoplasm or inner leaflet [82]. However, there is not much evidence for drug export by efflux pumps directly from the cytosol. The only experiment which would suggest drug export from the cytosol was performed with everted membrane vesicles from *E. coli*. Here, an accumulation of the membrane-impermeable taurocholate inside these vesicles could be observed when the transporter was activated by a pH gradient across the vesicle membrane [155]. Another *in vitro* experiment could detect the active extrusion of phospholipids from AcrB-loaded proteoliposomes [98]. But the question if the extruded molecules come from the outer or the inner leaflet of the membrane remains open.

Studies with the cation efflux pump of *R. metallidurans* CzcABC also support the idea that efflux pumps involving RND transporters export substances not from the cytosol, but from the periplasm. Kinetic measurements with the RND transporter CzcA revealed K_{50} values for Co^{2+}, Zn^{2+}, and Cd^{2+} in the low millimolar range [156, 157]. This is far above the intracellular concentration of free heavy metal ions [158, 159]. This means that the CzcABC efflux pump could not detoxify the cytoplasm, but there must exist other transporters with much higher affinity for removing heavy metal cations out of the cytosol. Transported in the periplasm, the cations could be fed in the RND transporters for efflux into the exterior.

8.5 Conclusion

Members of the channel-tunnel family are found in the genome of every Gram-negative bacterium sequenced so far. The involvement of efflux pumps in mediating drug resistance and in the type I secretion system exporting pathogenic factors shows the impact of this outer membrane protein on pathogenicity. The crystallization of the TolC channel-tunnel and the RND transporter AcrB within the last three years represent substantial progress in understanding these export processes. The question how substrates cross the periplasmic space could be answered by the remarkable channel-tunnel structure, which spans most of the intermembrane compartment. The mechanism of drug efflux was elucidated by the crystal structure of AcrB. The substrate-binding site in the cavity of the periplasmic extension is accessible from the periplasm. This suggests that at least efflux pumps involving RND transporters capture substances from the periplasm or from the outer leaflet of the cytoplasmic membrane. They can be considered as "periplasmic vacuum cleaners" in analogy with other multidrug transporters that capture their substrates from the inner leaflet of the cytoplasmic membrane, which are called "membrane vacuum cleaners" [160]. In principle, an efflux mechanism which captures and exports drugs before they reach the targets in the cytosol could work more efficiently than a mechanism which does not act until the drugs have entered the cell interior.

Whether efflux pumps involving MF or ABC transporters are also able to capture substrates from the periplasm remains unclear. It also needs to be shown if proteins of the novel transporter family PET are functional and operate as part of a tripartite efflux pump. The structure of the third protein involved in all these export apparatus, the accessory protein, is thus the missing link. It will be a future task to solve the structure of the protein acting as linker between the inner membrane transporter and the outer membrane channel-tunnel. This will help us to understand how the three proteins interact with each other to form a functional export apparatus. Regions which are important for interaction, as well as substrate-binding sites, are potential targets for novel pharmaceuticals. Inhibiting efflux pumps and type I secretion systems would help to disarm dangerous bacteria.

Acknowledgments

This work was supported by the Deutsche Forschungsgemeinschaft (Emmy Noether Programme, grant AN373/1-1).

References

1 H. Nikaido, Escherichia coli *and* Salmonella typhimurium*: Cellular and Molecular Biology.* ASM, Washington, DC, **1996**.
2 T. J. Beveridge, *J Bacteriol* **1999**, *181*, 4725–4733.
3 K. Postle, *Mol Microbiol* **1990**, *4*, 2019–2025.
4 J. B. Stock, B. Rauch, S. Roseman, *J Biol Chem* **1977**, *252*, 7850–7861.
5 P. Fekkes, A. J. M. Driessen, *Microbiol Mol Biol Rev* **1999**, *63*, 161–173.
6 E. H. Manting, A. J. M. Driessen, *Mol Microbiol* **2000**, *37*, 226–238.
7 D. G. Thanassi, C. Stathopoulos, K. Dodson, D. Geiger, S. J. Hultgren, *J Bacteriol* **2002**, *184*, 6260–6269.
8 U. W. Konninger, S. Hobbie, R. Benz, V. Braun, *Mol Microbiol* **1999**, *32*, 1212–1225.
9 I. R. Henderson, F. Navarro-Garcia, J. P. Nataro, *Trends Microbiol* **1998**, *6*, 370–378.
10 M. Sandkvist, *Mol Microbiol* **2001**, *40*, 271–283.
11 T. B. Cao, M. H. Saier, Jr, *Microbiology* **2001**, *147*, 3201–3214.
12 D. Nunn, *Trends Cell Biol* **1999**, *9*, 402–408.
13 C. J. Hueck, *Microbiol Mol Biol Rev* **1998**, *62*, 379–433.
14 G. M. Young, D. H. Schmiel, V. L. Miller, *Proc Natl Acad Sci USA* **1999**, *96*, 6456–6461.
15 I. T. Paulsen, J. Chen, K. E. Nelson, M. H. Saier, Jr, *J Mol Microbiol Biotechnol* **2001**, *3*, 145–150.
16 M. H. Saier, Jr, I. T. Paulsen, *Semin Cell Dev Biol* **2001**, *12*, 205–213.
17 S. S. Pao, I. T. Paulsen, M. H. Saier, Jr, *Microbiol Mol Biol Rev* **1998**, *62*, 1–34.
18 Y. J. Chung, C. Krueger, D. Metzgar, M. H. Saier, Jr, *J Bacteriol* **2001**, *183*, 1012–1021.
19 T. T. Tseng, K. S. Gratwick, J. Kollman, D. Park, D. H. Nies, A. Goffeau, M. H. Saier, Jr, *J Mol Microbiol Biotechnol* **1999**, *1*, 107–125.
20 M. H. Brown, I. T. Paulsen, R. A. Skurray, *Mol Microbiol* **1999**, *31*, 394–395.

21 P. Stanley, V. Koronakis, C. Hughes, *Mol Microbiol* **1991**, *5*, 2391–2403.

22 W. H. Bingle, J. F. Nomellini, J. Smit, *J Bacteriol* **1997**, *179*, 601–611.

23 I. Gentschev, H. Mollenkopf, Z. Sokolovic, J. Hess, S. H. E. Kaufmann, W. Goebel, *Gene* **1996**, *179*, 133–140.

24 H. P. Hahn, C. Hess, J. Gabelsberger, H. Domdey, B. U. von Specht, *FEMS Immun Med Microbiol* **1998**, *20*, 111–119.

25 J. L. Palacios, I. Zaror, P. Martinez, F. Uribe, P. Opazo, T. Socias, M. Gidekel, A. Venegas, *J Bacteriol* **2001**, *183*, 1346–1358.

26 J. M. Ghigo, C. Wandersman, *J Biol Chem* **1994**, *269*, 8979–8985.

27 F. Duong, A. Lazdunski, M. Murgier, *Mol Microbiol* **1996**, *21*, 459–470.

28 D. Hui, V. Ling, *Biochemistry* **2002**, *41*, 5333–5339.

29 N. Izadi-Pruneyre, N. Wolff, V. Redeker, C. Wandersman, M. Delepierre, A. Lecroisey, *Eur J Biochem* **1999**, *261*, 562–568.

30 P. Delepelaire, C. Wandersman, *EMBO J* **1998**, *17*, 936–944.

31 F. Zhang, Y. Yin, C. H. Arrowsmith, V. Ling, *Biochemistry* **1995**, *34*, 4193–4201.

32 N. Wolff, J. M. Ghigo, P. Delepelaire, C. Wandersman, M. Delepierre, *Biochemistry* **1994**, *33*, 6792–6801.

33 C. Wandersman, P. Delepelaire, *Proc Natl Acad Sci USA* **1990**, *87*, 4776–4780.

34 P. Glaser, H. Sakamoto, J. Bellalou, A. Ullmann, A. Danchin, *EMBO J* **1988**, *7*, 3997–4004.

35 F. Duong, C. Soscia, A. Lazdunski, M. Murgier, *Mol Microbiol* **1994**, *11*, 1117–1126.

36 S. A. Thompson, O. L. Shedd, K. C. Ray, M. H. Beins, J. P. Jorgensen, M. J. Blaser, *J Bacteriol* **1998**, *180*, 6450–6458.

37 E. Kawai, H. Akatsuka, A. Idei, T. Shibatani, K. Omori, *Mol Microbiol* **1998**, *27*, 941–952.

38 P. Awram, J. Smit, *J Bacteriol* **1998**, *180*, 3062–3069.

39 C. Finnie, N. M. Hartley, K. C. Findlay, J. A. Downie, *Mol Microbiol* **1997**, *25*, 135–146.

40 G. M. York, G. C. Walker, *Mol Microbiol* **1997**, *25*, 117–134.

41 C. Finnie, A. Zorreguieta, N. M. Hartley, J. A. Downie, *J Bacteriol* **1998**, *180*, 1691–1699.

42 S. M. Hinsa, M. Espinosa-Urgel, J. L. Ramos, G. A. O'Toole, *Mol Microbiol* **2003**, *49*, 905–918.

43 J. Hess, I. Gentschev, D. Miko, M. Welzel, C. Ladel, W. Goebel, S. H. E. Kaufmann, *Proc Natl Acad Sci USA* **1996**, *93*, 1458–1463.

44 B. D. Tzschaschel, C. A. Guzman, K. N. Timmis, V. deLorenzo, *Nature Biotechnol* **1996**, *14*, 765–769.

45 E. T. Ryan, J. R. Butterton, R. N. Smith, P. A. Carroll, T. I. Crean, S. B. Calderwood, *Infect Immun* **1997**, *65*, 2941–2949.

46 N. Orr, J. E. Galen, M. M. Levine, *Infect Immun* **1999**, *67*, 4290–4294.

47 J. Hess, L. Grode, J. Hellwig, P. Conradt, I. Gentschev, W. Goebel, C. Ladel, S. H. E. Kaufmann, *FEMS Immunol Med Microbiol* **2000**, *27*, 283–289.

48 H. J. Mollenkopf, D. Groine-Triebkorn, P. Andersen, J. Hess, S. H. E. Kaufmann, *Vaccine* **2001**, *19*, 4028–4035.

49 O. G. Gomez-Duarte, M. F. Pasetti, A. Santiago, M. B. Sztein, S. L. Hoffman, M. M. Levine, *Infect Immun* **2001**, *69*, 1192–1198.

50 L. S. Havarstein, H. Holo, I. F. Nes, *Microbiology* **1994**, *140*, 2383–2389.

51 R. Lagos, J. E. Villanueva, O. Monasterio, *J Bacteriol* **1999**, *181*, 212–217.

52 J. Michiels, G. Dirix, J. Vanderleyden, C. Xi, *Trends Microbiol* **2001**, *9*, 164–168.

53 L. Gilson, H. K. Mahanty, R. Kolter, *J Bacteriol* **1987**, *169*, 2466–2470.

54 J. O. Solbiati, M. Ciaccio, R. N. Farias, R. A. Salomon, *J Bacteriol* **1996**, *178*, 3661–3663.

55 M. J. Fath, L. H. Zhang, J. Rush, R. Kolter, *Biochemistry* **1994**, *33*, 6911–6917.

56 M. F. Azpiroz, E. Rodriguez, M. Lavina, *Antimicrob Agents Chemother* **2001**, *45*, 969–972.

57 T. R. Klaenhammer, *FEMS Microbiol Rev* **1993**, *12*, 39–86.

58 M. J. van Belkum, R. W. Worobo, M. E. Stiles, *Mol Microbiol* **1997**, *23*, 1293–1301.
59 M. Kleerebezem, L. E. N. Quadri, O. P. Kuipers, W. M. de Vos, *Mol Microbiol* **1997**, *24*, 895–904.
60 W. M. de Vos, O. P. Kuipers, d. M. van, Jr, R. J. Siezen, *Mol Microbiol* **1995**, *17*, 427–437.
61 E. Sablon, B. Contreras, E. Vandamme, *Adv Biochem Eng Biotechnol* **2000**, *68*, 21–60.
62 M. H. Saier, Jr, *Microbiol Mol Biol Rev* **2000**, *64*, 354–411.
63 G. Chang, C. B. Roth, *Science* **2001**, *293*, 1793–1800.
64 K. P. Locher, A. T. Lee, D. C. Rees, *Science* **2002**, *296*, 1091–1098.
65 L. Kranitz, H. Benabdelhak, C. Horn, M. A. Blight, I. B. Holland, L. Schmitt, *Acta Crystallogr D* **2002**, *58*, 539–541.
66 L. Schmitt, H. Benabdelhak, M. A. Blight, I. B. Holland, M. T. Stubbs, *J Mol Biol* **2003**, *330*, 333–342.
67 L. S. Havarstein, D. B. Diep, I. F. Nes, *Mol Microbiol* **1995**, *16*, 229–240.
68 M. J. Fath, R. C. Skvirsky, R. Kolter, *J Bacteriol* **1991**, *173*, 7549–7556.
69 H. Benabdelhak, S. Kiontke, C. Horn, R. Ernst, M. A. Blight, I. B. Holland, L. Schmitt, *J Mol Biol* **2003**, *327*, 1169–1179.
70 T. Thanabalu, E. Koronakis, C. Hughes, V. Koronakis, *EMBO J* **1998**, *17*, 6487–6496.
71 L. Balakrishnan, C. Hughes, V. Koronakis, *J Mol Biol* **2001**, *313*, 501–510.
72 I. Gentschev, W. Goebel, *Mol Gen Genet* **1992**, *232*, 40–48.
73 S. Schlor, A. Schmidt, E. Maier, R. Benz, W. Goebel, I. Gentschev, *Mol Gen Genet* **1997**, *256*, 306–319.
74 A. F. Neuwald, J. S. Liu, D. J. Lipman, C. E. Lawrence, *Nucleic Acids Res* **1997**, *25*, 1665–1677.
75 F. Dardel, A. L. Davis, E. D. Laue, R. N. Perham, *J Mol Biol* **1993**, *229*, 1037–1048.
76 J. D. Green, E. D. Laue, R. N. Perham, S. T. Ali, J. R. Guest, *J Mol Biol* **1995**, *248*, 328–343.
77 F. K. Athappilly, W. A. Hendrickson, *Structure* **1995**, *3*, 1407–1419.
78 J. W. Hwang, P. C. Tai, *Curr Microbiol* **1999**, *39*, 195–199.
79 R. Schulein, I. Gentschev, S. Schlor, R. Gross, W. Goebel, *Mol Gen Genet* **1994**, *245*, 203–211.
80 C. Andersen, *Rev Physiol Biochem Pharmacol* **2003**, *147*, 122–165.
81 K. Poole, *Curr Opin Microbiol* **2001**, *4*, 500–508.
82 S. Murakami, R. Nakashima, E. Yamashita, A. Yamaguchi, *Nature* **2002**, *419*, 587–593.
83 H. Nikaido, H. I. Zgurskaya, *J Mol Microbiol Biotechnol* **2001**, *3*, 215–218.
84 S. Eda, H. Yoneyama, T. Nakae, *Biochemistry* **2003**, *42*, 7238–7244.
85 E. W. Yu, G. McDermott, H. I. Zgurskaya, H. Nikaido, D. E. Koshland, Jr, *Science* **2003**, *300*, 976–980.
86 K. Nishino, A. Yamaguchi, *J Bacteriol* **2001**, *183*, 1455–1458.
87 J. Abramson, I. Smirnova, V. Kasho, G. Verner, H. R. Kaback, S. Iwata, *Science* **2003**, *301*, 610–615.
88 Y. Huang, M. J. Lemieux, J. Song, M. Auer, D. N. Wang, *Science* **2003**, *301*, 616–620.
89 N. Kobayashi, K. Nishino, A. Yamaguchi, *J Bacteriol* **2001**, *183*, 5639–5644.
90 N. Kobayashi, K. Nishino, T. Hirata, A. Yamaguchi, *FEBS Lett* **2003**, *546*, 241–246.
91 K. T. Harley, M. H. Saier, Jr, *J Mol Microbiol Biotechnol* **2000**, *2*, 195–198.
92 T. Dinh, I. T. Paulsen, M. H. Saier, Jr, *J Bacteriol* **1994**, *176*, 3825–3831.
93 J. M. Johnson, G. M. Church, *J Mol Biol* **1999**, *287*, 695–715.
94 H. I. Zgurskaya, H. Nikaido, *J Mol Biol* **1999**, *285*, 409–420.
95 H. Yoneyama, H. Maseda, H. Kamiguchi, T. Nakae, *J Biol Chem* **2000**, *275*, 4628–4634.
96 E. B. Tikhonova, Q. Wang, H. I. Zgurskaya, *J Bacteriol* **2002**, *184*, 6499–6507.
97 H. I. Zgurskaya, H. Nikaido, *J Bacteriol* **2000**, *182*, 4264–4267.
98 H. I. Zgurskaya, H. Nikaido, *Proc Natl Acad Sci USA* **1999**, *96*, 7190–7195.
99 A. J. Avila-Sakar, S. Misaghi, E. M. Wilson-Kubalek, K. H. Downing, H. Zgurskaya, H. Nikaido, E. Nogales, *J Struct Biol* **2001**, *136*, 81–88.

100 J. W. Hwang, X. T. Zhong, P. C. Tai, *J Bacteriol* **1997**, *179*, 6264–6270.
101 M. C. Sulavik, C. Houseweart, C. Cramer, N. Jiwani, N. Murgolo, J. Greene, B. DiDomenico, K. J. Shaw, G. H. Miller, R. Hare, G. Shimer, *Antimicrob Agents Chemother* **2001**, *45*, 1126–1136.
102 O. Lomovskaya, K. Lewis, *Proc Natl Acad Sci USA* **1992**, *89*, 8938–8942.
103 A. S. Jellen-Ritter, W. V. Kern, *Antimicrob Agents Chemother* **2001**, *45*, 1467–1472.
104 R. Aono, N. Tsukagoshi, M. Yamamoto, *J Bacteriol* **1998**, *180*, 938–944.
105 J. T. Jo, F. S. Brinkman, R. E. Hancock, *Antimicrob Agents Chemother* **2003**, *47*, 1101–1111.
106 R. Chuanchuen, C. T. Narasaki, H. P. Schweizer, *J Bacteriol* **2002**, *184*, 5036–5044.
107 N. Masuda, E. Sakagawa, S. Ohya, N. Gotoh, H. Tsujimoto, T. Nishino, *Antimicrob Agents Chemother* **2000**, *44*, 3322–3327.
108 A. Sharff, C. Fanutti, J. Y. Shi, C. Calladine, B. Luisi, *Eur J Biochem* **2001**, *268*, 5011–5026.
109 C. Andersen, C. Hughes, V. Koronakis, *EMBO Rep* **2000**, *1*, 313–318.
110 N. Gotoh, H. Tsujimoto, K. Poole, J. Yamagishi, T. Nishino, *Antimicrob Agents Chemother* **1995**, *39*, 2567–2569.
111 C. Rensing, T. Pribyl, D. H. Nies, *J Bacteriol* **1997**, *179*, 6871–6879.
112 V. Koronakis, A. Sharff, E. Koronakis, B. Luisi, C. Hughes, *Nature* **2000**, *405*, 914–919.
113 R. Gross, *Mol Microbiol* **1995**, *17*, 1219–1220.
114 X. Z. Li, K. Poole, *J Bacteriol* **2001**, *183*, 12–27.
115 K. K. Y. Wong, F. S. L. Brinkman, R. S. Benz, R. E. W. Hancock, *J Bacteriol* **2001**, *183*, 367–374.
116 J. A. Killian, G. von Heijne, *Trends Biochem Sci* **2000**, *25*, 429–434.
117 R. Koebnik, K. P. Locher, P. Van Gelder, *Mol Microbiol* **2000**, *37*, 239–253.
118 S. K. Buchanan, *Curr Opin Struct Biol* **1999**, *9*, 455–461.
119 A. Pautsch, G. E. Schulz, *Nat Struct Biol* **1998**, *5*, 1013–1017.
120 A. D. Ferguson, E. Hofmann, J. W. Coulton, K. Diederichs, W. Welte, *Science* **1998**, *282*, 2215–2220.
121 S. W. Cowan, T. Schirmer, G. Rummel, M. Steiert, R. Ghosh, R. A. Pauptit, J. N. Jansonius, J. P. Rosenbusch, *Nature* **1992**, *358*, 727–733.
122 T. Schirmer, T. A. Keller, Y. F. Wang, J. P. Rosenbusch, *Science* **1995**, *267*, 512–514.
123 R. Benz, A. Schmid, G. H. Vos-Scheperkeuter, *J Membr Biol* **1987**, *100*, 21–29.
124 N. Saint, K. L. Lou, C. Widmer, M. Luckey, T. Schirmer, J. P. Rosenbusch, *J Biol Chem* **1996**, *271*, 20676–20680.
125 K. Bauer, M. Struyve, D. Bosch, R. Benz, J. Tommassen, *J Biol Chem* **1989**, *264*, 16393–16398.
126 M. Jordy, C. Andersen, K. Schulein, T. Ferenci, R. Benz, *J Mol Biol* **1996**, *259*, 666–678.
127 V. Koronakis, C. Andersen, C. Hughes, *Curr Opin Struct Biol* **2001**, *11*, 403–407.
128 C. R. Calladine, A. Sharff, B. Luisi, *J Mol Biol* **2001**, *305*, 603–618.
129 F. H. C. Crick, *Acta Crystallogr* **1953**, *6*, 689–697.
130 R. Benz, E. Maier, I. Gentschev, *Zentralbl Bakteriol* **1993**, *278*, 187–196.
131 C. Andersen, E. Koronakis, C. Hughes, V. Koronakis, *Mol Microbiol* **2002**, *44*, 1131–1139.
132 C. Andersen, C. Hughes, V. Koronakis, *J Membr Biol* **2002**, *185*, 83–92.
133 R. Benz, *Bacterial Cell Wall*. Elsevier, Amsterdam, **1994**.
134 D. J. Muller, A. Engel, *J Mol Biol* **1999**, *285*, 1347–1351.
135 C. Andersen, B. Schiffler, A. Charbit, R. Benz, *J Biol Chem* **2002**, *277*, 41318–41325.
136 E. Yoshihara, H. Maseda, K. Saito, *Eur J Biochem* **2002**, *269*, 4738–4745.
137 J. Eswaran, C. Hughes, V. Koronakis, *J Mol Biol* **2003**, *327*, 309–315.
138 C. Andersen, E. Koronakis, E. Bokma, J. Eswaran, D. Humphreys, C. Hughes, V. Koronakis, *Proc Natl Acad Sci USA* **2002**, *99*, 11103–11108.
139 J. E. Vanwielink, J. A. Duine, *Trends Biochem Sci* **1990**, *15*, 136–137.

140 L. L. Graham, R. Harris, W. Villinger, T. J. Beveridge, *J Bacteriol* **1991**, *173*, 1623–1633.

141 J. Dubochet, A. W. McDowall, B. Menge, E. N. Schmid, K. G. Lickfeld, *J Bacteriol* **1983**, *155*, 381–390.

142 J. A. Hobot, E. Carlemalm, W. Villinger, E. Kellenberger, *J Bacteriol* **1984**, *160*, 143–152.

143 P. Stanley, L. C. Packman, V. Koronakis, C. Hughes, *Science* **1994**, *266*, 1992–1996.

144 N. Heveker, D. Bonnaffe, A. Ullmann, *J Biol Chem* **1994**, *269*, 32844–32847.

145 L. A. Fernandez, V. De Lorenzo, *Mol Microbiol* **2001**, *40*, 332–346.

146 J. G. Coote, *FEMS Microbiol Rev* **1992**, *88*, 137–162.

147 P. Sebo, D. Ladant, *Mol Microbiol* **1993**, *9*, 999–1009.

148 U. Baumann, S. Wu, K. M. Flaherty, D. B. McKay, *EMBO J* **1993**, *12*, 3357–3364.

149 S. Letoffe, C. Wandersman, *J Bacteriol* **1992**, *174*, 4920–4927.

150 M. A. Schumacher, M. C. Miller, S. Grkovic, M. H. Brown, R. A. Skurray, R. G. Brennan, *Science* **2001**, *294*, 2158–2163.

151 R. Edgar, E. Bibi, *EMBO J* **1999**, *18*, 822–832.

152 E. E. Zheleznova, P. Markham, R. Edgar, E. Bibi, A. A. Neyfakh, R. G. Brennan, *Trends Biochem Sci* **2000**, *25*, 39–43.

153 H. W. van Veen, A. Margolles, M. Muller, C. F. Higgins, W. N. Konings, *EMBO J* **2000**, *19*, 2503–2514.

154 S. Eda, H. Maseda, T. Nakae, *J Biol Chem* **2003**, *278*, 2085–2088.

155 D. G. Thanassi, L. W. Cheng, H. Nikaido, *J Bacteriol* **1997**, *179*, 2512–2518.

156 D. H. Nies, *J Bacteriol* **1995**, *177*, 2707–2712.

157 M. Goldberg, T. Pribyl, S. Juhnke, D. H. Nies, *J Biol Chem* **1999**, *274*, 26065–26070.

158 R. C. Fahey, W. C. Brown, W. B. Adams, M. B. Worsham, *J Bacteriol* **1978**, *133*, 1126–1129.

159 D. McLaggan, T. M. Logan, D. G. Lynn, W. Epstein, *J Bacteriol* **1990**, *172*, 3631–3636.

160 O. Lomovskaya, H. I. Zgurskaya, H. Nikaido, *Nat Biotechnol* **2002**, *20*, 1210–1212.

9
Structure–Function Relationships in Sugar-specific Porins

Tilman Schirmer

9.1
Introduction

The outer membrane of Gram-negative bacteria carries pore-forming integral membrane proteins that allow the passage of small solutes across this protective shield. They can be classified into general porins, which are reviewed in Chapter 2, and porins that exhibit solute specificity. Hereby, a pore is called specific with respect to a particular solute, when its translocation rate exhibits saturation upon increase of the solute concentration [1]. From such a behavior, the presence of a specific interaction of the solute with the pore protein can be inferred. Discrimination against other solutes may just be a consequence of this specific protein–solute interaction and not a property especially evolved to favor translocation of certain solutes over others.

In the simplest case, specific porins (P) can be described as enzymes that catalyze the translocation of a solute (S) across the membrane according to the kinetic scheme (one-site, two-barriers model [2]):

$$S_{out} + P \underset{k_{-1}}{\overset{k_1}{\rightleftarrows}} S{\cdot}P \underset{k_{-2}}{\overset{k_2}{\rightleftarrows}} P + S_{in} \tag{1}$$

In a detailed description, more than one binding site in the pore would have to be postulated which would result in the formation of a series of different intermediates (S·P) along the reaction path (Fig. 9.1). With solute present only outside, the net flux through one channel is given by (see, e. g. [3])

$$\Phi = k_{-2}\frac{[S_{out}]}{[S_{out}] + \dfrac{k_{-1}+k_{-2}}{k_1}} \tag{2}$$

At high substrate concentrations, $[S] >> (k_{-1} + k_{-2})/k_1$, the flux through the channel saturates with $\Phi_{max} = k_{-2}$. The channel gets clogged, which would be counter-

Bacterial and Eukaryotic Porins. Edited by Roland Benz

ISBN 3-527-30775-3

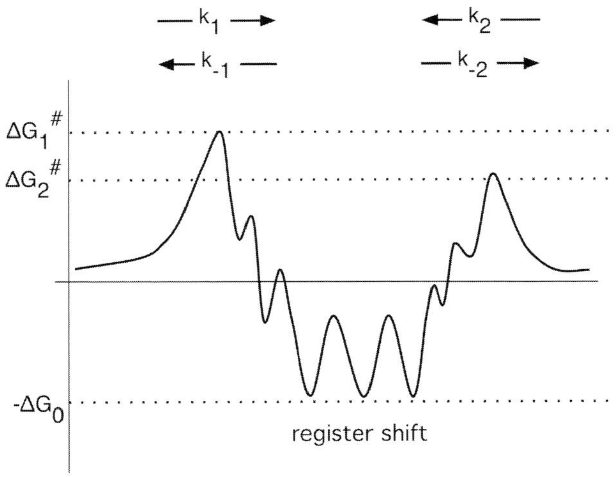

Figure 9.1 Kinetic model for maltopentaose translocation through LamB. The sugar can access the channel from both sides. The on-rates k_1 and k_2 correspond to the activation barriers $\Delta G_1^{\#}$ and $\Delta G_2^{\#}$, respectively. The off-rates, k_{-1} and k_{-2}, depend additionally on the (effective) depth of the binding site at the center of the channel, ΔG_0. The substructure in the energy profile is imposed by the modular nature of the polyglucose sugar. As there are three consecutive glucosyl-binding subsites, a maltodextrin molecule longer than maltotriose will find more than one energetically equivalent binding position [11]. These positions are related by register shifts of the oligosaccharide with respect to the channel.

productive (and is not the case for non-specific channels). However, at low substrate concentrations, i. e. under physiological conditions, specific channels excel. In this case, the translocation rate is given by:

$$\Phi = \frac{k_1 k_{-2}}{k_{-1} + k_{-2}} [S_{out}] = \frac{k_1}{1 + \dfrac{k_{-2}}{k_{-1}}} [S_{out}] \tag{3}$$

See also [4, 5]. It can be expected that the on- and off-rates has been maximized during evolution, since the structure of a specific channel can be fine-tuned to adapt to the specific substrate (see below). In contrast, the flux through non-specific channels is mainly governed by pore diameter and electrostatics [6]. It may be noted that, at low substrate concentration, the flux is independent of the depth of the energy well [7]. The presence of a binding site, however, is important to compensate for the loss of sugar hydration and may be important to reduce the heights of the kinetic barriers.

9.2
Maltoporin and Sucrose Porin

Maltoporin (LamB) from *Escherichia coli* is the prototype of a specific porin. It has initially been discovered as the receptor for λ phage [8] and later shown to be part of the maltose regulon. Maltoporin is indispensable for the uptake of maltodextrins (α1–4-linked polyglucose) longer than Glc_3 (which cannot pass through the general porins), but appears important also for the uptake of glucose at low concentrations [9].

The crystal structure (Figure 9.2) [10] shows that the framework of each monomer of the LamB trimer is an 18-stranded β-barrel (with a shear number of 22). Three of the connecting loops are folded inwardly and considerably constrict the barrel cross-section. At the channel lining, a contiguous path formed by six aromatic residues is found that extends from the vestibule through the channel narrowings to the periplasmic outlet. As demonstrated by the crystal structures of LamB saturated with various maltodextrins [11, 12] this path ("greasy slide") together with a number of charged residues ("ionic tracks") constitutes the sugar binding site (Figure 9.3). Its central location explains why sugar binding blocks ion current [2]. A guiding function of the greasy slide for maltodextrin translocation has been proposed early [10], involving hydrophobic interactions between the extracellular end of the greasy slide and the sugar.

Maltoporin binds sucrose, but does not permit its efficient translocation [13]. The crystal structure of the respective complex showed that the fructosyl moiety of sucrose is indeed too bulky to traverse the channel constriction [14]. In *Salmonella typhimurium*, a LamB homolog named sucrose porin (ScrY) has evolved for allowing translocation of sucrose. Its structure [15] is highly similar to that of maltoporin. The greasy slide is largely conserved, but the loop regions appear unrelated. A few residues are replaced at the channel constriction resulting in a larger cross-section and explaining why ScrY, in contrast to LamB, allows permeation of sucrose. ScrY carries an N-terminal 71 residues extension, which was not included in the construct used for structure determination.

With the detailed knowledge of the three-dimensional structures, it has now become possible to pose specific questions to get insight into the mechanism of these facilitated diffusion channels. There are two other channels of this class, aquaporin and the glycerol facilitator, for which crystal structures have been determined recently [16, 17]. This review will focus on LamB and ScrY, and summarize recent results obtained from site-directed mutagenesis, functional analysis of single porin molecules in artificial bilayers and molecular simulations. Older results and other aspects have been reviewed before [18–23].

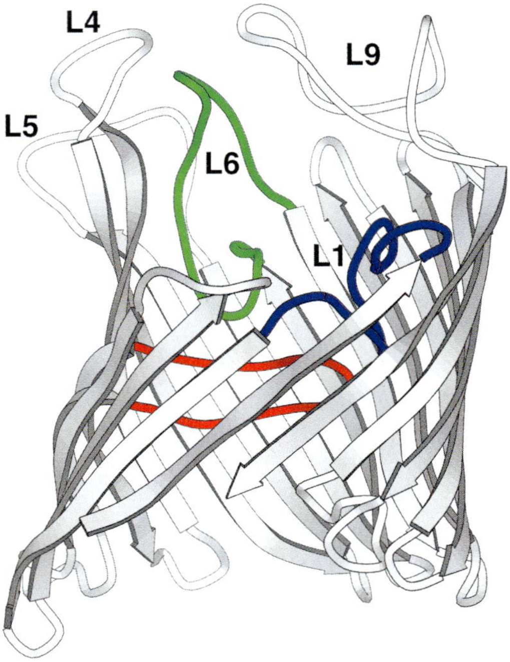

Figure 9.2 Schematic drawing of the maltoporin monomer (PDB code 1MAL [10]). The cell exterior is at the top and the periplasmic space is at the bottom. The area of the subunit involved in trimer contacts is facing the viewer. The 18 antiparallel β-strands are represented by arrows. Loops L1 (blue), L3 (red) and L6 (green) fold inward and constrict the channel. Loops L4–L6 and L9 form the protrusion that is the docking site for phage λ. This figure has been adapted with permission of the copyright holder from [10].

9.3
Probing Function by Site-directed Mutagenesis

9.3.1
Ionic Tracks

The channel lining of LamB has been probed extensively by mutagenesis. A large number of variants with residue replacements at the ionic tracks has been analyzed [24]. The study confirmed the role of the ionic tracks in sugar binding, but also demonstrated their importance for efficient *in vivo* sugar uptake. It is evident from the crystal structures [11] that the polar residues substitute effectively for the hydration shell of the substrate, while the sugar is passing through the narrow channel. Similarly, ion channels exhibit appropriate arrays of polar or charged residues to "solubilize" the translocating ion [25].

9.3.2
Greasy Slide

Alanine scanning mutagenesis was performed on the six aromatic residues of the greasy slide [7]. Not surprisingly, replacements at the center of the slide lowered the maltodextrin affinity, whereas for the others (W74A, W358A and F227A) virtually wild-type affinity was observed. This was expected from the knowledge of the

LamB–maltodextrin crystal structures [11]. The more interesting question of the study was whether shortening the greasy slide would affect the kinetics of sugar binding. Indeed, a 3- and 2-fold decreased on-rate was found for W74A and F227A, respectively, as determined by sugar-induced current-noise analysis [26]. The effect does not seem very large. It is, however, of physiological relevance as shown by the reduced *in vivo* sugar uptake rates [5]. That there was no indirect structural perturbation was shown for mutant W74A by crystal structure analysis. It can be concluded that the entire greasy slide is needed for efficient translocation with the terminal residues enhancing the kinetic rates, but not the sugar affinity.

9.3.3
Tyrosine 118

The role of residue Y118, located at the center of the channel opposite to the slide (Figure 9.3) has been probed extensively [4, 7, 27, 28]. Not surprisingly, most residue replacements at position 118 weaken maltodextrin binding. Mutants Y118F and Y118W, on the other hand, showed a 20- to 350-fold increase in the binding constant, probably due to a strengthening of the hydrophobic interaction. Indeed, the increase in binding affinity was mainly due to a strong decrease in the off-rate. Sugar binding and well-resolved kinetics do not necessarily imply sugar translocation. The sugar may leave the binding site towards the same side it had entered the channel. No translocation assays were performed for the above mutants, but the fact that they allow bacterial growth on maltodextrins as the sole carbon source was taken as evidence for their functionality [28].

9.3.4
Translocation Kinetics

The simple kinetic model of sugar translocation through LamB (Figure 9.1) involves two barriers that, in the initial studies, were assumed to be of equal height [4, 29]. Obviously, unidirectionally oriented channels are required to test this assumption. This was attempted [30] by adding the protein only to one side (*cis*) of the artificial bilayer membrane.

No significant difference in the apparent binding constants obtained upon addition of the sugar to the *cis* or *trans* compartment was observed. This still could mean that (asymmetric) channels were oriented randomly in the bilayer. Some loop-deletion mutants analyzed in the same study [30], however, exhibited asymmetry in sugar binding (only weak sugar binding from the *cis* side). In another experiment [31], the same group nicely showed that it was possible to invoke collapse of the extracellular loops at pH 4. Most relevant, the effect was seen only when lowering the pH in the *cis* compartment. Thus, both studies indicated that most of the LamB trimers had their extracellular side exposed to the *cis* compartment.

Different results were obtained by van Gelder [32] on wild-type LamB. λ Phage was employed to selectively prevent maltodextrin access from the extracellular side. Only when the sugar was added to the *trans* compartment this effect was ob-

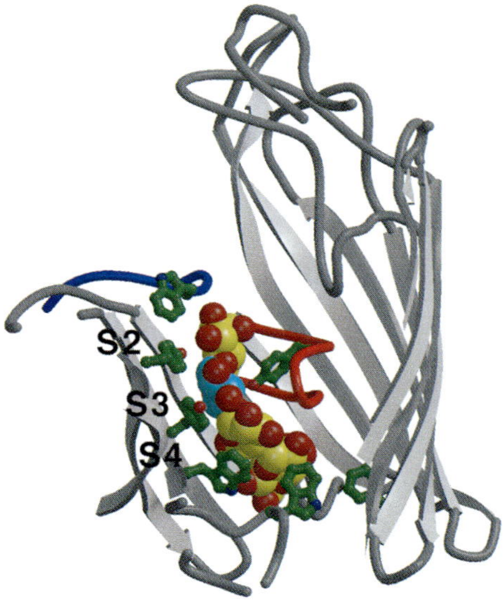

Figure 9.3 Structure of LamB with maltohexaose bound (PDB code 1MPO [11]). With respect to Figure 9.2, the monomer structure has been turned by about 60° around the vertical axis. Some strands in the front have been clipped for clarity. Residues of the greasy slide (from top to bottom: W74, Y41, Y6, W420, W358 and F227) are shown in stick representation. Y118 is found opposite the greasy slide (attached to loop L3, shown in red). The five ordered glucosyl moieties of the bound oligosaccharide are shown in space-filling representation. S2, S3 and S4 indicate the main glucosyl-binding subsites. This figure has been reproduced with permission of the copyright holder from [11].

served, implying that the extracellular loops were oriented towards *trans*. In this study, a transmembrane potential (positive at the *trans* side) was applied to facilitate insertion. The observed orientation may be due to the electrical dipole moment of LamB, which is negative at the extracellular side (as computed from the crystal structure). After removal of λ phage, addition of maltodextrins to either compartment revealed distinct apparent binding constants, reflecting a difference in the individual barrier heights. It was deduced that the on-rate from the extracellular side of the channel is larger by a factor of 2–3 compared to the rate from the periplasmic side, as also confirmed later directly by kinetic measurements [7].

Very clean and precise data were obtained recently on single trimeric maltoporin molecules inserted in artificial bilayers [33, 34]. This is obviously the most elegant and straightforward way to obtain unidirectional orientation. Clearly, the maltohexaose binding on-rate was found to be dependent on the side of sugar addition [34]. It was larger by a factor of 2.5–7 when the sugar was added to the *trans* side, which is consistent with the results obtained by van Gelder et al. Interestingly, sugar binding kinetics and affinity were strongly dependent on magnitude and sign of the applied transmembrane potential (which is needed to generate the ion current, which in turn is used to monitor sugar binding). This property has been overlooked so far and should be considered when comparing results of different studies. Since maltodextrin is not charged, the phenomenon probably implies that the external potential perturbs the structure of the sugar-binding site at the channel center; in particular, the arrangement of ionic track residues. The experimental data were fitted successfully with a model involving two channel states (one with high maltodextrin affinity and the other with no affinity) that are in voltage-dependent equilibrium [35]. Statistical evaluation of the time-resolved sugar-blocking events of single trimers [34] revealed that there is no cooperativity in sugar binding.

Most recently, it was shown [36] that a maltodextrin analog (Glc_6 modified at the reducing end with a bulky fluorophore) can bind to LamB only from the *cis* side. Because, as shown by crystallography [11], maltodextrins bind to the LamB channel with the non-reducing end pointing towards the periplasmic outlet, this observation would imply that the external loops of LamB face the *cis* compartment. Unfortunately, the polarity of the external potential used to stimulate membrane insertion was not reported; in a preceding study of the same group [33], the potential was positive at the *cis* side, i. e. opposite in polarity to that used in [32].

Taking the various studies together, it appears premature to come to a final conclusion regarding the orientation of LamB in artificial membranes and, thus, also regarding the assignment of the two distinct barriers to the extracellular or periplasmic side of the channel, respectively. In the biological context, this question appears of be of lesser relevance. The simple two-barrier, one-site model predicts that, at low (physiological) sugar concentration, the net flux across the outer membrane is the same, whether the larger barrier is at the entrance or at the outlet of the channel. [The formula for the flux, Equation [2], is symmetric with respect to indices 1 and 2. Note that $k_1k_{-2} = k_{-1}k_2 \sim \exp(-\Delta G_1^{\#} + \Delta G_0 - \Delta G_2^{\#})$; for definitions see Figure 9.1.] Only at close to saturating conditions would the location of the smaller barrier at the exit side confer an advantage over the reverse set-up [36].

9.3.5 Changing Substrate Specificity

The substrate range of LamB is restricted to maltodextrins and a few other sugars [13], whereas, as the name implies, sucrose porin catalyzes the efficient translocation of sucrose. The crystal structure of LamB in complex with sucrose [14] shows that the substrate is stuck at the channel center with Y118 preventing further trans-

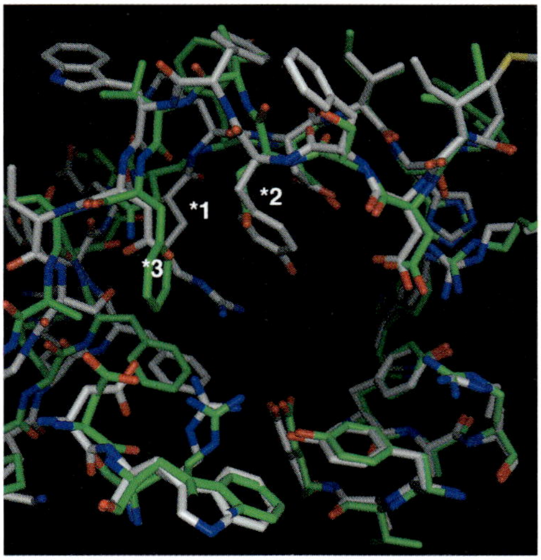

Figure 9.4 Superposition of the channel constrictions of LamB [10] and ScrY [15]. The PDB codes are 1MAL and 1OH2, respectively. The view is from the cell exterior along the channel direction. The greasy slide is visible at the lower right with W74 (LamB), the start of the slide, at the bottom of the picture. Most residues are conserved between the two channels with the notable exception of the residue replacements at site *1, R109 (LamB), N192 (ScrY, partly buried below L3); site *2, Y118 (LamB), D201 (ScrY); and site *3, D121 (LamB), F204 (ScrY). This figure has been reproduced with permission of the copyright holder from [5].

location. In ScrY, at the equivalent position, there is the much shorter side-chain of an aspartate (D201). Other residues at the channel constriction that differ are R109 and D121 of LamB, which are equivalent to N192 and F204 of ScrY (Figure 9.4). Are these differences responsible for the distinct specificity of the two porins? This question was addressed by mutating the ScrY residues to their LamB counterparts [37, 38] and vice versa [39].

The *in vivo* analysis of the ScrY triple mutant [37] indeed showed that the apparent K_m of the sucrose uptake apparatus was increased, demonstrating that translocation through ScrY became rate limiting. In contrast, the K_m of maltose uptake did not show much variation between strains expressing ScrY or LamB and was, in fact, also similar for the ScrY mutant. So, all these channels are probably not rate limiting for maltose uptake. The *in vitro* analysis of the ScrY mutants [38], however, revealed that all functional parameters (single-channel conductance, maltodextrin-binding rates and affinity) are quite different to LamB. This was attributed to the remaining differences in the channel structures.

In the reciprocal study [39], the crystal structure of one of the LamB mutants was also determined. As intended by design, the mutated side-chains 121 and 118 were at a very similar location as their counterparts in ScrY. The location and orientation of residue D109, however, was different from the corresponding ScrY residue. In fact, superposition revealed that in this region the backbone also adopts a different course (due to a one-residue insertion in ScrY). Functionally, as determined by *in vivo* sugar uptake and by liposome swelling, the triple mutant showed considerable sucrose permeation, although to a lesser degree than ScrY. Thus, the aim of the study was only partially achieved. The conclusion was that the channel constrictions of ScrY and LamB are too different to allow perfect conversion of substrate specificity by few residue replacements. In fact, a R109A/Y118A mutant that had the main obstructing side-chains pruned to alanine showed even better sucrose permeation [39]. It would be interesting to compare, on an absolute scale, sucrose and maltose permeation rates through ScrY and the various LamB mutants with the rate through a general porin such as OmpF. This could conclusively reveal whether the sugar porins with their greasy slides indeed confer an advantage for the translocation of small glycosyl substrates [9] or whether a sufficiently large channel is equally efficient.

There is still a controversy how efficiently maltodextrins can permeate ScrY. Maltodextrins bind to ScrY [40], Lamb$^-$, ScrY$^+$ bacteria can survive on Glc_4 as the sole carbon source [41] and the apparent K_m of maltose influx into whole cells is only about 2 times higher for bacteria expressing ScrY as opposed to those expressing LamB [37]. In contrast to all these findings, the translocation rate of maltose through ScrY is below the detection limit as determined by proteoliposome swelling [39, 42]. Therefore, translocation of maltose through ScrY seems to be suboptimal, but efficient enough to be not rate limiting *in vivo*.

9.3.6
N-terminal Domain of ScrY

Several studies have recently dealt with the N-terminal extension of ScrY. Ulmke et al. [37] showed that, in growth competition experiments, bacteria expressing a ScrY mutant lacking the N-terminal domain were disfavored, although the apparent transport K_m for sucrose was found unchanged with respect to wild-type ScrY. The difference was attributed to the different plasmid copy numbers in the strains employed and it was speculated that the N-terminal domain may confer increased stability to the channel. This was not supported by later thermal unfolding studies [43]. As suggested earlier [15], the N-terminal domain forms a trimeric coiled-coil. This can be concluded from CD, analytical ultracentrifuge and mass spectrometer measurements on recombinantly expressed or chemically synthesized peptide [43, 44].

The kinetics of maltodextrin binding is not significantly affected by deletion of the N-terminal domain [40], whereas single-channel ion conductance and cation/anion selectivity are altered [45]. This would be consistent with a largely unchanged channel structure, but with a changed electrostatic potential due to the missing charges of the N-terminal extension.

It has been speculated [15] that the N-terminal domain serves as a substitute for the missing sucrose-specific periplasmic binding protein. This notion is reinforced by the outcome of a recent study [43]. Equilibrium dialysis showed that sucrose binds strongly ($K_d = 0.1$ mM) to the isolated N-terminal domain. It is surprising that a coiled-coil domain performs so well in this task. In fact, the dissociation constant to the channel proper as determined by ion-current blocking is much larger ($K_d = 50$ mM [40]). Energetic downhill transfer of sucrose from the channel to the periplasmic extension may, thus, be ensured.

Furthermore, a hybrid consisting of the N-terminal domain of ScrY and LamB profoundly changed the translocation characteristics of sugars of the maltodextrin series [43]. Whereas, in wild-type LamB, the permeation rate (as measured by liposome swelling) decreases with increasing sugar length, the opposite is observed for the LamB hybrid, rendering this channel extremely efficient, e. g. for Glc_6. Therefore, one would think that this hybrid should confer an advantage over wild-type LamB. The authors point out that, *in vivo*, this is probably not the case due to the presence of the periplasmic maltose-binding protein. The analogous measurements have not yet been performed on ScrY. It would be very interesting to see whether wild-type ScrY is preferentially translocating longer maltodextrins.

9.3.7
Probing the Role of the External Loops

Viewed from the cell exterior, the LamB trimer confers the impression of a half open tulip with the three protrusions representing the petals. The protrusions are formed by loops L4, L5, L6 and L9, and probably prevent tipping over of the lipopolysaccharide molecules of the outer leaflet into the channel. As for other porins, the loops are the most variable segments when comparing related protein sequences. This is, of course, not unusual for a protein structure, because, during evolution, mutations can accumulate in loop regions without perturbing the fold or the function of the protein. In Gram-negative bacteria, the porin loops are about the only proteinaceous sites accessible from the outside. Therefore, their hypervariability may be important to evade recognition by harmful agents such as phages or antibodies. Indeed, it is in the protrusion where side-chain replacements are found in λ page-resistant mutants [43]. Notably, not all affected residues are at the protein surface, but some are buried at locations where the loops pack together [10]. Residue replacements at these positions most probably perturb indirectly the structure of the protrusions.

Quite early, the external loops were identified as such from topological studies [46]. Later they were mutagenized to probe their potential involvement in starch binding and maltodextrin translocation [47]. Loops L5, L6 and L9 were deleted one-by-one or in combination. Most of the mutants showed impaired function and a role of the loops in sugar binding was deduced. However, without knowing the three-dimensional structure of the deletion mutants cautious interpretation of the functional results is indicated. The authors point out quite rightly that deletion of any loop may destabilize the neighbors, up to the extent that they collapse and block the pore. *In vitro*, all deletion mutants showed strong maltodextrin binding [30]. Interestingly, it was shown for two of them that access to the binding site was indeed impeded from the extracellular side, in line with the interpretation that some of the remaining loops had collapsed.

In a more subtle approach, Charbit et. al [48] focused on the role of L9 by alanine-scanning mutagenesis of residues 371–379. In the native structure, residues K371, D373 and K375 are facing the membrane. It is hard to conceive how the latter residues can contribute to sugar binding and the modest reduction in the maltodextrin transport parameters is possibly not significant. Residues G378 and Y379, on the other hand, are part of the channel lining, but their replacement again showed only a moderate effect. The largest effect was observed for E374A and W376A. These residues are part of the L9–L1 interface; their replacement most probably results in indirect structural perturbation of the channel lining. Taken together, it appears to me that the postulated "primary" sugar-binding site, which would help to pre-orient the sugar prior to translocation, has not yet been identified.

9.4 Simulation of Maltoporin Function

LamB is the prototype of a facilitated diffusion channel. Based on the wealth of structural and functional information summarized in this review, the translocation process of maltodextrins can be described as follows. Upon diffusional encounter the oligosaccharide adheres to the start of the greasy slide by virtue of the hydrophobic interaction. Indeed, W74 is easily accessible within the vestibule of the trimeric channel. Unless the sugar detaches again, the following movement will be restricted along the hydrophobic patch of the greasy slide and, in effect, the long linear maltodextrin molecule will be steered into the narrow channel. The energy minimum will be found at the channel center, where the sugar hydroxyls can interact favorably with the ionic tracks. Rattling back and forth, the oligosaccharide will perform a number of register shifts with respect to the channel. The minimum energy path for such a register shift has been obtained recently by a molecular refinement procedure (conjugated peak refinement) [49]. Figure 9.5 shows a few snapshots of this movement. No large energy barriers were encountered, due to the smoothness of the greasy slide and the fact that the sugar hydroxyls can engage in a multitude of hydrogen bonding interactions throughout. Also, the protein matrix stays virtually unchanged during the process (Figure 9.5). Therefore, friction is kept low. Finally, the sugar can escape to one or the other side of the membrane. Net transport is achieved by the presence of the maltose-binding protein that keeps the concentration of free sugar in the periplasm very low, thus avoiding back transfer.

Sugar-specific porins and, in particular, LamB have proven to be very useful objects for the study of relationships between structure and guided diffusion channel function. It is remarkable that single trimers can now be analyzed routinely upon incorporation in artificial bilayers, permitting detailed evaluation of sugar binding kinetics and thermodynamics. The remaining open questions and discrepancies in the literature will be certainly resolved soon. That LamB may also be of practical use has been shown recently in a system which utilizes LamB for the ejection of λ phage DNA into artificial nanocontainers [50].

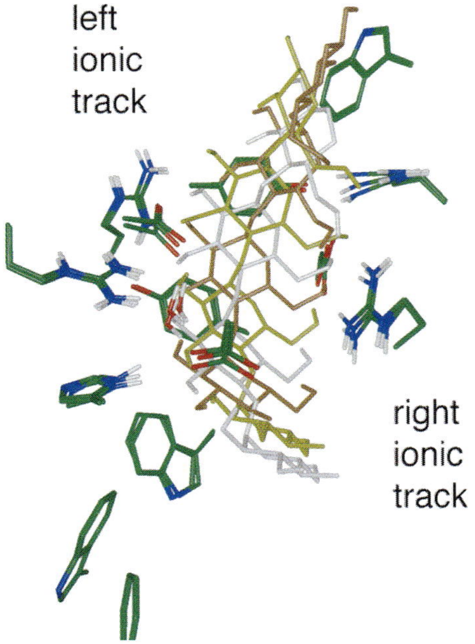

Figure 9.5 The register shift of maltohexaose. Superposition of three snapshots along the path. The picture is rotated with respect to the view in Figure 9.3 so that the greasy slide is in the back, and the ionic tracks are found to the left and right of the sugar. Note that the protein sidechains show only minor structural adaptation during the register shift. This figure has been reproduced with permission of the copyright holder from [49].

References

1 H. Nikaido, *Mol Microbiol* **1992**, *6*, 435–442.
2 R. Benz, A. Schmid, G. H. Vos-Scheperkeuter, *J Membr Biol* **1987**, *100*, 21–29.
3 A. Fersht, *Enzyme Structure and Mechanism*. Freeman, New York, **1985**.
4 M. Jordy, C. Andersen, K. Schulein, T. Ferenci, R. Benz, *J Mol Biol* **1996**, *259*, 666–678.
5 P. van Gelder, F. Dumas, I. Bartoldus, N. Saint, A. Prilipov, M. Winterhalter, Y. Wang, A. Philippsen, J. P. Rosenbusch, T. Schirmer, *J Bacteriol* **2002**, *184*, 2994–2999.
6 P. S. Phale, A. Philippsen, C. Widmer, V. P. Phale, J. P. Rosenbusch, T. Schirmer, *Biochemistry* **2001**, *40*, 6319–6325.
7 P. van Gelder, F. Dumas, I. Bartoldus, N. Saint, A. Prilipov, M. Winterhalter, Y. Wang, A. Philippsen, J. P. Rosenbusch, T. Schirmer, *J Bacteriol* **2002**, *184*, 2994–2999.
8 L. Randall-Hazelbauer, M. Schwartz, *J Bacteriol* **1973**, *116*, 1436–1446.
9 A. Death, L. Notley, T. Ferenci, *J Bacteriol* **1993**, *175*, 1475–1483.
10 T. Schirmer, T. A. Keller, Y.-F. Wang, J. P. Rosenbusch, *Science* **1995**, *267*, 512–514.
11 R. Dutzler, Y.-F. Wang, P. J. Rizkallah, J. P. Rosenbusch, T. Schirmer, *Structure* **1996**, *4*, 127–134.
12 J. E. W. Meyer, M. Hofnung, G. E. Schulz, *J Mol Biol* **1997**, *266*, 761–775.
13 M. Luckey, H. Nikaido, *Proc Natl Acad Sci USA* **1980**, *77*, 167–171.
14 Y. F. Wang, R. Dutzler, P. J. Rizkallah, J. P. Rosenbusch, T. Schirmer, *J Mol Biol* **1997**, *272*, 56–63.
15 D. Forst, W. Welte, T. Wacker, K. Diederichs, *Nat Struct Biol* **1998**, *5*, 37–46.
16 D. Fu, A. Libson, L. J. Miercke, C. Weitzman, P. Nollert, J. Krucinski, R. M. Stroud, *Science* **2000**, *290*, 481–486.
17 H. Sui, B. G. Han, J. K. Lee, P. Walian, B. K. Jap, *Nature* **2001**, *414*, 872–878.
18 G. E. Schulz, *Curr Opin Struct Biol* **1996**, *6*, 485–490.
19 W. Boos, H. Shuman, *Microbiol Mol Biol Rev* **1998**, *62*, 204–229.
20 T. Schirmer, *J Struct Biol* **1998**, *121*, 101–109.
21 R. Koebnik, K. P. Locher, P. van Gelder, *Mol Microbiol* **2000**, *37*, 239–253.
22 P. E. Klebba, *Res Microbiol* **2002**, *153*, 417–424.
23 A. Charbit, *Front Biosci* **2003**, *8*, S265–S274.
24 F. Dumas, R. Koebnik, M. Winterhalter, P. van Gelder, *J Biol Chem* **2000**, *275*, 19747–19751.
25 D. A. Doyle, J. M. Cabral, R. A. Pfuetzner, A. Kuo, J. M. Gulbis, S. L. Cohen, B. T. Chait, R. MacKinnon, *Science* **1998**, *280*, 69–77.
26 C. Andersen, M. Jordy, R. Benz, *J Gen Physiol* **1995**, *105*, 385–401.
27 F. Orlik, C. Andersen, R. Benz, *Biophys J* **2002**, *83*, 309–321.
28 F. Orlik, C. Andersen, R. Benz, *Biophys J* **2002**, *82*, 2466–2475.
29 R. Benz, A. Schmid, T. Nakae, G. H. Vos-Scherperkeuter, *J Bacteriol* **1986**, *165*, 978–986.
30 C. Andersen, C. Bachmeyer, H. Tauber, R. Benz, J. Wang, V. Michel, S. M. Newton, M. Hofnung, A. Charbit, *Mol Microbiol* **1999**, *32*, 851–867.
31 C. Andersen, B. Schiffler, A. Charbit, R. Benz, *J Biol Chem* **2002**, *277*, 41318–41325.
32 P. van Gelder, F. Dumas, J. P. Rosenbusch, M. Winterhalter, *Eur J Biochem* **2000**, *267*, 79–84.
33 S. M. Bezrukov, L. Kullman, M. Winterhalter, *FEBS Lett* **2000**, *476*, 224–228.
34 L. Kullman, M. Winterhalter, S. M. Bezrukov, *Biophys J* **2002**, *82*, 803–812.
35 G. Schwarz, C. Danelon, M. Winterhalter, *Biophys J* **2003**, *84*, 2990–2998.
36 C. Danelon, T. Brandot, M. Winterhalter, *J Biol Chem* **2003**, *278*, 35542–35551.
37 C. Ulmke, J. Kreth, J. W. Lengeler, W. Welte, K. Schmid, *J Bacteriol* **1999**, *181*, 1920–1923.

38 B. H. Kim, C. Andersen, J. Kreth, C. Ulmke, K. Schmid, R. Benz, *J Membr Biol* **2002**, *187*, 239–253.

39 P. van Gelder, R. Dutzler, F. Dumas, R. Koebnik, T. Schirmer, *Protein Eng* **2001**, *14*, 943–948.

40 C. Andersen, R. Cseh, K. Schulein, R. Benz, *J Membr Biol* **1998**, *164*, 263–274.

41 K. Schmid, R. Ebner, K. Jahreis, J. W. Lengeler, F. Titgemeyer, *Mol Microbiol* **1991**, *5*, 941–950.

42 C. Hardesty, C. Ferran, J. M. Dirienzo, *J Bacteriol* **1991**, *173*, 449–456.

43 F. Dumas, S. Frank, R. Koebnik, E. Maillet, A. Lustig, P. van Gelder, *J Mol Biol* **2000**, *300*, 687–695.

44 J. Michels, A. Geyer, V. Mocanu, W. Welte, A. L. Burlingame, M. Przybylski, *Protein Sci* **2002**, *11*, 1565–74.

45 K. Schülein, C. Andersen, R. Benz, *Mol Microbiol* **1995**, *17*, 757–767.

46 A. Charbit, J. Ronco, V. Michel, C. Werts, M. Hofnung, *J Bacteriol* **1991**, *173*, 262–275.

47 P. E. Klebba, M. Hofnung, A. Charbit, *EMBO J* **1994**, *13*, 4670–4675.

48 A. Charbit, J. Wang, V. Michel, M. Hofnung, *Mol Gen Genet* **1998**, *260*, 185–192.

49 R. Dutzler, T. Schirmer, M. Karplus, S. Fischer, *Structure* **2002**, *10*, 1273–1284.

50 A. Graff, M. Sauer, P. van Gelder, W. Meier, *Proc Natl Acad Sci USA* **2002**, *99*, 5064–5068.

10
Functional Reconstitution of Specific Porins

Roland Benz and Frank Orlik

10.1
Introduction

The cell envelope of Gram-negative bacteria consists of three distinct layers, the outer membrane, the periplasmic space containing the murein or the peptidoglycan layer and the inner membrane [1]. The inner membrane contains a typical phospholipid bilayer and represents an effective diffusion barrier for hydrophilic solutes. In addition to the respiration chain, it contains a large number of transport systems including the machineries for protein export and the secretion of substances harmful for the cells [2]. Murein represents a large heteropolymer from amino sugars and peptides that confers the rigidity of the cell envelope, the shape of the cell and protects it from osmotic lysis [3]. The amino sugars (*N*-acetylglucosaminyl-*N*-acetylmuramyl dimers) form long linear strands that surround the cells. They are covalently linked together between two muramyl residues by short tetrapeptides [4, 5]. Components of the outer membrane are either covalently bound to the peptidoglycan or may interact with it through ionic bridges in such a way that a tight network is produced.

The periplasmic space between the inner and outer membranes represents an additional cellular compartment that occupies between 5 and 20% of the total cell volume according to different estimations [6, 7]. It plays an important role in the physiology of Gram-negative bacteria because binding proteins for solutes such as maltose, phosphate and the β-lactamase activity are located there. The presence of the binding proteins is essential for vectorial uptake of nutrients across the outer and inner membranes because they confer the periplasmic space together with the inner membrane transport systems into a sink, which keeps the free concentration of the transported nutrients low. The periplasmic space is almost iso-osmotic with the cytoplasm (around 300 mosM). This means that the osmotic pressure, normally around 3.5 bar and at maximum about 7–8 bar [8], is maintained across the outer membrane and not across the inner membrane. The periplasmic space is strongly anionic compared to the external medium [7, 9]. This is partially due to anionic groups attached to the outer membrane; the

Bacterial and Eukaryotic Porins. Edited by Roland Benz

ISBN 3-527-30775-3

rest may be caused by the anionic membrane-derived oligosaccharides (MDOs) present in the periplasmic space maintaining part of the osmolarity. These oligosaccharides also contribute to the Donnan potential across the outer membrane, which can be as large as –80 mV (inside negative) in media of low ionic strength [7, 9].

The outer membrane of Gram-negative bacteria has special filter properties, which means that it also plays an important role in the physiology of these organisms. All nutrients or antibiotics, either hydrophilic or hydrophobic, have to cross this permeability barrier on their way into the periplasmic space and further on into the cell. Furthermore, the outer membrane contains proteins that allow the export of proteins such as the hemolysin of uropathogenic *Escherichia coli* strains and related RTX (Repeats in Toxin) toxins of other Gram-negative bacteria [10, 11] (see also Chapter 8). Other components of the outer membrane are parts of active export systems that, together with inner membrane transport systems, allow the secretion of harmful substances such as antibiotics out of the cells to keep their concentration inside the cells low. The active components of molecular sieving of the outer membrane are due to presence of a few major proteins called "porins", which form open channels in the outer membrane [12]. In addition to the open porin channels, the outer membrane of Gram-negative bacteria also contains receptors which represent gated channels that are closed under normal conditions [13] (see Chapters 11 and 12). Other porin-like structures of the outer membrane represent the channel-tunnel proteins of the TolC type [10, 11] (see Chapter 8).

Many known porins form transmembrane channels that sort solutes mainly according to their molecular masses and have little solute specificity, as discussed in Chapters 5 and 6. These porins represent general diffusion channels and Gram-negative bacteria contain at least one constituent general diffusion porin in the outer membrane. Other porins contain binding sites for certain classes of solutes. Their study *in vivo* and *in vitro* represents the topic of this chapter. Specific porins are often induced under special growth conditions together with components of an inner membrane uptake system, and enzymes necessary for degradation and fermentation of the solutes [14, 15]. Many of the uptake systems also possess periplasmic-binding proteins, which have a high affinity for the transported nutrients [14, 16]. The amino acid structures of the many porins known to date are deduced from of their amino acid or cDNA sequence. The sequences are not particularly hydrophobic, which means that the arrangement of the proteins in the secondary and tertiary structure is responsible for their function as membrane channels. Many porins are organized as trimers of three identical subunits of which each contains one channel [17, 18]. The three-dimensional (3-D) structures of the general diffusion porin from *Rhodobacter capsulatus* [17] and those of *Escherichia coli* [18] have been resolved from X-ray crystallography together with a variety of other general diffusion porins [19, 20]. These porins are organized as hollow cylinders formed by 16 antiparallel β-barrels (see Chapter 2). From the many specific porins characterized to date, only the 3-D structures of carbohydrate-specific porins from the LamB type are known [21–23]. The channel in these cases is also formed by trimers of three identical polypeptide subunits.

However, the single unit contains 18 antiparallel β-strands instead of the 16 found for the general diffusion pores.

The function of the porins can be studied in intact cells and *in vitro* reconstitution systems such as the liposome-swelling assay [24], measurement of the exclusion limit [12] or the lipid bilayer technique [25, 26]. Reconstitution experiments allow us to re-establish the function of the isolated porins. In addition, they are able to demonstrate the integrity of the pore-forming complex after the isolation process. Reconstitution experiments permit detailed studies of the channel properties, which is very often not possible using *in vivo* systems. Reconstituted systems contain only a few components and allow good control of the conditions. There also exist disadvantages of these studies because artifacts are possible and components may be missing that may be necessary for the full biological activity of the porin pores.

10.2
Isolation and Purification of Specific Porins

There exist different methods for the successful isolation of porins, and there is not much difference between the purification of general diffusion porins and specific porins with one exception (see below). Enterobacterial porins, including LamB are often tightly associated with the peptidoglycan layer [12, 27–29]. Their isolation is relatively simply because only a few outer membrane proteins are associated with the murein, whereas the others are lost during the washing of the cell envelope fraction with SDS-containing solutions. The SDS-insoluble material contains the peptidoglycan with the covalently bound lipoprotein and the peptidoglycan-associated proteins including OmpA and some porins. The porin can be separated from the peptidoglycan by the salt extraction method using the extraction of the murein with 1% SDS, 0.4 M NaCl and 5 mM EDTA at 37 °C [30]. Boiling of the porin–peptidoglycan complex for 5 min in SDS also leads to the release of the proteins, but it denatures the porins.

Porins from non-enteric bacteria are often not closely associated with the peptidoglycan layer. In these cases, the outer membrane has to be isolated first or the outer membrane fraction of the cell envelope has to be enriched by special treatment. The isolation of the outer membrane basically follows the procedure proposed first by Miura and Mizushima [31]. The cell envelope fraction is layered on top of a two-step sucrose gradient (70 and 54% [31]). After centrifugation in an ultracentrifuge for about 12 h at 80,000 *g* the outer membrane is in the pellet because of its higher density. Enrichment of the outer membrane in the cell envelope fraction can be achieved according to a rapid method for isolating detergent-insoluble outer membrane proteins from *Haemophilus* species [32]. For this, the cell envelope pellet is resuspended in 1% sodium *N*-lauryl sarcosinate, 10 mM HEPES, pH 7.4 and incubated for 30 min at room temperature to remove most soluble inner membrane components. The pellet of the subsequent centrifugation step is washed once in 10 mM HEPES, pH 7.4 and resuspended in 10 mM HEPES, pH 7.4. It contains most outer membrane components including the channel-forming activity, if there is any.

Porin purification can be achieved using chromatography across different columns. Most promising is fast protein liquid chromatography (FPLC) across affinity columns (e. g. ion-exchange columns such as Mono-Q, HiTrap-Q or Mono-SP) because the porins may adsorb to the column materials according to their net surface charge. For this procedure, the high ionic strength of the salt extraction solutions has first to be dialyzed against low ionic strength buffer. The protein is adsorbed to the column material, and the column is first washed with low ionic strength detergent buffer and then eluted with linear salt gradients ranging from 50 to 500 mM at a given pH. The salt and detergent content of the porins eluted at defined ionic strength can be decreased by another dialysis procedure, but this is not essential for the study of the pore-forming activity because the protein solutions are highly diluted in the membrane experiments. Chromatography across a starch column of carbohydrate-specific porins of the LamB type represents a simple and elegant way for their purification [33]. The porins bind tightly to the linear starch coupled to Sepharose 6B with the reducing end in advance. The protein is removed from the column material after a washing step with a buffer containing 20% maltose. For long-term stabilization it is possible to store the porins in a freezer at −20 °C or to lyophilize the protein solutions and to keep them frozen. Both procedures can often be applied without any loss of the pore-forming activity.

10.3
Reconstitution of Specific Porins in Lipid Bilayer Membranes

Several methods can be used for the study of the *in vitro* properties of specific porins. Some first reconstitution studies have been performed employing the liposome-swelling assay [24, 30]. Porin-containing multilamellar liposomes are formed from lipid and protein in a buffer containing a large molecular mass solute such as dextran or stachyose that is not able to penetrate the porin channels and is entrapped inside the liposomes. The liposomes are added under rapid mixing to an isotonic solution of a test solute. If this solute can penetrate the pores, the total concentration of solutes inside the liposomes increases because the large molecular solute is retained. The influx of the test solute occurs together with water because of the high water permeability of the lipid membranes. The liposome swelling, i. e. the influx of the test solute, can be detected by a measurement of the optical density, which is measured some time after the mixing [24]. The swelling rate is a measure for the penetration rate of the test solute through the porin channels. It is possible to compare the penetration rates of different solutes by using the same liposome preparation and to calculate the effective diameter of the porin channel from the initial swelling rates according to the theory of Renkin [34].

Many *in vitro* studies have been performed using lipid bilayers formed either according to the solvent-containing [25, 35, 36] or solvent-deficient technique [26, 37, 38] (see also Chapter 6). The cell apparatus used for membrane formation in both cases consists of Teflon chambers with thin walls separating two aqueous compartments. The Teflon divider between both compartments contains small circular holes. For membrane formation with solvents, the lipid is dissolved mainly in *n*-decane in a concentration of 1–2 % (w/v). The lipid solution is painted over the holes with diameters between 0.2 and 1 mm to form a thick colored lamella. The membrane experiments start after the lamella thins out and turns optically black in reflected light, because black membranes are much thinner than the wavelength of the visible light [39]. The formation of "solvent-free" or "solvent-depleted" lipid bilayer membranes occurs through the apposition of two lipid monolayers across small holes (diameter 10–100 μm) in the Teflon divider separating the two aqueous compartments [40, 41] (see also Chapter 6). Formation of solvent-depleted membranes cannot be controlled in reflected light because of their small surface area. The control of successful membrane formation (as opposed to the possibility that the small hole is simply plugged with lipid) is only possible on the basis of the measurement of the membrane capacitance [40, 41].

The lipid bilayer technique allows the very sensitive detection of current through the membrane down to 1 pA or less, which means that the conductance of a membrane channel can be measured with high accuracy because the bilayer itself has a very low conductance below 10 pS (corresponding to a resistance of more than 100 MΩ). Lipid bilayer membranes are, however, not very well suited for the study of fluxes of uncharged solutes. Lipid vesicles with a large surface to volume ratio allow much better access to this type of measurement. The lipid bilayer technique allows good access to the aqueous phases on both sides of the membrane. This means that the ionic composition on both sides of the membrane can be controlled with this method. It is possible to establish salt gradients across the membrane by the addition of concentrated salt solution to one side of the membrane or by exchange of the aqueous solution. Zero-current membrane potential measurements allow the measurement of the ionic selectivity of channels if the membrane contains a sufficient number of channels [42].

Figure 10.1 shows a typical reconstitution experiment using a solvent-containing membrane. The phosphate-specific porin OprP of *Pseudomonas aeruginosa* was added in final concentration of about 10 ng/ml to the aqueous phase on one side of a lipid bilayer membrane from diphytanoyl phosphatidylcholine/*n*-decane bathed in 0.1 M KCl. The addition of the specific porin results in a stepwise increase of the membrane conductance, which is typical for the reconstitution of both general diffusion pores and specific porins in model membranes. All steps had a single-channel conductance of about 160 pS. The single-channel events in the case of OprP were found to be fairly homogeneous, as has been observed for most specific porins, in contrast to general diffusion pores that very often show broad histograms [25, 35]. On the other hand, it is also possible that porin preparations may contain contaminants which cannot be detected in SDS–PAGE, but may be observed in lipid bilayer experiments because of its superior sensitivity. This means that it is possible to sort different porin channels using this method. It has to be noted that a similar sorting of channels is not possible by the liposome-swelling assay. Using the same approach, multichannel experiments (experi-

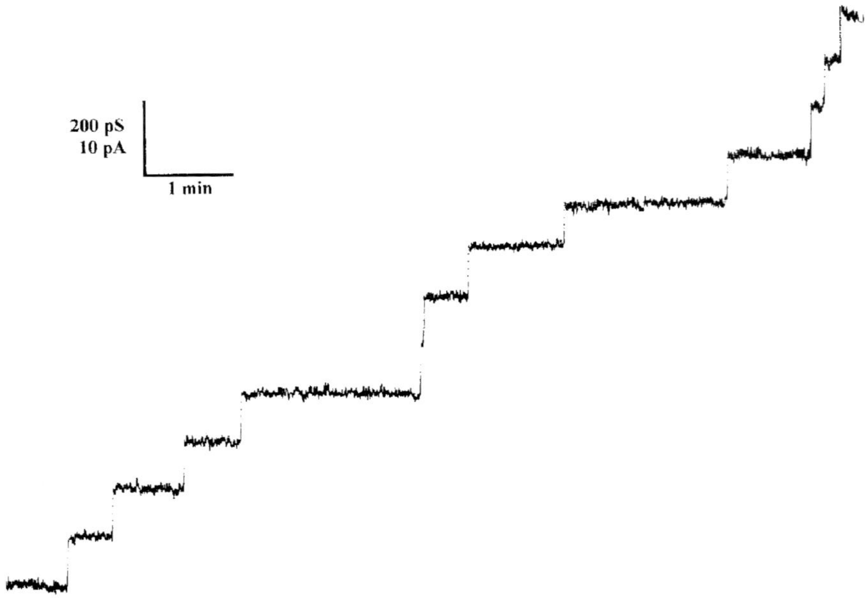

Figure 10.1 Stepwise increase of the membrane conductance (membrane current per unit voltage) in the presence of 50 ng/ml of the phosphate-specific OprP of *P. aeruginosa* in a 1 M KCl (pH 6) solution. The membrane was formed from diphytanoyl phosphatidylcholine/*n*-decane; $T = 25\,^{\circ}C$, $V_m = 50$ mV.

ments with a large number of channels) can be performed. A maximum of between 10^6 and 10^8 channels/cm^2 can be incorporated into lipid bilayer membranes. This means that the reconstitution of porin pores is not a rare event. On the other hand, the channel density in the outer membrane of Gram-negative bacteria is orders of magnitude above that obtained at maximum for lipid bilayers.

10.4
Analysis of Substrate Transport through Specific Porin Channels

10.4.1
Study of Ion Transport through the Phosphate-specific OprP of *P. aeruginosa*

The simple two-barrier, one-site model provides a good description of substrate transport through specific porins [43–46]. It assumes a binding site in the center of the specific porin channels. The rate constant k_1 describes the movement of the substrate molecules from the aqueous phase (concentration c) across the barrier to the central binding site, whereas the inverse motion is described by the rate constant k_{-1}. For simplicity, we assume symmetrical barriers with respect to the binding site, although some asymmetry has been found for the carbohydrate-specific LamB channel (maltoporin) of *Escherichia coli* [44, 45]. The stability constant of the binding between a substrate molecule and the binding site is $K = k_1/k_{-1}$. Furthermore, it is assumed that the channel is a single file channel – only one substrate molecule can bind to the binding site at a given time and that no substrate or other molecule, such as ions [43–46], can pass the channel when the binding site is occupied. This means that a substrate can enter the channel only when the binding site is free. The probability, p, that the binding site is occupied (identical concentrations on both sides) is given by:

$$p = \frac{Kc}{1 + Kc} \tag{1}$$

and that it is free by:

$$1 - p = \frac{1}{1 + Kc} \tag{2}$$

When a voltage is applied to the two-barrier one site-channel the barrier levels change (i. e. k_1 and k_{-1} have to be replaced by k_1', k_1'' and k_{-1}', k_{-1}'', respectively, because of the applied voltage) and a net flux of ions is observed. The net flux of ions, ϕ, through the channel under stationary conditions is equal to the net movement of ions across one barrier (for instance, across the left-hand side barrier) [46]:

$$\phi = k_1' c \frac{1}{1 + Kc} - k_{-1}'' \frac{Kc}{1 + Kc} \tag{3}$$

The rate constants k_1' and k_{-1}'' are multiplied by the probabilities that the binding site is free or occupied, respectively. These probabilities are independent of voltage when the top of the barriers is halfway between the binding site and the membrane surfaces. The current, I, through the channel as a function of an applied voltage is given by [46]:

$$I = e_0 \phi = e_0 \frac{Kc}{1 + Kc} (k_1'/K - k_2'') \tag{4}$$

Equation (4) can be used for the calculation of the single-channel conductance, $G^0(c)$, in the limits of small voltages ($V_m << 25$ mV):

$$G^0(c) = G^0_{max} \frac{Kc}{1 + Kc} \tag{5}$$

where $G_{max}{}^0$ is the maximum single-channel conductance of two-sites one-barrier channels at very high ion concentration:

$$G^0_{max} = \frac{e_0 F}{2RT} k_{-1} \tag{6}$$

The outer membrane of *P. aeruginosa* has very special sieving properties, which make this organism very resistant to most antibiotics (see Chapter 4). Part of these sieving properties arises from the limited permeability of the large protein F channel and of the many other outer membrane porins, which are probably all substrate specific and are expressed under special growth conditions [47, 48]. Protein D1 is part of the sugar uptake system and is induced in the presence of glucose [49]. Protein D2 probably plays an important role in the permeation of amino acids and peptides since it has been demonstrated that it contains a binding site for basic amino acids and peptides [50]. Under conditions of phosphate limitation, OprP is induced together with an alkaline phosphatase, a phosphate binding protein, and some inner membrane proteins [51]. The induced system therefore has some similarities with the pho system of the *Enterobacteriaceae* [52] although porin PhoE is besides its increased permeability for anions as compared to OmpF and OmpC of *E. coli* a general diffusion porin [53]. OprP and the closely related OprO (which is probably a channel for polyphosphate uptake [54, 55]) form highly anion-selective channels in lipid bilayer membranes, which are at least 100-times more permeable for chloride than for potassium [46, 51]. It is noteworthy that this *in vitro* system represents the only one suited for the study of this channel. Zero-current membrane potentials have Nernstian slopes (corresponding to $P_{cation}/P_{anion} << 1$) and indicate that the OprP channel is at least 1000 times more permeable for chloride than for potassium [46, 51]. Table 10.1 shows the analysis of the anion transport through the phosphate-specific OprP channel using the single-channel conductance data in different salt solutions with the same cation (potassium), but different anions. Different cations have no influence on the single-channel conductance and it is not a linear function of the bulk aqueous salt concentration and saturates at different concentrations of the various anions. Figure 10.2 shows the conductance of the OprP channel as functions of the chloride and phosphate concentrations at pH 6 [46]. The curve

Table 10.1 Average single-channel conductance, *G*, of OprP in different salt concentrations (the cation in all cases was potassium).

Anion	***G (pS)***						
	0.003 M	***0.01 M***	***0.03 M***	***0.10 M***	***0.30 M***	***1.0 M***	***3.0 M***
Chloride	17	49	102	159	245	260	270
Fluoride	–	17	50	136	240	415	466
Bromide	–	12	–	86	165	223	245
Iodide	–	–	4	13	29	64	–
Nitrite	–	10	37	83	153	208	243
Nitrate	–	–	7	24	49	94	–
Phosphate (pH 6)	9	7	9	10	9	10	–

The aqueous salt solutions contained 10 ng/ml OprP and less than 0.1 μg/ml SDS or Triton X-100. The pH of the aqueous salt solutions was 6 unless otherwise indicated. The membranes were formed from diphytanoyl phosphatidylcholine/*n*-decane. $T = 25\,°C$; $V_m = 50$ mV. (The data were taken from [46].)

for chloride could be fitted to Equation (5) with the following parameters: $G^0{}_{max}$ = 280 pS and K = 20 M^{-1}. A similar approach could not be used for phosphate because the single-channel conductance is independent of the phosphate concentration (see Table 10.1). Alternatively, it is possible to calculate the binding constant for phosphate from the inhibition of the chloride conductance since phosphate and chloride apparently compete for the same binding site of the one-site, two-barrier OprP channel [56].

The fit of the single-channel conductance as a function of the concentration of the different anions indicates that a one-site, two-barrier model provides a good fit of the experimental data. It means also that Equations (5) and (6) can be used to calculate the rate constants, k_{-1} and k_1, of anion transport through the OprP channel. They were calculated from the maximum single-channel conductance, $G^0{}_{max}$, and $K = k_1/k_{-1}$. The results of the calculation are included in Table 10.2. The rate constants for most of the anions are very similar. However, the on-rate constant for phosphate binding is considerable higher than that for the other anions, whereas the off-rate is much smaller. Furthermore, the OprP channel conducts much better at high salt concentration chloride than phosphate. This argues in principle against a phosphate-specific channel. However, phosphate limitation means that the phosphate concentration in the growth media is below 1 mM [51]. At these concentrations the channel offers a considerable advantage for the transport of phosphate as the following consideration demonstrates. The net flux of charged and uncharged solutes, ϕ, through the two-barrier, one-site channel (stability constant for solute binding, K) in the case of a gradient across the membrane (concentration on one side equal to zero, concentration on the other side equal c) and in the absence of a membrane potential is given by [45]:

$$\phi = \phi_{max} Kc/(Kc + 2) \tag{7}$$

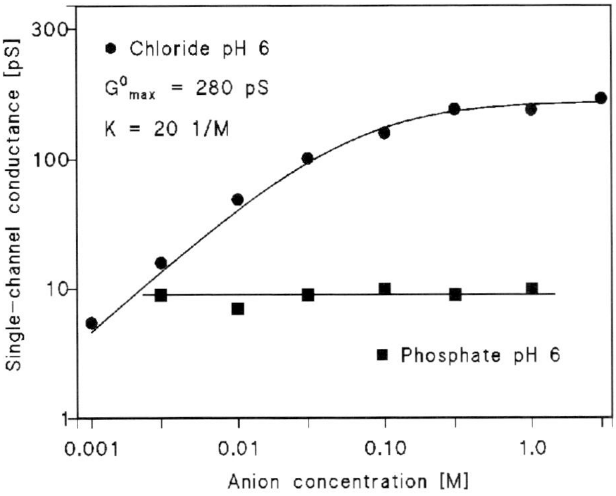

Figure 10.2 Single-channel conductance of the OprP channel of *P. aeruginosa* as a function of the KCl and phosphate concentration in the aqueous phase (pH 6). The membranes were formed from diphytanoyl phosphatidylcholine/*n*-decane; T = 20 °C; V_m = 50 mV. The solid lines were drawn using Equation (5) and the corresponding parameters given for chloride in Table 10.2. The data were taken from [46].

Table 10.2 Stability constants, K, maximum single channel conductance G^0_{max}, and rate constants k_1 and k_{-1} for the transport of different anions through the OprP channel of *P. aeruginosa* as calculated from the data of Table 10.1 using Equations (5) and (6) and $K = k_1/k_{-1}$

Anion	*K* (M^{-1})	G^0_{max} (pS)	k_1 [10^7 (Ms)$^{-1}$]	k_{-1} (10^7 s^{-1})	*r* (nm)
Fluoride	3.5	515	60	17	0.133
Chloride	20	280	180	9.0	0.181
Bromide	4.7	265	40	8.5	0.195
Iodide	1.3	110	4.6	3.5	0.216
Nitrite	4.6	263	39	8.4	–
Nitrate	2.0	140	9	4.5	0.198
Phosphate (pH 6)	11000	9	3500	0.28	–

where $\phi_{max} = k_{-1}$ is the maximum flux of solutes through the channel at very high concentrations (i. e. k_{-1} represents the turnover number of the channel similar to the turnover number of an enzyme at very high substrate concentration). The maximum permeability of a one-site, two-barrier channel is $k_1/2$ at very low solute concentration [45]. This means that the OprP channel has an about 20-fold higher permeability for phosphate than for chloride at very low ion concentration, which represents the *in vivo* situation. Figure 10.3 shows this situation in some more detail. It depicts the net flux of chloride and phosphate (pH 6) through the OprP-channel as functions of the corresponding anion concentration on one side of the channel (the concentration on the other side is set to zero, see above). The two lines were as calculated according to Equation (7) using the corresponding turnover numbers k_{-1} and the stability constants K taken from Table 10.2. The flux of phosphate is below 10 mM larger than that of chloride and is in the linear range of the flux versus concentration curve approximately 20 times higher. This is again consistent with the *in vivo* situation of phosphate starvation. The channel should have a high permeabil-

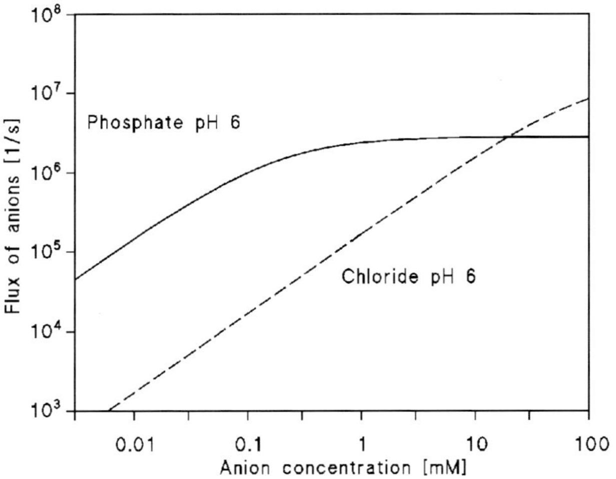

Figure 10.3 Flux of chloride and phosphate through one single OprP channel as a function of the corresponding anion concentration on one side of the channel. The concentration on the other side was set to zero. The flux was calculated using Equation (7) and the rate constants given in Table 10.2.

ity at very low external phosphate concentration, which together with the sink provided by the high-affinity phosphate-binding protein within the periplasmic space [51] should allow maximal phosphate transport across the outer membrane and also maximum presentation to the cytoplasmic membrane phosphate transport system. On the other hand, the phosphate flux saturates already at very low phosphate concentration (1 mM), whereas the saturation of chloride flux is at much higher chloride concentration. It is noteworthy that this does not contradict the view that OprP is a channel for phosphate uptake across the outer membrane of *P. aeruginosa*.

10.4.2
Evaluation of the Stability Constant for Binding of Neutral Solutes to the Binding Site inside Specific Porins

The transport of neutral solutes through substrate-specific porins has been studied in detail in recent years [44, 45, 57–60]. The minimum requirements for the description of solute transport are also given by the simple two-barrier, one site-model described above for the transport of phosphate and other anions through OprP. However, the approach is somewhat different because the transport of neutral solutes cannot be measured directly using the lipid bilayer method because of its poor resolution for flux measurements. Here, we use an alternative approach for the study of neutral solute transport on the basis of channel block for the movement of ions when a solute is bound to the binding site inside the specific porins and blocks the channel. The channel (given by P) is open when no substrate S is bound and closed when it is occupied by a substrate S to form the non-conducting substrate–channel complex PS:

$$P + S \underset{k_{-1}}{\overset{k_1 c}{\rightleftarrows}} PS \tag{8}$$

The probability, p, that the binding site is occupied by a substrate molecule is given by Equation (1), and that it is free and the channel conducts ions is given by Equation (2). The conductance, $G(c)$ [$= I(c)/V_m$], of a membrane containing many specific porins in the presence of a solute with the stability constant, K, and a solute concentration, c, is given by the probability that the binding site is free:

$$G(c) = \frac{G_{max}}{1 + Kc} \tag{9}$$

where G_{max} is the membrane conductance before the start of the solute addition to the aqueous phase when all channels are in the open configuration and $G(c)$ is the conductance at the solute concentration c. Equation (8) may also be written as:

$$\frac{G_{max} - G(c)}{G_{max}} = \frac{Kc}{1 + Kc} \tag{10}$$

This means that the stability constant for solute binding can be derived from titration experiments similar to that shown in Figure 10.4 for the titration of LamB-mediated membrane conductance with maltotetraose. It is possible to analyze the titration curve by using Lineweaver–Burke plots as has been shown in previous publications (see also Figure 10.5) [45, 46, 59]. The half-saturation constant, K_S, is given by the inverse stability constant $1/K$.

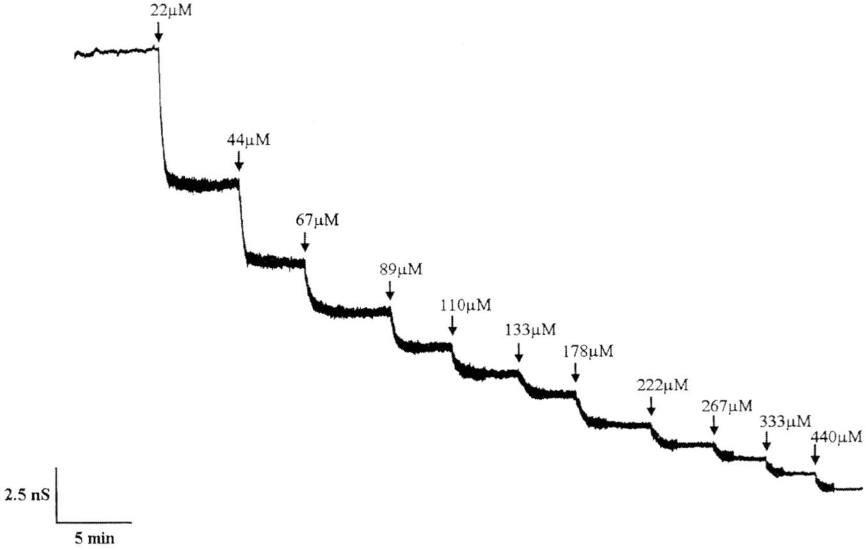

Figure 10.4 Titration of LamB-induced membrane conductance with maltotetraose. The membrane was formed from diphytanoyl phosphatidylcholine/*n*-decane. The aqueous phase contained 50 ng/ml LamB, 1 M KCl and maltotetraose at the concentrations shown at the top of the figure. The temperature was 25 °C and the applied voltage was 20 mV.

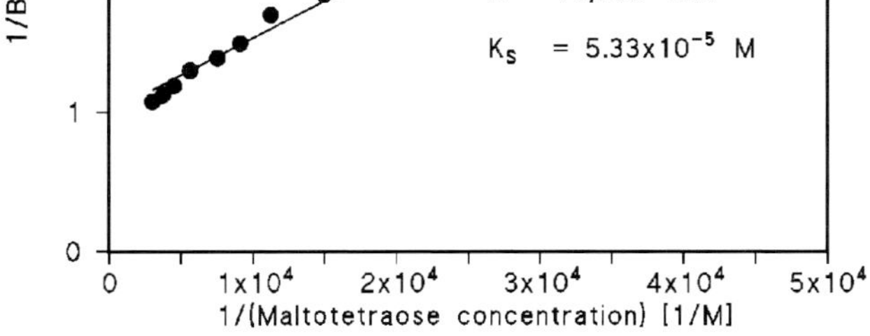

Figure 10.5 Lineweaver–Burke plot of the inhibition of LamB-induced membrane conductance by maltotetraose. The data were taken from Figure 10.4. The solid line was drawn according to Equation (10) using $K = 18{,}000\ M^{-1}$.

10.4.3
Investigation of Substrate-binding Kinetics using the Analysis of Current Fluctuations

It is not possible to derive the kinetic constants of substrate binding to the binding site inside the specific porins from the titration experiments. The reason for this is the high affinity of the substrates for the binding sites and the slow aqueous diffusion of the solutes through the unstirred layers on the surface of the membranes. An alternative approach was used to study the binding kinetics based on the microscopic reversibility of chemical reactions as described in the following. The numbers of open and closed channels in a membrane are coupled through a chemical reaction [Equation (8)]. Assuming that a lipid bilayer membrane contains 1000 solute-specific channels, 500 channels are blocked on average when the solute is added in a concentration equal to the half-saturation concentration K_S. At a given time the number of closed and open channels may vary from this number due to microscopic variations of the number of bound substrate molecules. The microscopic variations can be treated in the same manner as those caused by macroscopic disturbances such as concentration changes or temperature jumps. Their relaxation time, τ, is controlled by the chemical reaction rate $1/\tau$ of the second-order reaction between binding site and solute given in Equation (11) [61–63]:

$$\frac{1}{\tau} = 2\pi f_c = k_1 c + k_{-1} \tag{11}$$

Where f_c is the corner frequency of the power density spectrum, $S(f)$, which is given by a “Lorentzian” function [42, 43]:

$$S(f) = S_0/(1 + (f/f_c)^{\alpha}) \tag{12}$$

where a is the slope of the decay of the Lorentzian function (usually close to 2) and S_0 is the plateau value of the power density spectrum at small frequencies. It is given by [61–63]:

$$S_0 = 4Ni^2 p(1 - p)\tau \tag{13}$$

where N is the total number of channels (blocked and unblocked) within the membrane and i is the current through one single open channel. The membrane current, $I(c)$, is given by the number of open channels times the current through one single-channel and the probability that the channel is open:

$$I(c) = iN(1 - p) \tag{14}$$

Equations (13) and (14) can be used to calculate the single-channel conductance, g, of the channels from the applied membrane potential V_m and the current through one single channel:

$$g = \frac{i}{V_m} = \frac{S_0}{4V_m I(c)p\tau} \quad (15)$$

10.5 Study of Carbohydrate Binding to the Specific Porins of the LamB Family

10.5.1 LamB of *E. coli*

The genes coding for the malto-oligosaccharide uptake and degradation system are part of the maltose operon of enteric bacteria, which has two loci malA (74′) and malB (91′) [14]. malB also contains, in addition to the genes *malG, malF* and *malK*, responsible for inner membrane transport, and the gene coding for the periplasmic binding protein, *malE*, the gene, *lamB*, coding for an outer membrane protein, which is a carbohydrate-specific channel [44, 64, 65]. The carbohydrate-specific porins of enteric bacteria form a large family of proteins, which suggests that they have a common ancestor [66]. The Prototype of this family is LamB (maltoporin) of *E. coli*, also known as the receptor for phage λ, whose carbohydrate-binding properties are described here. LamB of *E. coli* K12, also called maltoporin, is a trimeric protein, which specifically facilitates the diffusion of maltose and malto-oligosaccharides through the bacterial outer membrane and acts as a non-specific porin for small hydrophilic molecules [44, 64, 65]. Each monomer consists of an 18-stranded antiparallel β-barrel, with nine surface-exposed loops (L1–L9) [18, 23]. Reconstitution experiments have shown that LamB has a defined substrate specificity and that it is able to discriminate between disaccharides of identical molecular mass, e. g. between sucrose and maltose [45, 57, 65]. Long-chain malto-oligosaccharides are able to block permeation of glucose and ions through LamB [64, 65]. This possibility allows a meaningful investigation of carbohydrate transport through LamB *in vitro*. In particular, it is possible to study the mechanism and kinetics of malto-oligosaccharide transport using the above-described formalism for a two-barrier, one-site model [44, 45]. This means it is possible to titrate the LamB-induced conductance with maltose, malto-oligosaccharides and other carbohydrates and to calculate the binding constant, *K*, of these solutes to the channel interior by using Equation (10).

Figure 10.4 shows the results of a titration experiment with this carbohydrate-specific porin. LamB was added in a concentration of 50 ng ml^{-1} to a black lipid bilayer membrane (BLM). Thirty minutes after the addition, the membrane conductance was almost stationary. Different concentrations of maltotetraose were added to the aqueous phase with stirring (arrows). The membrane conductance decreased as a function of the maltotetraose concentration (Figure 10.4). Using Equation (10), a binding constant of about 18,500 M^{-1} for maltotetraose to the binding site inside the channel can be calculated from the conductance decrease as a function of maltotetraose concentration (see Figure 10.5). Table 10.3 shows the results derived from similar titration experiments with a variety of carbohydrates and the

Table 10.3 Stability constants, *K*, for carbohydrate binding to the carbohydrate-specific LamB channel of *E. coli*.

Carbohydrate	*K* (M^{-1})	K_s (mM)	*Rate of permeation relative to maltose*[a]
D-Glucose	9.5	110	160
Maltose	100	10	100
Maltotriose	2,500	0.40	66
Maltotetraose	10,000	0.10	19
Maltopentaose	17,000	0.059	–
Maltohexaose	15,000	0.067	–
Maltoheptaose	15,000	0.067	2.5
Trehalose	46	22	76
Lactose	18	56	9
Sucrose	67	15	2.5

The data were derived from titration experiments similar to that shown in Figure 10.4 using Lineweaver–Burke plots [see Equation (10)] where the LamB-induced conductance was inhibited by binding of the carbohydrates to the channel. K_s is the half-saturation constant. The aqueous phase contained 1 M KCl and about 50 ng/ml LamB. $T = 25\,°C$; $V_m = 50$ mV. The data were taken from [45]. The relative rate of permeation was taken from [65].

[a]The LamB containing liposomes were prepared in 40 mM stachyose and diluted in 40 mM solution of the test solutes. The relative rate of permeation was derived from the swelling of the liposomes as measured from the decrease of the optical density of the liposomes at 500 nm [65].

relative rate of permeation of the same carbohydrates through LamB as measured in the liposome permeability assay [65]. The comparison of both data suggests that the kinetics of carbohydrate permeation plays an important role for their transport through LamB because the stability constants for binding of different disaccharides, such as sucrose and maltose, are similar despite largely different results of the liposome permeability assay *in vitro* (see Table 10.3) [65] and the *in vivo* situation [67]. The results indicate that it is possible to study the specificity of channels for neutral solutes in reconstitution experiments. Table 10.3 demonstrates that the stability constant for the binding of carbohydrates increased with the number of residues in a carbohydrate chain. For example, in the series glucose, maltose to maltopentaose, the binding constant increased about 1,800-fold, whereas no further increase occurred between five and seven glucose residues in the malto-oligosaccharides.

As described out above, the rate constants k_1 and k_{-1} for carbohydrate transport through LamB cannot be derived from the analysis of the sugar-induced block of the LamB channels because of the pure time resolution of this method. To measure the rate constant, the carbohydrate-induced current noise of the LamB channel was analyzed in multichannel experiments as is described above [68–70]. For this, a large number of LamB channels were reconstituted in a lipid bilayer membrane and the noise measurements could only be started when the conductance was absolutely stationary since a change of the membrane conductance has a large influence on the noise measurements. The power density spectrum of the

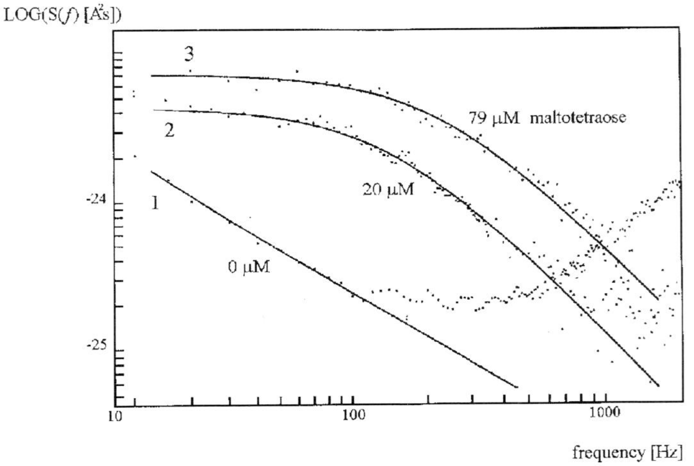

Figure 10.6 Power density spectra of maltotetraose-induced current noise of a diphytanoyl phosphatidylcholine/*n*-decane membrane containing 690 LamB channels. Trace 1 shows the control (1 M KCl). Trace 2: the aqueous phase contained 20 μM maltotetraose and the power density spectrum of Trace 1 was subtracted ($\tau = 1.1$ ms; $S_0 = 4.3 \times 10^{-24}$ A^2s). Trace 3: the aqueous phase contained 79 μM maltotetraose and the power density spectrum of Trace 1 was subtracted ($\tau = 0.69$ ms; $S_0 = 7.2 \times 10^{-24}$ A^2s). $V_m = 25$ mV. The number, *N*, of channels was calculated by dividing the membrane conductance through the single-channel conductance of LamB.

current noise was measured at low voltage (see Figure 10.6, trace 1). Then maltotetraose was added in a concentration of 20 μM and the power density spectrum was measured again and the background noise was subtracted (trace 2). At another concentration of maltotetraose ($c = 79$ μM) the power density spectrum (minus the background) corresponded to trace 3. The power density spectra of traces 2 and 3 are of the Lorentzian type expected for a random switch with unequal on and off probabilities [61, 68]. Both could be fitted to Equation (12) with sufficient accuracy. The corner frequency, f_c, increased with increasing malto-oligosaccharide concentration. The corner frequencies of the experiments shown in Figure 10.6 and of other maltotetraose concentrations (not shown) could be fitted reasonably well to Equation (11), which allowed the calculation of the rate constants of the chemical reaction [$k_1 = 8.9 \times 10^6$ $(Ms)^{-1}$ and $k_{-1} = 760$ s^{-1}].

It is possible to study the binding kinetics of a variety of carbohydrates, but not of maltose and glucose, to the binding site inside the LamB channel. The rate constants for the on- (k_1) and off-processes (k_{-1}) for the binding of these carbohydrates are given in Table 10.4 [69]. The data for glucose and maltose could not be derived from noise measurements because the corner frequency is outside the experimental range. They are estimated on the basis of the relative rates of permeation of glucose and maltose taken from Luckey and Nikaido [65], and using the binding constants of Table 10.3 and the following consideration valid for the transport of carbohydrates in the vesicle permeability assay. Substituting the experimental condi-

Table 10.4 Kinetics of carbohydrate-transport through the LamB-channel of *E. coli* as derived from noise analysis of sugar-induced block of ionic current through the channel.

Carbohydrate	k_1 [10^6 (Ms)$^{-1}$]	k_{-1} (s^{-1})	K (M^{-1})
Glucose	1.0	110,000	9.5
Maltose	0.8	8,000	100
Maltotriose	10	1,300	7,700
Maltotetraose	8.0	670	11,900
Maltopentaose	7.0	530	13,200
Maltohexaose	5.0	240	20,800
Maltoheptaose	4.0	120	33,000
Sucrose	0.004	50	80

The kinetic constants for glucose and maltose binding were derived from the data of the liposome-swelling assay using the stability constants of glucose and maltose binding [65]. The data were taken from [69].

tions of [65] ($c'' = 40$ mM, $c' = 0$) the flux, ϕ, of carbohydrates through LamB is given by [see Equation (7)]:

$$\Phi = k_1(40 \text{ mM})/[2{+}K(40 \text{ mM})] \quad (16)$$

This means that k_1 is given by:

$$k_1 = \phi[(50 \text{ M}^{-1}) + K] \quad (17)$$

Equation (17) can be used for the calculation of k_1 for glucose and maltose using the flux of other malto-oligosaccharides relative to them (see Table 10.3). The results shown in Table 10.4 suggest that the rate constants for the on-process are approximately the same for most of the malto-oligosaccharides. Substantial differences were observed for the movement of the sugars out of the channel (corresponding to k_{-1}), probably because the interaction between binding site and carbohydrate increases with the number of glucose residues, which means that the malto-oligosaccharides move forth and back many times until they can leave the channel to one or the other side [69].

Lipid bilayer experiments also allow us to study the "sideness" of the reconstitution of the LamB channels when the protein is added to only one side of the membrane. The channels are in the permanently open configuration at neutral pH. Similar to other Gram-negative bacterial porins, LamB shows almost complete channel closure when the pH on both sides of the membrane is lowered to pH 4 [71]. When LamB is added to only one side of the membrane, the *cis* side, and the pH was lowered at the *cis* or the *trans* side, the response to pH is asymmetric. This suggests preferential orientation of maltoporin channels and pH-dependent closure of only one side of the channel. Experiments with LamB mutants where the major external loops L4, L6 and/or L9 are deleted (see below) allow the identification of surface-exposed loops L4 and L6 as a cause for pH-mediated closure [71]. Obviously, the low pH reduces the interaction between the loops, and they collapse and close the channel. The pH dependence of LamB demonstrates that the channels insert into the lipid bilayer in a preferential orientation. About 70–80 % of the reconstituted channels are oriented with the extracellular entrance to the side to which the protein was added (the *cis* side) and with the periplasmic opening to the opposite side (the *trans* side). This means that the LamB channel has the opposite orientation in the reconstitution experiments as compared to the *in vivo* situation. The possibility to close channels which are oriented in the reverse direction by low pH at the *trans* side also allows the deduction of channel asymmetry with respect to carbohydrate-binding kinetics. Maltose binding is almost symmetric regarding the channel orientation at low voltage, which means that the on-rates differ only by a factor of 2 [71–73]. High voltages on the order of 100 mV and higher lead to a more substantial asymmetry of the on-rates, but the physiological meaning may be questionable because of the low Donnan potential across the outer membrane [73]. Interestingly, the sucrose binding to LamB is highly asymmetric, which is in some contrast to maltose binding [71].

10.5.1.1 Study of LamB (Maltoporin) Mutants

The study of LamB mutants yields interesting insight into the transport mechanism of carbohydrates through the channel. Important for carbohydrate binding are the components of the such called the "greasy slide", which represents aromatic side-chains distributed of the whole sequence of LamB and localized within the central hole of the monomeric channels (Y6, Y41, W74, Y118, F229, W358 and W420; see Chapter 9). The replacement of one or several of these amino acids by alanine in the LamB sequence results in a decrease of the stability constant for carbohydrate binding, which is caused by the decrease of the on-rate constant, k_1. The off-rate constant, k_{-1}, appears to be less affected [74]. Y118 is localized in the center of the LamB channel. This bulky amino acid plays an important role in ion and carbohydrate transport because it controls the central constriction of the channel (see Figure 10.7). The mutation of tyrosine 118 into alanine and other amino acids with small side-chains leads to a substantial increase of the single-channel conductance, which increases about 6-fold for Y118A (from 150 to 850 pS in 1 M KCl) as compared to wild-type [75]. The mutation of Y118 also causes strong effects on carbohydrate binding and transport. In general, the stability constant, K, decreases. However, the replacement of Y118 by phenylalanine or tryptophan results in an increase of K by factors of about 30 (Y118F) and about 140 (Y118W). This is caused by a small increase of the on-rate, k_1, and a dramatic decrease of the off-rate, k_{-1} (see Table 10.5). The bulky side-chains of the aromatic amino acids seem to interact with the carbohydrate in such a way that their movement through the channel is slowed down, which decreases the off-rate, k_{-1}.

The 3-D structure of LamB shows that the 18 antiparallel β-strands are connected by short turns at the periplasmic side (see Chapter 9), whereas the nine surface-exposed loops are much longer. These loops seem to interact with one another and to form a complicated architecture. Their function was studied in detail in a number of studies of deletion mutants *in vivo* and *in vitro* [76, 77]. Single deletions in either the L4 or the L6 surface loops abolishes maltoporin function, whereas the simultaneous deletion of these two loops (LamBΔL4+ΔL6) affects only moderately carbohydrate transport function and not sugar binding as determined by binding of LamB to a starch column. Single deletion of the fifth surface loop of LamB (LamBΔL5) also abolishes *in vivo* maltoporin function. The double mutants LamBΔL4+ΔL5 and LamBΔL5+ΔL6 are strongly impaired in maltoporin activity, whereas the triple-deletion mutant LamBΔL4+ΔL5+ΔL6 regained almost wild-type maltodextrin binding and transport.

The carbohydrate-binding properties of the different loop mutants were also studied in reconstitution experiments using the asymmetric and symmetric addition of the mutant porins to one or both sides of the lipid bilayer membrane. In the *in vitro* experiments, the carbohydrates were also added to one or both sides of the membranes [76]. The results of purification across the starch column and the asymmetric titration experiments suggest that carbohydrates enter the LamB channel from the cell surface side with the non-reducing end in advance. The deletion of one of the loops L4 (LamBΔL4) or L6 (LamBΔL6) results in a considerable reduction of carbohydrate binding from the external side, whereas it has only little influence on the carbohydrate binding from the periplasmic side. The deletion of loop L9v alone (LamBΔL9v), of two loops L4 and L6 (LamBΔL4+ΔL6), of three loops L4, L5 and L6 (LamBΔL4+ΔL5+ΔL6) or of L4, L6 and L9v (LamBΔL4+ΔL6+ΔL9v) had relatively little influence on the carbohydrate binding properties of the maltoporin channel in reconstitution experiments and the channel shows approximately similar binding properties for carbohydrate addition from both sides. The results of the *in vivo* and *in vitro* experiments suggest that the deletion of one or two surface-ex-

Table 10.5 Parameters of maltopentaose-induced transport noise in different LamB-mutants of *E. coli*[a] taken from [72].

Mutant	k_1 [10^6 $(Ms)^{-1}$]	k_{-1} (s^{-1})	K (10^3 M^{-1})
Wild-type	5.3	420	13
Y118A	1.5	970	1.8
Y118S	4.9	4,700	1.5
Y118I	2.4	3,100	0.84
Y118N	4.9	2,300	3.4
Y118H	5.0	4,700	1.2
Y118C	0.93	2,900	0.33
Y118D	5.8	2,900	1.9
Y118R	1.7	1,600	1.1
Y118W	16.5	9	1,800
Y118F	15.5	34	460

[a]The membranes were formed from diphytanoyl phosphatidylcholine/*n*-decane. The aqueous phase contained 1 M KCl and about 500 ng/ml LamB mutants. k_1 and k_{-1} were derived from a fit of the corner frequencies as a function of the maltose concentration [compare Equation (11)]. K is the stability constant for maltopentaose binding derived from the ratio k_1/k_{-1}. The parameters for wild-type LamB [69] and the Y118F mutant [70] are given for comparison.

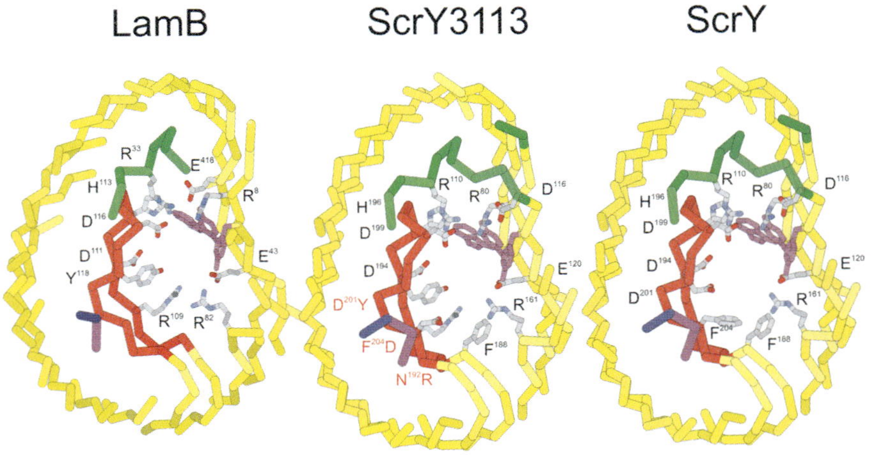

Figure 10.7 Cross-sections of the *E. coli* LamB monomer (maltoporin), the ScrY monomer (sucrose porin) and the ScrY3113 mutant. The panel shows loop 3 (in red), part of loop 1 (in blue) and the amino acid residues (denoted with their numbers from the mature N-terminal end) that are relevant for passage of carbohydrates and ions through the central constriction. The strands of the β-barrel cylinder of the porins are given in rose. The coordinates of maltoporin were taken from the crystallographic data of [21] and that of sucrose porin from [22]. The cross-section of the ScrY3113 mutant monomer was created by using the cross-section of ScrY and the replacement of the amino acids N192R, D201Y and F204D by using the program Hyperchem®.

posed loops leads to the collapse of their architecture, which leads to the plugging of the channel at its surface-exposed side. The removal of several loops, such as L4, L6 and L9v (LamBΔL4+ΔL6+ΔL9v) opens the channel again for carbohydrate transport *in vitro*, but not *in vivo*, probably because the O-antigen side-chains of lipopolysaccharide are also able to block the channel when the external loops are missing [76].

10.5.2
ScrY (Sucrose Porin) of Enteric Bacteria

Another member of the family of carbohydrate-specific porins is the sucrose-specific ScrY, which shares high sequence homologies with LamB [67, 79]. This outer membrane channel was discovered during the investigation of the plasmid-encoded metabolic pathway of sucrose in *Salmonella typhimurium*. Sucrose has, without the plasmid pUR400, only a small permeability through the outer membrane. Cells containing pUR400 are able to use sucrose as sole carbon source, which suggests that sucrose is better permeable through ScrY than through LamB [67, 78–80]. The 3-D structure of ScrY is known and shows an interesting homology to that of LamB [22]. Similar to LamB, the six aromatic residues that line the channel and form a path line, which extends from the floor of the vestibule through the constriction to the periplasmic space, are nearly conserved in ScrY. They form the so-called "greasy slide" (Y78, Y118, W151, F435 and W482 in the sequence of ScrY corresponding to Y6, Y41, W74, F229, W358 and W420 in that of LamB). On the other hand, the central constriction appears to be wider (see Figure 10.7). Consistent with this is the single-channel conductance of ScrY that is in the range of that observed for general diffusion porins [80]. The carbohydrate-specific porin function has been shown in electrophysiological measurements with ScrY reconstituted into planar lipid membranes and titration experiments similar to those with LamB, which means that the ionic current through the channel could also be blocked by addition of malto-oligosaccharides [80]. Liposome-swelling experiments show that transport of sucrose through ScrY [79] is more effective than through LamB. Noise experiments allow also for ScrY the evaluation of the on- and the off-rate constants for carbohydrate binding to the binding site inside the ScrY channel and of a mutant, in which the N-terminal 70 amino acids are deleted [81]. A combination of the results from noise measurements and those from liposome swelling allows also a meaningful comparison of the kinetics of carbohydrate transport through ScrY using the formalism described above for LamB [see Equations (16) and (17)]. Substituting the experimental conditions of [80] ($c' = 20$ mM, $c'' = 0$) the flux through ScrY is given by $\phi = k_1(20\ \text{mM})/[2 + K(20\ \text{mM})]$ or $k_1 = \phi[(100\ \text{M}^{-1}) + K]$. The calculation of k_1 allows a more quantitative description of the carbohydrate transport through ScrY by assuming that the flux, ϕ, of maltoheptaose is about 800 s^{-1} [82]. Table 10.6 shows the rate constants of carbohydrate binding to ScrY. The results are consistent with the *in vivo* situation, which demon-

Table 10.6 Rate constants for carbohydrate transport through the ScrY channel calculated using the relative rates of permeation of carbohydrates relative to that of sucrose[a].

Carbohydrate	*Relative rate of permeation*[a]	*Maximum flux through one single ScrY channel* (s^{-1})	*K* (M^{-1})	k_1 [10^3 $(Ms)^{-1}$]	k_{-1} (s^{-1})
Maltoheptaose	27	800	1,900	1,600	840
DL-Arabinose	109	3,233	9.1	350	38,800
D-Glucose	91	2,700	8.3	290	35,200
D-Fructose	149	4,420	1.9	450	237,000
Sucrose	100	2,970	20	360	17,800
Maltose	33	980	150	245 (430)	1,630 (4,800)
Lactose	33	980	17	115	6,700
Raffinose	19	560	640	414	648

[a] The on-rate constants, k_1, were calculated from Equation (10) using the flux of maltoheptaose (800 s^{-1}) through the ScrY channel under the conditions of the experiment of [79]. For the calculation of the rate constants for the individual carbohydrate it was assumed that its flux through ScrY relative to that of maltoheptaose is given by the relative rates of permeation taken from [79]. The stability constants, *K*, were taken from titration experiments [80] with the exception of maltoheptaose [82], k_{-1} was calculated from k_1/K. The numbers in brackets (for maltose) correspond to those obtained from noise measurements [82].

strates that ScrY can replace LamB for carbohydrate transport, but not vice versa for transport of sucrose [67]. Comparison of the parameters of maltose and sucrose transport demonstrates also that the on-rates are comparable, whereas the off-rate of sucrose binding to ScrY is 10 times higher than that for maltose, which represents a considerable advantage for the velocity of sucrose permeation through ScrY.

10.5.2.1 Study of Carbohydrate Binding to ScrY Mutants

A comparison of the 3-D structures of LamB and ScrY (see Figure 10.7) shows that the central constriction of the two channels formed by the external loop 3 is controlled by a number of different amino acids. The most prominent ones of this constriction are N192, D201 and F204, which correspond to R109, Y118 and D121 of LamB [83]. These amino acids were replaced by site-directed mutagenesis into those of LamB, which results in the ScrY single mutants ScrYN192R, ScrYD201Y and ScrYF204D, and the ScrY triple mutants ScrY3113 (N192R + D201Y + F204D) and ScrY3213, which also lacks amino acids 1–61 from the N-terminal end. Reconstitution experiments with lipid bilayer membranes demonstrate that all of these mutants are functional and have a higher single-channel conductance than LamB, which means that the mutation of these functional amino acids alone is not sufficient to transform ScrY in LamB and vice versa [83]. The carbohydrate-induced block of the channel function was also used for the study of current noise through the different mutant ScrY-channels. The analysis of the power density spectra allows the evaluation of the on- and off-rate constants (k_1 and k_{-1}) of carbohydrate binding to the binding site inside the channels. The results suggest that both on- and off-rate constants were affected by the mutations. Most of them show a substantial effect on carbohydrate binding kinetics, as Table 10.7 demonstrates. However, the results demonstrate that the kinetics of carbohydrate binding

Table 10.7 Parameters of maltopentaose-induced transport noise in different ScrY mutants[a] (data taken from [83]).

Mutant	k_1 [10^6 (Ms)$^{-1}$]	k_{-1} (s^{-1})	K (10^3 M^{-1})
ScrY	2.7	1,900	1.5
N192R	3.7	480	7.7
D201Y	3.7	680	5.4
F204D	3.7	770	4.8
3113	2.0	15	140
3213	1.9	11	170
LamB	5.3	420	13

[a] The membranes were formed from diphytanoyl phosphatidylcholine/n-decane. The aqueous phase contained 1 M KCl and about 500 ng/ml ScrY mutants. k_1 and k_{-1} were derived from a fit of the corner frequencies as a function of the maltopentaose concentration [compare Equation (11)]. K is the stability constant for maltopentaose binding derived from the ratio k_1/k_{-1}. The data represent the mean of at least three individual experiments. The parameters for wild-type ScrY [82] and LamB [69] are given for comparison.

to the ScrY3113 mutant is still different to that of LamB, suggesting that not only the amino acids of the central constriction, but also the general architecture of both channels has a substantial influence on carbohydrate transport [83].

10.6 Properties of the Cyclodextrin (CD)-specific Outer Membrane Porin CymA of *Klebsiella oxytoca*

Certain microorganism can live on cyclic dextrins as sole carbon source [15]. In the case of the Gram-negative bacterium *K. oxytoca* the utilization pathway has been studied in detail and it could be shown that the products of at least 10 genes, designated *cymA* to *cymJ*, are involved. Four of them are constituents of a periplasmic binding protein-dependent uptake system similar to the *mal* system of *E. coli* [14, 15], from which CymE (corresponding to MalE) is the binding protein [84], CymF (MalF) and CymG (MalG) are the integral membrane components, and CymD (MalK) the ATPase. α- and β-CDs are taken up through the outer membrane as intact entities [15, 85], and are linearized in the cytoplasm by the cyclodextrinase CymH. CymA, an outer membrane protein, which has a molecular mass of 39 kDa and is presumably active as a monomer, in contrast to the trimeric general diffusion pores and specific porins of *E. coli* outer membrane [84], is important for the transport of CDs across the cell wall. The primary sequence of CymA possesses a high content of antiparallel β-sheets, typical of outer membrane porins [85].

Reconstitution of CymA into lipid bilayers and conductance measurements demonstrate that it is able to form ion-permeable channels which do not appear in a step-like fashion as in other porins. Instead, they exhibit a substantial current noise probably caused by changes of channel size by protein breathing [86]. The binding of CDs to the channel resulted in its almost complete block for ion transport, similar as for other substrate-specific channels. This allows a detailed inves-

Table 10.8 Stability constants K (M^{-1}) of CD binding to CymA as derived from titration experiments.

Carbohydrate	*LamB E. coli*	*CymA K. oxytoca*			
		pH 6	*pH 5*	*pH 8*	*pH 6 (T = 6 °C)*
α-CD	ND	32,000	16,970	28,610	84,750
β-CD	ND	1,970	ND	ND	ND
γ-CD	ND	310	ND	ND	ND
Maltotriose	2,800	20	ND	ND	ND
Maltopentaose	14,000	75	ND	ND	ND
Maltoheptaose	17,000	91	ND	ND	ND

The membranes were formed of diphytanoyl phosphatidylcholine dissolved in *n*-decane. The aqueous solutions were unbuffered and had a pH of 6 unless otherwise indicated. The applied voltage was 20 mV and the temperature was 20 °C unless otherwise indicated. The data were taken from [86].

tigation of carbohydrate binding, and the stability constants for the binding of cyclic and linear carbohydrates to the binding site inside the channel can be calculated from titration experiments of the membrane conductance with the carbohydrates [86]. Highest stability constant was observed for α-CD followed by β-CD and γ-CD (see Table 10.8). Table 10.8 shows that linear malto-oligosaccharides also bind to CymA, but with much a smaller stability constant as compared to that measured for the CDs. The results suggest that the outer membrane protein CymA has a special role in the uptake of the bulky CDs. Although it is specific for dextrins, its structure is presumably completely different to that of the carbohydrate-specific porins of the LamB family.

10.7
Porin OmpP2 of *Haemophilus influenzae* is a Specific Porin for Nicotinamide-derived Nucleotide Substrates

The human pathogen *H. influenzae* has an absolute requirement for nicotinamide adenosine dinucleotide (NAD, factor V) because it lacks all biosynthetic enzymes necessary for *de novo* synthesis of this cofactor. Growth *in vitro* needs the presence of either NAD itself, or nicotinamide mononucleotide (NMN) or nicotinamide riboside (NR) [84]. *H. influenzae* contains a well-studied outer membrane porin, OmpP2, which acts in most cases as a general diffusion pore and shows an interesting diversity in different *H. influenzae* strains [35, 36]. Interestingly, in a recent study, substrate specificity of OmpP2 for NAD and NMN with a K_d of about 8 and 4 mM has been found, indicating that OmpP2 has also specific sieving properties [87]. This is very interesting because no interaction was detected for the nucleoside NR and other purine or pyrimidine nucleotide or nucleoside species (see Table 10.9). This means presumably that an intrinsic binding site within OmpP2 exists, which facilitates diffusion of these compounds across the outer membrane. The

Table 10.9 Stability constants *K* and half-saturation constants K_s for the binding of different nucleotide compounds to the OmpP2 channel of *H. influenzae*.

Salt concentration	*Compound*	*K* (M^{-1})	K_s (*mM*)	*Maximum percent inhibition of channel conductance*
1 M KCl	NMN	100	10.0	40
0.1 M KCl	NMN	250	4.0	55
	NAD	125	8.0	70
	NADH	<10	>100	
	NR	<10	>100	

[a]The membranes were formed from diphytanoyl phosphatidylcholine/*n*-decane. The aqueous phase contained either 1 or 0.1 M KCl and about 100 ng/ml OmpP2; $T = 20\,°C$. The stability constant *K* was derived from fits of the experimental data similar to that shown in Figure 10.5 using Lineweaver–Burke plots according to Equation (10). The data represent means of at least three titration experiments. Note that the ionic strength had a considerable influence on the stability constant of nucleotide compounds to OmpP2. The data were taken from [87].

binding site recognizes carbonyl and exposed phosphate groups. It may act as a facilitator for nicotinamide-based nucleotide transport, which supports and optimizes the utilization for the essential cofactor sources NAD or NMN for *H. influenzae*, but not for NR, which may use another outer membrane pathway for uptake into the periplasmic space [87].

10.8 Study of the Nucleoside-specific Tsx of *E. coli*

The outer membrane protein Tsx is the receptor of colicin K and bacteriophage T6 [88]. Synthesis of Tsx is co-regulated with the systems for nucleoside uptake and metabolism, suggesting that it plays an important role in the permeation of nucleosides across the outer membrane, but it is only a minor outer membrane protein under normal growth conditions. Mutants lacking Tsx are impaired in the uptake of several nucleosides [88–90]. Its specificity is very high because the rate of uptake of adenosine and thymidine is strongly reduced in these tsx^- mutants, whereas the uptake of cytidine was similar in tsx^+ and tsx^- strains. Furthermore, the uptake of desoxynucleosides is more strongly dependent on Tsx than that of the corresponding nucleosides [89]. On the other hand, Tsx plays no role in the uptake of the free bases or in the permeation of nucleoside monophosphates *in vivo* [89]. Similar to the situation described above for LamB and maltose uptake, transport by Tsx protein becomes rate limiting at nucleoside concentrations less than 1 μM [90]. Tsx is not murein associated like the outer membrane porins LamB, OmpF and OmpC. It seems to be susceptible to SDS, because pore-forming activity in reconstitution experiments and phage inactivation is lost in the presence of SDS [59]. Tsx of *E. coli* is not heat-modifiable as are other Gram-negative bacterial porins and OmpA because it has the same molecular mass of 28 kDa on SDS-PAGE independent on solubilization temperature [59], whereas its correct molecular mass is 31,418 Da [91]. This means that it remains unclear if the active Tsx channel is also an oligomer.

Reconstitution experiments with purified Tsx demonstrate that it forms small ion-permeable channels with the extremely small single-channel conductance of 10 pS in 1 M KCl [59]. These channels can be inhibited by the addition of nucleosides and desoxynucleosides to the membranes. Half-saturation constant can be calculated for the binding of the nucleosides and desoxynucleosides to Tsx from Lineweaver–Burke plots [see Equation (10)]. This indicates that Tsx forms a channel which contains a binding site for nucleosides and desoxynucleosides [92] (see Table 10.10). The data of Table 10.10 show that the stability constant for binding is highly dependent on the structure of the nucleosides and desoxynucleosides. Comparison of the structures of the different pyrimidines shows that thymine contains an additional methyl group in the 5 position as compared to uracil. This group obviously results in a larger stability constant for the binding, suggesting that hydrophobic interactions are also involved in nucleoside binding, which is in agreement with the extremely small single-channel conductance despite the fact that the bulky nucleosides and desoxynucleosides are permeable through the Tsx-channel.

Table 10.10 Stability constants K for the binding of nucleobases, nucleosides and deoxynucleosides to the Tsx channel (half-saturation constant $K_s = 1/K$).

Compound	*K* (M^{-1})	K_s (10^{-3} M)
Adenine	500	2.0
Adenosine	2,000	0.50
Deoxyadenosine	7,100	0.14
Guanine	ND	ND
Guanosine	1,000	1.0
Deoxyguanosine	3,100	0.32
Thymine	170	5.8
Thymidine	5,000	0.20
5′-Deoxythymidine	20,000	0.050
Cytosine	ND	ND
Cytidine	46	22
Deoxycytidine	100	10
Uracil	50	20
Uridine	1,900	0.54
Deoxyuridine	19,000	0.053

The data were taken from [93]. *K* was calculated from titration experiments similar to that shown for LamB (Figure 10.4).

Cytosine has an amino group in the 4 position instead of the carbonyl group as compared to uracil. This results in a considerable decrease of the binding affinity of this compound, probably because of the loss of one hydrogen bound between an OH group of the protein and the carbonyl or because of the increased hydrophilic nature of the whole molecule (see Table 10.10). Similar considerations also apply to a comparison of the binding affinities of adenine and guanine [92]. The binding between Tsx and purines and pyrimidines is stabilized if a ribofuranose is bound to the 9 position in the case of the purines or to the 1 position in the case of the pyrimidines. This increase of the stability constant could in principle be caused by the formation of additional hydrogen bounds. However, the removal of the hydroxyl group in the 2′ position (desoxynucleosides) resulted in an even larger stability constant, which makes it rather unlikely that hydrogen bonds are the reason for the larger stability constants, which again suggests that hydrophobic interactions between binding site and nucleosides and desoxynucleosides are involved in binding. Saturation of the flux of nucleosides and through the Tsx channel is expected for increasing concentrations of nucleosides and desoxynucleosides, similar to the situation in other specific porin channels [92].

10.9 Conclusions

The outer membrane of Gram-negative bacteria contains several different pathways for the permeation of solutes. The permeation of vitamins and siderophores occurs through TonB-dependent highly specified receptors that are closed in the absence of substrates and energy (see Chapter 11). The general diffusion pores form more or less structured holes with defined exclusion limits and little solute specificity. They provide an pathway through the outer membrane with defined exclusion limits that is always open. The flux of substrates through general diffusion pores is a linear function of the concentration. Nevertheless, it is possible that general diffusion pores also have a certain solute specificity. The phosphate starvation-inducible PhoE porin of enteric bacteria is a general diffusion pore, but because of its anion selectivity provides a considerable advantage for the permeation of phosphate and polyphosphate across the outer membrane without any phosphate-specific binding site because its permeability for phosphate is about 7 times higher than that of OmpF and about 25 times higher than that of OmpC under otherwise identical conditions [53]. Specific porins represent another class of permeability pathway. They form highly structured channels and contain binding sites for classes of specific solutes. The presence of binding sites inside the specific porins leads to the saturation of solute flux through the channels above the half-saturation concentration $1/K$. However, the presence of a binding site offers a considerable advantage for the flux of solutes across the outer membrane at small external solute concentrations, the *in vivo* situation, at which the general diffusion pores are

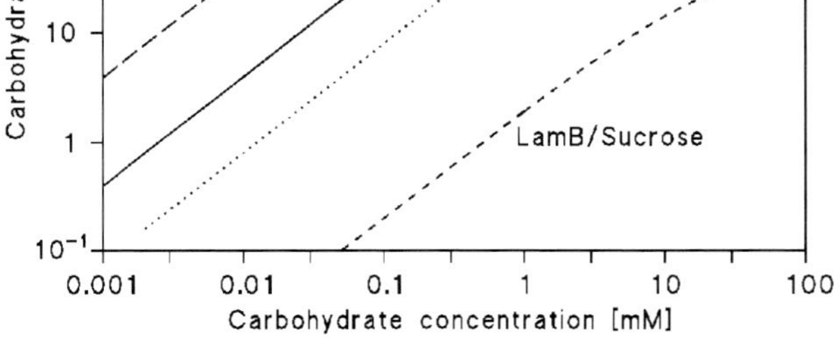

Figure 10.8 Flux of maltose, sucrose and maltotetraose through one single LamB channel as a function of the corresponding sugar concentration on one side of the channel. The concentration on the other side was set to zero. The flux was calculated using Equation (7) and the rate constants given in Table 10.4. The flux of maltose through OmpF was calculated on the basis on its dimensions [18] and using the Renkin correction factor [34].

rate limiting [94]. The reason for this is the increase of the effective solute concentration inside the channel, similar to the increase of substrate concentration near the binding site of an enzyme.

Specific porins have their maximum substrate permeability in the linear range of the flux-concentration relationship [see Equation (7)]. The maximum permeability is given by the on-rate, $k_1/2$, of substrate binding to the binding site. At high concentration the substrate flux saturates. It is given by the rate constant k_{-1}, which represents the maximum turnover number of substrates through a one-site, two-barrier channel. The comparison of substrate fluxes through OmpF and LamB shown in Figure 10.8 demonstrates again the advantage of a binding site for the maximum scavenging of substrates at small substrate concentration. It shows that the flux through a general diffusion pore can exceed that through a specific porin at high substrate concentration. The important advantage of the study of specific porins in reconstituted systems is the knowledge of some of their structures at atomic resolution, which allows the understanding of substrate–channel interaction at molecular level. The study of specific porins and its mutants represents an exciting object for the investigation of substrate translocation through channels reconstituted in model membranes.

Acknowledgements

The authors thank Christian Andersen for helpful discussions. The financial support of the author's own research by the Deutsche Forschungsgemeinschaft (Project Be 865/10, "Specific Porins") is gratefully acknowledged.

References

1 J.-M. Ghuisen, R. Hakenbeck (Eds), *Bacterial Cell Wall*. Elsevier, Amsterdam, **1994**.
2 G. Winkelmann (Ed.), *Microbial Transport Systems*. Wiley-VCH, Weinheim, **2001**.
3 K. D. Young, *Mol Microbiol* **2003**, *49*, 571–580.
4 K. Lazar, S. Walker, *Curr Opin Chem Biol* **2002**, *6*, 786–793.
5 J. van Heijenoort, In: *Bacterial Cell Wall*, J.-M. Ghuisen, R. Hakenbeck (Eds). Elsevier, Amsterdam, **1994**, pp. 39–54.
6 H. Nikaido, M. Vaara, *Microbiol Rev* **1985**, *49*, 1–32.
7 J. B. Stock, B. Rauch, S. Roseman, *J Biol Chem* **1977**, *252*, 7850–7861.
8 T. J. Beveridge, *Int Rev Cytol* **1981**, *72*, 229–317.
9 K. Sen, J. Hellman, H. Nikaido, *J Biol Chem* **1988**, *263*, 1182–1187.
10 V. Koronakis, A. Sharff, E. Koronakis, B. Luisi, C. Hughes, *Nature* **2000**, *405*, 914–919.
11 C. Andersen, C. Hughes, V. Koronakis, *EMBO Rep* **2000**, *1*, 313–318.
12 T. Nakae, *Biochem Biophys Res Commun* **1975**, *64*, 1224–1230.
13 H. Nikaido, *J Biol Chem* **1994**, *269*, 3905–3908.

14 M. Schwartz, In: Escherichia coli *and* Salmonella typhimurium, *Vol. 2*, F. C. Neidhardt (Ed.). ASM, Washington DC, **1987**, pp. 1482–1502.
15 G. Fiedler, M. Pajatsch, A. Böck, *J Mol Biol* **1996**, *256*, 279–291.
16 K. Poole, R. E. W. Hancock, *Eur J Biochem* **1984**, *144*, 607–612.
17 M. S. Weiss, U. Abele, J. Weckesser, W. Welte, E. Schiltz, G. E. Schultz, *Science* **1991**, *254*, 1627–1630.
18 S. W. Cowan, T. Schirmer, G. Rummel, M. Steiert, R. Gosh, R. A. Pauptit, J. N. Jansonius, J. P. Rosenbusch, *Nature* **1992**, *356*, 727–733.
19 A. Kreusch, G. E. Schulz, *J Mol Biol* **1994**, *243*, 891–905.
20 K. Zeth, K. Diederichs, W. Welte, H. Engelhardt, *Structure* **2000**, *8*, 981–992.
21 T. Schirmer, T. A. Keller, Y. F. Wang, J. P. Rosenbusch, *Science* **1995**, *267*, 512–514.
22 D. Forst, W. Welte, T. Wacker, K. Diederichs, *Nature Struct Biol* **1998**, *5*, 37–46.
23 J. E. Meyer, M. Hofnung, G. E. Schulz, *J Mol Biol* **1997**, *266*, 761–775.
24 H. Nikaido, E. Y. Rosenberg, *J Gen Physiol* **1981**, 77, 121–135.
25 R. Benz, K. Janko, W. Boos, P. Läuger, *Biochim Biophys Acta* **1978**, *511*, 305–319.
26 H. Schindler, J. P. Rosenbusch, *Proc Natl Acad Sci USA* **1978**, 75, 3751–3755.
27 Y. Hasegawa, H. Yamada, S. Mizushima, *J Biochem* **1976**, *80*, 1401–1409.
28 K. Nakamura, S. Mizushima, *J Biochem* **1976**, *80*, 1411–1422.
29 B. Lugtenberg, L. Van Alphen, *Biochim Biophys Acta* **1983**, *737*, 51–115.
30 H. Nikaido, *Methods Enzymol* **1983**, *97*, 85–100.
31 T. Miura, S. Mizushima, *Biochim Biophys Acta* **1968**, *150*, 159–164.
32 G. M. Carlone, M. L. Thomas, H. S Rumschlag, F. O. Sottnek, *J Clin Microbiol* **1986**, *24*, 330–332.
33 T. Ferenci, K.-S. Lee, *J Bacteriol* **1989**, *171*, 855–861.
34 E. M. Renkin, *J Gen Physiol* **1954**, *38*, 225–253.
35 V. Vachon, R. Laprade, J. W. Coulton, *Biochim Biophys Acta* **1986**, *861*, 74–82.
36 R. Srikumar, D. Dahan, M. F. Gras, M. J. H. Ratcliffe, L. van Alphen, J. W. Coulton, *J Bacteriol* **1992**, *174*, 4007–4016.
37 B. Dargent, W. Hofmann, F. Pattus, J. P. Rosenbusch, *EMBO J* **1986**, *5*, 773–778.
38 J. H. Lakey, *FEBS Lett* **1987**, *211*, 1–4.
39 J. P. Dilger, R. Benz, *J Membr Biol* **1985**, *85*, 181–189.
40 M. Montal, P. Mueller, *Proc Natl Acad Sci USA* **1972**, *69*, 3561–3566.
41 R. Benz, O. Fröhlich, P. Läuger, M. Montal, *Biochim Biophys Acta* **1975**, *394*, 323–334.
42 R. Benz, A. Schmid, R. E. W. Hancock, *J Bacteriol* **1985**, *162*, 722–727.
43 P. Läuger, *Biochim Biophys Acta* **1973**, *300*, 423–441.
44 R. Benz, A. Schmid, T. Nakae, G. Vos-Scheperkeuter *J Bacteriol* **1986**, *165*, 978–986.
45 R. Benz, A. Schmid, G. H. Vos-Scheperkeuter *J Membr Biol* **1987**, *100*, 21–29.
46 R. Benz, R. E. W. Hancock, *J Gen Physiol* **1987**, *89*, 275–295.
47 F. Yoshimura, H. Nikaido, *J Bacteriol* **1982**, *152*, 636–642.
48 J. Trias, E. Y. Rosenberg, H. Nikaido, *Biochim Biophys Acta* **1988**, *938*, 493–496.
49 R. E. W. Hancock, A. M. Carey, *FEMS Microbiol Lett* **1980**, *8*, 105–109.
50 J. Trias, H. Nikaido, *J Biol Chem* **1990**, *265*, 15680–15684.
51 R. E. W. Hancock, K. Poole, R. Benz, *J Bacteriol* **1982**, *150*, 730–738.
52 B. L. Wanner, *J Cell Biochem* **1993**, *51*, 47–54.
53 K. Bauer, P. van der Ley, R. Benz, J. Tommassen, *J Biol Chem* **1988**, *263*, 13046–13053.
54 R. J. Siehnel, C. Egli, R. E. W. Hancock, *Mol Microbiol* **1992**, *6*, 2319–2326.
55 R. E. W. Hancock, C. Egli, R. Benz, R. J. Siehnel, *J Bacteriol* **1992**, *174*, 471–476.
56 R. Benz, C. Egli, R. E. W. Hancock, *Biochim Biophys Acta* **1993**, *1149*, 224–230.
57 B. Dargent, A. Charbit, M. Hofnung, F. Pattus *J Mol Biol* **1988**, *201*, 497–506.

58 K. Schülein, R. Benz, *Mol Microbiol* **1990**, *4*, 625–632.
59 C. Maier, E. Bremer, A. Schmid, R. Benz, *J Biol Chem* **1988**, *263*, 2493–2499.
60 C. Andersen, B. Rak, R. Benz, *Mol Microbiol* **1999**, *31*, 499–510.
61 F. Conti, I. Wanke, *Q Rev Biophys* **1975**, *8*, 451–506.
62 B. Lindemann, W. van Driessche, *Science* **1977**, *195*, 292–294.
63 L. J. De Felice, *Introduction to Membrane Noise*. Plenum Press, New York, **1981**.
64 T. Ferenci, M. Schwentorat, S. Ullrich, J. Vilmart, *J Bacteriol* **1980**, *142*, 521–526.
65 M. Luckey, H. Nikaido, *Proc Natl Acad Sci US* **1980**, *77*, 167–171.
66 M. A. Bloch, C. Desaymard, *J Bacteriol* **1985**, *163*, 106–110.
67 K. Schmid, R. Ebner, K. Jahreis, J. W. Lengeler, F. Titgemeyer, *Mol Microbiol* **1991**, *5*, 941–950.
68 S. Nekolla, C. Andersen, R. Benz, *Biophys J* **1994**, *86*, 1388–1397.
69 C. Andersen, M. Jordy, R. Benz, *J Gen Physiol* **1995**, *105*, 385–401.
70 M. Jordy, C. Andersen, K. Schülein, T. Ferenci, R. Benz, *J Mol Biol* **1996**, *259*, 666–678.
71 C. Andersen, B. Schiffler, A. Charbit, R. Benz, *J Biol Chem* **2002**, *277*, 41318–41325.
72 F. Orlik, C. Andersen, R. Benz, *Biophys J* **2002**, *83*, 309–321.
73 L. Kullman, M. Winterhalter, S. M. Bezrukov, *Biophys J* **2002**, *82*, 803–812.
74 P. Van Gelder, F. Dumas, I. Bartoldus, N. Saint, A. Prilipov, M. Winterhalter, Y. Wang, A. Philippsen, J. P. Rosenbusch, T. Schirmer, *J Bacteriol* **2000**, *184*, 2994–2999.
75 F. Orlik, C. Andersen, R. Benz, *Biophys J* **2002**, *82*, 2466–2475.
76 C. Andersen, C. Bachmeyer, H. Täuber, R. Benz, J. Wang, V. Michel, S. M. C. Newton, M. Hofnung, A. Charbit, *Mol Microbiol* **1999**, *32*, 851–867.
77 A. Charbit, C. Andersen, J. Wang, B. Schiffler, V. Michel, R. Benz, M. Hofnung, *Mol Microbiol* **2000**, *35*, 777–790.
78 K. Schmid, R. Ebner, J. Altenbuchner, R. Schmitt, J. W. Lengeler, *Mol Microbiol* **1988**, *2*, 1–8.
79 C. Hardesty, C. Ferran, J. M. DiRienzo, *J Bacteriol* **1991**, *173*, 449–456.
80 K. Schülein, K. Schmid, R. Benz, *Mol Microbiol* **1991**, *5*, 2233–2241.
81 K. Schülein, C. Andersen, R. Benz, *Mol Microbiol* **1995**, *17*, 757–767.
82 C. Andersen, R. Cseh, K. Schülein, R. Benz, *J Membr Biol* **1998**, *164*, 263–274.
83 B.-H. Kim, C. Andersen, J. Kreth, C. Ulmke, K. Schmid, R. Benz, *J Membr Biol* **2002**, *187*, 239–253.
84 M. Pajatsch, M. Gerhart, R. Peist, R. Horlacher, W. Boos, A. Böck, *J Bacteriol* **1998**, *180*, 2630–2635.
85 M. Pajatsch, C. Andersen, A. Mathes, A. Böck, R. Benz, H. Engelhardt, *J Biol Chem* **1999**, *274*, 25159–25166.
86 F. Orlik, C. Andersen, C. Danelon, M. Winterhalter, M. Pajatsch, A. Böck, R. Benz, *Biophys J* **2003**, *85*, 876–885.
87 C. Andersen, E. Maier, G. Kemmer, J. Blass, A.-K. Hilpert, R. Benz, J. Reidl, *J Biol Chem* **2003**, *85*, 876–885.
88 N. M. Evans, D. D. Smith, A. J. Wicken, *J Med Microbiol* **1974**, *7*, 359–365.
89 P. A. Manning, A. P. Pugsley, P. Reeves, *J Mol Biol* **1977**, *116*, 285–300.
90 H. J. Krieger-Brauer, V. Braun, *Arch Microbiol* **1980**, *124*, 33–242.
91 A. Munch-Petersen, B. Mygind, A. Nicolaisen, N. J. Pihl, *J Biol Chem* **1979**, *254*, 3730–3737.
92 E. Bremer, P. Gerlach, A. Middendorf, *J Bacteriol* **1988**, *170*, 108–116.
93 R. Benz, A. Schmid, C. Maier, E. Bremer, *Eur J Biochem* **1988**, *176*, 699–705.
94 I. C. West, M. G. P. Page, *J Theor Biol* **1984**, *110*, 11–19.

11
Energy-coupled Outer Membrane Iron Transporters

Volkmar Braun and Michael Braun

11.1
Common Features of Outer Membrane Iron Transporters

Energy-coupled outer membrane transport gained special interest because of the lack of an energy source in the outer membrane. Energy is derived from the cytoplasmic membrane, and the question is how and in what form energy gets into the outer membrane. A subclass of the transporters also serves as signalers of a novel mechanism of gene transcription control. Moreover, crystal structure analysis of transporters revealed a novel structure that has not been seen in any other protein.

Import of hydrophilic substrates across the outer membrane of Gram-negative bacteria is determined by proteins which form channels through which compounds pass by diffusion, facilitated diffusion or energy-consuming active transport. Diffusion takes place without recognition of the substrate by the protein. Facilitated diffusion involves substrate-specific binding to the protein. Active transport is associated with tight and stereospecific binding of the substrates to specific protein sites from where the substrates are released by an energy-consuming step, followed by opening of the channel and further substrate translocation. The three mechanisms are ideally fulfilled in the permeation of various substrates across the outer membrane. This chapter deals with the energy-consuming active transport.

Stereospecific binding can be determined by systematically changing the substrate structure and by introducing site-specific mutations in the active site of the transporter. Mutant analysis may be guided by crystal structures, three of which have been determined for iron transporters. Crystal structure analysis reveals structural changes upon binding of substrates if transporters are crystallized in the presence and absence of substrates or inhibitors. However, the dynamics of structural changes during substrate translocation and elicited by energy input are not revealed by crystals. Crystal analysis of individual proteins fails to show energy-dependent structural changes, especially in the mechanistically complex energization of outer membrane transport where energy consumed in the outer membrane is provided by the cytoplasmic membrane. The entire protein complex that achieves energization and transport would have to be crystallized, and specific ex-

Bacterial and Eukaryotic Porins. Edited by Roland Benz

ISBN 3-527-30775-3

perimental methods to energize the complex would have to be devised. Until this can be achieved, indirect methods must be applied to measure structural changes during transport into cells, such as fluorescence spectroscopy and electron spin resonance spectroscopy.

11.1.1 Energy Coupling of Transport

The energy-coupled outer membrane transport proteins not only serve as substrate transporters but also function as phage and colicin receptors. The distinction between the transporter and receptor functions are, on the one hand, semantic because proteins at cell surfaces that serve as binding sites for viruses and protein toxins are generally designated receptors. On the other hand, the distinction also points to functional differences because no transport activity of the receptors has been demonstrated for any of the biopolymers, phage DNA and protein toxins. However, the receptor activities are, with a few exceptions, not passive, but rather require energy input of the same kind as found for the transport activities [1, 2]. The energy is derived from the proton motive force of the cytoplasmic membrane. Three proteins have been identified which are involved in energy transfer from the cytoplasmic membrane into the outer membrane. The TonB and ExbD proteins are anchored through their N-proximal portions in the cytoplasmic membrane, and the rest of the proteins are located in the periplasm. The ExbB protein traverses the cytoplasmic membrane 3 times and most of the protein is locate in the cytoplasm. The TonB protein directly interacts with the outer membrane transporters as shown by mutations in a conserved region at the N-terminus of the transporters, designated the TonB box, which inactivate the transporters. These mutations are suppressed in *Escherichia. coli* by mutations in glutamine 160 of TonB (Q160) which when converted to leucine or lysine partially restore the energy and TonB-dependent transport and receptor activity of the outer membrane protein mutants in the TonB box [3–5]. In addition, TonB can be cross-linked to transporters with formaldehyde [6, 7], and *in vivo* by spontaneous disulfide cross-linking between cysteine residues introduced into the TonB box and cysteine residues introduced into region 160 of TonB [7, 8]. The receptor and transporter functions are abolished in mutants in the TonB box, region 160 of TonB, the transmembrane regions and various other regions of TonB, ExbB and ExbD, and in energy-deprived cells.

Current thinking proposes that TonB alters its conformation in response to the proton gradient (proton motive force) across the cytoplasmic membrane. In this energized conformation TonB interacts with outer membrane transporters and causes, similar to an allosteric regulatory protein, a structural change in the transporters that alters the structure of binding sites of the Fe^{3+} complexes. The Fe^{3+} siderophores and heme are released from the binding sites, and a channel is opened through which the iron complexes enter the periplasm. Opening of the channel most likely is also an energy-consuming transport step.

Since the binding sites of the iron chelates are above the cell surface, the structural transition provoked by TonB must traverse through the entire transporters from the periplasm to the surface-exposed regions. For example, the exchange rate of pyoverdin by Fe^{3+} pyoverdin at the FpvA transporter of *Pseudomonas aeruginosa* occurs much faster at TonB-coupled FpvA than on isolated FpvA [9]. Furthermore, release of heme from the heme-binding proteins and iron from the iron-binding proteins also requires a TonB-triggered conformational change. For example, the hemoglobin-binding capacity of a *Neisseria meningitidis tonB* mutant or a de-energized *tonB* wild-type treated with 10 μM carbonylcyanide *m*-chlorophenylhydrazone (CCCP) is 5- to 20-fold lower than that of the energized wild-type [10]. In contrast, Fe^{3+} transferrin is much more tightly bound to the TbpAB transporter of a *tonB* mutant than to a *tonB* wild-type [11]. Transmembrane transmission of structural changes in transporters elicited by TonB must also be claimed for binding of the phages T1 and ϕ80 to loop 4 of *E. coli* FhuA [12] which requires TonB and energized cells [1].

11.1.2
Iron Sources

Outer membrane iron transporters transport ferric siderophores, heme or iron ions delivered by transferrin and lactoferrin from their eukaryotic hosts (Figure 11.1) [13–16]. Since under oxic conditions and at pH 7 iron forms insoluble Fe^{3+} hydroxy polymers, all aerobically growing cells have to complex iron. To achieve that, microbes synthesize low-molecular-weight compounds, designated siderophores, which are secreted and form very strong Fe^{3+}complexes in the order of 10^{-20} μM [17] in the growth medium from where they are transported into the cells by highly specific transport systems. Eukaryotes bind iron to transferrin and lactoferrin, incorporate Fe^{2+} into heme which is bound to hemoglobin, and store iron intracellularly as ferritin. Bacteria use these iron sources of their hosts and Gram-negative bacteria synthesize outer membrane transporters which bind specifically Fe^{3+} siderophores, heme, hemoglobin, hemoglobin/haptoglobin, heme/hemopexin, transferrin or lactoferrin (Figure 11.1). Energy-coupled transport by outer membrane proteins fulfills the purpose by first concentrating the scarce Fe^{3+} siderophores in the bacteria's ambience at the cell surface by extremely tight binding to the outer membrane transporters (K_d values in the nanomolar range), followed by energy-driven transport into the periplasm. The Fe^{3+} siderophores are also too large (in the order of 750 Da) to diffuse through the *E. coli* K-12 porins. In the case of iron bound to host proteins, iron must be mobilized from transferrin and lactoferrin, and heme must be taken out of hemoglobin and hemopexin by the outer membrane transporters at the bacterial surfaces. Bacteria do not take up proteins with the exception of certain toxins (bacteriocins, colicins).

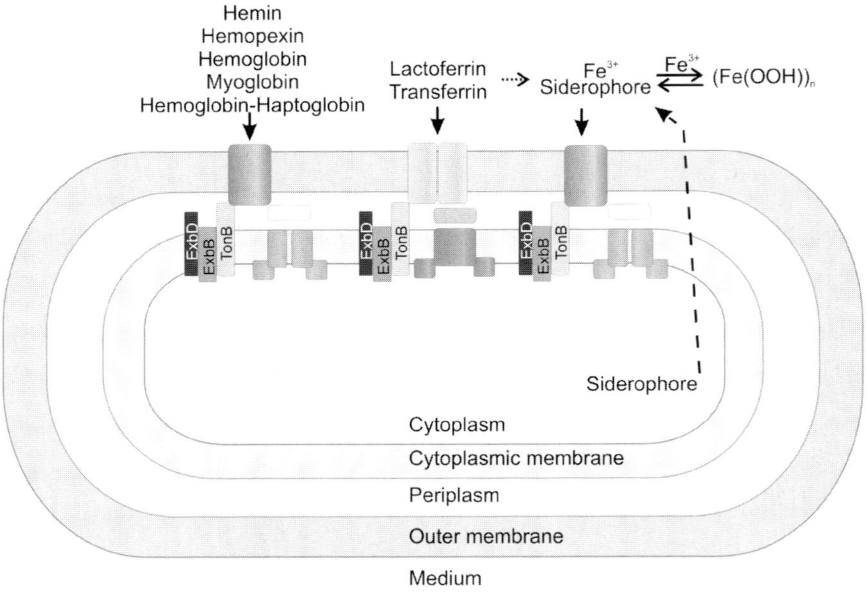

Figure 11.1 Illustration of iron transport systems into Gram-negative bacteria. Iron is delivered as heme, heme bound to hemopexin, heme incorporated into hemoglobin or hemoglobin bound to haptoglobin, complexed by transferrin, lactoferrin or siderophores. Siderophores complex iron in equilibrium with $(FeOOH)_n$ transferrin or lactoferrin. TonB of the TonB–ExbB–ExbD complex, whose stoichiometry is unknown, interacts with the outer membrane transport proteins. Heme, Fe^{3+} or Fe^{3+} siderophores bind to binding proteins in the periplasm from where they are transported across the cytoplasmic membrane by permeases energized by ATPases associated with the cytoplasmic membrane. Siderophores are synthesized by the cells or used from other bacteria and fungi.

11.1.3
Regulation

Under iron-replete growth conditions only low amounts of the iron transporters are synthesized. Transcription of the transporter genes is repressed through the Fur protein which when loaded with Fe^{2+} functions as a repressor that binds to the promoters and inhibits transcription. Under iron-deplete conditions, to which bacteria are usually exposed under natural conditions, Fur is unloaded, does not function as a repressor and transcription takes place. The number of transporter molecules can become very high, close to 100 000 copies per cell. The copy number of outer membrane proteins is much higher than the copy number of the transport proteins across the cytoplasmic membrane. The reason may be the requirement of the outer membrane transporters to extract the scarce iron complexes from the medium in addition to transport. Nevertheless, transport across the outer membrane seems to be the rate-limiting step, for two reasons: (i) most transporters are not loaded with the respective iron complex and (ii) most transporters are not connected to TonB since the number of TonB molecules is only in the order of 2000 per cell [18].

11.1.4
Transport across the Cytoplasmic Membrane

After transport across the outer membrane, iron, Fe^{3+} siderophores and heme are bound to binding proteins in the periplasm [13–15, 19, 20]. The binding proteins deliver iron and the iron complexes to transport proteins (permeases) which are integrated in the cytoplasmic membrane. One or two polypeptides form the permeases [21]. Energy for transport across the cytoplasmic membrane is derived from ATP through proteins which are attached at the cytoplasmic side to the permeases. These proteins contain typical Walker ATP binding motifs and hydrolyze ATP [22]. Detailed knowledge of the structure and arrangement of the permease, the associated ATPase, the structure of the binding protein, and how it may interact with the permease has recently been obtained through determination of the crystal structure of the BtuC permease associated with the BtuD ATPase and the separately crystallized BtuF binding protein [23, 24]. This is the first and only completely determined crystal structure of a bacterial ABC transporter.

11.2
Crystal Structures of Energy-coupled Outer Membrane Transport Proteins

To date, crystal structures of four transport proteins have been determined: FhuA [25, 26], FecA [27], FepA [28] and BtuB [29] (Figure 11.2). They display the same basic structure composed of a β-barrel formed by 22 antiparallel β-strands and a globular domain composed of a four-stranded β-sheet with four interspersed α-helices. The β-sheet of the globular domain forms a platform which is conserved among the four crystal structures. The globular domain is at the N-proximal end and is inserted from the periplasmic side into the β-barrel. The globular domain completely closes the channel of the β-barrel and for this reason was designated the cork, plug or hatch. With the exception of BtuB, the very N-terminal end, residues 1–18 of FhuA, residues 1–10 of FepA and residues 1–79 of FecA, are not observed in the crystal structure because they are flexible. The TonB box is located in the flexible region and, as will be discussed later, flexibility may have functional implications.

The β-strands are tilted relative to the barrel axis, in FhuA about 45°. A co-crystal of FhuA with a lipopolysaccharide (LPS) molecule [25] allows exact positioning of FhuA in the outer membrane (Figure 11.2) since it is known that the fatty acid residues of LPS form part of the outer leaflet of the lipid bilayer. The localization of FhuA is representative of the localization of the other outer membrane transporters. The loops that connect the β-strands protrude beyond the lipid bilayer; the most prominent L4 loop of FhuA 34 Å from the cell surface. More than half of

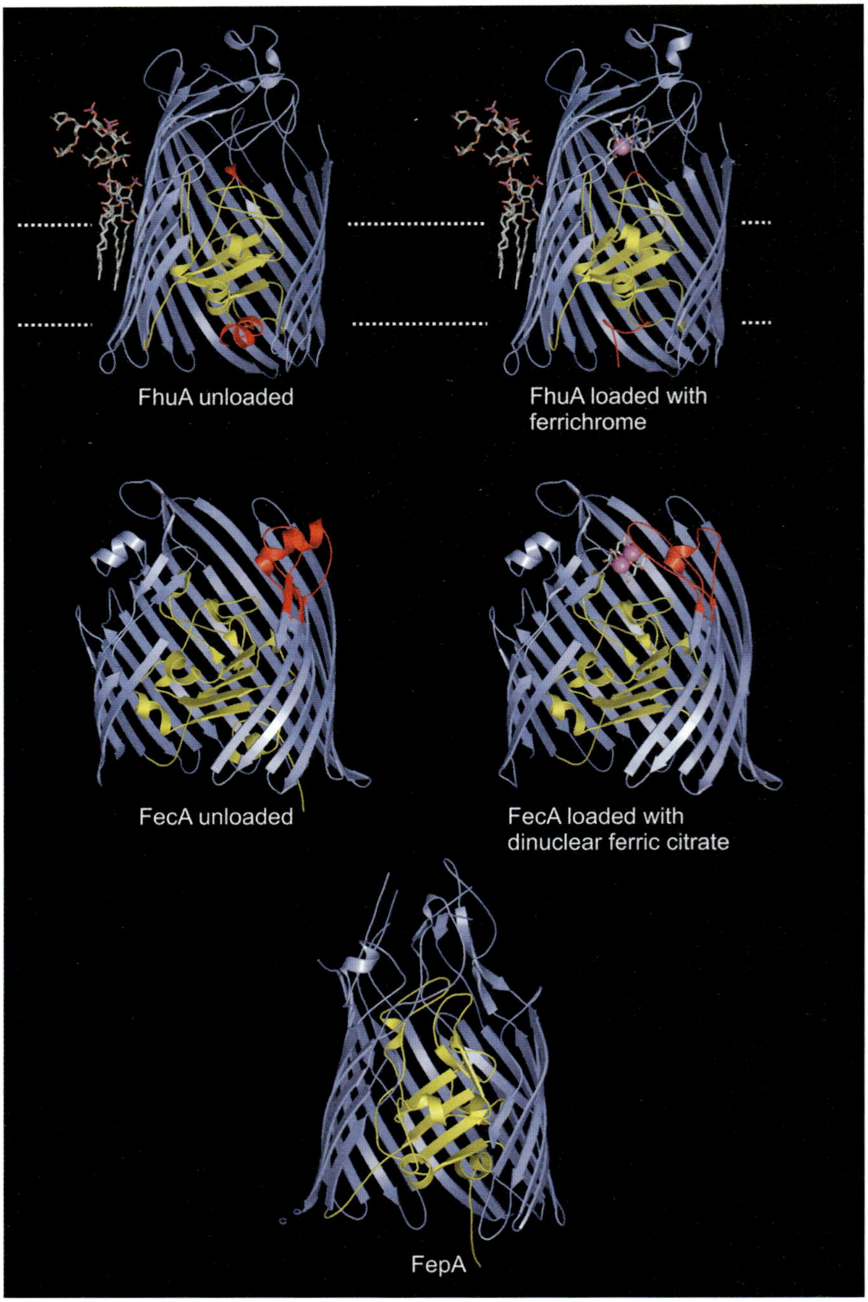

Figure 11.2 Crystal structures of outer membrane transporters. Note the large structural transitions that occur in FhuA and FecA upon binding of ferrichrome and ferric citrate, respectively, indicated by the red colors. No clear distinction could be made between unloaded FepA and FepA loaded with ferric enterobactin since the crystals cracked upon incubation with ferric enterobactin [28, 31]. Lipopolysaccharide was co-crystallized with FhuA [25] and defines the location of FhuA in the outer membrane.

the transporters are not embedded in the outer membrane, but extend above the cell surface. The ligand binding sites are located above the cell surface, e. g. 20 Å in FhuA, and can be reached by substrates through water-filled channels in the transporters. It is likely that substrates weakly bind consecutively to amino acid side-chains located along the channels until they reach their high-affinity binding sites with K_d values in the nanomolar range.

Detailed discussions of the various structures are found in [30,31].

11.2.1
FhuA Transporter and Receptor

11.2.1.1 The Transport Activity of FhuA

Ferrichrome is considered as the natural substrate of the FhuA transporter although it is not synthesized by any of the FhuA-containing bacteria, but by certain fungi. Ferrichrome is a cyclic hexapeptide composed of three consecutive residues of glycine and three consecutive residues of N^5-hydroxy-N^5-acetyl-ornithine. Natural variants of ferrichrome, ferricrocin and ferrichrysin, in which one or two glycine residues are replaced by serine residues, are transported as well as ferrichrome. Ferricrocin is bound by 10 residues, four of the cork and six of the barrel, seven of them aromatic side-chains. Binding of ferricrocin [25] or ferrichrome [26] to FhuA leaves the barrel conformation largely unchanged. However, in the cork domain rotations by 90° of side-chain E98 and by 50° of Q100 and translations of E98 by 1.5 Å towards the barrel wall and of Q100 by 1.8 Å towards ferricrocin are observed. Numerous smaller shifts in the cork occur. A much larger transition occurs at the periplasmic side of the FhuA molecule where a short helix is converted into a coil, and residues E19, S20 and W22 move 17 Å. It is thought that this large structural transition is required for productive interaction of FhuA with energized TonB. This assumption is supported by the finding that prevention of the 17-Å translation by fixing the cork to the barrel through disulfide bridges between residues T27C (threonine replaced by cysteine) and P533C abolishes ferrichrome transport [32].

The I9P and V11D mutations in the TonB box of FhuA and its deletion (residues 7–11 of the mature protein) abolish TonB-dependent FhuA activities, but retain the TonB-independent infection by phage T5 [32]. This result indicates the important role the TonB box plays in FhuA activity, as was originally derived from the suppression of the above point mutations by mutations Q160L and Q160K in TonB [5]. Further proof that these two regions interact with each other comes from the spontaneous *in vivo* formation of disulfide links between I9C and V11C and TonB (Q160C) (F. Endriß and V. Braun, unpublished results). If conformational energy of TonB is used to change the conformation of FhuA such that it functions as a transporter and if this conformational change is mediated through interaction of TonB with the TonB box, one would expect that the amino acid sequence that connects the TonB box with the cork is very important. The conformational change would have to be conveyed to the cork and the barrel to change the stereochemistry of the ferrichrome binding site and to move the cork inside the barrel to open a channel. However, deletion of residues 13–22 and 24–31 which include the switch helix (residues 24–29), and duplication of residues 23–30 only partially reduce FhuA activities [32].

Deletion of the entire cork (residues 5–160) converts FhuA into an open channel that facilitates diffusion of ferrichrome, antibiotics and maltodextrins through the outer membrane. FhuAΔ5–160 incorporated into artificial lipid bilayers increases the conductance for KCl and other solutes, but does not from stable channels with clearly discernable single channel conductance steps [33]. In a complete chromosomal FhuA deletion strain plasmid encoded FhuAΔ5–160 exerts none of the FhuA activities [34]. However, in strains which synthesize chromosomally encoded FhuA inactivated by multiple mutations in the barrel, or a FhuA N-proximal fragment of 357 residues, FhuAΔ5–160 display high transport and receptor activities [34, 35]. These activities are the result of an unusual complementation. Co-synthesis of one of these FhuA derivatives with FhuAΔ5–160 leads to the incorporation of the missing cork derived from the FhuA mutant derivatives [34]. It is not clear how this happens – whether the cork of the inactive FhuA does not incorporate into the barrel to which it is covalently linked, but instead is incorporated into co-synthesized FhuAΔ5–160, or whether it is proteolytically released and then incorporated into FhuAΔ5–160. For steric reasons the latter mechanism is more likely because it is difficult to envisage how a cork still bound to a barrel is incorporated into another barrel. In addition, the cork and the barrel form different structural domains, and it is frequently observed that the linkage between domains is less structured and subject to proteolytic cleavage. The N-terminal FhuA 1–357 fragment, which actually contains 27 additional residues due to the out-of-frame mutation after residue 357, may be proteolytically trimmed prior to its incorporation into FhuAΔ5–160 or incorporated as such with the region beyond residue 160 extending into the periplasm. Such an arrangement is feasible since the linkage between residue 160 of the cork and 161 of the barrel is exposed to the periplasm. However, it is more likely that most or all of the extra sequence is unfolded and cut off by periplasmic proteases. Separate synthesis of the cork and FhuAΔ5–160 results in an active FhuA, provided both proteins are endowed with a signal sequence showing that assembly of complete FhuA by the separately synthesized cork and barrel occurs outside the cytoplasm in the periplasm or while the FhuA barrel is incorporated into the outer membrane [34]. These results suggest that in wild-type FhuA the barrel folds and the cork is subsequently incorporated in the barrel. The cork is proteolytically degraded in the periplasm, but not in the cytoplasm, which indicates that after secretion across the cytoplasmic membrane, accompanied by the release of the signal sequence, the cork does not assume a conformation that is resistant to proteolysis. A study on the conformation of the separately synthesized cork of FepA by nuclear magnetic resonance and circular dichroism reveals an essentially unfolded protein [36]. The FhuA and FepA cork studies suggest that the corks gain their conformation during incorporation into the channel. This kind of protein folding differs from folding of many proteins whereby the N-proximal region starts to fold while synthesis of the polypeptide proceeds on the ribosome.

Corks of other strains fit into the barrel of *E. coli* FhuA and vice versa [37]. Corks of *E. coli*, *Salmonella enterica*, *Salmonella paratyphi* B and *Pantoea agglomerans* can mutually be exchanged by genetic means resulting in hybrid proteins which transport ferrichrome with rates that range between 20 and 100 % of the homologous proteins. The hybrid proteins also display wild-type phage and colicin receptor activities [37]. However, there are exceptions. The cork of *P. agglomerans* does not assemble with the *S. enterica* barrel to yield a functional protein and infection by phage ES18 only occurs through *S. enterica* FhuA. As mentioned above, separately synthesized cork of *E. coli* FhuA incorporates into the barrel of *E. coli* FhuA, but the corks of FecA and FepA do not yield an active protein [34].

In contrast to FecA (see Section 11.2.2.1) no surface loop movements are observed upon binding of ferrichrome. It is possible that crystal forces prevent movement although movement is not observed in two distinct crystal forms [25, 26]. However, labeling of FhuA with fluorescein maleimide at the introduced cysteine 336 of loop 4 reveals fluorescence quenching in the presence of ferrichrome, indicating movement of loop 4 [38].

11.2.1.2 Substrate Specificity of the *E. coli* FhuA Transporter

Ferrichrome and the structurally closely related ferricrocin and ferrichrysin are transported at similar rates by FhuA. Some other ferric hydroxamate siderophores also support growth via FhuA, but transport rates were not determined, among them phenylferricrocin [39] for which a crystal structure has been determined [40]. The ferricrocin moiety is bound as ferricrocin and the phenyl group is bound by two aromatic amino acids of the cork (Y87, F115). The antibiotic albomycin is the most prominent and most extensively studied ferrichrome derivative that is transported by FhuA into the periplasm and the FhuCDB proteins into the cytoplasm. Since cells actively take up albomycin and are killed (suicide), the minimal inhibitory concentration of albomycin is 100-fold less than that of ampicillin, one of the most active β-lactam antibiotics against *E. coli* which enters the periplasm by diffusion. As ferrichrome, albomycin is composed of three N^5-acetyl-N^5-hydroxy-L-ornithine residues which complex Fe^{3+}. It lacks the three glycine residues of ferrichrome and thus does not form a cyclic hexapeptide. Instead, it contains a single serine residue to which is bound an antibiotically active moiety, in essence a thioribosylpyrimidine group that inhibits Ser-t-RNA synthetase [41]. The antibiotic moiety must be released by peptidases from the iron hydroxamate carrier to kill cells. Peptidase mutants are albomycin resistant as are mutants in the Fhu transport system. After cleavage, the iron carrier is secreted while the antibiotic remains in the cytoplasm [42, 43].

To see how albomycin fits into the FhuA transport protein, co-crystals were prepared and their crystal structure was determined [40]. The iron carrier occupies the same position as ferrichrome. The antibiotic forms hydrogen bonds and van der Waals contacts with numerous side-chain residues of the extracellular pocket, extracellular loops and barrel strands. Albomycin assumes two conformations in the crystal, an extended and a compact form, which is made possible by the flexibility of the amino acyl linker connecting the iron carrier to the antibiotic.

Another antibiotic, a synthetic rifamycin derivative, designated CGP 4832, is also transported by FhuA into the periplasm, but not further by FhuCDB into the cytoplasm [44]. Active transport into the periplasm as opposed to diffusion of rifamycin across the outer membrane decreases the minimal inhibitory concentration of CGP 4832 200-fold. The chemical structure of CGP 4832 has nothing in common with ferrichrome. It was, therefore, of interest to determine the crystal structure of CGP 4832 bound to FhuA [45]. CGP 4832 shares with ferricrocin all 10 amino acid side-chains of FhuA except one and binds to eight additional side-chains. High-affinity binding occurs through the *N*-methyl-3-piperidyl-acetoxyacetyl moieties by which CGP 4832 differs from rifamycin. In contrast to ferrichrome, ferricrocin, phenylferricrocin and albomycin, CGP 4832 binding does not unwind the switch helix accompanied by the large translation of residues E19 and W22. However, the relative increase of the B factors of the C_α atoms of the first 21 amino acids indicates destabilization of the switch helix. As discussed in Section 11.2.1.1, fixation of the cork to the barrel in this region abolishes FhuA activity. On the other hand, deletion of the switch helix and further mutations in this region do not inactivate FhuA, suggesting that FhuA contacts TonB and the cork can move.

Transport of albomycin demonstrates that rather bulky side-chains can be added to ferrichrome without abolishment of Fhu-mediated transport. This finding indicates that synthetic linkage of antibiotics which slowly penetrate into bacteria to ferrichrome may strongly increase the antibiotic activity. CGP 4832 demonstrates how by pure chance a synthetic derivative fits the binding stereochemistry of FhuA and is transported. The results with CGP 4832 suggest that if a compound is properly bound to FhuA it is also transported. Co-crystals of albomycin were also obtained with FhuD, the transporter in the periplasm following FhuA [46]. In the FhuD crystal albomycin is bound as ferrichrome and the antibiotic moiety sticks out of the protein into the ambient, indicating that there is no stereospecific binding of the antibiotic.

11.2.1.3 The Receptor Activity of FhuA

E. coli cells that synthesize FhuA are infected by the phages T1, T5, ϕ80 and UC-1, and are sensitive to colicin M, a toxic protein with a molecular weight of 27000 that is encoded on certain plasmids and kills sensitive *E. coli* cells. FhuA serves as receptor for these biopolymers. It is unclear whether FhuA also serves as a transporter since for the uptake of colicin M the cork would have to be expelled completely to provide a channel with a sufficient diameter to accommodate unfolded colicin M. The phage-binding sites were identified by competitive peptide mapping. Synthetic hexapeptides which cover residues 316–356 interfere with infection of *E. coli* cells by phages T1, T5 and ϕ80 [12]. Two of the regions extending from residues 316 to 325 and 331 to 340 cover loop 4, of which the former specifically inhibits infection by ϕ80. A third region common to T1 and ϕ80 infection (residues 347–353) is located in the β8 strand where the D348 deletion is also situated which strongly affects FhuA activity. The peptides do not interfere only with phage binding. As shown with phage T5, peptide binding triggers release of phage DNA accompanied by strong alterations in the phage heads and loss of the phage tails [12]. This process is temperature dependent as many more phages are inactivated at 37 than 20 °C.

Models for phage infection must take into account how far the phage tails penetrate into the cells. The model shown in Figure 11.3 demonstrates the dimensions of phage T5 and FhuA. The model is based on studies with isolated FhuA reconstituted into lipid bilayers. Isolated FhuA adsorbs T5 which releases its DNA, a finding that goes back to 1952 [47]. The pb5 protein that recognizes FhuA is not located at the tip of the tail fiber, as one might expect, but at its distal end [48]. Isolated pb5 forms a very stable equimolar complex with FhuA [49], and inhibits infection by T5 and ϕ80 as well as ferrichrome uptake. The tail fiber made up of the pb2 protein traverses the lipid bilayer of the proteoliposome and undergoes a major conformational change in that it shrinks from 50 to 23 nm while the diameter increases from 2 to 4 nm [50]. A channel is formed through which ferrichrome, trapped in the proteoliposomes, is released [51, 52]. Although formation of a channel upon binding of T5 to FhuA suggests a channel within FhuA, this has not been proven. The channel is sufficiently large (2.5 nm width) that phage DNA could get through if T5 only binds to the FhuA surface. Demonstration that the tip of the phage tail is within the proteoliposome favors a model as shown in Figure 11.3. Penetration of the phage tail through the outer membrane and the periplasm would also protect phage DNA from being degraded by periplasmic nucleases.

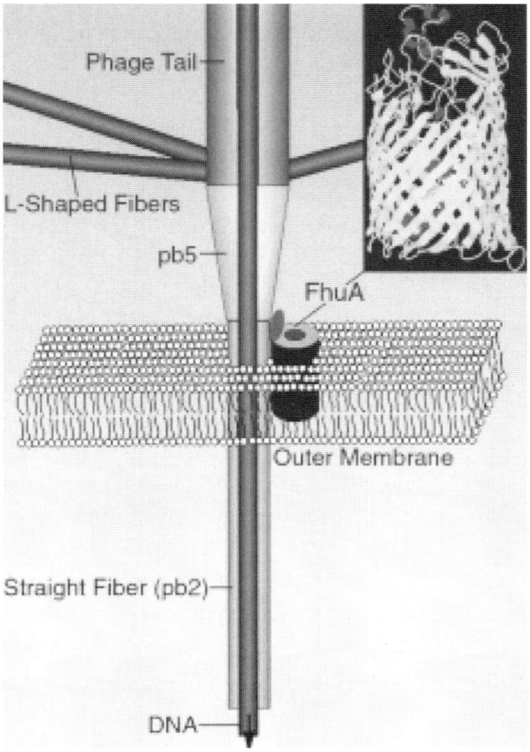

Figure 11.3 Illustration of the dimensions of the phage T5 tail and the FhuA receptor used for phage infection (insert FhuA crystal structure). Reproduced from [50], with permission.

11.2.1.4 Analysis of Previously Isolated Mutants in the Light of the FhuA Crystal Structure

Prior to the determination of the FhuA crystal structure many FhuA mutants resulting in amino acid replacements, insertions and deletions were created to obtain insights into FhuA function. Deletion of amino acids 236–243 and 236–248 results in transport-inactive derivatives which, however, still function as receptors [53]. These deletions cover residues Y244 and W246, which serve as binding sites for the transported substrates and antibiotics. Insertion of a tetrapeptide after residue 241 still shows residual transport activity and partial sensitivity to colicin M but full sensitivity to the phages [54]. Tetrapeptide insertions after residues 59, 69, 70, 82, 108, 128, 134 and 135 of the cork region abolish or nearly abolish all FhuA activities [20]. These insertions most likely change the conformation and folding of the cork. Insertion of 4 amino acids after residue 20 [55] and 16 amino acids after residues 5 and 23 [54] retains all FhuA activities. These insertions are located in or close to the flexible region and may not affect conformation because there is no fixed conformation. They agree with the results obtained with the deletion and duplication derivatives in this region guided by the crystal structure, discussed in Section 11.2.1.1.

In contrast to cork insertions, most insertions of four and 16 residues in the barrel do not abolish FhuA activities. Even those which are located in the β-strands have no effect or only reduce FhuA activities [54, 55]. Probably the inserted peptides bulge from the β-strands and leave the binding frame of the anti-parallel β-strands largely intact. Insertions in surface loops and periplasmic turns do not inactivate FhuA. Notable exceptions are insertions after residue 321 which do not affect ferrichrome transport but strongly reduce phage infection [54–56]. This result agrees with the identification of the phage binding site in loop 4 by competitive peptide mapping (Section 11.2.1.3).

Deletions of FhuA fragments are less tolerated than insertions. Deletion of residues 322–336 strongly reduces phage infection, but largely retains ferrichrome transport [57], as may be expected from removal of part of the phage-binding site. Deletion of D348 inactivates the receptor function for the TonB-dependent phages T1 and ϕ80, sensitivity to phage T5 is reduced 10-fold, and ferrichrome transport activity is strongly reduced [39]. This deletion is located in the β8-strand well above the outer membrane and may disturb formation of the β-pleated sheet.

Residues R93 and R133 of the cork are close enough and oriented such that they can form salt bridges to residues E522 and E571 of the β-barrel. These residues contribute to the fixation of the cork in the β-barrel and possibility serve as a cross-talk device between the two domains. They are strictly conserved in TonB-dependent outer membrane transport proteins. Replacement of the arginine residues reduce FhuA activity most (42 % of the wild-type activity) when substituted by proline, which changes the conformation of the polypeptide. Replacement by leucine retains FhuA transport activity fully and replacement by glutamate nearly completely (93 %). The latter result is particularly remarkable since E93 should repulse E522. In other repulsion constructs, FhuA(R93 E522R) and FhuA(R93 R133 E571R), no transport activity remains (0 and 1 %), but this is also the case with

FhuA(R93L E522R) (1%) and FhuA(R93L E571R) (5%), where no repulsion occurs. Repulsion of two residues may not be sufficient to destabilize FhuA since approximately 60 hydrogen bonds and seven salt bridges are predicted to be involved in the interaction between the cork and the β-barrel [26]. However, the strict conservation of these residues suggests a function in the communication between the cork and the β-barrel beyond the mere structural role. The arginine residues at positions 522 and 571 may disrupt this communication. The glutamate residues are not essential since their replacement by alanine results in highly transport-active FhuA derivatives. FhuA(E522R) and FhuA(E571R) also fail to confer sensitivity to the phages and sensitivity to colicin M is reduced 1000-fold. Since infection by phage T5 is also impaired, the arginine residues do not, or do not only, disrupt the functional response of FhuA to TonB.

11.2.2 FecA Transporter and Signaler

11.2.2.1 Transport Activity of FecA

The crystal structure of FecA reveals for the first time an idea about the gating mechanism of outer membrane transporters. Upon binding of dinuclear ferric citrate, (Fe^{3+} citrate)$_2$, the large surface loops 7 and 8 move 11 and 15 Å, respectively, such that the extracellular pocket through which ferric citrate gains access to the ligand-binding site is closed. Ferric citrate can no longer escape to the extracellular milieu and is only translocated into the periplasm. However, transport requires additional structural changes in the barrel and the cork so that a channel is opened to the periplasm. The structural changes are thought to be elicited by the proton motive force of the cytoplasmic membrane, mediated by the TonB, ExbB and ExbD proteins. Interaction of FecA with TonB is shown by *in vivo* disulfide bond formation between cysteines genetically inserted into the TonB box of FecA and region 160 of TonB [7]. These results also prove that the TonB box of FecA is not located at the N-terminus, as is the case for most outer membrane transporters and all TonB-dependent colicins, but at residues 80–84 of the mature protein. The mutation D80C in the TonB box strongly reduces transport and signaling activity of FecA (Section 11.2.2.2), whereas L82P and V84G and deletion of the TonB box eliminate both activities. However, the TonB box of FecA can be replaced by the TonB box of FhuA and FepA, yielding a functional FecA. Mutation Y163C in TonB inactivates TonB-related FecA functions, but does not affect the BtuB vitamin B_{12} transporter activity [7, 8].

The N-proximal portion of residues 1–79 is not observed in the crystal structure. It yields no discernible electron density because of its flexibility. The TonB box of FecA (residues 80–84) is seen, but disappears in the crystal structure after binding of ferric citrate which is taken as evidence that it becomes flexible. FecA, like FhuA, contains a switch helix which unwinds upon binding of ferric citrate. Ferric citrate elicits numerous additional small changes in FecA.

Ferric citrate probably moves along amino acid residues in the water-filled surface pocket of FecA until it is firmly bound by 11 residues (T138, Q176, Q178, S180, R365, L369, R380, R438, Q570, D573 and N721; A. D. Ferguson, personal communication). Loops 7 and 8 close the entrance, and the cork moves such that a channel is formed between the surface pocket and the periplasmic pocket through which ferric citrate diffuses, presumably along amino acid side-chains. When it arrives in the periplasm, it is bound to the FecB protein. Iron, but not citrate, is transported into the cytoplasm; ferric citrate must dissociate in the periplasm which should occur at FecB. In fact, ferric iron as well as ferric citrate protect isolated FecB from degradation by added proteases (Braun and Herrmann, unpublished results). Iron is then transported by the FecBC permease energized by the FecE ATPase.

11.2.2.2 Signaler Function of FecA

FecA not only functions as a ferric citrate transporter, but also as a signal donor for transcription initiation of the *fecABCDE* transport genes [58]. Binding of ferric citrate to FecA without further transport into the periplasm triggers a signal that in the cytoplasm induces transcription of the transport genes. The trigger may be caused by the structural changes in FecA upon binding of ferric citrate. FecA contains an extra N-proximal portion (residues 1–79) that is not found in outer membrane transport proteins that only transport ferric siderophores and are not involved in transcription regulation. This extra segment functions in signaling, but is dispensable for transport, as shown by its deletion [59]. The signal from FecA is received by FecR which interacts with its periplasmic C-proximal end with the periplasmic N-proximal end of FecA [60]. FecR is a transmembrane protein that transduces the signal across the cytoplasmic membrane. In the cytoplasm, the N-proximal end of FecR interacts with the FecI protein which is converted into a σ factor that binds to the RNA polymerase which initiates transcription at the promoter upstream of the *fecA* gene. *fecA* is highly transcribed, which guarantees in uninduced cells a minimal level of FecA able to respond to ferric citrate and initiate *fec* transport gene transcription.

Binding of ferric citrate to FecA is not sufficient to initiate transcription. Energy from the cytoplasmic membrane mediated by the TonB, ExbB and ExbD proteins is also required. Apparently, ferric citrate-induced structural changes in FecA are not sufficient to generate the transcription initiation signal but additional structural changes through energy input are required.

Transcription initiation mechanisms similar to the FecAIR signaling cascade were demonstrated in *Pseudomonas aeruginosa*, *Pseudomonas putida*, *Bordetella pertussis*, *Bordetella avium*, *Bordetella bronchiseptica* and *Ralstonia solanacearum* [60–62]. The outer membrane proteins transport Fe^{3+} siderophores or heme; they contain an N-terminal segment as in FecA and they initiate transcription via FecIR homologs. Exceptionally, in *R. solanacearum*, the outer membrane protein is not involved in iron transport and does not respond to iron complexes for transcription initiation, but is triggered by contact to plant cells in which it elicits the hypersensitive response. None of these systems has been characterized with regard to the extent of the ferric citrate regulatory device which forms the paradigm of this type of regulatory mechanism.

11.2.3
FepA Transporter and Receptor

11.2.3.1 FepA Transport Activity for Fe^{3+} Enterobactin and Receptor Activity for Colicin B

The FepA outer membrane protein transports ferric enterobactin, a cyclic trimer of dihydroxybenzoyl serine, and serves as a receptor for colicins B and D. It has been studied in *E. coli* K-12, but anti-FepA antibody cross-reacting proteins also occur in a variety of other Gram-negative bacteria such as *Shigella flexneri, Salmonella typhimurium, Citrobacer, Edwardsiella, Enterobacter, Haemophilus, Hafnia, Morganella, Neisseria, Proteus, Providencia, Serratia* and *Yersinia*, most of which also transport ferric enterobactin [63]. The crystal structure of FepA seems to be an intermediate between Fe^{3+} enterobactin-loaded and ligand-free structures [28]. Structure determination was performed with a crystal briefly soaked with Fe^{3+} enterobactin. Longer incubation cracked the crystal so that the binding site of Fe^{3+} enterobactin and structural alterations in FepA upon ligand binding could not be discerned. However, a number of physicochemical studies reveal structural alterations upon binding and transport of ferric enterobactin. Spin-labeled FepA at residues E280C and E310C show structural changes upon addition of Fe^{3+} enterobactin [64]. During Fe^{3+} enterobactin transport the spin label at E280C indicates a series of structural changes [65]. Fluorescein maleimide bound to cysteines introduced at positions 271 and 397 exhibits a reduced fluorescence upon binding of ferric enterobactin and reveals a K_d of 0.2 nM which agrees with the K_d determined by binding of radiolabeled ferric enterobactin [66]. C271 is located in loop L3 and C397 in loop L5 which both move and position the fluorescence reporter molecule into another environment.

Transport of ferric enterobactin bound to FepA across the outer membrane can be followed by the recovery of fluorescence during removal of Fe^{3+} enterobactin from FepA. Transport is inhibited by the ionophore CCCP and the electron transport inhibitors cyanide and azide. Arsenate at rather high concentrations (10 mM), which is used to interfere with ATP-driven transport, partially inhibits FepA-mediated transport. The latter result is unexpected if only the proton motive force drives transport across the outer membrane. It is possible that the lower rate of depletion of FepA from ferric enterobactin indicated by the fluorescence increase results from a lower transport rate across the cytoplasmic membrane which, as an ATP-driven transport, is sensitive to arsenate. Ferric enterobactin-induced loop movements are also revealed by cross-linking studies with ethylene glycolbis(sulfosuccinimidylsuccinate) [67]. K483 in loop 7 is cross-linked with K332 in loop 5 and to the OmpA and OmpF/C proteins. Ferric enterobactin reduces cross-linking to K332 and abolishes cross-linking to the Omp proteins. The movement of K483 involves at least 15 Å, which is similar to the ferric citrate-induced 11 Å movement of L7 in FecA [27].

Assuming that aromatic amino acid residues may be important for binding of enterobactin, aromatic amino acids of FepA were replaced by alanine residues [68]. Of eight residues examined, only Y260A located in loop 3 strongly reduce

(100-fold) the affinity for ferric enterobactin. In other cases where binding is reduced, FepA carries double mutations. Three of the replaced arginine residues are located in loops 3 and 4, two in β6 and β7, and one in the periplasmic turn β6/7. Most, but not all, of the binding mutants also show reduced transport. The data suggest that some of the arginine residues serve as primary binding sites along which ferric enterobactin reaches the final binding site. This interpretation is consistent with the observation made with fluorescence studies which suggest two sequential decay processes upon binding of ferric enterobactin [69]. Random polymerase chain reaction mutagenesis reveals two functionally important FepA regions, the TonB box and extracellular loops of the cork and the β-barrel [70]. The TonB box mutants I14S, I14P, V16D and V16L do not transport ferric enterobactin, show nearly normal binding, and are resistant to colicin B. Other mutations, R75 and R126 of the cork, form salt bridges to E511 and E567 of the β-barrel which are disrupted by the R74L, R74P and R126 replacements, and for this reason may strongly reduce ferric enterobactin transport. The FepA mutant E567A displays a 3-fold higher K_m and a 5-fold lower V_{max} of ferric enterobactin transport than the FepA wild-type, whereas the transport activity of E511Q is less affected [71] (see also Section 11.2.1.1 in which equivalent mutations are described in FhuA).

11.3 Other Fe^{3+} Siderophore Transporters

11.3.1 The Outer Membrane Protein FpvA of *P. aeruginosa* Transports Fe^{3+} Pyoverdin

Isolation of the FpvA protein with bound iron-free pyoverdin [72] was a surprise since only iron-loaded siderophores usually bind with high affinity to transporters because they frequently assume a conformation which differs from the unloaded form. *In vitro*, both pyoverdin and Fe^{3+} pyoverdin bind equally well to isolated FpvA; however, *in vivo*, Fe^{3+} pyoverdin binds with a 17-fold higher affinity than pyoverdin [73]. In contrast to Fe^{3+} pyoverdin, pyoverdin is not transported and replaced at FpvA by Fe^{3+} pyoverdin with fast kinetics that reflect transport kinetics. Displacement of pyoverdin by Fe^{3+} pyoverdin requires TonB in contrast to binding of pyoverdin, which is TonB independent. It is an interesting, but unsolved question how FpvA distinguishes between Fe^{3+} pyoverdin which is transported and pyoverdin which is not transported.

Like FecA, FpvA contains an N-terminal extension which serves to induce transcription of the pyoverdin synthase gene *pvdD* [74]. Removal of the N-terminal extension abolishes transcription induction, but retains pyoverdin transport. Regulation of pyoverdin synthesis by a transport component coordinates formation of the siderophore with its usage. Pyoverdin synthesis is not directly regulated by iron since the biosynthesis genes do not contain a Fur box. Rather the Fe^{2+} Fur repressible regulatory gene *pvdS* encodes an ECF σ factor which regulates transcription of pyoverdin biosynthesis genes, but not the *pvdA* transporter gene. The latter is regulated by a second ECF σ factor, FpvI, which is controlled by the anti-σ factor FpvR. Transcription of the divergently transcribed *fpvI* and *fpvR* genes is repressed by Fe^{2+} Fur which binds to the DNA between *fpvI* and *fpvR*. FpvR also functions as an anti-σ factor of PvdS. According to the experimentally demonstrated regulation of the ferric citrate transport system, it is assumed that binding of Fe^{2+} pyoverdin to FpvA triggers a structural change in FpvA that leads in the periplasm to the interaction of its N-proximal region with FpvR, which in turn leads to inactivation of FpvR as an anti-σ factor, resulting in PvdS and FpvI σ factor activity. PvdS initiates transcription of a number of genes, including regulatory genes for exotoxin A, *prpL* that encodes an endoproteinase, and *aprA* that encodes an alkaline metalloproteinase [75].

11.3.2 The IroN Protein Transports Salmochelin

An *iro* locus involved in iron supply was identified in most *Salmonella enterica* serovars [76] and in extraintestinal *E. coli* strains located on the pathogenicity island III located on the chromosome [77, 78]. The *iro* locus contains genes which encode proteins involved in Fe^{3+} salmochelin transport, salmochelin biosynthesis and secretion. IroN transports a number of Fe^{3+} catecholates across the outer membrane; in particular, salmochelin, a glucosylated derivative of enterobactin [79].

11.3.3 FyuA Transports Fe^{3+} Yersiniabactin

Yersinia enterocolitica, *Yersinia pestis* and *Yersinia pseudotuberculosis* synthesize the siderophore yersiniabactin [80]. The biosynthesis genes and the Fe^{3+} yersiniabactin transport genes are usually encoded on pathogenicity islands which also occur in pathogenic *E. coli* and *Salmonella* strains [81–83]. Horizontal gene transfer of the yersiniabactin gene cluster as such or as part of the pathogenicity island among *Enterobacteriaceae* is considered to be the reason for its wide distribution. Yersiniabactin is secreted, incorporates Fe^{3+} and the iron complex is transported across the outer membrane by the FyuA protein. FyuA also serves as receptor of pesticin, a protein toxin synthesized by *Y. pestis* that hydrolyzes murein (muramidase) [84].

11.4 Outer Membrane Proteins that Transport Heme

TonB-dependent heme transport across the outer membrane has been demonstrated in over 20 Gram-negative bacteria [85–87]. The heme transporters consist of one outer membrane protein such as found in *Y. enterocolitica* [88], *Y. pestis*, *E. coli*, *Serratia marcescens*, *Shigella dysenteriae*, *P. aeruginosa*, *Haemophilus influenzae*, *Helicobacter pylori* or two proteins contained in *Neisseria gonorrhoeae* and *Neis-*

seria meningitidis. The heme transporters may be further subdivided into transporters which bind heme, heme bound to hemopexin, heme incorporated into hemoglobin and hemoglobin bound to haptoglobin (Figure 11.1). It is thought that heme is released from the heme proteins at the transporters and transported into the periplasm. Transport of heme across the cytoplasmic membrane is mediated by ABC transporters. Heme is then acquired as such and incorporated into bacterial heme proteins or the iron is released, presumably by a heme oxidase, as has been shown for *N. meningitidis* [89], *P. aeruginosa* [90] and the Gram-positive *Corynebacterium diphtheriae* [91].

In cases of two proteins that form the outer membrane transporters, both proteins are required to bind hemoglobin [10]. HpuB of *N. meningitidis* represents the TonB-coupled transporter and HpuA is a lipoprotein. HpuA in intact cells is degraded by trypsin to a 21-kDa stable product indicating its surface exposure. In the absence of HpuB, HpuA is further degraded. HpuB is completely degraded by trypsin and the time course of degradation is influenced by HpuA. This and other data suggest that both proteins are associated with each other and together form the hemoglobin-binding site. HpuB shows an increased trypsin sensitivity in CCCP-treated wild-type cells and *tonB* mutants, indicating conformation changes when uncoupled from TonB and the proton motive force.

Cells may contain more than one heme transporter, for example PhuR and HasR in *P. aeruginosa* [92], and HutA, HutR and HasR in *Vibrio cholerae* [93]. Both species also synthesize two distinct TonB proteins which are differently involved in transport of heme provided as heme or hemoglobin.

In *S. marcescens*, a protein designated hemophore is released into the culture medium that mobilizes heme from hemoglobin. The heme loaded HasA hemophore binds to the outer membrane transporter HasR to which it delivers the heme. Transfer of heme from HasA to HasR requires TonB since it does not occur in a *tonB* mutant [94]. HasA is required for heme uptake at low heme concentrations (below 0.5 μM), whereas HasR is essential for heme uptake at all physiological heme concentrations [85, 95]. If heme is bound to hemopexin, HasA is required for heme transport. A hemophore is also formed by *H. influenzae* (HuxA), *P. aeruginosa* (HasAp) and *Pseudomonas fluorescens* [96].

The *hasA* gene is preceded by the *hasR* gene of which both are preceded by the *hasI hasS* regulatory genes. *hasI* and *hasR* display sequence homology to the *fecI fecR* regulatory genes discussed in Section 11.2.2. In fact, binding of heme-loaded HasA to HasR initiates transcription of the *hasA* and *hasR* genes, as has been demonstrated for binding of ferric citrate to FecA. The HasI protein is essential for transcription initiation as is FecI, but in contrast to FecR, HasS is not required for HasI activity but rather inhibits HasI [95]. As in the *fec* system, *hasIS* and *hasR* transcription are regulated by iron via the Fur repressor. Of the three genes located downstream of *hasA*, two encode proteins required for HasA secretion and a third gene, *hasB*, encodes a TonB homolog. The TonB homolog is not specifically required for heme-promoted transcription regulation and heme import, since only double mutants in *tonB* and *hasB* are deficient in heme uptake, and a *hasB* mutant transcribes *hasR* with wild-type levels.

Another means to gain access to heme of hemoglobin is extracellular proteolytic degradation of hemoglobin which has been demonstrated for *E. coli* [97] and *Porphyromonas gingivalis* [98]. The protease of the latter organism degrades hemoglobin, hemopexin, haptoglobin and transferrin.

A heme transport system was also found in *Bradyrizobium japonicum*, a nitrogen-fixing symbiont of soybeans [99]. The transport system is not essential for nitrogen fixation, but may serve to deliver heme in environments outside plants. The outer membrane transporter HmuR is energy-coupled to TonB, ExbB and ExbD. Transcription of *hmuR* is regulated by two regulators, Fur and Irr, and there is evidence for a third regulator.

11.5 Outer Membrane Proteins that Transport Iron Delivered as Transferrin and Lactoferrin

Bacteria like *N. gonorrhoeae, N. meningitidis, H. influenzae, Moraxella catarrhalis, Actinobacilus pleuropnomonia* and *Pasteurella multicoda* use iron contained in the major extracellular iron proteins of their hosts, transferrin and lactoferrin. With a few exceptions, the outer membrane transporters (receptors) consist of two proteins, named TbpA and TbpB for the transferrin receptors, and LbpA and LbpB for the lactoferrin receptors. Transferrin and lactoferrin are structurally and functionally very similar, as are the bacterial receptors. Therefore, in the following the transferrin receptors will be discussed as they were studied in more detail [100] and findings on the transferrin receptors apply equally to the lactoferrin receptors. TbpA and TbpB are exposed at the cell surface. Both proteins can bind separately transferrin, but binding is more efficient when both proteins act in concert. In contrast to TbpA, TbpB preferentially binds iron-loaded transferrin. TbpA is the TonB-dependent iron transporter, whereas TbpB is a lipoprotein and not essential for iron transport. TbpA contains a TonB box, residues 14–18 in *N. gonorrhoeae* FA19 in which the mutation I16P inactivates iron transport. Wild-type TbpA can be co-purified by affinity chromatography with TbpB which fails with a TbpA TonB box mutant [101]. Interaction of TbpA and TbpB is also indicated by protection of TbpB against protease degradation in the presence of TbpA energized by TonB [11]. TbpA and TonB are co-immunoprecipitated in the absence of transferrin, which resembles the findings with the *E. coli* FhuA, FepA, FecA and BtuB transporters that interact with TonB in the absence of their ligands, but interact more strongly in the presence of ligands. It is likely that removal of Fe^{3+} from transferrin requires the proton motive force of the cytoplasmic membrane mediated by TonB since in a *tonB* mutant transferrin is not released from TbpA [11].

Guided by the crystal structures of FhuA, FepA, FecA and BtuB, a topology model for TbpA was proposed which consists of 22 antiparallel β-strands and a globular domain that encompasses the N-proximal 163 residues [102]. This model is supported by deletion of the predicted surface loops 4 and 5, which render TbpA incapable of binding transferrin [102], and by insertion of an epitope into loops 2, 3, 5, 7 and 10, which affects transferrin binding to different extents [100].

It is not known how Fe^{3+} is released from transferrin and how Fe^{3+} is transported across the outer membrane by TbpA. It is conceivable that Fe^{3+} diffuses along amino acid side-chains of the β-barrel and the globular domain until it reaches the periplasm where it binds to the FbpA protein which delivers it to the ABC transporter composed of the FbpB permease and the FbpC ATPase [20].

11.6 Perspectives

The crystal structures of the outer membrane transporters do not reveal how these proteins function. However, they provide valuable suggestions of how these proteins may function. These suggestions can be studied experimentally by isolating mutant proteins that are defect in ligand binding and ligand transport. To derive a dynamic model of transporters in action, biophysical techniques must be employed such as various fluorescence methods with reporter molecules placed at appropriate sites of the transporters. Open questions are: How are the ligands released from their binding sites at the transporters? How are they translocated through the channels of the β-barrels? How are the channels opened? The answers to these questions are intimately linked to the mode of action of the TonB protein, and the ExbB and ExbD proteins. What is the stoichiometry of the three proteins in the predicted complex and how large is the complex they form? How does the complex sense the proton motive force of the cytoplasmic membrane and how does the complex react? What is the structural meaning of an energized TonB, how is the proposed potential energy stored in TonB conveyed to the transporters and what does it elicit in the transporters? For iron and heme incorporated in host proteins it is unknown whether and how TonB affects mobilization of iron and heme from the host proteins and binding to the transporters. What is the role of the lipoproteins in composite heme and iron transporters? Why are there lipoproteins in some heme transporters and only single TonB-coupled proteins in other transporters? Is iron released from transferrin and lactoferrin transported as Fe^{3+} ion or as chelate complexed by a ligand? Do the receptors for phages and colicins function as nucleic acid and protein transporters or pass these biopolymers along the outer surface of the β-barrels across the outer membrane? If phage tails insert into the outer membrane do they go through the interior of the β-barrels which would disrupt the β-barrels or do they penetrate the outer membrane apart from the transport proteins which only serve as initial binding sites? For which purpose do colicins, which use TonB-coupled transport proteins, contain a TonB box which in addition to the TonB box of the transporters is required to kill cells? The genome sequences of some organisms predict many more TonB-coupled transporters, 28 in *P. aeruginosa* and 67 in *Caulobacter crescentus*, than are required to transport iron and vitamin B_{12}, up to now the only characterized transporters. The data suggest that large variety of substrates are transported by TonB-coupled transporters and the reason may be their very low concentrations in the bacteria's habitat. The list of principal questions is long – much longer than the list of answers!

Acknowledgments

We thank Robert Olisa for reading the manuscript. This work was supported by the Deutsche Forschungsgemeinschaft (Forschergruppe "Bakterielle Zellhülle: Synthese, Funktion und Wirkort", BR330/19-1, 2, 3, BR330 20-1) and the Fonds der Chemischen Industrie.

References

1 R. E. Hancock, V. Braun, *J Bacteriol* **1976**, *125*, 409–415.
2 C. Bradbeer, *J Bacteriol* **1993**, *175*, 3146–3150.
3 K. J. Heller, R. J. Kadner, K. Günther, *Gene* **1988**, *64*, 147–153.
4 P. E. Bell, C. D. Nau, J. T. Brown, J. Konisky, R. J. Kadner, *J Bacteriol* **1990**, *172*, 3826–3829.
5 H. Schöffler, V. Braun, *Mol Gen Genet* **1989**, *217*, 378–383.
6 G. S. Moeck, J. W. Coulton, K. Postle, *J Biol Chem* **1997**, *272*, 28391–28397.
7 M. Ogierman, V. Braun, *J Bacteriol* **2003**, *185*, 1870–1885.
8 N. Cadieux, R. J. Kadner, *Proc Natl Acad Sci USA* **1999**, *96*, 10673–10678.
9 J. Schalk, C. Hennard, C. Dugave, K. Poole, M. A. Abdallah, F. Pattus, *Mol Microbiol* **2001**, *39*, 351–360.
10 K. H. Rohde, A. F. Gillaspy, M. D. Hatfield, L. A. Lewis, D. W. Dyer, *Mol Microbiol* **2002**, *43*, 335–354.
11 C. N. Cornelissen, J. E. Anderson, P. F. Sparling, *Mol Microbiol* **1997**, *26*, 25–35.
12 H. Killmann, G. Videnov, G. Jung, H. Schwarz, V. Braun, *J Bacteriol* **1995**, *177*, 694–698.
13 V. Braun, K. Hantke, W. Köster, In: A. Sigel, H. Sigel (Eds), *Metal Ions in Biological Systems, 35 Iron Transport and Storage in Microorganisms.* Marcel Dekker, New York, **1998**, pp. 67–145.
14 V. Braun, K. Hantke, In: D. M. Templeton (Ed.), *Molecular and Cellular Iron Transport.* Marcel Dekker, New York, **2002**, pp. 395–426.
15 V. Braun, K. Hantke, In: G. Winkelmann (Ed.), *Microbiol Transport Systems.* Wiley-VCH, Weinheim, **2001**, pp. 289–312.
16 L. Letellier, M. Santamaria, *Mini Rev Med Chem* **2002**, *2*, 343–351.
17 G. B. Wong, M. J. Kappel, K. N. Raymond, B. Matzanke, G. Winkelmann, *J Am Chem Soc* **1983**, *105*, 810–815.
18 P. I. Higgs, R. A. Larsen, K. Postle, *Mol Microbiol* **2002**, *44*, 271–281.
19 T. E. Clarke, L. W. Tari, H. J. Vogel, *Curr Topics Med Chem* **2001**, *1*, 7–30.
20 T. A. Mietzner, S. B. Tencza, P. Adhikari, K. G. Vaughan, A. J. Nowalk, *Curr Topics Microbiol Immunol* **1998**, *225*, 113–135.
21 W. Köster, *Res Microbiol* **2001**, *152*, 291–301.
22 A. L. Davidson, *J Bacteriol* **2002**, *184*, 1225–1233.
23 K. P. Locher, A. T. Lee, D. C. Rees, *Science* **2002**, *296*, 1091–1098.
24 E. L. Borths, K. P. Locher, A. T. Lee, *Proc Natl Acad Sci USA* **2002**, *99*, 16642–16647.

25 A. D. Ferguson, E. Hofmann, J. W. Coulton, K. Diederichs, W. Welte, *Science* **1998**, *282*, 2215–2220.
26 K. P. Locher, B. Rees, R. Koebnik, A. Mitschler, L. Moulinier, J. P. Rosenbusch, D. Moras, *Cell* **1998**, *95*, 771–778.
27 A. D. Ferguson, R. Chakraborty, B. S. Smith, L. Esser, D. van der Helm, J. Deisenhofer, *Science* **2002**, *295*, 1715–1719.
28 S. K. Buchanan, B. S. Smith, L. Venkatramani, D. Xia, L. Esser, M. Palnitkar, R. Chakraborty, D. van der Helm, J. Deisenhofer, *Nat Struct Biol* **1999**, *6*, 56–63.
29 D. P. Chimento, A. K. Mohanty, R. J. Kadner, M. C. Wiener, *Nat Struct Biol* **2003**, *10*, 394–401.
30 A. D. Ferguson, J. W. Coulton, K. Diederichs, W. Welte, In: A. Messerschmidt, R. Huber, T. Poulos, K. Wieghardt (Eds), *Handbook of Metalloproteins*. Wiley, Chichester, **2001**, pp. 834–849.
31 D. van der Helm, R, Chakraborty, In: G. Winkelmann (Ed.), *Microbiol Transport Systems*. Wiley-VCH, Weinheim, **2001**, pp. 261–287.
32 F. Endriß, M. Braun, H. Killmann, V. Braun, *J Bacteriol* **2003**, *185*, 4683–4692.
33 M. Braun, H. Killmann, E. Maier, R. Benz, V. Braun, *Eur J Biochem* **2002**, *269*, 4948–4959.
34 M. Braun, F. Endriß, H. Killmann, V. Braun, *J Bacteriol* **2003**, *185*, 5508–5518.
35 M. Braun, H. Killmann, V. Braun, *Mol Microbiol* **1999**, *33*, 1037–1049.
36 K. C. Usher, E. Ozkan, K. H. Gardner, J. Deisenhofer, *Proc Natl Acad Sci USA* **2001**, *98*, 10676–10681.
37 H. Killmann, M. Braun, C. Herrmann, V. Braun, *J Bacteriol* **2001**, *183*, 3476–3487.
38 C. Bös, D. Lorenzen, V. Braun, *J Bacteriol* **1998**, *180*, 605–613.
39 H. Killmann, V. Braun, *J Bacteriol* **1992**, *174*, 3479–3486.
40 A. D. Ferguson, V. Braun, H. P. Fiedler, J. W. Coulton, K. Diederichs, W. Welte, *Protein Sci* **2000**, *9*, 956–963.
41 A. L. Stefanska, M. Fulston, C. S. Houge-Frydrych, J. J. Jones, S. R. Warr, *J Antibiot* **2000**, *53*, 1346–1353.
42 A. Hartmann, H. P. Fiedler, V. Braun, *Eur J Biochem* **1979**, *99*, 517–524.
43 V. Braun, K. Günthner, K. Hantke, L. Zimmermann, *J Bacteriol* **1983**, *156*, 308–315.
44 A. P. Pugsley, W. Zimmerman, W. Wehrli, *J Gen Microbiol* **1987**, *133*, 3505–3511.
45 A. D. Ferguson, J. Kodding, G. Walker, C. Bös, J. W. Coulton, K. Diederichs, V. Braun, W. Welte, *Structure* **2001**, *9*, 707–716.
46 T. E. Clarke, V. Braun, G. Winkelmann, L. W. Tari, H. J. Vogel, *J Biol Chem* **2002**, *277*, 13966–13972.
47 H. Frank, M. L. Zarnitz, W. Weidel, *Z Naturforsch* **1963**, *18*, 281–284.
48 K. J. Heller, H. Schwarz, *J Bacteriol* **1985**, *162*, 621–625.
49 L. Plancon, C. Janmot, M. le Maire, M. Desmadril, M. Bonhivers, L. Letellier, P. Boulanger, *J Mol Biol* **2002**, *318*, 557–569.
50 J. Böhm, O. Lambert, A. S. Frangakis, L. Letellier, W. Baumeister, J. L. Rigaud, *Curr Biol* **2001**, *11*, 1168–1175.
51 L. Letellier, K. P. Locher, L. Plancon, J. P. Rosenbusch, *J Biol Chem* **1997**, *272*, 1448–1451.
52 L. Plancon, M. Chami, L. Letellier, *J Biol Chem* **1997**, *272*, 16868–16872.
53 H. Killmann, C. Herrmann, H. Wolff, V. Braun, *J Bacteriol* **1998**, *180*, 3845–3852.
54 R. Koebnik, V. Braun, *J Bacteriol* **1993**, *175*, 826–839.
55 G. Carmel, D. Hellstern, D. Henning, J. W. Coulton, *J Bacteriol* **1990**, *172*, 1861–1869.
56 G. S. Moeck, B. S, Bazzaz, M. F. Gras, T. S. Ravi, M. J. Ratcliffe, J. W. Coulton, *J Bacteriol* **1994**, *176*, 4250–4259.
57 H. Killmann, R. Benz, V. Braun, *J Bacteriol* **1996**, *178*, 6913–6920.
58 V. Braun, *Arch Microbiol* **1997**, *167*, 325–331.
59 I. Kim, A. Stiefel, S. Plantör, A. Angerer, V. Braun, *Mol Microbiol* **1997**, *23*, 333–344.
60 V. Braun, S. Mahren, M. Ogierman, *Curr Opin Microbiol* **2003**, *6*, 173–180.

61 P. Visca, L. Leoni, M. J. Wilson, I. L. Lamont, *Mol Microbiol* **2002**, *45*, 1177–1190.

62 J. L. Ramos, M. T. Gallegos, S. Marques, M. I. Ramos-Gonzalez, M. Espinosa-Urgel, A. Segura, *Curr Opin Microbiol* **2001**, *4*,166–171.

63 J. M. Rutz, T. Abdullah, S. P. Singh, V. I. Kalve, P. E. Klebba, *J Bacteriol* **1991**, *173*, 5964–5974.

64 J. Liu, J. M. Rutz, P. E. Klebba, J. B. Feix, *Biochemistry* **1994**, *33*, 13274–13283.

65 X. Jiang, M. A. Payne, Z. Cao, S. B. Foster, J. B. Feix, S. M Newton, P. E. Klebba, *Science* **1997**, *276*, 1261–1264.

66 Z. Cao, P. Warfel, S. M. Newton, P. E. Klebba, *J Biol Chem* **2003**, *278*, 1022–1028.

67 D. C. Scott, S. M. Newton, P. E. Klebba, *J Bacteriol* **2002**, *184*, 4906–4911.

68 Z. Cao, Z. Qi, C. Sprencel, S. M. Newton, P. E. Klebba, *Mol Microbiol* **2000**, *37*, 1306–1317.

69 M. A. Payne, J. D. Igo, Z. Cao, S. B. Foster, S. M. Newton, P. E. Klebba, *J Biol Chem* **1997**, *272*, 21950–21955.

70 T. J. Barnard, M. E. Watson, Jr, M. A. McIntosh, *Mol Microbiol* **2001**, 41, 527–536.

71 R. Chakraborty, E. A. Lemke, Z. Cao, P. E. Klebba, D. van der Helm, *BioMetals* **2003**, *16*, 507–518.

72 I. J. Schalk, P. Kyslik, D. Prome, A. van Dorsselaer, K. Poole, M. A. Abdallah, F. Pattus, *Biochemistry* **1999**, *38*, 9357–9365.

73 I. J. Schalk, C. Hennard, C. Dugave, K. Poole, M. A. Abdallah, F. Pattus, *Mol Microbiol* **2001**, *39*, 351–360.

74 J. Shen, A. Meldrum, K. Poole, *J Bacteriol* **2002**, *184*, 3268–3275.

75 P. A. Beare, R. J. For, L. W. Martin, I. L. Lamont, *Mol Microbiol* **2003**, *47*, 195–207.

76 A. J. Bäumler, T. L. Norris, T. Lasco, W. Voight, R. Reissbrodt, W. Rabsch, F. Heffron, *J Bacteriol* **1998**, *180*, 1446–1453.

77 U. Dobrindt, G. Blum-Oehler, T. Hartsch, G. Gottschalk, E. Z. Ron, R. Funfstuck, J. Hacker, *Infect Immun* **2001**, *69*, 4248–4256.

78 L. J. Sorsa, S. Dufke, J. Heesemann, S. Schubert, *Infect Immun* **2003**, *71*, 3285–3293.

79 K. Hantke, G. Nicholson, W. Rabsch, G. Winkelmann, *Proc Natl Acad Sci USA* **2003**, *100*, 3677–3682.

80 H. Haag, K. Hantke, H. Drechsel, I. Stojiljkovic, G. Jung, H. Zähner, *J Gen Microbiol* **1993**, *139*, 2159–2165.

81 A. Rakin, C. Noelting, S. Schubert, J. Heesemann, *Infect Immun* **1999**, *67*, 5265–5274.

82 D. Brem, C. Pelludat, A. Rakin, C. A. Jacobi, J. Heesemann, *Microbiology* **2001**, *147*, 1115–1127.

83 T. A. Oelschlaeger, D. Zhang, S. Schubert, E. Carniel, W. Rabsch, H. Karch, J. Hacker, *J Bacteriol* **2003**, *185*, 1107–1111.

84 W. Vollmer, H. Pilsl, K. Hantke, J. V. Höltje, V. Braun, *J Bacteriol* **1997**, *179*, 1580–1583.

85 C. Wandersman, I. Stojiljkovic, *Curr Opin Microbiol* **2000**, *3*, 215–220.

86 C. A. Genco, D. W. Dixon, *Mol Microbiol* **2001**, *39*, 1–11.

87 E. E. Wyckoff, D. Duncan, A. G. Torres, M. Mills, K. Maase, S. M. Payne, *Mol Microbiol* **1998**, *28*, 1139–1152.

88 I. Stojiljkovic, K. Hantke, *Mol Microbiol* **1994**, *13*, 719–732.

89 W. Zhu, A. Wilks, I. Stojiljkovic, *J Bacteriol* **2000**, *182*, 6783–6790.

90 M. Ratcliff, W. Zhu, R. Deshmukh, A. Wilks, I. Stojiljkovic, *J Bacteriol* **2001**, *183*, 6394–6403.

91 G. C. Chu, K. Katakura, X. Zhang, T. Yoshida, M. Ikeda-Saito, *J Biol Chem* **1999**, *274*, 21319–21325.

92 U. A. Ochsner, Z. Johnson, M. L. Vasil, *Microbiology* **2000**, *146*, 185–198.

93 A. R. Mey, S. M. Payne, *Mol Microbiol* **2001**, *42*, 835–849.

94 S. Letoffe, F. Nato, M. E. Goldberg, C. Wandersman, *Mol Microbiol* **1999**, *33*, 546–555.

95 M. S. Rossi, A. Paquelin, J. M. Ghigo, C. Wandersman, *Mol Microbiol* **2003**, *48*, 1467–1480.

96 S. Letoffe, K. Omori, C. Wandersman, *J Bacteriol* **2000**, *182*, 4401–4405.

97 B. R. Otto, S. J. van Dooren, J. H. Nuijens, J. Luirink, B. Oudega, *J Exp Med* **1998**, *188*, 1091–1103.
98 A. Sroka, M. Sztukowska, J. Potempa, J. Travis, C. A. Genco, *J Bacteriol* **2001**, *183*, 5609–5616.
99 A. Nienaber, H. Hennecke, H. M. Fischer, *Mol Microbiol* **2001**, *41*, 787–800.
100 C. N. Cornelissen, *Front Biosci* **2003**, *8*, 836–847.
101 C. D. Kenney, C. N. Cornelissen, *J Bacteriol* **2002**, *184*, 6138–6145.
102 I. C. Boulton, M. K. Yost, J. E. Anderson, C. N. Cornelissen, *Infect Immun* **2000**, *68*, 6988–6996.

12
Structural and Functional Aspects of the Vitamin B_{12} Receptor BtuB

Robert J. Kadner, David P. Chimento and Nathalie Cadieux

12.1
Introduction

12.1.1
Overview

The β-barrel structure forms the transmembrane domain of porins and most, if not all, other integral proteins of the bacterial outer membrane. Porins, as described in the other chapters of this book, form diffusion channels which allow downhill movement of solutes. The preference of porins for the substrates able to pass through them is related to the molecule's size, electrostatic charge and hydrophobic character. Some porins display a high degree of substrate preference, as in the cases of LamB maltoporin, which enhances passage of maltose and maltodextrins [1], and of *Pseudomonas aeruginosa* porins OprD and OprB, which facilitate the passage of amino acids and carbohydrates, respectively [2]. The structures of the porins and their channel properties are described in most of the chapters in this book.

This chapter deals with a sophisticated specialization of the β-barrel structure, which converts the basic features of a porin channel into a transporter that mediates high-affinity, substrate-specific, energy-dependent uptake. In this family of transporters the porin channel is blocked by insertion of a tethered protein domain which contributes to formation of the substrate-binding site. This central globular domain has been called the cork or plug. We prefer to call it the hatch, which connotes a greater degree of movement and specific recognition than the other terms. Because these transporters are in the outer membrane, they are separated from the forms of metabolic energy used to directly drive the cell's other active transporters, i. e. the ATP pool, or the electrochemical ion gradients of protons, sodium or phosphate. Energy for outer membrane transport activity is provided by interaction with a *trans*-periplasmic protein called TonB, whose activity is somehow coupled to the proton motive force across the cytoplasmic membrane. These transport systems are termed TonB-dependent transporters (TBDT); other designations are TonB-gated porins or ligand-gated porins.

Bacterial and Eukaryotic Porins. Edited by Roland Benz

ISBN 3-527-30775-3

TBDT systems have been found only in Gram-negative bacteria, since they use a component targeted to the outer membrane. They are not present in all Gram-negative bacteria, whose physiology does not require the uptake of the TBDT substrates. Their absence from mitochondria and chloroplasts is not surprising because the substrates are not required or are supplied in abundance in the eukaryotic cytoplasm. The number of TBDT homologs present in a bacterial species can be quite high, e.g. 35 are found in the *P. aeruginosa* genome sequence [2] and 65 are encoded in *Caulobacter crescentus*. There are two types of substrates for the known TBDTs. Most important for the cell is ferric iron chelated to various carriers, including siderophores, heme and host proteins that bind iron (e.g. transferrin and lactoferrin) or heme (e.g. hemoglobin) (some reviews in [3, 4]). As described in Chapter 11 on siderophore transporters, ferric iron is practically insoluble at neutral or basic pH in aerobic conditions and its levels in these growth conditions are far too low to support growth. Thus, special mechanisms are needed to solubilize iron and carry it to the bacterial cell.

The other substrates for known TBDTs are cobalamins, such as vitamin B_{12}, and related corrin derivatives (generically designated here as Cbl). The Btu transport system responsible for Cbl uptake is the topic of this chapter. Our understanding of these transport systems has been greatly enhanced by the recent crystallographic determination of the structures of several TBDT, including the Cbl transporter BtuB. In addition, the Btu system provides added structural insights about the BtuCDF protein complex involved in the ATP-dependent transport of Cbl across the cytoplasmic membrane [5, 6]. This chapter describes the current status of the structure–function analysis of the Btu transport system, with emphasis on the comparison with the structure and mechanism of the other TBDTs, and an informed interpretation of previous genetic studies on transport-defective mutants.

12.1.2
Cbl Uptake and Utilization

Vitamin B_{12}, or cyano-Cbl (CN-Cbl), is the largest and most complex enzyme cofactor and participates in a variety of enzymatic processes (reviewed in [7]). The structure is shown in Figure 12.1. The corrin ring is related to the porphyrin tetrapyrrole ring or heme group, and is large, planar and rigid. The central metal atom is cobalt, present in several oxidation states. The ability to form carbon–cobalt bonds is a key feature of many Cbl-dependent reactions. Unlike heme derivatives, Cbls carry equatorial ligands to the central metal atom. The lower ligand is typically 2,3-dimethyl-benzimidazole (DMB) linked to the corrin ring via an aminopropanol-phosphate-ribose linkage. The upper ligand can be cyano, in the case of the drugstore form, or methyl, adenosyl or water in the intracellular forms. These equatorial ligands, along with the axial propionamide and acetamide groups, make Cbl a very large molecule to be transported across cell membranes.

Cbls are produced by many bacteria and Archaea, and are required by animals, but not by plants. Cbl-dependent reactions include some ribonucleotide reductases, diol dehydratases for glycerol and propanediol, methylmalonyl-CoA mutase, etha-

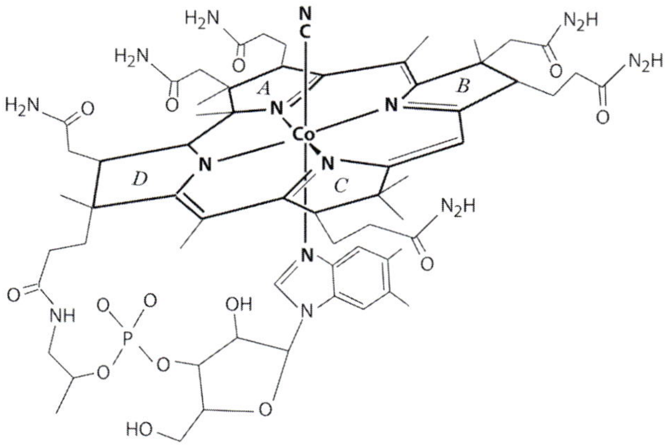

Figure 12.1 Structure of vitamin B_{12} (CN-Cbl). The corrin ring is large and inflexible, and the axial and equatorial substitutions further increase the size and bulk.

nolamine ammonia lyase, and several one-carbon transfer reactions including the synthesis of acetate or methane from CO_2. In the bacterium *Escherichia coli*, the three Cbl-dependent reactions are the transfer by MetH of a methyl group to homocysteine to form methionine, the ethanolamine ammonia lyase reaction catalyzed by the Eut complex and a step in the synthesis of the hypermodified base queuine in some tRNAs. *Salmonella enterica* adds the Cbl-dependent reactions of

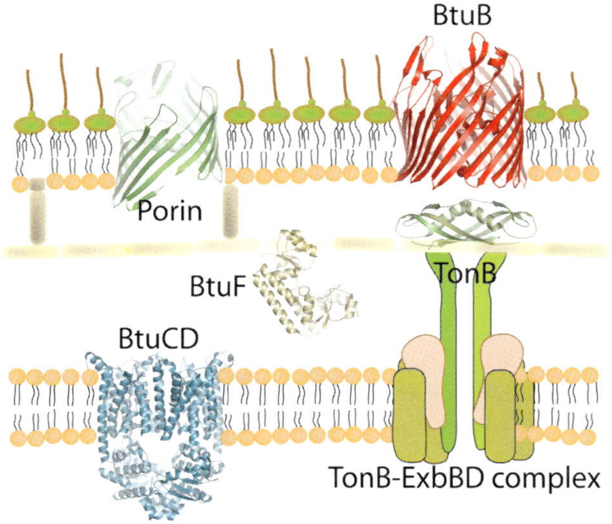

Figure 12.2 Schematic version of the components involved in CN-Cbl uptake, showing the structures available at present. The outer membrane is an asymmetric bilayer with lipopolysaccharide in the outer leaf and phospholipids in the inner. The structure of outer membrane proteins porin and BtuB are composed of β-barrels. BtuF is the periplasmic Cbl-binding protein and BtuCDs form a complex in the cytoplasmic membrane. The structure of the C-terminal domain of TonB is shown, but the structures of the rest of TonB or of its accessory proteins ExbB and ExbD in the cytoplasmic membrane are not known yet.

propanediol catabolism, as well as a *de novo* Cbl biosynthetic system. Many other enteric and environmental bacteria make Cbl by similar pathways.

Cells of *E. coli* grow well in the absence of Cbl. Conditions which impose a requirement for Cbl are useful for genetic analysis of Cbl transport and include the use of ethanolamine as the sole nitrogen or carbon source, or when a *metE* mutation inactivates the Cbl-independent methionine synthase isoenzyme. These two conditions confer requirements for different levels of Cbl transport. Maximal growth of a *metE* mutant needs as few as 25 molecules per cell and these cells grow well at Cbl concentrations of 0.1–0.5 nM. Because the growth response in this methionine-replacement assay requires a very low level of transport activity, it is useful only to detect the presence of low transport activities. In contrast, growth on ethanolamine requires hundreds to thousands of molecules of Ado-Cbl per cell and thus reflects higher levels of transport; maximal growth requires 50–100 nM CN-Cbl. The transport assay using radiolabeled [^{57}Co]CN-Cbl measures uptake of thousands of molecules per cell and usually correlates well with the growth response in the ethanolamine assay, but not in the methionine assay. Thus, Cbl is very effectively accumulated from the medium.

12.1.3
Transport Components

The uptake of Cbl into *E. coli* cells uses two independent energy-dependent transport processes. Like the uptake of iron and iron–chelate complexes, these serial uptake mechanisms employ a TBDT in the outer membrane which binds its substrates with high affinity. Transport across the outer membrane requires input of metabolic energy mediated by the action of the trans-periplasmic energy-coupling protein TonB in association with accessory cytoplasmic membrane proteins ExbB and ExbD. For Cbl uptake via BtuB and for a few of the iron–siderophore uptake systems, the substrate is accumulated in the periplasm to a much higher concentration than in the medium. For other outer membrane transporters, substrate accumulation across the outer membrane is not apparent. In all cases, however, there is no indication for energy-independent transport through these outer membrane transporters, although the potential of their passage through porin channels exists and likely is a factor at high substrate supply. As described in detail below, the structures and potential mechanisms of the TBDTs reveal generally similar features, but with substrate-specific characteristics. The TBDTs for iron–siderophore complexes are discussed in Chapter 11.

Figure 12.2 (see p. 239) provides a schematic of the components involved in Cbl uptake and their structures. Transport of Cbl across the cytoplasmic membrane occurs via an ATP-driven, binding protein-dependent periplasmic permease. The components are the periplasmic Cbl-binding protein BtuF, the transmembrane permease component BtuC and the peripheral ATP-binding cassette (ABC) protein BtuD [8–10]. The structures of the BtuCD complex and of BtuF have been described recently, and are consistent with the operation of these proteins in a hetero-pentameric complex as occurs in the well-studied permeases for maltose, oligopeptides and histidine [5, 6]. The elucidation of the structure of the BtuCD complex provides the first view of the widespread and physiologically important ABC transporter family, perhaps the largest family of proteins in nature. The transport of iron–siderophore complexes and of free iron across the cytoplasmic membrane use closely related periplasmic permeases, but they have different combinations of the component proteins.

Mutants defective in the Btu transport components or of TonB function show loss of detectable transport of CN-Cbl into the cytoplasm [11, 12]. This defect in transport activity does not prevent use of elevated concentrations of Cbl in the growth assays, but the mutants require at least 4-log higher levels of supplied Cbl. Mutants defective in the cytoplasmic membrane transport components also require elevated Cbl supply, but the extent of the defect is complicated by the frequent appearance of suppressor variants [8]. The nature of the suppressors are yet unknown. Double mutants defective in both the outer membrane and cytoplasmic membrane transport processes lack any detectable Cbl uptake.

Of interest to considerations of horizontal gene transfer of transport activities, the genes for all components of each iron transport system are typically linked, allowing all to be transferred in a single event. In contrast, in the enteric bacteria at least, the genes required to form an intact transporter are scattered, with *btuB* and *btuF* each far from the *btuCD* genes. This gene organization makes it very unlikely that an intact Cbl-transport system can be acquired, since acquisition of any one set of genes will not result in a selectable function.

12.1.4 Colicins and Phages

It has long been recognized that various lethal agents, such as phages, bacteriocins, microcins and lantibiotics, employ and require specific bacterial cell-surface components as their receptor for entry into the cell (reviewed in [13–15]. BtuB is required for binding, entry and action of colicin A, and the nine proteins in the E colicins, as well as for bacteriophage BF23. The colicins A and E are proteins of around 60 kDa and have various mechanisms of cell killing. Some form voltage-gated channels in the cytoplasmic membrane, others enter the cytoplasm and block macromolecular synthesis by acting as nucleases active on DNA or on ribosomal RNA. Entry of these colicins does not require function of TonB, but needs a separate translocation system that has no apparent effect on Cbl uptake.

The binding and action of the colicins and phage BF23 are competitively inhibited by Cbl. This situation has been suggested to be an example of molecular parasitism or mimicry, in which the infectious agent has a binding epitope that resembles the molecular surface of the physiological transport substrate [16]. It is also possible that the competitive interactions do not reflect binding of the substrates to the same BtuB surface, but that they bind to distinct surface regions, in such a way that the binding of one substrate alters the protein surface to prevent binding of another substrate. To describe the interaction of the various substrates with BtuB will require detailed identification of the binding sites.

12.2 BtuB Structure

The *btuB* gene was initially identified as the site of mutations conferring resistance to phage BF23 and the E colicins (gene *bfe*). It was renamed *btuB* upon recognition of its physiological role in vitamin B_{12} uptake. BtuB is one of the first TBDT proteins whose gene was sequenced [17]. Its natural low abundance of around 200 molecules per cell is strongly decreased when cells are grown with sufficient Cbl. It is interesting that *btuB* regulation appears to occur by binding of Ado-Cbl to the *btuB* mRNA in a manner that restricts access to ribosome binding [18, 19].

BtuB is an integral outer membrane protein of 594 amino acids and is synthesized with a 20-amino-acid signal sequence which is removed during translocation across the cytoplasmic membrane. BtuB can be expressed to high levels from plasmid vectors and can be purified by detergent extraction of the outer membrane proteins, followed by several ion-exchange fractionation steps in the presence of detergents [20–22]. The purified protein forms crystals suitable for structure determination [20]. Although high-resolution diffraction data were obtained from crystals of the native BtuB protein, efforts to obtain isomorphous heavy-atom derivatives to provide phasing information were unsuccessful. A variant protein with six additional methionines in predicted transmembrane strands was made [23]. The Se-methionine-containing version of this variant of BtuB with eight Met residues allowed solution of its structure by single wavelength anomalous dispersion. Molecular replacement revealed the structure of the native protein in magnesium-containing buffer at 2.0 Å, and of the protein with bound calcium or calcium + CN-Cbl at 3.3 and 3.1 Å, respectively. Comparison of these structures with the known structures of other TBDT has clearly revealed common and specific features.

12.2.1 Shared Structural Features

Structural information has now been presented for four TBDT: ferric enterobactin transporter FepA [24], ferrichrome transporter FhuA [25, 26], ferric dicitrate transporter FecA [27] and Cbl transporter BtuB [23]. All TBDT share a conserved two-domain organization in which a transmembrane β-barrel encloses a globular domain, which fills the channel of the barrel. Figure 12.3 presents the BtuB structure with one face of the barrel cut away to reveal the internal hatch domain. Comparison of the properties of the domains and their interface, and of the conformational changes associated with substrate binding, suggest models for the overall transport process. The β-barrel serves as the bushing that directs insertion of the protein into the outer membrane, and the embedded hatch domain mediates substrate recognition and energy coupling. The new structural insights provide exciting prospects for understanding how energy coupling and active transport occur.

Gram-negative bacteria have two membranes and a number of protein translocation and export systems. An unresolved question concerns what features of a protein determine its specific insertion into the cytoplasmic membrane or the outer

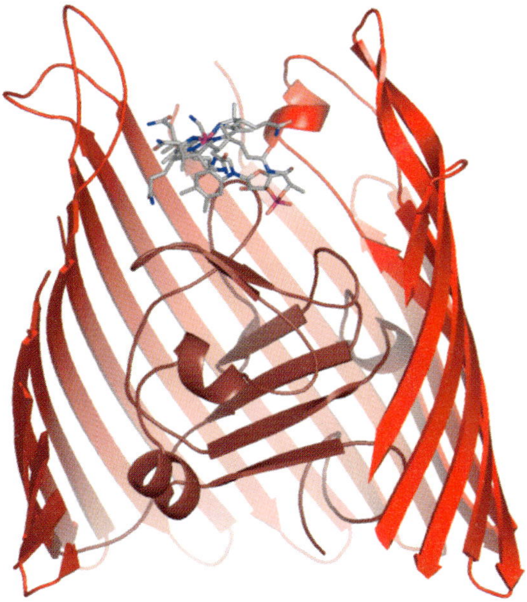

Figure 12.3 Structure of BtuB with barrel strands 16–22 cut away to reveal the internal domain. The substrate CN-Cbl, in color, is bound well above the membrane surface.

membrane. Outer membrane proteins can be readily reconstituted into phospholipid vesicles just like cytoplasmic membrane proteins. In nature, the outer membrane proteins are exclusively found in the outer membrane and cytoplasmic membrane proteins in the cytoplasmic membrane. One expects that insertion of a channel-forming outer membrane porin in the cytoplasmic membrane would be lethal for the cell. Hence insertion of the β-barrel structure in the cytoplasmic membrane must be avoided.

12.2.1.1 The β-barrel

The C-terminal β-barrel domains of the TBDT, despite their different molecular weights, are comprised of 22 antiparallel amphipathic β strands. The side-chains of barrel residues alternate between exposure to the membrane bilayer and exposure to the barrel interior. The outward-facing residues are generally non-polar, but contain numerous polar residues. As found in many membrane proteins, an aromatic collar rings the membrane bilayer interface. The transmembrane β-strands are tilted relative to the membrane normal by 30–40°, so that at least nine residues are needed to span the bilayer in a β-strand. The transmembrane orientation and the location of the bilayer center has been identified in BtuB by electron paramagnetic resonance (EPR) analysis [28]. These results showed that the periplasmic turns are located in or close to the bilayer interface. On the external side of the membrane, all of the β strands extend at least to the outer bilayer interface and many extend beyond it. The interior face of the barrel is quite polar.

The TBDT barrels have the same number of transmembrane strands and the same shear number of 24, and thus similar pore areas. Nonetheless, the dimensions of the barrels differ in their ellipticity. BtuB and FepA are more circular in projected dimension than are FhuA and FepA. One factor in this difference is that the latter two proteins possess a short helical region near the N-terminus, called the switch helix. The switch helix is unfolded in the substrate-bound state leading to a substantial translation across the periplasmic opening of the residues proximal to it, including the Ton box (see below). The switch helix makes contacts with the barrel wall and thus may be responsible for the more elliptical shape of those proteins. The absence of the switch helix in FepA and BtuB suggests that it is not needed for structural integrity. Likewise, its unfolding is not needed for the transmembrane signaling process involving the Ton box, but may serve to exaggerate the range of movement of the N-terminus in proteins, such as FecA, whose extended N-terminal region directs transcriptional regulation [29].

12.2.1.2 Periplasmic Turns

Like most other outer membrane proteins, the periplasmic turns of BtuB are quite short, two to four residues in length. Although the turns are positioned near the membrane interface, it is unlikely that they are completely buried because many contain polar, especially negatively charged residues. For BtuB, five of 10 turns have a net negative charge. Since the membrane interface is also negatively charged, it is unlikely that the turns are completely buried in the membrane.

This consideration is relevant in light of claims that the barrel domain of FhuA and FepA, in the absence of the hatch domain, was able to mediate TonB-dependent transport function. These findings were interpreted to indicate that TonB can interact and function with the barrel itself, suggesting that it recognizes and interacts with the turns on the barrel. There was other indirect evidence that TonB recognizes multiple elements on the TBDT [30]. Recent findings indicate that the apparent TonB-dependent function of the FepA and FhuA barrels is not an accurate interpretation (see Section 12.5.2). Nonetheless, it is still possible that the periplasmic turns of BtuB and its relatives are involved in transport function or TonB recognition.

12.2.1.3 External Loops

In contrast to the short periplasmic turns, the extracellular loops are much longer and varied in length. For BtuB the size of the loops ranges from nine to 19 residues. BtuB is much shorter than the other TBDT and the difference lies almost exclusively in the size of the external loops.

In the ferric citrate transporter FecA the external loops completely close over the barrel lumen when substrate is bound [27]. This occlusion of the central channel provoked the interesting model that the extracellular loops fold over the barrel opening so that the substrate can be transported without leaving open a large porin channel [31]. For BtuB, the shorter loops make it unlikely that their closure can completely block the barrel against passage of molecules from the medium. In all of the structures of BtuB, at least one of the loops is disordered. In the apo-form three loops are disordered for almost their entire length, and, as described further below, become fully ordered in the holo form (+ Ca + CN-Cbl). The fifth loop, joining strands 9 and 10, is ordered in the apo form, but is disordered in the holo form. Even if one takes all loops in their most ordered state, it is clear that they do not close over the barrel opening enough to prevent passage of small molecules. We propose that the closure of the loops serves not to block the barrel from passage of unrelated substrates, but to trap the bound substrate.

12.2.1.4 The Hatch Domain

The globular hatch domain has a very conserved structure. A four-stranded β-sheet is its hydrophobic core and this core has a virtually identical structure in all four proteins. The strands are connected by several short helical stretches and two apical substrate-binding loops. Although one apical loop moves substantially upon binding of substrate, most other substrate-induced movements are fairly subtle. The hatch contains the majority of the sequences conserved among TBDTs, many of which form sharp turns between structural elements.

There are numerous polar interactions holding the hatch in the barrel. The substantial number of hydrogen bonds and salt bridges has been taken as evidence that pulling of the hatch out of the barrel during transport is unlikely [32]. On the other hand, the hatch–barrel interface is remarkably polar with no large hydrophobic patches. Furthermore, about half of the polar interactions are bridged by water molecules. These transient interactions could allow substantial movement of the hatch.

12.2.2 Calcium Binding

An unusual feature of Cbl binding to BtuB is its requirement for calcium. This requirement was first noted by Clive Bradbeer [33], but might not have been widely accepted for several reasons. The apparent affinity of Btu for calcium to support Cbl binding was less than 1 μM, which was below the level of calcium associated with the outer membrane. Hence, calcium chelators were needed to deplete the calcium level sufficiently to reveal its requirement for Cbl binding. The recognition that calcium chelation perturbs the outer membrane structure and composition raised concerns that the requirement for calcium was for membrane integrity rather than Cbl binding. Several mutants altered near residue 50 of BtuB displayed a lower apparent affinity for calcium, so that the dependence for calcium in Cbl binding could be seen without need for titration with chelators. Unfortunately, all the mutations that decreased the affinity for calcium also rendered BtuB labile for proteolysis [33, 34]. Lastly, there was no indication of a requirement for calcium for uptake of iron siderophores.

Binding of CN-Cbl by BtuB crystals required replacement of the magnesium ions in the buffer with calcium ions [20]. Single wavelength anomalous dispersion of the crystal soaked in Ca + CN-Cbl, carried out at the cobalt K-edge, produced maps with three anomalous difference density peaks. The single large peak was identified as the CN-Cbl cobalt atom based on size and the presence of additional difference density clearly outlining the molecule shape of CN-Cbl. Two additional smaller peaks were identified as calcium atoms, based on size and coordination behavior. The same two atoms were present in the crystal of BtuB with calcium. Thus, calcium is required for binding of Cbl. The calcium-binding residues are missing in the calcium-independent iron transporters, but they are highly conserved in the close orthologs of BtuB.

In the absence of calcium, loops 2, 3 and 4 are mostly disordered. The binding of calcium atoms by residues in these loops induces complete ordering of loop 2, and partial ordering of loops 3 and 4. Binding of CN-Cbl into the substrate-binding site

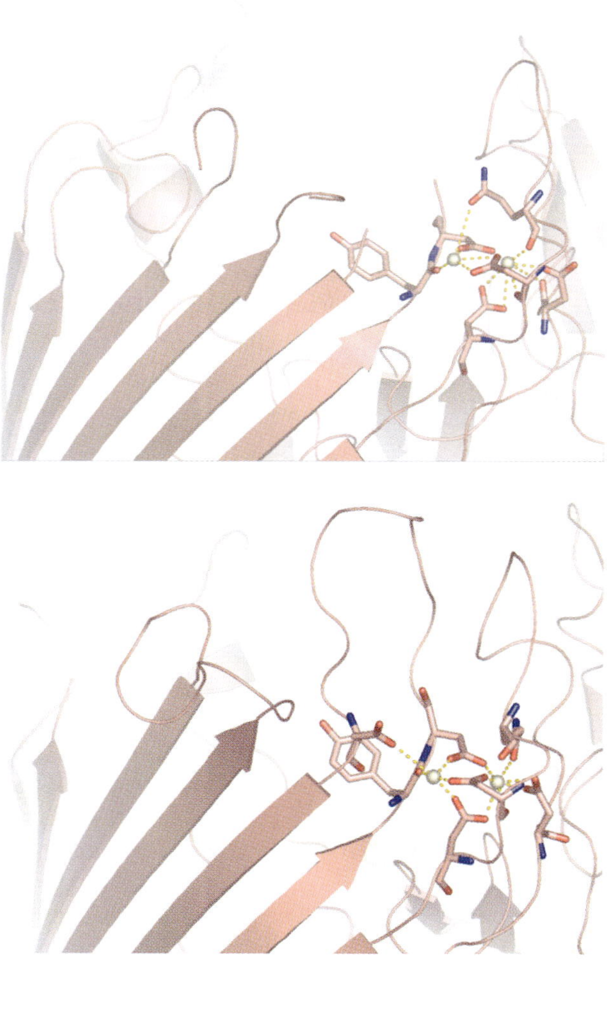

Figure 12.4 Structure of the calcium-binding region of BtuB. (Top panel) Calcium binding in the absence of CN-Cbl; most of loops 3 and 4 are disordered. (Bottom panel) Calcium binding in the presence of CN-Cbl; loops 2–4 are fully ordered, but loop 5 is disordered.

causes complete ordering of the three loops and the disordering of loop 5. Prior to substrate binding, loop 5 rests close to the CN-Cbl binding pocket and is likely displaced to accommodate substrate and the other loop movements. The two calcium atoms bind in an aspartate cage structure that is formed mainly from residues contributed by these three loops, This calcium coordination scheme is unlike the usual calcium-binding EF-hand, but occurs in several unrelated proteins, including thermolysin, pentraxin, serum amyloid protein and the C2 domain of numerous eukaryotic signaling molecules.

As shown in Figure 12.4, the coordination chemistry around the calcium atoms differs in the two structures. In the absence of CN-Cbl, Ca-1 is held by a bidentate complex to the carboxylate of Asp-179 and unidentate complex to Asp-193, Asp-195 and Asp-230. Additional ligands of Ca-1 are the side-chain of Gln-191 and the backbone carbonyl of Asp-195, totaling eight protein oxygen contacts. In contrast, Ca-2 contacts five BtuB ligands, the carboxyls of Asp-193, Asp-195, Asp-230 and Asp-241, and the carbonyl of Tyr-239. Remaining ligands to Ca-2 are likely to be bound water molecules. Reflecting the apparently weaker binding contacts of Ca-2 relative to Ca-1, the crystallographic B-factor of Ca-2, which is a measure of the motion of the atom in the crystal, is 15 $Å^2$ greater than that of Ca-1.

When CN-Cbl is also present, the calcium binding sites undergo a remarkable change. In the holo structure, the rearrangement of the calcium-binding geometry is accompanied by full ordering of the three loops. A couple of amino acids switch ligands from one calcium to the other. Asp-241 moves by 5.8 Å to enter the coordination complex and replace Gln-191 to contact Ca-2. Inferring from the geometry of the calcium-chelation cage, we expect that the affinity for Ca-1 may not change much when Cbl binds, but that Ca-2 is more tightly bound, supported by their now very similar B factors. The ordering of loops 2–4 by calcium and substrate forms part of the Cbl-binding surface that clamps down onto the bound Cbl and may trap it.

12.2.3 Cbl-binding Surfaces

BtuB binds CN-Cbl somewhat like a hand holding a plate from its edge, with one face of the corrin ring nestled into a tight pocket. The lower part of the binding site is comprised of the two apical loops rising from the hatch domain which packs against the DMB and ribose-phosphate linker. Upon binding Cbl, apex-2 twists and moves upward by as much as 6 Å toward the Cbl. Residues from barrel loops 2–4 become fully ordered and clamp down onto the top of the corrin ring. The equatorial acetamide and propionamide substituents on the corrin rings A and D insert into small pockets, and make polar and van der Waal contacts with residues in the hatch and barrel domains. Substituents on the B ring are not inserted into pockets, but make polar contacts to residues from both domains. Ring C is not contacted. Thus, the CN-Cbl molecule binds in a defined orientation and is held by 11 hydrogen bonds from 10 residues with another 20 residues in close contact. CN-Cbl is bound in the base-on form, unlike its binding to some enzymes where a protein His residue replaces the DMB as ligand to cobalt.

12.2.3.1 Comparison to Iron–Siderophore Binding Surfaces

Although the Cbl-binding site is specifically designed for recognition of Cbl derivatives, the substrate-binding sites in FhuA and FecA are located in the same place relative to the hatch apices. These three proteins bind ligands of different size and chemical character, and use entirely different sets of protein residues for substrate recognition. Although each protein possesses substrate-specific residues, their placement within the hatch shows a striking similarity. These different binding sites are constructed at essentially the same position of the protein backbone. This is an intriguing example of protein evolution, whereby identical protein architecture is used, but the choice of side-chain residues and protein flexibility allows it to recognize entirely different substrates. The binding sites recognize their substrate through complementary physical properties. Ferric dicitrate fits snugly into an arginine-rich, polar and basic pocket in FecA, whereas ferrichrome binds into a hydrophobic pocket in FhuA. The common features are the landing pad on the hatch apices and the use of some loop residues from the barrel as a clamp.

12.2.4 The Ton Box

The Ton box is a short 6- or 8-amino-acid segment located near the N-terminus of most TBDT. This sequence is fairly conserved, yet has substantial variability. Indeed, most residues can be changed to a variety of side-chain types without serious or even noticeable change in TonB function [35, 36]. However, certain changes such as Leu-8 to Pro or Val-10 to Gly or Pro in BtuB, or V10P in Cir, or I9D in FhuA [30, 35, 37] result in a phenotype indicative of uncoupling from TonB recognition or action. Transporters carrying these mutations are impaired or defective in all TonB-dependent activities, but are fully functional for all TonB-independent activities, and have normal stability and amount in the outer membrane. Mutations changing Gln-160 of TonB to Leu, Lys or Pro restored some level of BtuB function in an allele-specific manner to the Ton box mutants [30, 38]. These results were interpreted that the Ton box region interacts with TonB in a manner that depended on backbone conformation, but not side-chain recognition.

Recently, Cadieux et al. showed that of Pro substitutions throughout the Ton box, only those at positions 8 and 10 abrogated response to TonB [36]. Thus, it is not the polypeptide chain conformation of the entire Ton box that matters, but specifically these two positions. Second, the amount of Cbl transported under conditions of suppression of a Ton box mutation yielded a value of less that one Cbl per BtuB molecule. Thus, suppression may allow the mutant BtuB protein to transport only one molecule of Cbl, but not complete a transport cycle.

12.2.4.1 Interaction of the Ton Box and TonB

The genetic results suggested that the Ton box and TonB interact. Support for this conclusion came from formaldehyde-mediated cross-linking experiments from the Postle group [39]. Heterodimer formation between TonB and FepA or FhuA was stimulated by the presence of the cognate substrate and was prevented by the presence of a TonB-uncoupling mutation in the Ton box [40, 41]. Although formaldehyde cross-linking can indicate proximity between two proteins, it does not identify the interacting amino acid residues.

Site-directed disulfide cross-linking has been useful in various settings to identify the ability of pairs of protein residues to come into close proximity. Cadieux and Kadner [42] adapted this approach to the BtuB–TonB interaction. Single Cys residues were introduced into each position of the BtuB Ton box and into the region around Gln-160 of TonB, the site of Ton box suppressors. Co-expression of pairs of Cys-containing proteins in *E. coli* cells resulted in the formation of disulfide-linked BtuB-TonB heterodimers, as well as BtuB homodimers and several multimeric species that contained TonB, but not BtuB. Multimer formation required the presence of the introduced Cys residue and they dissociated back to monomer size by disulfide reduction.

Only a few pairs of Cys positions in BtuB and TonB gave rise to BtuB–TonB heterodimers [36, 42]. Of the 42 tested pairs of combinations of BtuB (Cys at positions 6–12) and TonB (Cys at positions 159–164), only 15 pairs (36 %) gave rise to detectable heterodimers and only six pairs reacted efficiently. For most pairs the level of cross-linking was markedly increased by the presence of CN-Cbl for 15 min before cell harvest. The observed degree of stimulation by Cbl is a marked underestimate because of the very long time available for unstimulated cross-linking to occur prior to Cbl addition. Similar cross-linking of TonB to the Ton box of FecA occurred, but without stimulation by citrate [43].

The influence of the uncoupling L8P and V10P substitutions in the BtuB Ton box on cross-linking behavior was determined. The assumption was that Ton box mutations would interfere with association with TonB and decrease cross-linking. In fact, both uncoupling mutations resulted in a dramatic increase in the number of pairs of Cys residues that became linked. In the presence of L8P, 19 of the 36 pairs of Cys-containing proteins showed cross-linking. Positions that did not cross-link with the wild-type sequence, such as T7C and V9C showed substantial linking to TonB when L8P was present. Conversely, cross-linking from BtuB A13C to TonB V163C or P164C, which occurred in the parental situation, was lost with L8P. The BtuB V10P uncoupling mutation, which has the same phenotype as L8P, also expanded the range of cross-linking. Here, 25 of the 36 pairs (70 %) formed heterodimers. Cross-linking between proteins when BtuB L8P was present was stimulated by CN-Cbl to a similar degree as in the wild-type situation, whereas proteins with BtuB V10P gave high cross-linking regardless of the presence of CN-Cbl [36].

These cross-linking results indicate that the BtuB Ton box comes near the Q160 region of TonB when CN-Cbl is bound. To look at possible changes in the conformation of the Ton box, Cadieux [60] examined the reaction of the Cys residues in the BtuB Ton box and flanking positions using a sulfhydryl-reactive maleimide coupled to biotin through a cyclohexyl-butane linker (BMCC). Extent of the reaction was detected by probing electrophoretically separated proteins with horseradish peroxidase-labeled Neutravidin, which binds biotin. Although the chemiluminescence detection assay is not quantitative, it provides an adequate indication of the accessibility of the BtuB residues for labeling. Residues 1–6 are highly available for labeling and are not affected by the addition of CN-Cbl. Starting at residue 7, the extent of labeling decreases progressively in a manner that fits with the solvent-accessible area deduced from the protein structure. Labeling of all of the Ton box positions increases to the level of the freely accessible position 6 when CN-Cbl is present. Thus, substrate binding alters the conformation of the Ton box to increase its availability for reaction with BMCC and with TonB. Residues beyond the Ton box showed generally lower labeling and varying responses to substrates. A biophysical method to investigate the nature of these conformational transitions was provided by site-directed spin labeling.

12.3 BtuB Dynamics

Changes in the dynamics or environment of a protein segment can be investigated by several biophysical techniques, such as fluorescence or nuclear magnetic resonance (NMR) spectrometry. EPR spectrometry, in combination with site-directed spin labeling, has been applied to address the substrate-induced changes in TBDT.

12.3.1 Site-directed Spin Labeling

A spin probe can be attached to specific sites on a protein, usually by derivatization of a cysteine residue. Its EPR spectrum provides several forms of information about the dynamics and environment of that region. It is particularly useful for analysis of membrane proteins [44]. The usual spin probe has a nitroxide moiety with an unpaired electron, and is roughly the size and polarity of a large amino acid. Its incorporation often has no obvious effect on the structure of the protein, but perturbations can occur. The line shape of the EPR X-band spectrum reflects the free volume available for the probe to relax its electron spin orientation relative to the imposed magnetic field. A probe that is free to rotate, such as on the surface of a protein, gives sharp spectral peaks, whereas a probe whose motion is restricted, due to burial in the protein interior or involvement in tertiary contact with other parts of the protein, gives broadened peaks. A probe attached to a residue in a secondary structure, such as an α-helix, has less mobility than a probe on a disorganized region, even if both are on the protein surface. Thus, the EPR spectrum can reveal changes in the environment of a probe upon an experimental manipulation.

A second application of EPR is in measurement of the depth of a spin probe in a membrane bilayer. In a power saturation protocol, the amount of microwave power the probe can absorb becomes saturated when it overwhelms the ability of the probe to relax through thermal motion and decay to the environment. Addition of collisional quenching agents with unpaired electrons provides another mode for spin relaxation and thereby increases the amount of microwave power needed for saturation of the probe. If two quenchers with different lipid solubility are used, their effect on the power saturation value reflects their concentration in the vicinity of the probe. Comparison of the power saturation values for the quenchers, oxygen, which is soluble in lipid, and nickel salt, which is poorly soluble in lipid, provides a direct measure of the depth of the probe in the bilayer. This technique was used to scan a pair of strands of BtuB and revealed the bilayer mid-point. The identification of the two residues in the middle of the bilayer by this method agreed well with their positions opposite each other and equidistant from the aromatic collars in the crystal structure.

E. coli outer membrane proteins contain few free sulfhydryl groups. Exposure of intact cells to Cys-reactive reagents (spin probe or BMCC) results in very little labeling of envelope proteins unless a protein is expressed with a novel Cys substitution. This lack of much background reactivity allows accessibility and dynamics measurements to be carried out in intact cells or outer membrane. These properties can be compared between the native membrane, proteoliposomes reconstituted with the purified protein or the protein in detergent micelles. The EPR spectra of spin probes attached to several regions of BtuB in different membrane environments revealed that parts of the protein respond differently to changes in the environment.

12.3.2 Substrate-induced Changes in the Ton Box

Spin probe was attached to Cys residues placed along successive positions in the N-terminal region of BtuB (residues 1–17 and 25–30) [45–47]. This region includes the Ton box from residues 6 to 12. The spectra indicated a striking pattern in which some positions appeared to be highly mobile and nearby residues were highly restricted. The initial interpretation was that this region of BtuB folded in a way that some side-chains were in contact with another part of the protein, giving restricted mobility, and other resides were open to the medium, giving high probe mobility. However, positions 9–12, which showed high mobility, were poorly labeled by BMCC and had very low solvent accessibility in the crystal structure. This raised the possibility that those residues are buried in the native structure and the attachment of the spin probe interferes with their docking into this native state. Those residues appeared to be highly mobile because the entire region of the Ton box was prevented from folding into its native structure. Several lines of evidence support this view. When a phenylalanine residue was placed at position 11 (T11F) to produce a side-chain comparable in size to the spin probe, a spin probe at position 7, which is normally highly restricted, became mobile. This result is interpreted that the T11F substitution also prevents docking of the Ton box region [47]. Secondly, a spin probe which has greatly reduced rotation around its internal bonds and is more reflective of protein backbone motion, showed identical behavior as the usual spin probe with respect to the residues of low and high mobility. It is unlikely that two adjacent peptide bonds could display such great differences in their movement. Thus, certain positions were sensitive to the addition of the spin probe and showed probe-induced undocking.

The behavior of spin probe attached to the positions of restricted mobility showed a very striking effect when CN-Cbl was present. The entire Ton box displayed high mobility, typical of the flexible undocked state. This undocking was mainly restricted to the Ton box region. Upstream residues 2–5 were already fairly mobile, although their mobility increased further when CN-Cbl was bound. Downstream residues 14–17 were fairly restricted and showed little change in EPR spectrum in the presence of CN-Cbl. Thus, it seems that the Ton box undergoes a transition from a docked conformation to an extended and highly mobile one when CN-Cbl binds. These finding are consistent with the CN-Cbl-induced increase in BMCC labeling and cross-linking to TonB.

The Ton box appears to undergo this transition readily. The L8P and V10P uncoupling mutations resulted in a highly mobile state of the Ton box. Even removal of BtuB from the membrane environment into detergent micelles resulted in undocking, which was reversed upon reconstitution into a phospholipid environment in proteoliposomes.

Recent findings [60, 61] show that the Ton box has a decreased probability of undocking when in the presence of another BtuB substrate, the receptor-binding domain of colicin E3. With this substrate, the Ton box shows lower mobility in EPR analysis and a lower degree of labeling with BMCC. The colicin domain could even promote re-docking of the Ton box that was released in detergent micelles. These findings suggest that the two substrates, whose binding is competitive, may bind to similar regions on the outside face of BtuB, but trigger a different conformational transition of the Ton box.

12.3.3
Transmembrane Region and Barrel Dynamics

Site-directed spin-labeling was applied to a stretch of 29 consecutive residues that form two transmembrane segments and the intervening periplasmic turn in the barrel domain [28]. Power saturation in the presence of collisional quenching agents oxygen and Ni-acetylacetonate showed the expected alternating exposure of successive residues to the membrane bilayer and the barrel interior. The depth in the bilayer of the residues facing the membrane fit well with the subsequently determined crystal structure. In POPC bilayers, spin label mobility indicated that backbone flexibility increased toward the periplasmic side of the barrel. This conclusion fits with the distribution of crystallographic B-factors, reflecting atomic motion, along this particular β-hairpin (D. Chimento and M. Wiener, personal communication). This gradient of greater flexibility toward the cell exterior is not seen for other strands of BtuB. In detergent-solubilized POPC/OG mixed micelles, the outward-facing residues show increased mobility relative to the bilayer environment, whereas the inward-facing residues showed little change. Thus, the structure remains intact in detergent solution, but undergoes moderate changes in flexibility.

12.3.4
Comparison to Crystal Structure

When the crystal structure became available, it provided a clear and convincing verification of the predictions of BtuB topology and the role of calcium in Cbl binding and transport. There was one striking difference from the behavior predicted from EPR and cross-linking experiments. The Ton box remained stubbornly visible in all three structures, with only a rotation of the last visible residue at position 6. There was no indication of the extension of the Ton box toward the periplasmic opening as predicted from the EPR and cross-linking or of any major movement across the opening as expected from the structures of FecA and FhuA. This discrepancy is a matter of great interest and could result from the crystallization conditions.

12.4
Revisiting Old Data

The new structural information and comparisons of the four TBDT provide opportunities to design experiments to test models of function and to improve the reliability of sequence annotation. In addition, it is worthwhile to interpret previous data in light of the structures.

12.4.1
phoA Fusions

The construction of *phoA* gene fusions encoding a hybrid protein with N-terminal sequences from the target gene linked to the mature form of alkaline phosphatase was widely used for analysis of protein localization and transmembrane topology. Alkaline phosphatase is active and stable only when its three disulfide bonds form. Their oxidation normally occurs after the protein is secreted to the periplasm. High phosphatase activity is observed only if the sequence of the fusion protein specifies translocation of the PhoA moiety to the periplasm, such as an N-terminal signal sequence for a periplasmic or outer membrane protein, or an odd number of transmembrane segments for a cytoplasmic membrane protein. If the PhoA moiety is retained in the cytoplasm, it has low activity and stability.

This approach was applied to several TBDT. The localization of PhoA fusions to FhuA [48] and FepA [49] depended on the length of the outer membrane protein that was present. When the fusion had less that 88 residues from FhuA or 178 residues of FepA, the fusion was found in the periplasm. Fusions with more than 180 residues of FhuA or 227 residues of FepA were associated with the outer membrane. These results were interpreted to indicate that the region associated with the change in cellular location contained a membrane-association determinant. It is intriguing that the transition between periplasmic and outer membrane localization occurs when the fusion contains an intact hatch domain. However, insufficient information is available to decide whether the determinant of outer membrane location is the intact hatch or the additional presence of a transmembrane hairpin.

If a transmembrane hairpin were sufficient to target the fusion to the outer membrane, then BtuB–PhoA fusions of similar length should show the same behavior. In fact, BtuB–PhoA fusions are not targeted to the outer membrane until at least 400 BtuB residues were present [50]. One explanation of these results is that a hatch can occupy a barrel even if they are not covalently tethered. The different behavior of the PhoA fusions to the different TBDT might reflect the different amounts of the intact proteins in the outer membrane. The amount of BtuB is much lower than that of FhuA and FepA. The hatches of the two iron proteins could insert into their cognate barrels, whereas the amount of BtuB barrels is too low to accommodate the overexpressed level of BtuB–PhoA. The concept that a hatch can readily slip out of its own barrel and into another needs exploration.

12.4.2 Behavior of In-frame Deletions

Before the structural information was available, it was a common practice to generate and analyze in-frame deletions of segments of the TBDT. Unfortunately, the vast majority of these deletions removed parts of transmembrane segments. The fact that most of these mutants were not stably incorporated in the outer membrane [51] is indicative of the derangement of structure they incurred. Some deletions were expressed in the outer membrane, but had substantial effects on outer membrane integrity as revealed by hypersensitivity to drugs, dyes and detergents. These results were often interpreted as the formation of a permanently opened channel and, in fact, exhibited such behavior in black lipid films. However, the protein structures suggest that deletions which remove one or more TMS might not affect permeability properties unless they prevented insertion of the hatch domain. The resultant empty barrel could display high non-specific permeability. The ability of a deletion variant to insert into the outer membrane would probably require a very fortuitous deletion of an even number of transmembrane segments. Although no rules can be formulated to predict the nature of a TMS and its orientation in the outer membrane, it is likely that most deletions would not be an exact dissection. In-frame deletions in BtuB showed that removal of part of the hatch usually resulted in loss of transport function but the protein could be recovered in the outer membrane, although often in decreased amount. Only a few of the deletions of barrel sequences were inserted in the outer membrane and they produced truncated BtuB proteins deleted for one-half or for three transmembrane strands, making simple explanations of their behavior specious.

12.5
Myths and Models about TonB-dependent Transport Mechanism

The structural information allows us to draw some predictions regarding the mechanism of TBDT and to discount some previous claims regarding their behavior. Here, we discuss prior interpretations of TBDT action and properties.

12.5.1
"Ligand-gated Pores"?

When the concepts of porin structure and function were applied to other outer membrane proteins, there arose the concept that TBDT were ligand-gated porins [52, 53]. It is now clear that ligand binding results in substantial conformational transitions in the outer loops and these somehow are associated with conformational changes in the Ton box on the periplasmic face. These substrate-induced movements are seen by fluorescence [54], cross-linking, spin-labeling and EPR spectrometry, and crystallography. Somehow, the notion arose that these changes in structure upon ligand binding resulted in opening of the central channel to allow substrate passage. To our knowledge there is no evidence that substrate binding opens a channel or is sufficient for its own entry. In the three proteins where the structures of the substrate-bound and -free forms can be compared, there is no indication of movement of the hatch domain out of the barrel or even to form a channel. If anything, substrate binding results in increased order of the loops and a narrowing of the pore opening. This structural information fits with the genetic findings that substrate passage across the outer membrane is totally dependent on TonB function.

12.5.2
The Barrel is Sufficient for TonB-dependent Transport?

Genetic evidence indicated that substitutions in the Ton box region of numerous TBDT could result in loss of TonB-coupled transport activities, but many substitutions had no detectable effect. Mutations affecting other parts of the TBDT have not yet been found to display a similar TonB-uncoupled phenotype. The structure of TBDT revealed that the hatch domain forms part of the substrate-binding surfaces. Cross-linking experiments showed that the Ton box interacts with TonB and that the sequence of the Ton box influences the interaction [36, 41, 55].

However, several claims were made that the hatch domain and the Ton box, in particular, are not required for TonB-dependent transport [56–58]. The primary evidence was provided by in-frame deletions truncated for the entire hatch and Ton box, e. g. the deletion of residues 5–160 in FepA [58]. The hatch-less variants of FhuA and FepA had a markedly reduced affinity for their cognate substrate, but their uptake rates were only decreased about 2-fold relative to the wild-type protein. These findings led to the obvious conclusion that the hatch and Ton box were irrelevant, and that TonB must interact with the turns on the periplasmic surface of the TBDT.

It has recently become clear that these analyses were flawed because the *fepA* and *fhuA* mutant host strains used for expression of the hatch-less barrels, although themselves lacking function of that transporter, still expressed a substantial but incomplete portion of the respective protein, including the entire hatch domain. When other mutants that produced no hatch were used for expression, the hatch-less proteins were completely inactive for any TonB-dependent functions [59]. These new results invalidate the previous claims about the irrelevance of the hatch domain, but also call into question most genetic studies of TBDT that used uncharacterized or incomplete deletions of the chromosomal allele. However, these results also indicate that separate expression of a hatch and barrel domain can result in their assembly into a functional transporter. If a hatch can insert into a barrel, perhaps it can do so in a reversible manner.

12.5.3
The Hatch Stays in the Barrel?

Although not in the category of myth or unwarranted supposition, claims were made from examination of the structures of FhuA [25, 32] that the hatches were tightly held in place in the barrel by more than 40 hydrogen bonds or salt bridges. These multiple bonds must make it difficult or impossible for the hatch to be removed from the barrel during the transport cycle because of the high energy cost of breaking these interactions.

Certainly the hatch–barrel interface is quite large and contains many polar interactions. Does this mean that it is held tightly in place and cannot move? The answer is not yet available, but some aspects are known. Perhaps most important is the recognition that the inner face of the barrel and its interface with the hatch is generally highly polar. There are no large regions of hydrophobic patches holding the two domains together. Water bridges mediate at least half of the interactions between domain. These are expected to be more mobile and transient than hydrophobic interfaces. The recognition that barrel and hatch can be separately expressed and still insert into the other with restoration of function support this possibility. Whether the hatch is removed from the barrel during transport, or only rearranges to let the substrate pass, remains to be shown.

12.6
Conclusions

The wealth of structural information about the TBDT has allowed explanation of many earlier findings and also raises many questions about transport mechanism that remain to be tested. Key questions remain concerning the mechanism of energy coupling and of TonB action, and of the movement of the hatch domain during transport.

Acknowledgments

The authors appreciate insightful discussion with Michael Wiener, David Cafiso and Gail Fanucci. Work in the authors' laboratory was supported by NIH research grant GM19078.

References

1 Schirmer T, Keller TA, Wang YF, Rosenbusch JP. *Science* **1995**, *267*, 512–514.
2 Hancock RE, Brinkman FS. *Annu Rev Microbiol* **2002**, *56*, 17–38.
3 Braun V, Braun M. *Curr Opin Microbiol* **2002**, *5*, 194–201.
4 Ratledge C, Dover LG. *Annu Rev Microbiol* **2000**, *54*, 881–941.
5 Locher KP, Lee AT, Rees DC. *Science* **2002**, *296*, 1091–1098.
6 Borths EL, Locher KP, Lee AT, Rees DC. *Proc Natl Acad Sci USA* **2002**, *99*, 16642–16647.
7 Roth JR, Lawrence JG, Bobik TA. *Annu Rev Microbiol* **1996**, *50*, 137–181.
8 Cadieux N, Bradbeer C, Reeger-Schneider E, Koester W, Mohanty AK, Wiener MC, Kadner RJ. *J Bacteriol* **2002**, *184*, 706–717.
9 DeVeaux LC, Clevenson DS, Bradbeer C, Kadner RJ. *J Bacteriol* **1986**, *167*, 920–927.
10 Freidrich MJ, DeVeaux LC, Kadner RJ. *J Bacteriol* **1986**, *167*, 928–934.
11 Bassford PJ, Jr, Bradbeer C, Kadner RJ, Schnaitman CA. *J Bacteriol* **1976**, *128*, 242–247.
12 Bassford PJ, Jr, Kadner RJ. *J Bacteriol* **1977**, *132*, 796–805.
13 James R, Kleanthous C, Moore GR. *Microbiology* **1996**, *142*, 1569–1580.
14 Lazdunski CJ, Bouveret E, Rigal A, Journet L, Lloubes R, Benedetti H. *J Bacteriol* **1998**, *180*, 4993–5002.
15 Riley MA, Wertz JE. *Annu Rev Microbiol* **2002**, *56*, 117–137.
16 Cao Z, Klebba PE. *Biochimie* **2002**, *84*, 399–412.
17 Heller K, Kadner RJ. *J Bacteriol* **1985**, *161*, 904–908.
18 Nou X, Kadner RJ. *J Bacteriol* **1998**, *180*, 6719–6728.
19 Nou X, Kadner RJ. *Proc Natl Acad Sci USA* **2000**, *97*, 7190–7195.
20 Chimento DP, Mohanty AK, Kadner RJ, Wiener MC. *Acta Crystallogr D* **2003**, *59*, 509–511.
21 Imajoh S, Ohno-Iwashita Y, Imahori K. *J Biol Chem* **1982**, *257*, 6481–6487.
22 Taylor R, Burgner JW, Clifton J, Cramer WA. *J Biol Chem* **1998**, *273*, 31113–31118.
23 Chimento DP, Mohanty AK, Kadner RJ, Wiener MC. *Nature Struct Biol* **2003**, *10*, 394–401.
24 Buchanan SK, Smith BS, Venkatramani L, Xia D, Esser L, Palnitkar M, Chakraborty R, van der Helm D, Deisenhofer J. *Nature Struct Biol* **1999**, *6*, 56–63.
25 Ferguson AD, Hofmann E, Coulton JW, Diederichs K, Welte W. *Science* **1998**, *282*, 2215–2220.
26 Locher KP, Rees B, Koebnik R, Mitschler A, Moulinier L, Rosenbusch JP, Moras D. *Cell* **1998**, *95*, 771–778.
27 Ferguson AD, Chakraborty R, Smith BS, Esser L, van der Helm D, Deisenhofer J. *Science* **2002**, *295*, 1715–1719.
28 Fanucci GE, Cadieux N, Piedmont CA, Kadner RJ, Cafiso DS. *Biochemistry* **2002**, *41*, 11543–11551.
29 Enz S, Mahren S, Stroeher UW, Braun V. *J Bacteriol* **2000**, *182*, 637–646.
30 Bell PE, Nau CD, Brown JT, Konisky J, Kadner RJ. *J Bacteriol* **1990**, *172*, 3826–3829.
31 Postle K. *Science* **2002**, *295*, 1658–1659.

32 Ferguson AD, Deisenhofer J. *Biochim Biophys Acta* **2002**, *1565*, 318–332.
33 Bradbeer C, Gudmundsdottir A. *J Bacteriol* **1990**, *172*, 4919–4926.
34 Gudmundsdottir A, Bradbeer C, Kadner RJ. *J Biol chem* **1988**, *263*, 14224–14230.
35 Gudmundsdottir A, Bell PE, Lundrigan MD, Bradbeer C, Kadner RJ. *J Bacteriol* **1989**, *171*, 6526–6533.
36 Cadieux N, Bradbeer C, Kadner RJ. *J Bacteriol* **2001**, *182*, 5954–5961.
37 Schoffler H, Braun V. *Mol Gen Genet* **1989**, *217*, 378–383.
38 Heller KJ, Kadner RJ, Gunther K. *Gene* **1988**, *64*, 147–153.
39 Skare JT, Ahmer BMM, Seachord CL, Darveau RP, Postle K. *J Biol Chem* **1993**, *268*, 16302–16308.
40 Larsen RA, Thomas MG, Postle K. *Mol Microbiol* **1999**, *31*, 1809–1824.
41 Moeck GS, Coulton JW, Postle K. *J Biol Chem* **1997**, *272*, 28391–28397.
42 Cadieux N, Kadner RJ. *Proc Natl Acad Sci USA* **1999**, *96*, 10673–10678.
43 Ogierman M, Braun V. *J Bacteriol* **2003**, *185*, 1870–1885.
44 Hubbell WL, Cafiso DS, Alenbach C. *Nature Struct Biol* **2000**, *7*, 735–739.
45 Merianos HJ, Cadieux N, Lin CH, Kadner RJ, Cafiso DS. *Nature Struct Biol* **2000**, *7*, 205–209.
46 Coggshall KA, Cadieux N, Piedmont C, Kadner RJ, Cafiso DS. *Biochemistry* **2001**, *40*, 13964–13971.
47 Fanucci GE, Coggshall KA, Cadieux N, Kim M, Kadner RJ, Cafiso DS. *Biochemistry* **2003**, *42*, 1391–1400.
48 Coulton JW, Reid GK, Campana A. *J Bacteriol* **1988**, *170*, 2267–2275.
49 Murphy CK, Klebba PE. *J Bacteriol* **1989**, *171*, 5894–5900.
50 Koester W, Gudmundsdottir A, Lundrigan MD, Seiffert A, Kadner RJ. *J Bacteriol* **1991**, *173*, 5639–5647.
51 Lathrop JT, Wei B, Touchie GA, Kadner RJ. *J Bacteriol* **1995**, *177*, 6810–6819.
52 Klebba PE, Rutz JM, Liu J, Murphy CK. *J Bioenerget Biomembr* **1993**, *25*, 603–611.
53 Klebba PE, Newton SMC. *Curr Opin Microbiol* **1998**, *1*, 238–248.
54 Scott DC, Newton SM, Klebba PE. *J Bacteriol* **2002**, *184*, 4906–4911.
55 Larsen RA, Foster-Hartnett D, McIntosh MA, Postle K. *J Bacteriol* **1997**, *179*, 3213–3221.
56 Braun M, Killmann H, Braun V. *Mol Microbiol* **1999**, *33*, 1037–1049.
57 Killmann H, Herrman C, Torun A, Jung G, Braun V. *Microbiol* **2002**, *148*, 3497–3509.
58 Scott DC, Cao Z, Qi Z, Bauler M, Igo JD, Newton SM, Klebba PE. *J Biol Chem* **2001**, *276*, 13025–13033.
59 Vakharia H, Postle K. *J Bacteriol* **2002**, *184*, 5508–5512.
60 Cadieux N, Phan PG, Cafiso DS, Kadner RJ *Proc Natl Acad Sci USA* **2003**, *100*, 10688–10693.
61 Fanucci GE, Cadieux N, Kader RJ, Cafiso DS *Proc Natl Acad Sci USA* **2003**, *100*, 11382–11387.

13
Structure and Function of Mitochondrial (Eukaryotic) Porins

Roland Benz

13.1
Introduction

The matrix space of mitochondria is, similar to the situation in Gram-negative bacteria, surrounded by two membranes. The mitochondrial inner membrane is especially rich in proteins. Important for mitochondrial function are the enzymes of the respiration chain and oxidative phosphorylation and the many highly specific carriers for the substrates of mitochondria [1, 2]. In contrast to this, the mitochondrial outer membrane is far less substrate specific and acts as a molecular filter for hydrophilic solutes since it is freely permeable to a variety of small hydrophilic solutes up to a well-defined exclusion limit of about 3000–5000 Da. Molecules with higher molecular masses are retained [3, 4]. The well-defined exclusion limit indicates that the mitochondrial outer membranes contain defined passages for substrates. A considerable part of its permeability properties is caused by the presence of general diffusion pores, called mitochondrial porins or voltage-dependent anion-selective channels (VDACs) [5–9]. Schein et al. [5] were the first to notice that a channel-forming component was present in mitochondria, but not in other membrane fractions, of *Paramecium aurelia*. Because of the voltage dependence and the anion selectivity of the channel they used the name VDAC. It is obvious that the anion selectivity of mitochondrial porin was overestimated at the beginning because the selectivity ratio of the channel was estimated to be more than 7 for chloride over potassium ions, while more recent data suggest that it is close to 2 at neutral pH [8, 9]. Mitochondrial porins were isolated from a variety of mitochondria from different organisms and tissues [8–15], and there exists emerging evidence that similar proteins are also present in cytoplasmic membranes of eukaryotic cells with yet unknown function [16–18]. They form a family of slightly basic proteins with molecular masses around 30 kDa [11, 13, 15, 16, 19, 20–23]. The genes of many mitochondrial porins are known to date. They are encoded in the nucleus without a leader sequence and are synthesized on cytoplasmic ribosomes. They are targeted to mitochondria as chaperone-bound species, recognized by the outer membrane general import complex (TOM) and then inserted into the outer membrane [24, 25].

Bacterial and Eukaryotic Porins. Edited by Roland Benz

ISBN 3-527-30775-3

The eukaryotic genomes may contain several genes and pseudogenes coding for porins (see Chapter 15). Mitochondria contain mostly isoform 1 (VDAC1, porin 31HL), but the role of the other two isoforms in eukaryotic cells was also studied (see Chapter 16). Experiments with water soluble porin from *Neurospora crassa* and the composition of isolated mammalian porin suggest that the channel-forming unit is a complex composed of probably only one polypeptide chain and sterols [26–28]. The amino acid composition exhibits a high polarity similar to bacterial porins, although the channel-forming unit is a membrane protein that contains many membrane-spanning segments. The exact structure of the pore is still a matter of debate because has not been crystallized to date. Secondary structure predictions, electron microscopic analyses of two-dimensional arrays and circular dichroism (CD) spectra of mitochondrial porins suggest that they contain large amounts of β-sheet structure with no indication of hydrophobic transmembrane α-helical regions, which makes it likely that bacterial and mitochondrial porins share common structural motifs [29–40]. So far it is not clear if this has to do with the evolution of mitochondria or if it simply reflects that both membrane channels have water-soluble precursors.

Reconstitution experiments with planar lipid bilayers and liposomes [5–15] define mitochondrial porins as pore-forming components which have a high permeability. According to these studies, the mitochondrial outer membrane pore has a diameter of about 2–3 nm in the open state, which agrees well with the electron microscopic analysis of mitochondrial porin *in vitro* and *in vivo* [32–39]. The open channel appears to be wide and water filled. It is slightly anion selective for salts composed of cations and anions of the same aqueous mobility such as potassium and chloride [8, 9], but may change selectivity when the anion has a smaller aqueous mobility than the cation. For membrane voltages larger than about 30 mV, the channel closes in distinct steps in several closed states. These closed states have a reduced permeability towards hydrophilic solutes and a completely different selectivity than the open state [12, 41–43]. In addition to increasing membrane potential, polyanions such as König's polyanion (a copolymer of molecular mass 10 kDa from methacrylate, maleate and styrene in a 1:2:3 proportion [44]) or dextran sulfate result in channel closure *in vivo* and *in vitro* [41, 42, 45]. In particular, adenosine triphosphate (ATP) and adenosine diphosphate (ADP) are not permeable through the closed states of mitochondrial porin [41, 42, 46].

The role of the mitochondrial outer membrane in the physiology and metabolism of mitochondria was underestimated for a long time. It was described as a simple barrier that only prevented the destruction of the mitochondrial inner membrane during osmotic swelling. More recent papers gave some insight into the function of the mitochondrial outer membrane and the role of the porin in mitochondrial metabolism. Porin is the hexokinase-binding protein and glycerokinase also binds to this protein [47, 48]. Cross-linking studies of the mitochondrial outer membrane of yeast revealed that the pore associates or interacts with a 14-kDa protein that was identified as glutathione transferase [49]. Subfractionation of the mitochondrial membranes by sucrose density centrifugation allowed the isolation of contact sites between inner and outer membranes. These contact sites are especially enriched in hexokinase-binding capacity [50]. A further interesting aspect of mitochondrial porin is its possible role in the mitochondrial permeability transition pore (PTP) and its part in cell signaling (see Chapter 16). Furthermore, there exists a strong indication that porin is involved in apoptosis because of its interaction with pro-apoptotic and anti-apoptotic factors such as Bax and Bcl-2 [51–53] (see also Chapter 14). Furthermore, it shows negative regulation by C-Ref kinase [54]. This may indicate that the porin pore plays a crucial role for the interaction of mitochondria with cellular components.

13.2 Reconstitution of Mitochondrial Porins (VDACs) in Model Membranes

13.2.1 Isolation and Purification of Mitochondrial Porins

Several different methods were established for the successful isolation and purification of mitochondrial porins. All start from mitochondria that are obtained by differential centrifugation or density gradient centrifugation of homogenized cellular material [55, 56]. It has to be noted, however, that both preparations also contain other cellular material [57]. For preparation of the mitochondrial outer membrane, the mitochondria are exposed to a swelling/shrinking procedure [58]. Mitochondria are first exposed to 10 mM phosphate followed by addition of 0.45 M sucrose. Centrifugation at 10000 g yields the mitoplasts in the pellet that also contain most of the contact sites between the inner and outer membrane. The outer membrane is in the supernatant. Porin can be isolated from the outer membrane following a standard protocol by using neutral detergents [8]. A higher yield of porin may be obtained following an alternative method starting from whole mitochondria [27]. The cell organelles are extracted with neutral detergents such as Genapol X-80, *N,N*-dimethyldodecylamine-*N*-oxide (LDAO) or Triton X-100. The extract is passed through a dry hydroxyapatite (HTP) column where most of the proteins bind. Porin appears just after the void volume of the column because it is deeply buried in lipid-detergent micelles and does not interact with the column material when Triton X-100 or Genapol X-80 are employed [27]. The use of LDAO instead of Genapol or Triton obviously leads to smaller micelles that bind to HTP and allow a normal elution protocol [27]. The use of ionic detergents such as cholate, desoxycholate and sodium dodecylsulfate (SDS) during the isolation and purification procedure seems to destroy the structure of the channel-forming complex, and leads to its inactivity in reconstitution experiments.

13.2.2 Heterologous Expression of Eukaryotic Porins

It is possible to create a water-soluble form of mitochondrial porin from *N. crassa* [26] and other organisms [59]. Water-soluble porin is inactive in reconstitution experiments and does not form ion-permeable channels. However, when the porin is treated with Triton or Genapol in the presence of sterols, channel formation is regained with properties that cannot be distinguished from its detergent-solubilized form. It is possible that the detergents act as some kind of chaperone. Careful analyses of the sterol requirements suggested that cholesterol leads to highest activation, whereas the addition of epicholesterol did not influence the low channel-forming ability of water-soluble porin [59]. The possibility to refold mitochondrial porin allows an interesting insight in to the structure–function relationship of porin mutants because the genes of many of the mitochondrial porins are known to date. For heterologous expression of mitochondrial porins in *Escherichia coli*, the porin-coding sequences have to be cloned in an expression vector that contains for easy purification of the expressed protein a DNA sequence coding for an additional six histidines (His-tag) upstream of the polycloning site [31, 60, 61]. This His-tag allows the purification of the overexpressed protein by Ni-NTA affinity chromatography. The expressed protein is found within inclusion bodies and needs to be solved in 1% SDS or a high concentration of urea. After chromatography across a Ni-NTA column in SDS or urea the protein is renatured by an overnight dialysis against a 1% Genapol or Triton X-100 solution and by the addition of a cholesterol suspension in the same detergent.

13.2.3
Reconstitution Methods

The properties of the channels formed by mitochondrial porins (VDACs) were studied in different model membranes. Mitochondrial porins from mung bean [7] and rat liver [62] were investigated in lipid vesicles. The reconstitution of rat liver porin into the vesicles made them permeable for [^{14}C]sucrose, but not for [^{3}H]dextran, which suggested that a specific pathway and not a leak is introduced into the vesicles. The exclusion limit of the channel from the outer membrane of mung bean mitochondria was measured in similar experiments using radioactively labeled solutes of different molecular mass to be between 1 and 10 kDa, probably between 3 and 6 kDa. An alternative method was used for the study of porin from *N. crassa* mitochondria [63]. Porin-containing liposomes were transferred into a hypertonic solution containing a test solute. First, the liposomes shrink and can only swell when the test solute is permeable through the channels. The results of experiments using carbohydrates and polyethylene glycols (PEG) of different molecular masses suggest that PEGs with molecular masses up to about 6.8 kDa are permeable through the mitochondrial pore. From these results, the diameter of the mitochondrial pore was calculated to be around 3–4 nm [63]. It is noteworthy that this type of vesicle-swelling assay is not as precise as the swelling method used previously for bacterial porins [64]. The reason for this is that the method used for the study of the channel diameter of bacterial porins is based on the relative rate of permeation of many different solutes through the channel, which depends on the channel diameter according to the Renkin equation [65, 66].

Many reconstitution studies with mitochondrial porins were performed using the planar lipid bilayer technique (see Chapters 6 and 10). The reconstitution is performed as follows. Purified porin is added at low concentration (10 ng/ml to 1 μg/ml) to the aqueous phase bathing black lipid bilayer membranes formed according to the two different methods, which yield more or less similar results. After the addition of the porins to a preformed membrane, the membrane current starts to increase in a stepwise fashion after an initial lag of about 2 min (see Figure 13.1 for the reconstitution of mitochondrial porin from *N. crassa*). This process indicates the insertion of ion-permeable channels into the membrane [6, 8]. The number of channels formed is dependent on time and concentration of protein. In general, the membrane conductance increased for about 15–20 min. After that time, the membrane conductance (i. e. the current per unit voltage) increased at much slower rate. When the rate of conductance increase was relatively slow (as compared with the initial one) it was shown for different mitochondrial porins that the membrane conductance was a linear function of the protein concentration up to porin concentrations of about 1 μg/ml [8, 9]. The results are consistent with the assumption that the protein samples contain preformed channels.

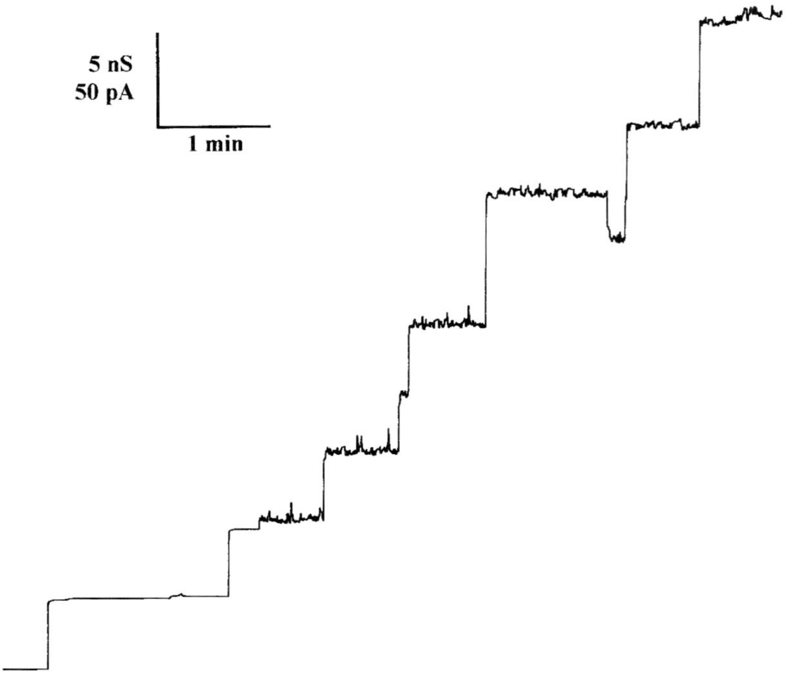

Figure 13.1 Stepwise increase of the membrane current (given in pA) after the addition of *N. crassa* porin to a black lipid bilayer membrane given as a function of time. The aqueous phase contained 10 ng/ml protein and 1 M KCl. The membrane was formed from diphytanoyl phosphatidylcholine/*n*-decane. The voltage applied was 10 mV; $T = 25\,°C$.

13.3
Characterization of the Pore-forming Properties of Eukaryotic Porins

13.3.1
Single-channel Analysis of the Mitochondrial Pore in the Open State

Mitochondrial porins from different eukaryotic cells form ion-permeable channels in lipid bilayer membranes – each step in Figure 13.1 reflects the insertion of one conductive unit (i. e. of one channel into the membrane). These conductance steps are caused by the reconstitution of channels since they are not observed when only the detergents Triton X-100, Genapol X-80 or LDAO are added to the aqueous phase. Figure 13.1 demonstrates that most of the conductance steps are directed upwards and closing steps are only rarely observed at small transmembrane potentials of about 10–20 mV. The most frequent value for the single-channel conductance of mitochondrial porins in 1 M KCl (the conditions of Figure 13.1) was about 4–5 nS (see Figure 13.2 for a histogram of conductance fluctuations observed with *N. crassa* porin). Only a limited number of smaller steps were observed under these conditions (about 25 % of the total number of conductance steps). It should be noted that smaller steps are also found for other mitochondrial porins, including those from yeast [12], rat liver [8] and a number of cells from other tissues including mammalian cells [10, 11, 13, 15, 19]. They may be explained as substates of the pore with the exception of porin from *Paramecium* mitochondria, where only small pores with a single-channel conductance around 2.5 nS were observed in 1 M KCl [43]. This result may indicate that the mitochondrial outer membrane pore does not only switch to substates, but that it may also exist in two different stable conformations. The channels formed by eukaryotic porins are wide and water filled. This is the result of measurements with salts composed of different anions and cations. Even large organic anions and cations such as $Tris^+$ and $HEPES^-$, were found to be permeable through the open state of mitochondrial por-

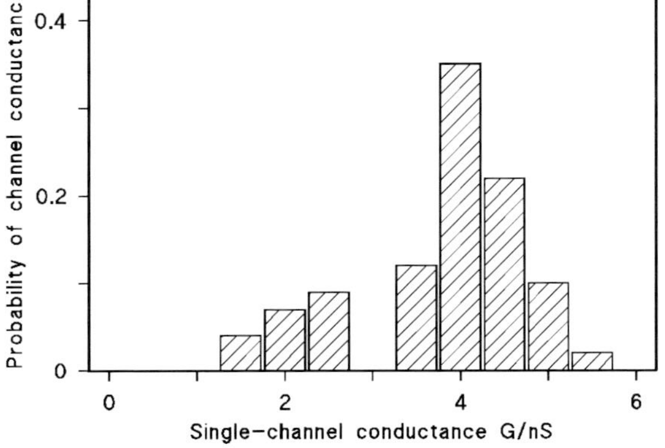

Figure 13.2 Histogram of conductance fluctuations observed with membranes of diphytanoyl phosphatidylcholine/*n*-decane in the presence of *N. crassa* porin. *P*(*G*) is the probability for the occurrence of a conductance step with a certain single-channel conductance (given in nS). The aqueous phase contained 1 M KCl. The voltage applied was 10 mV. The mean value of all upward directed steps was 4.2 nS for the right-side maximum and 2.5 nS for the left-side maximum (in total 187 single events); $T = 25\,°C$.

Table 13.1 Average single-channel conductance of porin 31HL in different salt solutions.

Salt	*c (M)*	*G (nS)*
KCl	0.01	0.05
	0.03	0.15
	0.1	0.45
	0.3	1.3
	1.0	4.3
	3.0	11
LiCl	1	3.2
K-acetate	1	1.5
Tris–HCl	0.5	1.5
Tris–HEPES	0.5	0.18
K-MES	0.5	0.70

The solutions contained 5–10 ng/ml porin 31HL and less than 0.1 μg/ml of the detergent NP-40; the pH was between 6.0 and 7.0. The membranes were made of diphytanoyl phosphatidylcholine/*n*-decane; $T = 25\,°C$; $V_m = 10$ mV. *G* was determined by recording at least 70 conductance steps and averaging over the right-hand maximum (see Figure 13.2). *c* is the concentration of the salt solution. Taken from [67]

ins (see Table 13.1). The single-channel conductance of these salts is a linear function of the specific conductivity of the bulk aqueous phase at small membrane potentials [67].

13.3.2 Eukaryotic Porins are Voltage Gated

The current recording shown in Figure 13.1 demonstrates that closing events represent only a minor fraction of the recordings at 10 mV transmembrane potential. At larger voltages, beginning with about 20–25 mV, the number of closing events increases. They are always smaller than the initial on-steps, which indicates that the pore switches to ion-permeable substates at voltages higher than 20 mV. An experiment of this type is shown in Figure 13.3. Transmembrane potentials of +50 and –50 mV were applied to a membrane containing first three (upper trace) and then four (lower trace) mitochondrial pores from *Paramecium*. The arrow indicates the reconstitution of the fourth channel, which closes after about 0.8 s. The membrane conductance decays in defined steps to smaller values. Similar experiments were also performed with membranes containing many mitochondrial porins, such as shown in Figure 13.4. Voltages of –40 and +40 mV were applied to a diphytanoyl phosphatidylcholine/*n*-decane membrane containing about 150 mitochondrial porin channels from *Paramecium*. In this case, the closing steps cannot be resolved because of the large number of pores in the membrane. The current decays in a single exponential function (see Figure 13.4). The results of this and additional experiments at different voltages allowed the evaluation of the voltage dependence. The steady-state conductance showed a bell-shaped curve as a function of the applied voltage when the conductance at a given voltage $G(V)$

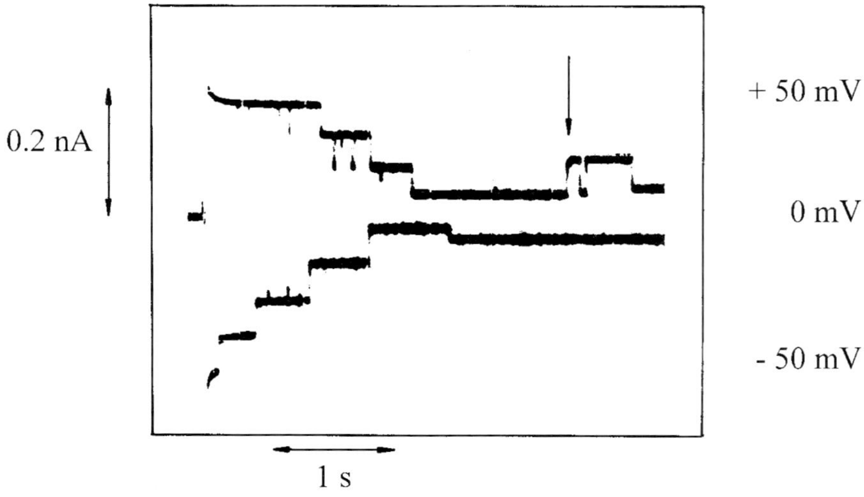

Figure 13.3 Voltage-dependence of *Paramecium* porin. First a voltage of +50 mV was applied to a membrane containing three channels (upper trace). During the experiment a fourth channel was reconstituted into the membrane (arrow), which closed after 0.8 s. Then, −50 mV was applied to the same membrane containing four channels (lower trace). The porin channels closed in defined steps. The membrane was formed from diphytanoyl phosphatidylcholine/*n*-decane. The aqueous phase contained 0.5 M Tris–HCl (pH 7.2); $T = 25\,°C$.

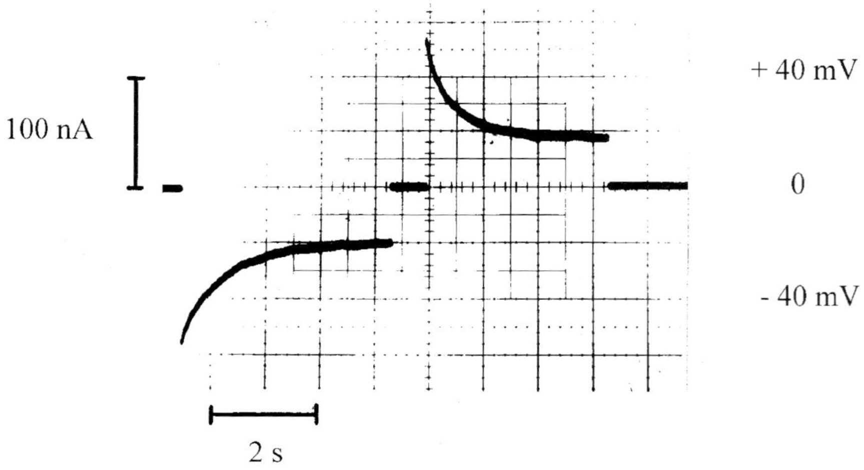

Figure 13.4 Voltage-dependence of *Paramecium* porin in a mutichannel experiment. About 150 porin pores were incorporated in a membrane from 1 % diphytanoyl phosphatidylcholine/*n*-decane. The voltage across the membrane was switched to −40 mV (with respect to the *cis* side, the side of the addition of 500 ng/ml protein) and then to +40 mV. The channels switched to substates of the open state in a single exponential curve. The aqueous phase contained 1 M KCl (pH 6); $T = 25\,°C$.

divided by G_0 at zero potential is plotted as a function of membrane voltage. Figure 13.5 shows the results for *Paramecium* porin and three different salts (KCl, K-MES and Tris–HCl, all at pH 7.2). The results differ somewhat for different salts, although the voltage-dependence is approximately the same. This has to do with the selectivity of the open channel as compared to that of the closed state (see below). The data given in Figure 13.5 could be analyzed as proposed by Schein et al. [5], based on an earlier study of excitability-inducing material (EIM) [68] by calculating the ratio of the number of open, N_o, to closed channels, N_c, from the data of the bell-shaped curves of Figure 13.5:

$$N_o/N_c = (G - G_{min})/(G_0 - G) \qquad (1)$$

where G is in this equation the conductance at a given membrane potential V_m, G_0 and G_{min} are the conductance at zero voltage and very high potentials, respectively. The open to closed ratio of the channels, N_o/N_c, is also given by:

$$N_o/N_c = \exp[-nF(V_m - V_0)/RT] \qquad (2)$$

where F (Faraday's constant), R (gas constant) and T (absolute temperature) are standard symbols, n is the number of gating charges moving through the entire transmembrane potential gradient for channel gating (i. e. a measure for the strength of the interaction between the electric field and the open channel), and

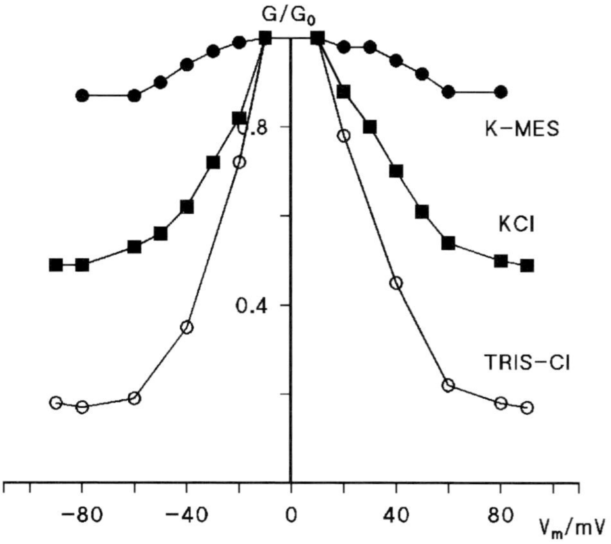

Figure 13.5 Ratio of the conductance, G, at a given voltage, V_m, divided by the conductance, G_0, at 5 mV as a function of the voltage. The aqueous phase contained either 0.5 M KCl, 0.5 M K-MES or 0.5 M Tris–HCl (pH in all cases 7.2). The *cis* side contained 50 ng/ml *Paramecium* porin. The sign of the voltage is given with respect to the *cis* side, the side of the addition of porin.

V_0 is the potential at which 50 % of the total number of channels are in the closed configuration (i. e. $N_o/N_c = 1$).

A semilogarithmic plot of the data given in Figure 13.5 shows that they could be fitted to a straight line with a slope of 13 mV for an e-fold change of V_m. This result suggests that the number of charges involved in the gating process is approximately 2 in the case of *Paramecium* porin (see Figure 13.6). The distribution of open and closed channels, N_o/N_c, follows a Boltzmann distribution. This means that (2) allows also the calculation of the energy difference for channel gating according to [68]:

$$N_o/N_c = \exp[-W(V_m)/RT] \tag{3}$$

$W(V_m)$ is the voltage-dependent energy difference of one mole channels between the open and the closed state. A comparison of (2) and (3) shows that $W(V_m) = nF(V_m - V_0)$. The energy needed for channel closure, nFV_0, calculated from the data of Figures 13.5 and 13.6 is approximately 7.7 kJ mol^{-1}, an energy that is not very high. It is considerably below the energy of one hydrogen bond, which means that channel gating is a low-energy process.

The time constant τ of the single exponential relaxation process shown in Figure 13.4 decreases with increasing voltage. Interestingly, the time constants could be

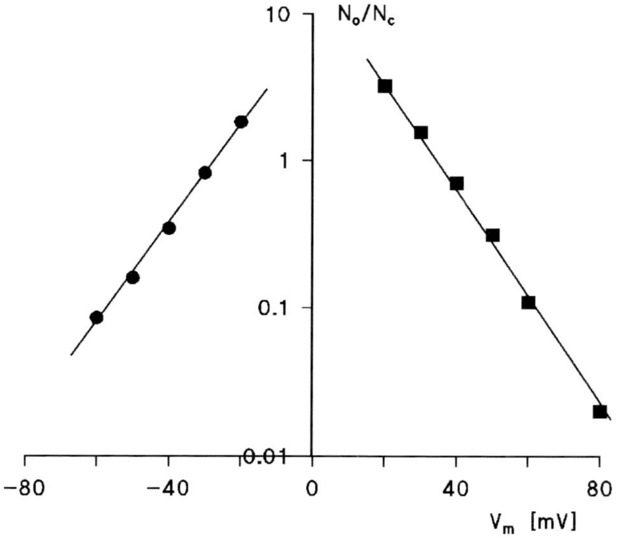

Figure 13.6 Semilogarithmic plot of the ratio, N_o/N_c as a function of the transmembrane potential V_m. The data were taken from Figure 13.5. The closed circles indicate the measurements with positive voltages, V_m, while the squares show measurements with negative potentials (with respect to the *cis* side, the side of the addition of the protein). The slope of the straight lines is such that an e-fold change of N_o/N_c is produced by a change in V_m of 13 mV. The midpoint potential of the N_o/N_c distribution (i. e. $N_o = N_c$) is 35 mV for positive potentials (circles) and –28 mV for negative potentials (squares). Only data for KCl are shown.

fitted to a similar formalism as given in (2) for the ratio N_o/N_c, which means that a semilogarithmic plot of the time constants versus voltage yields a straight line [43]. The slope of this line corresponded to an e-fold decrease of the time constant τ for an increase of the voltage by about 13 mV under the conditions of Figure 13.4. This result suggested again that the number of gating charges involved in channel gating is about 2, which agreed satisfactorily with the number of gating charges derived from the plot of N_o/N_c (see Figure 13.5). The time constant of the switching of the pores from the "closed" to the "open" state could not be followed for mitochondrial porins within the time resolution of the experimental instrumentation (about 1 ms). Figure 13.7 shows an experiment in which the voltage across a lipid bilayer membrane containing about 30 yeast porins was first set to 70 mV. After the exponential decay of the membrane current, the voltage was decreased to 20 mV (arrow), but no relaxation was observed (see Figure 13.7). This result indicated largely different reaction rates for the closing and opening of the mitochondrial porins.

All eukaryotic porins studied to date are voltage dependent. Nevertheless, the threshold voltage for complete channel closure is somewhat different for porins from different organisms. Whereas 35 mV is sufficient to decrease the initial conductance of yeast porin to 50 % [12], about 90 mV has to be applied to membranes containing porin from rat brain to obtain the same effect [10]. Semilogarithmic plots of the ratio N_o/N_c showed that both V_0 and n were different in these cases. These results indicated that the number of charges involved in the gating process varies for mitochondrial porins from different organisms.

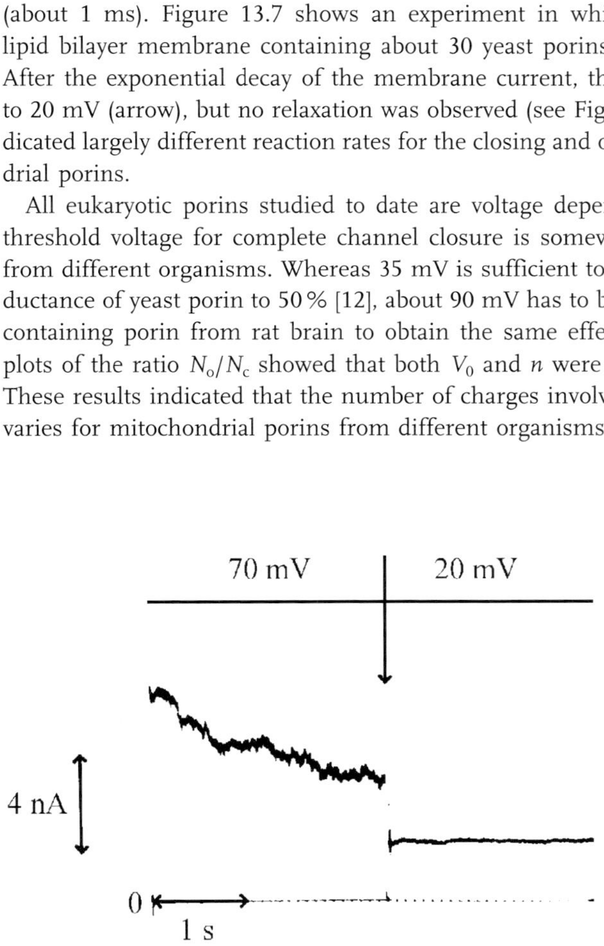

Figure 13.7 Voltage dependence of the open and closed states of yeast porin. The voltage across a diphytanoyl phosphatidylcholine/*n*-decane membrane containing about 30 channels was first set to 70 mV at the *cis* side of the membrane. After the relaxation of the membrane current (time constant about 1.5 s), the voltage was decreased to 20 mV (arrow). The opening of the channels was very fast (less than 50 ms) and could not be followed at the time scale of the experiment. The aqueous phase contained 1 M KCl and 20 ng/ml yeast porin at the *cis* side of the membrane; $T = 25\,°C$.

Table 13.2 Average single-channel conductance of the open and closed states of yeast [12] and human [67] porins in different 0.5 M salt solutions.

Salt	*Open state (nS)*	*Closed state (nS)*
Yeast porin		
KCl	2.3	1.3
K-MES	0.95	0.65
Tris–HCl	1.5	0.30
Human porin (porin 31 HL)		
KCl	2.4	1.4
K-MES	0.70	0.65
Tris–HCl	1.5	0.30

The pH of the aqueous salt solutions was adjusted to 7.2. The protein concentration was between 5 and 10 ng/ml; $V_m = 30$ mV, $T = 25$ °C. The single-channel conductance of the closed state was calculated by subtracting the conductance of the closing events from the conductance of the initial opening of the pores.

13.3.3
Single-channel Conductance of the Closed States

The closed state of eukaryotic porins has a reduced permeability for ions. Single-channel conductance experiments allow the evaluation of the conductance of the closed state at higher membrane potentials of 30 or 40 mV. At these voltages the open state of the channels has only a limited lifetime because of the voltage dependence of the open state. It is possible to evaluate the single-channel conductance of the closed state by subtracting the conductance of the closing events from those of the open state. Table 13.2 shows the results of these type of measurements obtained for three different salts and two types of porin, from yeast [12] and VDAC1 from human cells (porin 31HL [67]). The single-channel conductance of the closed state of the pore was considerably smaller for Tris–HCl than for K-MES, despite a similar aqueous mobility of K^+ and Cl^-. This result suggested that the closed state(s) of mitochondrial porins is cation selective.

13.3.4
Selectivity of the Open and Closed State of Eukaryotic Porins

The data of Table 13.1 suggests that ions move through the open state of mitochondrial porin in the same way they move in the bulk aqueous phase [6, 10, 12, 18, 21, 67]. Interestingly, the pores also exhibit a certain specificity for charged solutes because the single-channel conductance in potassium acetate is somewhat smaller than that in LiCl despite the same aqueous mobility of lithium ions as compared to acetate (see Table 13.1). This means that the mitochondrial porin is selective although it is a wide and water-filled channel in the open state. Experiments with lipid bilayer membranes under zero-current conditions and an externally applied concentration gradient, c''/c', across the membrane allow the evaluation of the ionic selectivity. From the membrane potential, V_m, caused by the preferential movement of one sort of ions through the channel, the ratio P_{cation}/P_{anion} of the permeabilities for cations and anions can be calculated using the Goldman–Hodgkin–Katz equation [69].

Table 13.3 shows the zero-current membrane potentials and the permeability ratios for different mitochondrial and eukaryotic porins [porin 31HL (human), rat liver, yeast and *Paramecium*] in potassium chloride, potassium acetate and lithium chloride. It is obvious from the data of Table 13.3 that the asymmetry potential and the ion selectivity of mitochondrial porins are dependent on the combination of different anions and cations. The different porin channels are slightly anion selective (ratio $P_{cation}/P_{anion} = 2{:}1$) for potassium and chloride, which have approximately the same aqueous mobility. For the combination of the less mobile lithium ion (because of its larger hydration shell [70]) with chloride, the anion selectivity increases. Since the ions move within the channel in a similar way as in bulk aqueous phase, the channel becomes cation selective for potassium acetate because of the smaller mobility of acetate as compared with potassium ions. This result represents another support for the assumption of the mitochondrial pore as a wide water-filled channel in the "open" state.

The open state of all mitochondrial porins characterized so far is slightly anion-selective for salts composed of equally mobile cations and anions such as KCl (see above). The mitochondrial porins switch to closed states when the transmembrane voltage exceeds 20–30 mV. These states definitely have a reduced ion permeability as the bell-shaped curves of Figure 13.5 clearly indicate. The ion selectivity of the closed states cannot be measured using zero-current membrane potential measure-

Table 13.3 Zero-current membrane potentials, V_m, of membranes from diphytanoyl phosphatidylcholine/*n*-decane in the presence of rat liver [8], yeast [12] and *Paramecium* [43] porins measured for a 10-fold gradient of different salts[a]

Salt	*V_m (mV)*	*P_{anion}/P_{cation}*
Rat liver		
KCl (pH 6)	−11	1.7
LiCl (pH 6)	−24	3.4
K acetate(pH 7)	+14	0.50
Yeast		
KCl (pH 6)	− 7	1.4
LiCl (pH 6)	−20	2.6
K acetate (pH 7)	+14	0.50
Paramecium		
KCl (pH 6)	−11	1.7
LiCl (pH 6)	−24	3.4
K acetate (pH 7)	+14	0.50

[a] V_m is defined as the potential of the dilute side (10 mM) relative to that of the concentrated side (100 mM); the temperature was 25 °C. P_{anion}/P_{cation} was calculated from the Goldman–Hodgkin–Katz equation [69].

ments with pores in the closed state. However, the results of Figure 13.5 and Table 13.2 suggest that the closed state is cation selective since for the combination K-MES (a mobile cation combined with a less mobile anion) the conductance in the open and in the closed states differs only little. The difference between open state and closed state is more substantial for Tris–HCl (a mobile anion combined with a less mobile cation). This result suggests indeed that the channel is cation selective in the closed state.

13.4
Inhibition of the Mitochondrial Pore

König et al. [44, 71] showed that an amphiphilic, synthetic polyanion (a copolymer of 10 kDa molecular mass of methacrylate, maleate and styrene in a 1:2:3 proportion) inhibits different carriers in the mitochondrial inner membrane and the ATPase. It is possible that these effects have nothing to do with a direct interaction between the polyanion and inner membrane carriers. Moreover, it seems that the polyanion binds to mitochondrial porin and shifts its voltage dependence in a defined way, thus closing the channel when the membrane voltage has a negative sign at the *cis* side, the side of the addition of the polyanion [41, 42, 46, 72]. For a positive potential at the *cis* side, the channel is always in its open configuration

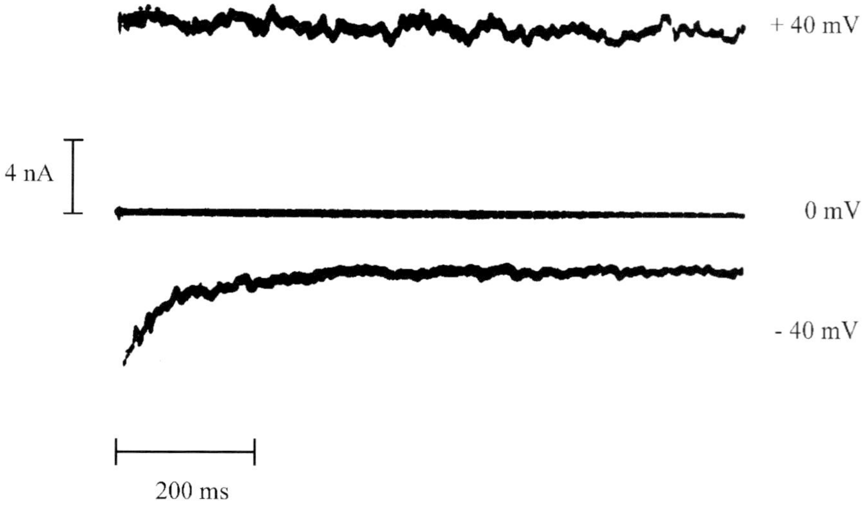

Figure 13.8 Asymmetric current response of rat liver porin after application of membrane potentials of different polarity. The membrane was made of asolectin/*n*-decane. The concentration of the porin was 100 ng/ml in the 1 M KCl solution. The *cis* side contained in addition 1 µg/ml polyanion. A membrane potential of +40 mV (upper trace) and –40 mV (lower trace) was applied to the membrane (referred to as the *cis* side) 40 min after the addition of the porin and 10 min after that of the polyanion. Note that the membrane current only decreased if the *cis* side was negative; $T = 25\,°C$.

even for voltages up to 120 mV and higher. The effect of the polyanion on mitochondrial porin from rat liver is demonstrated in Figure 13.8. First the voltage-dependence of reconstituted rat liver porin was studied and showed the typical bell-shaped curve. The polyanion was added at a concentration of 100 ng/ml polyanion to one side of the membrane (the *cis* side). When the voltage had a negative sign on the *cis* side, the membrane current started to decrease at smaller voltages than without the polyanion. A membrane potential of –10 mV was already sufficient to close the channels. Higher voltages (–40 mV; see Figure 13.8) resulted in a more rapid closure of rat liver porin (relaxation time constant smaller than 100 ms), whereas +40 mV had no effect on porin-induced membrane conductance. Even voltages with magnitudes up to +120 mV did not result in any current decrease, which means that the polyanion stabilized the pore in the open state if the sign of the transmembrane potential was positive on the *cis* side. The addition of the polyanion to both sides resulted in a bell-shaped curve because the channels closed at positive and negative potentials. However, the parameters of channel closure were completely different to those without polyanion, i. e. n was much larger (about 4–5) and V_0 was smaller than 10 mV. The experimental data suggest that there does not exist a "sideness" for the interaction of the polyanion with porin, which means that the polyanion can interact with the gate(s) from both sides of the channel. Furthermore, the asymmetric effect of the polyanion on rat liver porin was stable during the whole lifetime of the membranes (up to several hours), indicating that the polyanion could not penetrate the channel.

The interaction of the polyanion with mitochondrial porins from different sources also allows us to measure the conductance of the closed state when the polyanion was added at a concentration of 100 ng/ml to the side of the membrane with negative polarity (the *cis* side). The single-channel conductance was about half that of the open state measured at low voltage, which means that the closed channels are still permeable for ions, which makes a direct polyanion-induced block of the channel rather unlikely [41, 42, 45, 46, 72]. Table 13.4 shows the effect of polyanion on the single-channel conductance obtained with salts other than KCl. The reduction of the single-channel conductance caused by the polyanion is especially large when a mobile anion (e. g. Cl^-) is combined with a less mobile cation (e. g.

Table 13.4 Average single-channel conductance of the open and the polyanion-induced closed state of rat liver porin in different 0.5 M salt solutions [42].

Salt	*Open state (nS)*	*Closed state (nS)*
KCl	2.2	1.2
LiCl	1.8	0.40
K acetate	1.1	0.85
K-MES	0.88	0.74
Tris–HCl	1.5	0.25

The pH of the aqueous salt solutions was adjusted to 7.2. The membrane voltage was 10 mV at the *cis* side; $T = 250$ C. The aqueous phase contained in the measurements of the closed state 0.1 µg/ml polyanion added to the *cis* side.

Tris$^+$). In this case, the single-channel conductance was reduced to less than 10% of the open state value. The effect of the polyanion on combinations of mobile cations with less mobile anions (e.g. on K-MES) was very small. These results are consistent with the assumption that the slightly anion-selective channel in the open state becomes highly cation selective in the closed state (see also Figure 13.5). This was also the result of selectivity measurements in the presence of polyanion. The ratio P_{cation}/P_{anion} for rat liver porin in selectivity measurements without polyanion is approximately 0.6 for KCl [8]. For a polyanion concentration of 15 μg/ml on both sides of the membrane, P_{cation}/P_{anion} increases to about 10, which suggests that chloride has a very small permeability through the polyanion-mediated closed state of rat liver porin [42].

Channel blockage by the polyanion can be compared to the situation in whole mitochondria in which polyanions added to the outside (presumably the negative side of the outer membrane) are able to close the channel and inhibit the passage of anionic metabolites across the mitochondrial outer membrane, including ADP and ATP, thus blocking the mitochondrial metabolism [41, 42, 46]. Other polyanions such as dextran sulfate have a similar effect on the voltage dependence of mitochondrial porin, but these compounds need a considerably larger concentration for channel closure than the polyanion [45]. In addition to the different polyanions, the action of a protein component from mitochondria was studied in mitochondria and in reconstituted systems [45]. This "modulator" exhibits a similar action on the porin channel as the polyanion, which means that it disturbs nucleotide exchange across the outer membrane and shifts the voltage dependence of the channel in a dose-dependent way. Although its action is known, it has not been possible to identify the modulator. It is presumably a polypeptide since it is susceptible to pronase, i.e. its action on porin is destroyed by pronase treatment [45, 73].

13.5 Structure of the Channel formed by Eukaryotic Porins

13.5.1 Primary Structure of Eukaryotic Porins

The primary structure of eukaryotic porins from many organisms, including humans, mouse, fruit fly, plants, yeast, fungi and *Dictyostelium*, is known to date. With the exception of porin 31HL (HVDAC1 from humans), all were derived from their cDNA sequences (see also Chapters 15 and 16) [13, 21, 22, 74–77]. Porin 31HL is known from its amino acid sequence [78]. Despite many variations of single amino acids, all sequences are related to one another and have an approximate length of about 280 amino acids. Figure 13.9 shows a comparison of the amino acid sequences of human (HVDAC1, porin 31HL) [78], *Drosophila melanogaster* [74], potato [76], *Saccharomyces cerevisiae* [21] and *N. crassa* [22] mitochondrial porins. Despite the clear homology between the different sequences, only 15 (out of about 280) amino acid residues are strictly conserved in all five porins (marked by points in Figure 13.9). The most remarkable identity in all five porins is the GLK triplet near amino acid 90, which may play an essential, but unknown, role in eukaryotic porin function. However, it is not the place of ATP binding [79]. Otherwise, the identities between the five porins are rather small in large stretches of the aligned porin sequences. These results suggest that the possible β-barrel structure of mitochondrial porins (see below) tolerates extensive amino acid variations without substantial alterations in the secondary structure and of the function of a voltage-dependent membrane channel, which was found for all the mitochondrial porins. Similarly, all porins form wide water-filled channels with an average single-

```
                           ●   ●  ●
Porin 31HL   AVPPTYADLGKSARDVFTK-GYGFGLIKLDLKTKSENGLEFTSSGSANTET 50
   DM-VDAC   MAPPSYSDLGKQARDIFSK-GYNFGLWKLDLKTKTSSGIEFNTAGHSNQES 50
    P-VDAC MGKGPGLYTEIGKKARDLLYKD-YQ-SDHKFSITTYSPTGVVITSSGSKKGD- 50
   SC-VDAC   MSPPVYSDISRNINDLLNKDFYHATPAAFDVQTTTANGIKFSLKAKQPVKD 51
   NC-VDAC   MAVPAFSDIAKSANDLLNKDFYHLAAGTIEVKSNTPNNVAFKVTGK-STHD 50

                                        ● ●          ●●●
Porin 31HL TKVTGSLETKYRWTEYGLTFTEKWNTDNTLGTEITVEDQLARGLKLTFDSSFS 103
   DM-VDAC GKVFGSLETKYKVKDYGLTLTEKWNTDNTLFTEVAVQDQLLEGLKLSLEGNFA 103
    P-VDAC -LFLADVNTQLKNKN--VTTDIKVDTNSNLFTTITV-DEAAPGLK--TILSF- 96
   SC-VDAC GPLSTNVEAKLNDKQTGLGLTQGWSNTNNLQTKLEFA-NLTPGLKNELITSLT 103
   NC-VDAC KVTSGALEGKFTDKPNGLTVTQTWNTANALETKVEMADNLAKGLKAEGIFSFL 103

                                          ●               ●
Porin 31HL PNTGKKNAKIKTGYKREHINLGCDMDFDIAGPSIRGALVLGYEGWLAGYQMNF 156
   DM-VDAC PQSGNKNGKFKVAYGHENVKADSDVNIDLKGPLINASAVLGYQGWLAGYQTAF 156
    P-VDAC RVPDQRSGKLEVQYLHDYAGICTSVGLTAN-PIVNFSGVVGTNIIALGTDVSF 148
   SC-VDAC P-GVAKSAVLNTTFTQPFFTARGAFDLCLKSPTFVGDLTMAHEGIVGGAEFGY 155
   NC-VDAC PATNARGAKFNLHFKQSNFHGRAFFDL-LKGPTANIDAIVGHEGFLAGASAGY 155

Porin 31HL ETAKSRVTQSNFAVGYKTDEFQL-HTNVNDGTEFGGSIYQKVNKKLETAVNLA 208
   DM-VDAC DTQQSKLTTNNFALGYTTKDFVL-HTAVNDGQEFSGSIFQRTSDKLDVGVQLS 208
    P-VDAC DTKTGDFTKCNAGLSFTNADLVASLNLNNKGDNLTASYYHTVSPLTSTAVGAE 201
   SC-VDAC DISAGSISRYAMALSYFAKDYSLGAT-LNNEQITTVDFFQNVNAFLQVGAKAT 207
   NC-VDAC DVQKAAITGYSAAVGYHAPTYSAAITATDNLSVFSASYYHKVNSQVEAGSKAT 208

                  ●             ●
Porin 31HL WT-AGNSNTR--FGIAAKYQIDPDACFSAKVNNSSLIGLGYTQTLKPGIKLTL 258
   DM-VDAC WA-SGTSNTK--FAIGAKYQLDDDASVRAKVNNASQVGLGYQQKLRDGVTLTL 258
    P-VDAC VNHSFSTNENI-ITVGTQHRLDPLTSVKARINNFGKASALLQHEWRPKSLFTV 253
   SC-VDAC MN-CKLPNSNVNIEFATRYLPDASSQVKAKVSDSGIVTLAYKQLLRPGVTLGV 259
   NC-VDAC WN-SKTGNT-VGLEVATKYRIDPVSFVKGKINDRGVAAIAYNVLLREGVTLGV 259

               ●         ● ●
Porin 31HL SALLDGKNVNAGGHKLGLGLEFQ                                281
   DM-VDAC STLVDGKNFNAGGHKIGVGLELE                                281
    P-VDAC SGEVDTKSVDK-GAKFGLALALK                                275
   SC-VDAC GSSFDALKLSEPVHKLGWSLSFD                                282
   NC-VDAC GASFDTQKLDQATHKVGTSFTFE                                282
```

Figure 13.9 Comparison of the amino acid sequences of human (HVDAC1, porin31HL) [78], *D. melanogaster* [74], potato [76], *S. cerevisiae* [21] and *N. crassa* [22] mitochondrial porins using BLAST. Three or more identical amino acids are shaded. Points indicate identical amino acids in all five sequences.

channel conductance of about 4 nS in 1 M KCl, that are slightly cation selective in the open state and highly cation-selective in the voltage-dependent closed state.

No identity exists between primary sequences of mitochondrial porins and those of bacterial porins, especially of phototrophic bacteria, which are related to the ancestor of mitochondria. On the other hand, the structure and function of bacterial porins as channels are clearly similar to those of the mitochondrial porins, since they both contain predominantly antiparallel β-barrel stave structures, which form hollow cylinders in the outer membrane and they both have similar molecular masses. Hydropathicity plots indicate that the primary sequence of mitochondrial porins is predominantly hydrophilic (46–50% polar residues). This clearly distinguishes them from the ion channels in nerve and muscle tissue with their typical α-helical structure, although the p-elements of potassium and similar ion channels contain an approximately 20-amino-acid stretch that lines the channel and is organized similar to a β-sheet structure [80]. Eukaryotic porins contain only an about 19-amino-acid stretch at the N-terminal end that may form an α-helical region. The function of the amphipathic α-helical structure is still a matter of debate, and it is suggested that it has to do with channel gating, protein targeting or it is even a part of the channel itself.

13.5.2
Secondary Structure of Mitochondrial Porins

The secondary structure of most mitochondrial porins was analyzed using computer-based prediction programs. The simplest possibility to identify amphipathic β-strands is the method proposed by Vogel and Jähnig [81]. It is based on the observation that every second amino acid in a sided β-strand is either on average hydrophobic or hydrophilic. This means that one side of a β-strand is hydrophilic and faces the channel interior; the other side is hydrophobic and faces the lipid side chains. In addition to this simple approach, more sophisticated methods are also possible, such as secondary structure predictions based on a neural network-based predictor [82], which is suited to find β-sheets along protein sequences and is trained to recognize the topography of known bacterial β-strands. An example for this treatment is given in Figure 13.10 for *N. crassa* porin [40]. Within its primary sequence, 16 amphipathic β-strands (indicated by shaded boxes) can be recognized by the method. All studies of porin structure agree that the eukaryotic porins contain a high degree of antiparallel β-sheets as the main structural feature [13, 21–23, 29, 30, 37, 40]. The results of the secondary structure predictions are supported by CD measurements of eukaryotic porins. Native and recombinant porin from *N. crassa* shows in these experiments a high degree of β-sheet structure [31, 61, 83]. In general, the recombinant porins seem to be a valuable model system for the study of eukaryotic porins by spectroscopic methods, in which high amounts of protein are needed. CD spectroscopy of root plastid porin was performed to determine the secondary structure of the protein under different conditions. It has a high degree of β-sheet structure in the non-ionic detergent Genapol X-80 and in lipid vesicles. Although the presence of sterols is essential for recon-

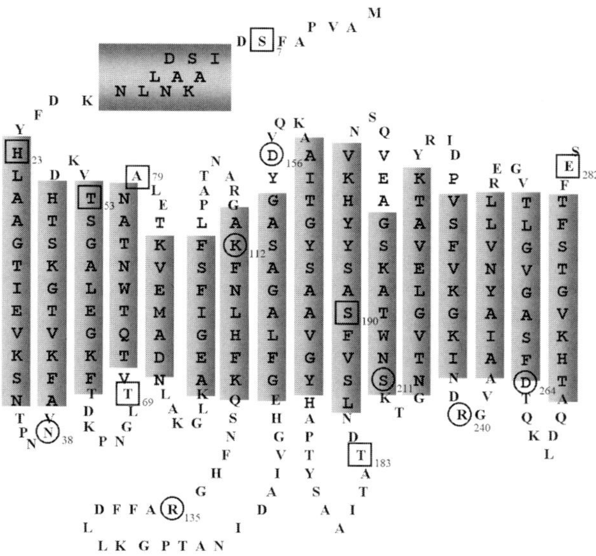

Figure 13.10 Model of the arrangement of *N. crassa* porin in the membrane according to [40]. The 16 putative β-strands are shaded in grey. Taken from [40] with permission.

stitution of eukaryotic porins in lipid bilayer membranes, the presence of sterols in the experiments with lipid vesicles does not change the CD spectra [31]. The more polar detergent SDS induces a large amount of α-helix structure in the protein. Similar results were obtained from experiments with two plant porin isoforms using attenuated total reflection Fourier transform infrared spectroscopy (ATR-FTIR) [77]. Here, the β-sheet content of the eukaryotic porins was about 50 %. Furthermore, the arrangement of the porins as detected by ATR-FTIR was very similar to that of OmpF of *E. coli*.

13.5.3
Structure of the Channel-forming Unit

Two different models of the channel-forming unit were derived from secondary structure predictions [23, 28–30, 37, 40, 83], site-directed mutagenesis [75] and deletion of part of the primary sequence of mitochondrial porins [60]. One model assumes that the channel is composed of either 12 (yeast [75]) or 13 (*N. crassa*[84]) β-strands and the amphipathic N-terminal α-helix as part of the channel. This means that in the yeast model C- and N-termini of porin are on different sides of the membrane. It is, however, more likely that both are on the same side as the surface of the mitochondrial outer membrane is similar to the situation in Gram-negative bacteria, where they are at the periplasmic side of the outer membrane (see Chapters 2 and 9).According to the model, the amphipathic N-terminal α-helix resides in the lumen wall as part of channel gating [28, 37, 75, 84]. Its displacement by voltage should result in some sort of structural rearrangement of the pore, which leads to partial channel closure. Our own experiments with ΔN mutants of *N. crassa* porin (ΔN2–12porin and ΔN3–20porin) suggest that the N-terminus itself seems to be important for overall channel stability, rather than being directly involved in gating. The gating characteristics of the ΔN channels were only slightly different from those of the wild-type channels, indicating that the structures responsible for the gating process are still intact in these mutants [60]. This means that the N-terminus probably does not reside in the lumen wall of the mitochondrial outer membrane channel.

The second model assumes that the channel is formed by 16 antiparallel membrane-spanning β-strands, similar to the arrangement for general diffusion pores in the bacterial outer membranes [29, 30, 40, 83, 85]. In this case, the amphipathic α-helix at the N-terminal end is localized within the surface of the membrane and exposed to the aqueous phase because it represents a major antigenic determinant of mitochondrial porins [16, 29, 67, 86]. These results suggest that the N-terminus is indeed localized in the surface of the membrane, most probably on the external surface of mitochondria; however, its possible role in channel gating remains open, but it may play a major role in channel stability (see below). Treatment of intact and broken bovine heart mitochondria with different proteases identified different cleavage sites of the pore-forming unit exposed to the aqueous phase and this was unchanged whether intact or broken mitochondria were exposed to the proteases [29]. The cleavage sites are probably localized within surface-exposed loops of the protein. The C-terminal end may also be exposed to the external surface, but it is not accessible to proteases because only a few amino acids may be localized outside the membrane [29, 30, 40]. The C-terminus of *N. crassa* porin is clearly organized in antiparallel β-strands [40, 60]. The mutant that lacks 15 amino acids at the C-terminal (ΔC269–283porin) showed the same channel-forming activity as wild-type porin. It had a smaller single-channel conductance than wild-type protein, indicating a smaller channel size. However, its other biophysical properties, in particular the voltage dependence, were identical to those of wild-type porin [60].

Figure 13.11 shows a model for *N. crassa* porin derived from neural network-based predictors [40, 82]. The result of the prediction was compared to bacterial β-sheet proteins and used to align the primary sequence in coordinates of a three-dimensional (3-D) model. It is very similar to that proposed for yeast porin using the same formalism and has 16 antiparallel β-strands similar to the general diffusion porins of Gram-negative bacteria [40, 85]. The model of *N. crassa* porin is shown in Figure 13.11 (A from the side and B from the top). N- and C-termini are localized on the surface of mitochondria. The β-strands of *N. crassa* porin are tilted by about 40° to 60° similar to those of general diffusion pores (see Chapter 2). The loops connecting two β-strands are short on the external surface of mitochondria and small at the opposite side. The hole is slightly oval and has a diameter of about 2.5–3 nm. The position of the amphipathic α-helical structure (not shown in Figure 13.11) presumably represents a major problem for the validity of the model because it should somehow be involved in the stability of the pore as experiments with N-terminal deletion porins suggest. A simple, hollow β-stranded cylinder is presumably not stable in a membrane because of the lateral pressure in a lipid bilayer. It may result in a variation of its size as was shown for FhuA when the cork inside the β-barrel cylinder was removed (see Chapter 11). All known β-stranded membrane channels are somehow stabilized in their structure. Loop 3 stabilizes the bacterial porin channels (see Chapters 2 and 9). The α-hemolysin channel of *Staphylococcus aureus* is stabilized by the huge mushroom-like protrusion on the *cis* side of the membrane [87] and the receptors are stabilized by the cork (see Chapters 11 and 12). A similar stabilizing element is also missing in both models for eukaryotic porins.

The model with 16 membrane-spanning β-strands appears to be more likely than that containing 13 (12) β-strands and the amphipathic N-terminus, although the latter model cannot be ruled out completely. The reasons for this can be summarized as follows:

- The porin channel is the primordial outer membrane channel and existed already when mitochondria evolved. Its overall structure may have been conserved, although there does not exist any homology between the primary structures of mitochondrial and bacterial porins. It has to be noted that homology between different bacterial porins can be very low, but the channel structure is well preserved.
- The amphipathic α-helix at the N-terminal end is the major antigenic determinant of eukaryotic porins, which means that it is probably not hidden inside the membrane.
- The results of experiments in which channel properties are changed by site-directed mutagenesis or by chemical modifications have to be taken with some care since the change of a single amino acid may induce a change that is related to a completely different region of the channel.

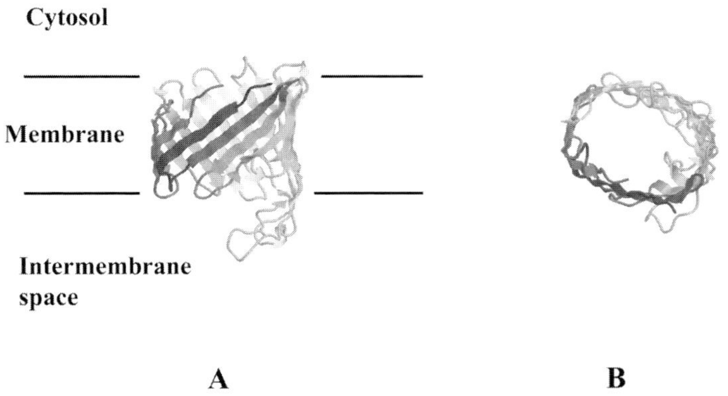

Figure 13.11 Schematic view of the 3-D structure of the *N. crassa* porin. Note that the location of the N-terminus is not given. The coordinates were kindly provided by Rita Casadio and Vito De Pinto. (A) Side view. (B) View from the mitochondrial surface-exposed side.

13.5.4
Are Sterols Involved in the Formation of the Channel-forming Unit?

Mitochondrial inner and outer membranes have different lipid compositions. The inner membrane contains cardiolipin, whereas sterols of different kind are exclusively found in the outer membrane dependent on the organism. Purified *N. crassa* porin contains the sterol ergosterol [9]. Water-soluble mitochondrial porin from *N. crassa* that lost its associated lipids and sterols needs sterols for functional reconstitution into lipid bilayer membranes [26, 59, 88]. The detergent Triton X-100 plays a critical chaperone-like role in this process, similar to as has been found for the folding of bacterial porins [89]. The role of different sterols was studied in functional reconstitution of water-soluble porins from *Dictyostelium discoideum, Paramecium,* plant and rat liver [59, 88]. The water-soluble porins regain their channel-forming ability after pre-incubation of the polypeptides with sterols in the presence of detergents. Some of the sterols can replace one another for channel formation, but influence channel gating [59]. Interestingly, a sterol content of five cholesterol molecules per polypeptide unit was calculated in pig heart porin purified using the detergent LDAO [27]. Sterols are probably not tightly bound to the polypeptide. The removal of sterols from the channel-forming complex may be responsible for the substantial loss of channel-forming activity during purification of mitochondrial porins.

13.5.5
Electron Microscopic Analysis of *N. crassa* Porin

Three-dimensional crystals of mitochondrial porin have not been obtained to date, which means that its tertiary folding structure is still not yet clear. However, it was possible to study 2-D crystals of fungal [32–38] and recombinant human porin [39] using electron microscopic analysis. Porin from *N. crassa* mitochondria crystallizes in periodic polymorphs with well-defined lattice parameters when outer membranes are subjected to phospholipase A2 treatment [34]. According to the Fourier-filtered electron microscopic images of the crystalline arrays, the mitochondrial pore is a cylinder normal to the membrane plane with an outer diameter of about 3.8 nm for the hydrocarbon cylinder (the diameter of the polypeptide backbone is about 3.8 nm) and an inner diameter of about 2.5 nm [35, 36]. The crystalline arrays are composed of subunits that contain six porin channels. The lattice geometry responds to a number of external compounds. In particular, polyanion binds to the porin and leads to a substantial contraction of the lattice [35]. Besides polyanion, apocytochrome *c* [36] and the targeting sequence of an imported cytochrome oxidase subunit [90] seem to bind to the periphery of the porin channel. More recently, 2-D crystals of *N. crassa* porin were analyzed using single-particle approaches [38]. The results look promising, but it is not known yet if these approaches will lead to an improved resolution of porin structure. Human porin expressed in *E. coli* and reconstituted using detergent dialysis also forms 2-D crystals [39]. The data derived from electron microscopic analysis of these crystals agree basically with those derived from 2-D crystals of *N. crassa* porin, but two neighboring molecules have opposite directions, which means that the 2-D crystals of human porin are not suitable for high-resolution imaging.

13.6 Conclusions

Eukaryotic porins form large water-filled channels in the "open" state with diameters around 2.5–3 nm that are slightly anion selective. The channels appear to be voltage gated starting at about 20–30 mV and switch to ion-permeable closed states that are highly cation selective. Eukaryotic porins represent the major permeability pathway of mitochondrial outer membranes, although porin-deficient mitochondria also exhibit oxidative phosphorylation [91–93]. In addition to mitochondrial localization, there exists emerging evidence that other cellular compartments also contain eukaryotic porins with a yet unknown function. The first indication was the extramitochondrial location of porin 31HL, which is localized in the cytoplasmic membrane of a transformed human B cell line [16]. A similar channel is localized in astrocytic plasma membranes [17]. Later, more evidence was found for such location because a 36-kDa porin-like protein from rat brain co-purified with the γ-aminobutyric acid/benzodiazepine receptor protein [18]. It is possible that the channel from the astrocytic plasma membrane and the 36-kDa protein from rat brain relate to the "maxi" chloride channels present in the cell membrane of a number of eukaryotic cells. In excised patches, the channel has a maximum open probability at zero voltage, similar to the situation in lipid bilayer experiments.

The electrophysiological characterization of mitochondrial porin suggests that the permeability properties of the mitochondrial outer membrane could be voltage controlled. The generation of a voltage across the mitochondrial outer membrane is still open and it is not clear if the close apposition of inner and outer membranes *in situ* mitochondria is sufficient for electric coupling of both membranes and for channel gating. Furthermore the exact mechanism of channel gating and the residues responsible for selectivity in the different states are completely open. In general, there exists some knowledge about the secondary structure of the eukaryotic porins. However, the precise 3-D structure, including the exact number of membrane-spanning β-strands and the diameter, are still unknown. Crystallographic data are badly needed to give clear answers to these questions and possibly also the localization of sterols in the channel-forming unit. Another open problem is the assembly of the porin channel. It is unlikely that it (i. e. the β-strands) forms inside a membrane since eukaryotic porins are polar proteins and are synthesized as water-soluble precursors outside membranes. This means that the channel is formed chaperone-promoted on the surface of the mitochondria at a receptor site (Tom 20) from polypeptide and sterols [25]. After formation the channel inserts into the membrane using the general import pathway (TOM). So far it is not clear if it requires special insertion sites in the mitochondrial outer membrane or if the insertion process is similar to that in lipid bilayer membranes. In any case, it is likely that the channel is preformed before insertion. The pre-formation of channels in the aqueous phase or on the membrane surface is probably also the reason why lipid bilayer membranes represent a powerful tool for the study of mitochondrial and bacterial porins.

Acknowledgments

The generous supply of Figures 13.10 and 13.11 by Rita Casadio and Vito De Pinto is gratefully acknowledged. I would like to thank Dieter Brdiczka, Vito De Pinto and Friedrich P. Thinnes for the fruitful and excellent collaboration during many years, and my collaborators Otto Ludwig, Birgit Popp, Elke Maier and Angela Schmid for their excellent work with outer membrane porins on which this review is largely based. My own research was supported by the Deutsche Forschungsgemeinschaft (Project Be865/10) and by the Fonds der Chemischen Industrie.

References

1 F. L. Hoch, *Biochim Biophys Acta* **1992**, *1113*, 71.
2 L. Palmieri, M. J. Runswick, G. Fiermonte, J. E. Walker, F. Palmieri, *J Bioenerg Biomembr* **2000**, *32*, 67.
3 R. L. O'Brien, G. Brierly, *J Biol Chem* **1965**, *240*, 4527.
4 E. Pfaff, M. Klingenberg, E. Ritt, W. Vogell, *Eur J Biochem* **1968**, *5*, 222.
5 S. J. Schein, M. Colombini, A. Finkelstein, *J Membr Biol* **1976**, *30*, 99.
6 M. Colombini, *Nature* **1979**, *279*, 643.
7 L. S. Zalman, H. Nikaido, Y. Kagawa, *J Biol Chem* **1980**, *225*, 1771.
8 N. Roos, R. Benz, D. Brdiczka, *Biochim Biophys Acta* **1982**, *686*, 204.
9 H. Freitag, W. Neupert, R. Benz, *Eur J Biochem* **1982**, *123*, 629.
10 O. Ludwig, V. De Pinto, F. Palmieri, R. Benz, *Biochim Biophys Acta* **1986**, *860*, 268.
11 V. De Pinto, O. Ludwig, J. Krause, R. Benz, F. Palmieri, *Biochim Biophys Acta* **1985**, *894*, 109.
12 O. Ludwig, J. Krause, R. Hay, R Benz, *Eur Biophys J* **1988**, *15*, 269.
13 H. Troll, D. Malchow, A. Müller-Taubenberger, B. Humbel, F. Lottspeich, G. Gerisch, R. Benz, *J Biol Chem* **1992**, *267*, 21072.
14 D. Smack, M. Colombini, *Plant Physiol* **1985**, *79*, 1094.
15 A. Schmid, S. Krömer, H. W. Heldt, R. Benz, *Biochim Biophys Acta* **1992**, *1112*, 174.
16 F. P. Thinnes, H. Götz, H. Kayser, R. Benz, W. E. Schmidt, H. D. Kratzin, N. Hilschmann, *Biol Chem Hoppe-Seyler* **1989**, *370*, 1253.
17 R. Dermietzel, H. Theng-Khing, R. Buettner, A. Hofer, E. Dotzler, M. Kremer, R. Deutzmann, F. P. Thinnes, G. I. Fishman, D. C. Spray, D. Siemen, *Proc Natl Acad Sci USA* **1994**, *91*, 499.
18 M. H. Bureau, M. Khrestchatisky, M. A. Heeren, E. B. Zambrowicz, H. Kim, T. M. Grisar, M. Colombini, A. J. Tobin, R. W. Olsen, *J Biol Chem* **1992**, *267*, 8679.
19 J. A. Aljamal, G. Genchi, V. De Pinto, L. Stefanazzi, R. Benz, F. Palmieri, *Plant Physiol* **1993**, *102*, 615.

20 A. Blumenthal, K. Kahn, O. Beja, E. Galun, M. Colombini, A. Breiman, *Plant Physiol* **1993**, *101*, 579.

21 K. Mihara, R. Sato, *EMBO J* **1985**, *4*, 769.

22 R. Kleene, N. Pfanner, R. Pfaller, T. A. Link, W. Sebald, W. Neupert, M. Tropschug, *EMBO J* **1987**, *6*, 2627.

23 M. Forte, H. R. Guy, C. A. Mannella, *J Bioenerg Biomembr* **1987**, *19*, 341.

24 H. Freitag, M. Janes, W. Neupert, *Eur J Biochem* **1982** *126*, 197.

25 T. Krimmer, D. Rapaport, M. T. Ryan, C. Meisinger, C. K. Kassenbrock, E. Blachly-Dyson, M. G. Douglas, W. Neupert, F. E. Nargang, N. Pfanner, *J Cell Biol* **2001**, *152*, 289.

26 R. Pfaller, H. Freitag, M. A. Harmey, R. Benz, W. Neupert, *J Biol Chem* **1985**, *260*, 8188.

27 V. De Pinto, R. Benz, F. Palmieri, *Eur J Biochem* **1989**, *183*, 179.

28 S. Peng, E. Blachly-Dyson, M. Colombini, M. Forte, *J Bioenerg Biomembr* **1992**, *24*, 27.

29 V. De Pinto, G. Prezioso, F. P. Thinnes, T. A. Link, F. Palmieri, *Biochemistry* **1991**, *30*, 10191.

30 R. Benz, *Biochim Biophys Acta* **1994**, *1197*, 167.

31 B. Popp, S. Gebauer, K. Fischer, U. I. Flügge, R. Benz, *Biochemistry* **1997**, *36*, 2844.

32 C. A. Mannella, A. Ribeiro, J. Frank, *Biophys J* **1986**, *49*, 307.

33 C. A. Mannella, X. W. Guo, B. Cognon, *FEBS Lett* **1989**, *253*, 231.

34 C. A. Mannella, *Science* **1984**, *224*, 165.

35 C. A. Mannella, X. W. Guo, *Biophys J* **1990**, *57*, 23.

36 C. A. Mannella, A. Ribeiro, J. Frank, *Biophys J* **1987**, *51*, 221.

37 C. A. Mannella, J. *J Struct Biol* **1998**, *121*, 207.

38 A. Verschoor, W. F. Tivo, C. A. Mannella, *J Struct Biol* **2001**, *133*, 254.

39 M. Dolder, K. Zeth, P. Tittmann, H. Gross, W. Welte, T. Wallimann, *J Struct Biol* **1999**, *127*, 64.

40 R. Casadio, I. Jacoboni, A. Messina, V. De Pinto, *FEBS Lett* **2002**, *520*, 1.

41 R. Benz, L. Wojtczak, W. Bosch, D. Brdiczka, *FEBS Lett* **1988**, *231*, 75.

42 R. Benz, M. Kottke, D. Brdiczka, *Biochim Biophys Acta* **1990**, *1022*, 313.

43 O. Ludwig, R. Benz, J. E. Schultz, *Biochim Biophys Acta* **1989**, *978*, 319.

44 T. König, B. Kocsis, L. Mészáros, K. Nahm, S. Zoltán, I. Horváth, *Biochim Biophys Acta* **1977**, *462*, 380.

45 M. Colombini, M. J. Holden, P. Mangan, In: *Ion Carriers of Mitochondrial Membranes*, A. Azzi, K. A. Nalecz, M. J. Nalecz, L. Wojtczak (Eds). Springer, New York, **1989**, pp. 215–224.

46 M. Colombini, C. L. Yeung, J. Tung, T. König, *Biochim Biophys Acta* **1987**, *905*, 279.

47 C. Fiek, R. Benz, N. Roos, D. Brdiczka, *Biochim Biophys Acta* **1982**, *688*, 429.

48 A. K. Östlund, U. Göhring, J. Krause, D. Brdiczka, *Biochem Med* **1983**, *30*, 231.

49 J. Krause, R. Hay, C. Kowollik, D. Brdiczka, *Biochim Biophys Acta* **1986**, *860*, 690.

50 D. Brdiczka, *Biochim Biophys Acta* **1991**, *1071*, 291.

51 E. H. Cheng, T. V. Sheiko, J. K. Fisher, W. J. Craigen, S. J. Korsmeyer, *Science* **2003**, *301*, 513.

52 A. Godbole, J. Varghese, A. Sarin, M. K. Mathew, *Biochim Biophys Acta* **2003**, *1642*, 87.

53 Y. Tsujimoto, S. Shimizu, *Biochimie* **2002**, *84*, 187.

54 V. Le Mellay, J. Troppmair, R. Benz, U. R. Rapp, *BMC Cell Biol* **2002**, 3, 14.

55 P. L. Pedersen, J. W. Greenawalt, B. Reynafaeje, J. Hullihen, G. L. Decker, J. W. Soper, E. Bustamente, *Methods Cell Biol* **1978**, *20*, 411.

56 R. Hovius, K. Lambrechts, K. Nicolay, B. de Kruiff, *Biochim Biophys Acta* **1990**, *1021*, 217.

57 K. K. Arora, D. M. Parry, P. L. Pedersen, *J Bioenerg Biomembr* **1992**, *24*, 47.

58 G. L. Sottocasa, B. Kuylenstierna, L. Ernster, A. Bergstrand, *J Cell Biol* **1967**, *32*, 415.

59 B. Popp, A. Schmid, R. Benz, *Biochemistry* **1995**, *34*, 3352.

60 B. Popp, D. A. Court, R. Benz, W. Neupert, R. Lill, *J Biol Chem* **1996**, *271*, 13593.

61 D. A. Koppel, K. W. Kinnally, P. Masters, M. Forte, E. Blachly-Dyson, C. A.

Mannella, *J Biol Chem* **1998**, *273*, 13794.
62 M. Lindén, M. Gellerfors, B. D. Nelson, *Biochem J* **1982**, *208*, 77.
63 M. Colombini, *J Membr Biol* **1980**, *53*, 79.
64 H. Nikaido, *Methods Enzymol* **1983**, *97*, 85.
65 E. M. Renkin, *J Gen Physiol* **1954**, *38*, 225.
66 H. Nikaido, E. Y. Rosenberg, *J Gen Physiol* **1981**, *77*, 121.
67 R. Benz, E. Maier, F. P. Thinnes, H. Götz, N. Hilschmann, *Biol Chem Hoppe-Seyler* **1992**, *373*, 295.
68 G. Ehrenstein, H. Lecar, R. Nossal, *J Gen Physiol* **1970**, *55*, 119.
69 R. Benz, K. Janko, P. Läuger, *Biochim Biophys Acta* **1979**, *551*, 238.
70 G. W. Castellan, In: *Physical Chemistry*. Addison-Wesley, Reading, MA, **1983**, pp. 769–784.
71 T. König, I. Stipani, I. Horváth, F. Palmieri. *J Bioenerg Biomembr* **1982**, *14*, 297.
72 V. De Pinto, R. Benz, C. Caggese, F. Palmieri, *Biochim Biophys Acta* **1989**, *987*, 1.
73 M. Liu, M. Colombini, *J Bioenerg Biomembr* **1992**, *24*, 41.
74 A. Messina, M. Neri, F. Perosa, C. Caggese, M. Marino, R. Caizzi, V. De Pinto, *FEBS Lett* **1996**, *384*, 9.
75 E. Blachly-Dyson, S. Peng, M. Colombini, M. Forte, *Science* **1990**, *247*, 1233.
76 L. Heins, H. Mentzel, A. Schmid, R. Benz, U. K. Schmitz, *J Biol Chem* **1994**, *269*, 26402.
77 H. Albrecht, E. Goormaghtigh, J.-M. Ruysshaert, F. Homblé, *J Biol Chem* 2000, *275*, 40992.
78 H. Kayser, H. D. Kratzin, F. P. Thinnes, H. Götz, W. E. Schmidt, K. Eckart, N. Hilschmann, *Biol Chem Hoppe-Seyler* **1989**, *370*, 1265.
79 G. Runke, E. Maier, J. D. O'Neil, R. Benz, D. A. Court, *J Bioenerg Biomembr* **2000** *32*, 563.
80 R. MacKinnon, *FEBS Lett* **2003**, *27*, 62.
81 H. Vogel, F. Jähnig, *J Mol Biol* **1986**, *190*, 191.
82 L. Shao, K. W. Kinnally, C. A. Mannella, *Biophys J* **1996**, *71*, 778.
82 I. Jacoboni, P. L. Martelli, P. Fariselli, V. De Pinto, R. Casadio, *Protein Sci* **2001**, *10*, 779.
83 D. C. Bay, D. A. Court, *Biochem Cell Biol* **2002**, *80*, 551.
84 J. Song, C. Midson, E. Blachly-Dyson, M. Forte, M. Colombini, *J Biol Chem* **1998**, *273*, 24406.
85 S. W. Cowan, T. Schirmer, G. Rummel, M. Steiert, R. Gosh, R. A. Pauptit, J. N. Jansonius, J. P. Rosenbusch, *Nature* **1992**, *356*, 727.
86 D. Babel, G. Walter, H. Götz, F. P. Thinnes, L. Jürgens, U. König, N. Hilschmann, *Biol Chem Hoppe-Seyler* **1991**, *372*, 1027.
87 L. Song, M. R. Hobaugh, C. Shustak, S. Cheley, H. Bayley, J. E. Gouaux, *Science* **1996**, *274*, 1859.
88 F. Carbonara, B. Popp, A. Schmid, V. Iacobazzi, G. Genchi, F. Palmieri, R. Benz, *J Bioenerg Biomembr* **1996**, *28*, 181.
89 K. Sen, H. Nikaido, *Proc Natl Acad Sci USA* **1990**, *87*, 743.
90 C. A. Mannella, X. W. Guo, J. Dias, *J Bioenerg Biomembr* **1992**, 24, 55.
91 M. Dihanich, K. Suda, G. Schatz, *EMBO J* **1987**, *6*, 723.
92 J. Michejda, X. J. Guo, G. J.-M. Lauquin, *Biochem Biophys Res Commun* **1990**, *171*, 354.
93 R. Benz, A. Schmid, M. Dihanich, *J Bioenerg Biomembr* **1989**, *21*, 439.

14
Mitochondrial Porins in Mammals: Insights into Functional Roles from Mutant Mice and Cells

Keltoum Anflous and William J. Craigen

14.1
Introduction

Voltage-dependent anion channels (VDACs) are pore-forming proteins found in the mitochondrial outer membrane of all eukaryotes [1, 2]. VDACs have been commonly referred to as "mitochondrial porins" by analogy to the porins of the outer membrane of Gram-negative bacteria [3]. VDAC is a voltage-gated protein and, *in vitro*, the membrane potential is the primary factor in determining the different conductance states. Secondary and tertiary structure analyses predict that the VDAC channel adopts a conserved β-barrel conformation formed by 16 β-strands reminiscent of bacterial porins [4–7]. The N-terminal globular α-helix and putative flexible loop regions are predicted to face the cytoplasmic compartment. The existence of large anion-selective channels fits well with a major function of mitochondria – energy transduction. Although VDACs likely play a central role in the regulated flux of metabolites across the mitochondrial outer membrane, their exact cellular roles are not well understood.

The existence of three related, but distinct, isoforms suggests that each have distinct functions. Although VDACs are structurally and electrophysiologically highly conserved across species, the specific function of each isoform remains unknown. It has been reported that VDACs associate with the adenine nucleotide translocator at the contact sites between the mitochondrial inner and outer membranes [8]. VDACs may serve as a binding site for cytosolic hexokinase, providing the enzyme with preferential access to ATP derived from oxidative phosphorylation [9]. VDACs are also believed to participate in the phosphocreatine circuit that links sites of energy production to sites of energy consumption, leading to the compartmentalization of high-energy metabolites [10]. Although it has been reported that VDACs interact with both anti-apoptotic and pro-apoptotic Bcl-2 family members, whether it is the opening or the closure of VDACs that contributes to the onset of apoptosis (e. g. mitochondrial swelling and cytochrome *c* release) is still a matter of considerable debate [11, 12].

Bacterial and Eukaryotic Porins. Edited by Roland Benz

ISBN 3-527-30775-3

In higher eukaryotes, three VDAC isoforms have been characterized, and designated VDAC1, VDAC2 and VDAC3. Prior studies on VDACs, either *in vitro* or in reconstituted systems, typically do not take into account the multiple interactions of VDACs with different cellular components (e. g. cytosolic enzymes, cytoskeleton). Furthermore, these studies often do not distinguish between VDAC isoforms. The present chapter summarizes recent reports that investigate VDAC-deficient mice and cell lines in order to shed light on the physiological function of mouse VDAC1, VDAC2 and VDAC3 in the context of the cell and the organism. Unresolved questions remain concerning whether there are VDACs in different cellular compartments, the nature of isoform-specific transport and any developmental or pathophysiological roles for VDACs.

14.2
Channel Activity

When VDAC proteins are inserted into planar phospholipid membranes, incremental increases in membrane conductance, clustered at about 4.5 nS, can be recorded. VDACs are slightly anion selective at low voltage and switch to substates characterized by lower conductance at a voltage above 20 mV [1, 3, 31]. The substates of VDACs are cation selective [32]. The characterization of these different substates has been studied by using polyanionic compounds, both in intact mitochondria and *in vitro* lipid bilayers [33]. The structural changes by which VDAC goes from a state of high permeability to one of low permeability are large. The pore diameter that accompanies the "closed" and "open" states is estimated to change from 0.8–0.9 to 1.2–1.5 nm. This is consistent with significant changes in VDAC permeability to the metabolites that cross the mitochondrial outer membrane, e. g. pyruvate, succinate, $MgADP^{-}$ and $MgATP^{2-}$. The induction of impermeable substates in intact mitochondria would be predicted to lead to a complete inhibition of intermembrane kinases such as creatine kinase and adenylate kinase, which due to their location are excluded from the external ATP pool. It has been proposed that the N-terminus of VDAC regulates the conductance state of the protein [34]. Upon expression of mouse VDAC isoforms in yeast cells missing the major yeast VDAC gene (*por1*), isolated mitochondria exhibit large differences in their outer membrane permeability to NADH, depending on which mouse VDAC isoform is present [35]. Electrophysiological studies of purified proteins inserted into artificial bilayers also reveal differences in channel properties amongst the mouse isoforms. While VDAC1 displays the prototypic characteristics of VDAC activity that are highly conserved amongst species, VDAC2 shows two possible conductance states, whereas VDAC3 does not gate well even at high membrane potential. These results indicate that the different mammalian isoforms may provide different cellular functions.

14.3
The Genetics of Mammalian VDACs

The genes encoding VDACs have been cloned and sequenced from a number of species, including human [13, 14], mouse [15, 16], rat [17, 18], bovine [19], *Drosophila melanogaster* [20], *Saccharomyces cerevisiae* [21], wheat [22] and rice [23]. In mammals, three VDAC genes have been isolated and characterized: VDAC1, VDAC2 and VDAC3. In humans and mice the three isoforms are each encoded by separate autosomal genes (Figure 14.1). There are also numerous intron-less pseudogenes for each isoform, one of which in the mouse arose recently enough to still contain an open reading frame. The three VDAC isoforms belong to a single gene family that arose by gene duplication and divergence [24]. Phylogenetic analysis indicates that VDAC1 is the more primordial of the vertebrate VDAC genes, suggesting that as the multiple isoforms arose from gene duplication events. VDAC3 diverged first from the primordial VDAC, with VDAC2 arising more recently. An additional VDAC3 isoform is synthesized via alternative splicing of a 3-nucleotide exon (ATG) [18, 25]. The exon introduces a methionine 39 amino acids downstream of the N-terminus of the polypeptide (Figure 14.2). Expression of this alternative form appears limited to brain, heart and skeletal muscle. The functional significance of this alternative splicing remains unknown, but its presence in multiple species presumably reflects some highly conserved function. Both mouse VDAC3 isoforms are able to partially rescue the temperature-sensitive

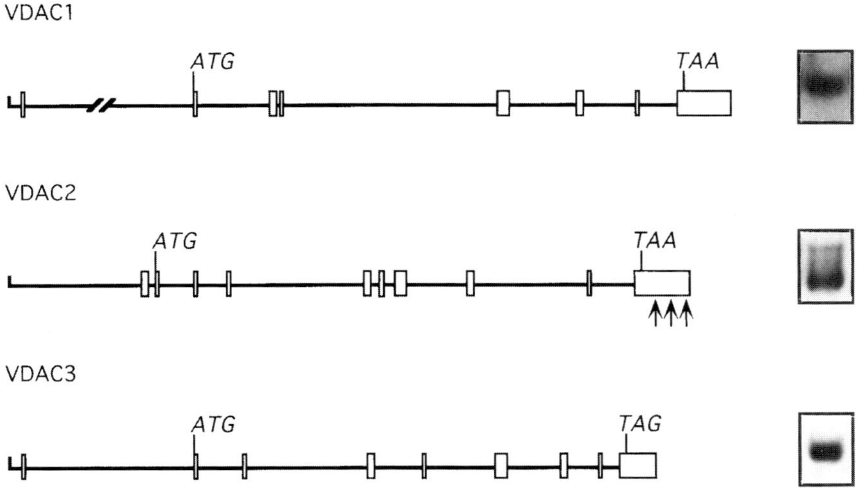

Figure 14.1 Diagram of the genomic organization of the three mouse VDAC genes (drawn to scale). Both VDAC1 and VDAC3 are encoded in nine exons, while VDAC2 is encoded in 10 exons. The relative sizes and locations of the different exons are indicated by open boxes. The translation start and stop sites for each gene are indicated. A Northern blot of each gene using ES cell RNA is shown. Note the presence of multiple VDAC2 transcripts in comparison to VDAC1 and VDAC3 (from [25]).

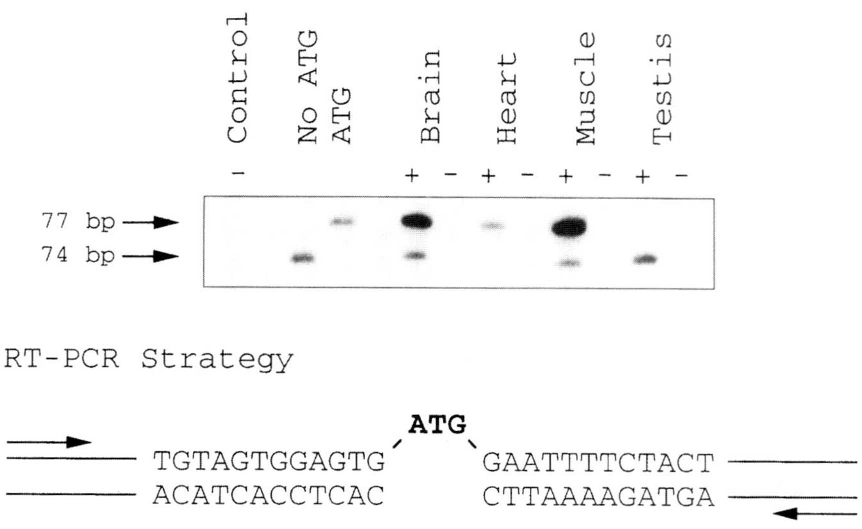

Figure 14.2 VDAC3 mRNA contains a 3-bp insertion. Oligonucleotides were designed to flank the site of the ATG codon insertion and tissues were examined for expression of the alternative isoform. The alternate VDAC3 transcript with the ATG insertion is only found in brain, heart and muscle. The control lane contains no DNA, while the second and third lanes are cDNAs in plasmids that lack or contain the extra ATG. Each PCR was carried out with (+) or without (–) reverse transcriptase (from [24]).

phenotype of VDAC-deficient (*ΔPor1*) yeast, indicating that in yeast the proteins appear to act similarly. However, the biochemical basis for this complementation remains unknown. Interestingly, deletion of the N-terminal 39 residues from mouse VDAC3 results in a truncated VDAC3 that can, in contrast to the full-length protein, fully rescue the mutant yeast strain. This suggests that the N-terminal region of VDAC3 has some regulatory property.

The human and mouse VDAC amino acid sequences share around 25 % similarity with *Saccharomyces* and *Neurospora* VDAC sequences. However, predictions of the secondary structures demonstrate a greater degree of conservation, with the hydrophobicity plots of each VDAC being almost super-imposable. Using radiation hybrid analysis, the human genes encoding VDAC1 and VDAC2 have been assigned to chromosomes 5q31 and 10q22, respectively [26]. The mouse genes encoding VADC1, VDAC2 and VDAC3 were mapped using an interspecies backcross panel to the proximal regions of mouse chromosomes 11, 14 and 8, respectively [15, 16], and maintain the same syntenic relationships to adjacent genes as the human genes. Each gene is composed of 9 or 10 exons contained within less than 15 kb of DNA (Figure 14.1). The 5′-flanking region of each VDAC isoform contains multiple putative transcription factor binding sites [24]. Commonly, genes expressed in a constitutive fashion are transcribed from promoters that contain at least one Sp1 binding site. Mouse VDAC1 contains two Sp1 sites, VDAC2 contains 10 Sp1 sites and VDAC3 five Sp1 sites. A sterol repressor element 1 binding site is also located in the promoter region of VDAC1 and VDAC2, which suggests a role for sterols in the regulation of VDAC1 and VDAC2 expression; however, there is no direct evidence for this currently. Interestingly, the VDAC2 gene contains three polyadenylation sites in the 3′ region. Each site is used to varying degrees, with the most proximal site being the most common site of transcription terminal. The role of these three distinct transcripts is unknown, but may reflect a post-transcriptional mechanism for regulating protein levels.

14.4 Generation of mammalian VDAC mutants

Embryonic stem (ES) cell lines lacking each VDAC isoform were generated by homologous recombination [27]. The strategy used to target the VDAC1 gene was designed to delete exons 2–5. Gene targeting of the VDAC2 gene was designed to delete the promoter and first two exons, including the predicted start codon. Gene targeting of the VDAC3 gene deleted the last four of the nine exons of the VDAC3 gene. All three mutations create a null allele. In each cell line the remaining wild-type allele was subsequently disrupted to generate VDAC-deficient cell lines. Based upon the significant loss of cytochrome *c* oxidase (COX) activity in yeast lacking *por1*, the cell lines were examined for loss of respiration and any reduction in respiratory chain activity. ES cells lacking each isoform are viable, but exhibit around 30% reduction in oxygen consumption whether coupled or not. VDAC1- and VDAC2-deficient cells exhibit a partial reduction in COX activity, whereas VDAC3-deficient cells have normal COX activity. These results indicate that VDACs are not essential for cell viability and it was speculated that reduced respiration in part reflects decreased outer membrane permeability for small metabolites necessary for oxidative phosphorylation.

VDAC1- and VDAC3-deficient ES cells have been used to generate mouse chimeras and germline-transmitting mice [28–30]. VDAC1 mutant mice are born at a reduced frequency, depending on the mouse strain background. For example, VDAC1 deficiency when bred onto the C57Bl6 strain is almost completely lethal, whereas on a CD1 background mutant mice appear at almost expected frequencies. Thus, this lethal trait is subject to strain-specific modifier gene effects and offers the opportunity to identify those genes that interact with VDAC1. Double-mutant mice (VDAC1/3) have also been obtained by intercrossing VDAC1 and VDAC3 heterozygous mice. Although double-mutant mice also exhibit a partial *in utero* lethality and are growth retarded, the mice are viable and, therefore, offer the possibility of studying the role of VDACs in cellular metabolism in intact organisms. Using double-mutant mice allows for the investigation of a single channel function in the absence of other isoforms.

14.5
The Role of the Mitochondrial Outer Membrane in Compartmentalization of High-energy Metabolites

The exchange of metabolites between the matrix and the cytosol is a potential site of regulation of mitochondrial function that involves transport across both mitochondrial membranes. Under a variety of conditions, the mitochondrial outer membrane has been shown to limit the rate of metabolite flux. In parallel, compounds that close VDACs greatly inhibit mitochondrial function by restricting the flux of adenine nucleotides into the mitochondrial intermembrane space [36–40]. It has been reported that pyridine dinucleotides decrease the permeability of the mitochondrial outer membrane [40–41]. By using intact mitochondria from potato tubers, it was demonstrated that β-NADH reduced the permeability of the outer membrane by a factor of 6 in a concentration-dependent manner. Although less pronounced, NADPH exerts the same effect on the outer membrane. In a reconstituted system, both NADH and NADPH induce the closure of VDAC channels. The authors commented that the results obtained *in vitro* may not reflect the situation *in vivo*. However, these observations suggest a new pathway for cross-talk between mitochondrial and cytoplasmic energy production and may explain an aspect of the Crabtree effect (i. e. inhibition of respiration after the addition of glucose or other hexoses that are capable of being phosphorylated by hexokinase). In two separate reports, Rostovtseva and Colombini demonstrated that VDAC channels isolated from mitochondria of *N. crassa* and reconstituted into planar phospholipid membranes mediate and gate the flow of ATP in their open state [42, 43]. Channel closure results in 50% reduction in ion conductance and ATP flux is almost completely blocked. Hence, VDACs can potentially regulate the movement of ATP between the cytosol and the mitochondrial spaces, and therefore may participate in controlling ATP flux through the mitochondrial outer membrane.

Evidence that the regulation of mitochondrial outer membrane permeability may play a role in coupled respiration and cell survival has been shown by the defect in ATP/ADP exchange across the mitochondrial outer membrane upon induction of apoptosis following growth factor withdrawal. This was associated with the accumulation of creatine phosphate in the intermembrane space, and subsequent loss of outer membrane integrity, cytochrome *c* release and apoptosis [44]. This reduction in the outer membrane permeability correlates with the changes in conductance properties that accompany closure of VDACs. In a subsequent report, Vander-Heiden et al. also demonstrated that the anti-apoptotic protein Bcl-x_L is able to restore metabolite exchange across the outer membrane without inducing the loss of cytochrome *c* from the intermembrane space, presumably by maintaining VDACs in an open configuration [11]. In a separate report, Le Mellay et al. have shown that the oncogenic protein c-Raf forms a complex with VDAC *in vivo* and blocks reconstitution of VDAC channels in planar bilayer membranes *in vitro* [45]. The authors proposed that this interaction might be responsible for the Raf-induced inhibition of cytochrome *c* release from mitochondria upon growth factor withdrawal. They further speculated that c-Raf kinase-induced VDAC inhibition might regulate the metabolic function of mitochondria and mediate the switch to aerobic glycolysis that is common to cancer cells. It was also proposed that the action of c-Raf and Bcl-2 in blocking cytochrome *c* release upon binding to VDAC might be mediated by two distinct mechanisms – pore reduction versus pore closure.

VDACs have been identified as the mitochondrial outer membrane binding sites for cytosolic kinases, e. g. glycerol kinase and hexokinase [46, 47]. It is believed that this interaction facilitates access of kinases to ATP and overcomes the restriction that the mitochondrial outer membrane exerts on the permeability for the small metabolites [48]. The mitochondrial isoforms of creatine kinase (mi-CK) and adenylate kinase, located in the intermembrane space, have been also reported to interact with VDACs [49, 50]. There is some evidence to suggest that this interaction occurs at the contact sites between the inner and outer mitochondrial membranes [49, 51]. The functional coupling of mi-CK and VDACs to mitochondrial respiration has been extensively characterized by using the technique of selectively permeabilized skinned fibers prepared from heart and skeletal muscle [52–55]. Studies using this technique, unlike studies using isolated mitochondria, preserve the *in vivo* structural relationships between mitochondria and the cytoskeleton, and examine the total population of mitochondria. The functional coupling of mi-CK to mitochondrial respiration can be determined in the presence of creatine by measuring the mitochondrial sensitivity for $MgADP^-$, i. e. the apparent K_m for $MgADP^-$ (K_m[ADP]). The K_m[ADP] is high in skinned fibers in comparison to isolated mitochondria, presumably reflecting the low *in vivo* permeability of the mitochondrial outer membrane for $MgADP^-$. In the presence of creatine, mi-CK generates phosphocreatine and ADP from newly synthesized ATP generated in the matrix. In this scenario, creatine enters mitochondria through VDAC, phosphocreatine is the high-energy compound that exits mitochondria through VDAC, while ADP is recycled back to the matrix via the adenine nucleotide translocase (ANT) to generate more ATP, leading to a shift in K_m[ADP] toward lower values. Hence, this functional coupling increases the apparent concentration of ADP in close proximity of ANT, which compensates for the low *in vivo* permeability of the mitochondrial outer membrane for ADP [38, 56]. The K_m[ADP] potentially represents an adaptable mechanism for mitochondrial regulation of cell metabolism. A 3-fold decrease in the K_m[ADP] in skinned cardiac fibers in rats fed a creatine analog that competitively inhibits creatine transport across the plasma membrane has been reported [57]. This decrease in K_m[ADP] is thought to compensate for the reduced cellular level of creatine and phosphocreatine that would presumably affect the phosphocreatine/creatine shuttle. Mi-CK-deficient mice also exhibit a decreased K_m[ADP] in cardiac and oxidative skeletal muscle fibers [53]. This suggests that if the phosphocreatine/creatine shuttle is impaired, ADP rather than creatine plays the main role of diffusible phosphate acceptor in the cells.

Recently, it has been demonstrated that mitochondria from different rat tissues possess creatine transport activity [58]. This result has been supported by immunoblotting and *in situ* immunofluorescence that localizes a putative creatine transporter to the mitochondrial inner membrane. To address the question of whether VDAC1 constitutes the main pathway for ADP and creatine transport across the mitochondrial outer membrane we used skinned fibers from VDAC1-deficient mice. We reported an alteration in the mitochondrial outer membrane permeability for ADP in different muscle types in the absence of VDAC1 [29]. In cardiac muscle and a glycolytic skeletal muscle (gastrocnemius) we demonstrated an increase in the K_m[ADP], which supports the hypothesis that VDAC1 constitutes a pathway for ADP across the mitochondrial outer membrane. In contrast, an oxidative skeletal muscle (soleus) exhibited a decrease in both the K_m[ADP] and V_{max}, which suggests that, in the soleus, the absence of VDAC1 affects the properties of both mitochondrial membranes. These physiologic changes are accompanied by striking ultrastructural changes in both subsarcolemmal and intermyofibrillar mitochondria. The cristae become very compact and the mitochondria enlarge many fold (Figure 14.3). The muscle-type-specific changes also point to the complexity of the role(s) played by VDAC1 in different muscle types. However, the creatine effect on respiration in VDAC1-deficient muscles remained unchanged (Figure 14.4). This result suggests that the functional coupling of mi-CK to respiration does not require VDAC1 and that VDAC1 is not required for creatine diffusion across the mitochondrial outer membrane. Whether the persistence of the functional coupling of mi-CK to respiration in VDAC1-deficient mice is due to functional redundancy among VDAC isoforms or to increased activity of mi-CK needs further study. Similar results have been obtained with VDAC3-deficient cardiac skinned fibers (K. Anflous et al., unpublished results).

To gain more insight into the physiologic consequence of altered permeability, VDAC-deficient mice have been subjected to exercise-induced stress testing. Mice lacking both VDAC1 and VDAC3 show a significant reduction in time to fatigue compared to control littermates (K. Anflous et al., unpublished results). It is interesting that the single mutants do not show a reduction in the time to fatigue, despite a partial respiratory chain defect (K. Anflous et al., unpublished results). The question of whether the exercise intolerance in double-mutant mice is due to the absence of some compensatory mechanisms that may exist in single mutant

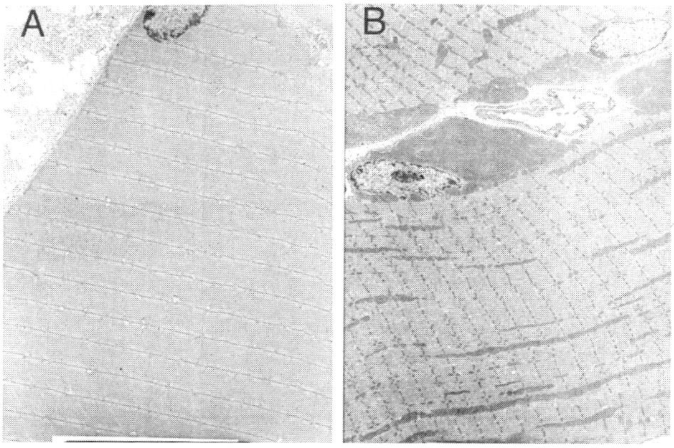

Figure 14.3 Electron microscopic study showing mitochondria from wild-type (A) and VDAC1-deficient (B) gastrocnemius muscle (from [29]).

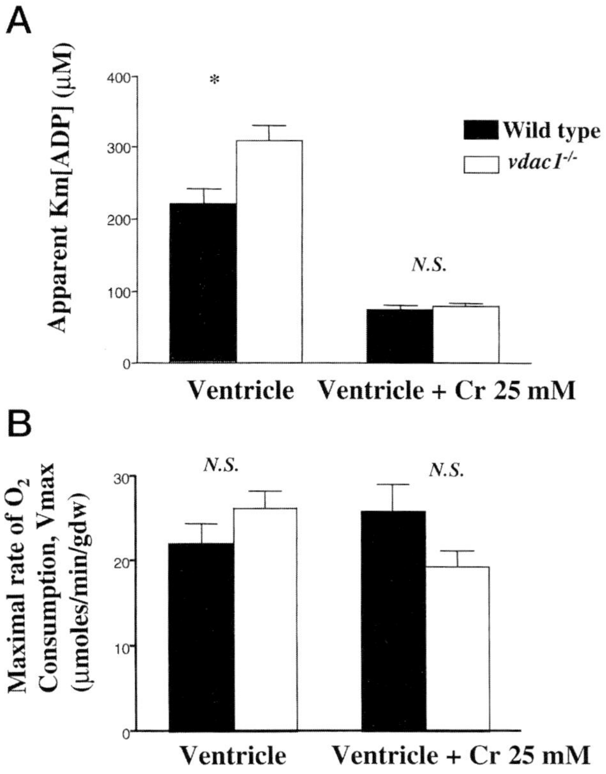

Figure 14.4 (A) Apparent K_m[ADP] in saponin-skinned fibers prepared from wild-type (black bars) and VDAC1-deficient (white bars) ventricular muscle in the absence and presence of 25 mM creatine (Cr). [* $p < 0.05$; N. S., no significant difference]. (B) V_{max} in saponin-skinned fibers prepared from wild-type (black bars) and VDAC1-deficient (white bars) (from [29]).

mice needs further study. The fact that VDAC-deficient hearts do not show any sign of hypertrophy, despite structurally aberrant mitochondria (Figure 14.5), likely reflects a compensatory mechanism, e. g. alterations in the expression of some key metabolic genes, to switch metabolic substrate utilization. Indeed, it has been suggested that metabolic switches are a prerequisite for the successful adaptation of the heart to an altered work load [59]. Measuring substrate oxidation in perfused control and VDAC-deficient hearts will address this question and preliminary results indicate that metabolic adaptation occurs, with a switch from fatty acid oxidation to glycolytic metabolism (K. Anflous et al, unpublished observations). Recently, it has been reported that COX IVaH-deficient mice have impaired left ventricular filling or diastolic dysfunction under maximal cardiac load despite a normal cellular ATP content [60]. Therefore, evaluation of myocardial systolic and diastolic performance under different working conditions in VDAC-deficient mice will give more insight to whether the heart develops subtle phenotypes as a consequence of alterations in the permeability of the mitochondrial outer membrane.

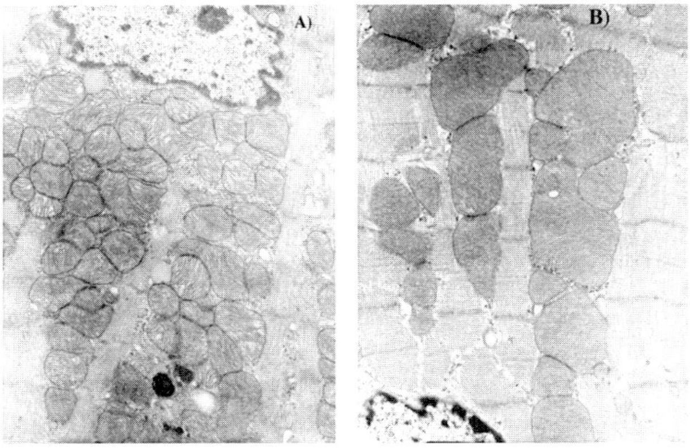

Figure 14.5 Electron microscopic study showing intermyofibrillar population of mitochondria from wild-type (A) and VDAC1-deficient (B) ventricular muscle (from [29]).

14.6 VDAC–Cytoskeletal Interactions

Sampson et al. [28] reported that mice lacking VDAC3 are healthy, but males are infertile. Infertility affects as many as 5–10% of men, making it one of the more common disorders in humans [61]. Sperm motility is a major determinant of fertility in men, and poor sperm motility is a significant cause of male infertility [62]. The structure of sperm is highly organized, especially the sperm axoneme. Epididymal sperm axonemes have a 9+2 microtubule doublet arrangement typical of cilia. Axonemes can be oriented by bisecting the axoneme through the longitudinal columns and two associated microtubules (doublets 3 and 8), with three doublets on one side and four on the other side. Each microtubule doublet has a corresponding outer dense fiber, all of which are morphologically distinct. Two of the outer dense fibers that are associated with microtubules 3 and 8 terminate within the principal piece and form the longitudinal columns of the fibrous sheath that partition the axoneme into two unequal compartments.

VDAC3-deficient males exhibit normal copulatory behavior, but no pregnancies were observed in a large number of matings. VDAC3-deficient males have normal numbers of sperm per epididymis in comparison to wild-type and heterozygous males and show no significant differences in the size, weight or histologic features of testes. However, whereas in wild-type and heterozygous groups around 70% of the sperm were categorized as motile, only 17% of VDAC3-deficient sperm were categorized as motile. When viewed by electron microscopy, 68% of VDAC3-deficient epididymal sperm axonemes (versus 9% of wild-type axonemes) in cross-section demonstrated some structural aberration, most commonly loss of one outer doublet from the normal 9+2 microtubules doublet arrangement (Figure 14.6). In the majority of VDAC3-deficient axonemes, the missing doublet corresponds to the last of the four doublets (doublet 7), reflecting a single recurring defect in the axonemal structure. It has been proposed that loss of doublets 4–7 represent sliding of microtubules during attempted motility, with extrusion of half of the axoneme [63]. Images through the midpiece of epididymal sperm demonstrate the same defect in doublet 7. Cross-sections through the distal principal piece of testicular sperm only occasionally revealed structural abnormalities, suggesting that the structural abnormality found in epididymal sperm represents instability of the axoneme rather than a defect in assembly. Furthermore, the normal structures found in spermatids within the testis suggest that the defect develops with maturation of sperm in the transition from the testes to the epididymis. Electron microscopy of spermatids in the testes revealed enlarged and abnormally shaped mitochondria along the midpiece, similar to that seen in cardiac and skeletal muscle from VDAC1-deficient mice [29].

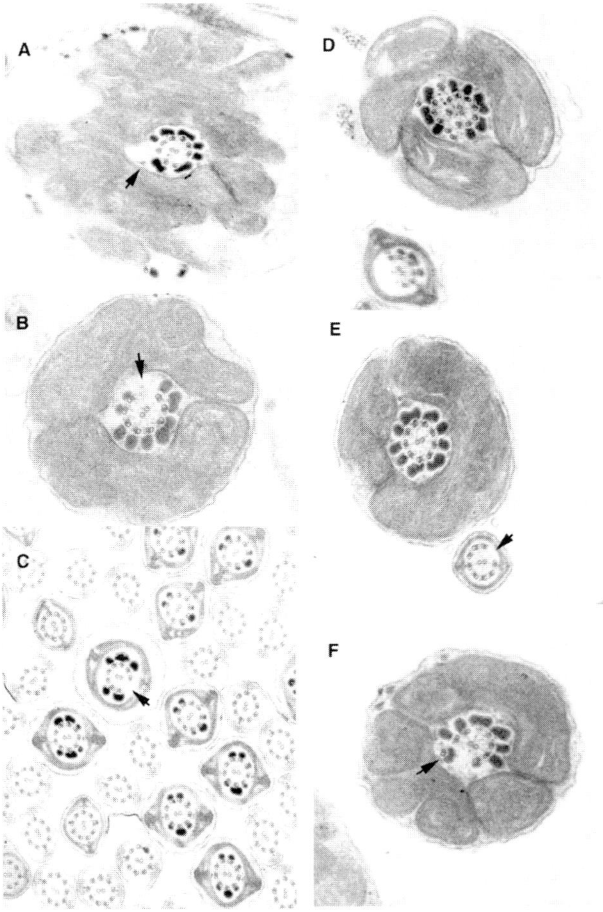

Figure 14.6 Axonemal instability in the absence of VDAC3. Electron micrographs of the midpiece of VDAC3-deficient sperm (×13000). Different images (A, B and D–F) of microtubule doublet loss in the epipdymal sperm. Image (C) is from testicular sperm, with only one abnormal axoneme identified. In image (F) the outer dense fiber is still associated with the doublet (from [28]).

Cilia from VDAC3-deficient tracheal epithelial have a normal 9+2 structural arrangement and normal ciliary movement when examined by dark field microscopy. However, when sections were dissected 2 mm below the vocal cords and examined by scanning electron microscopy, reduced numbers of ciliated cells were consistently observed, although no increase in respiratory pathology has been seen in the mice. This may reflect a developmental effect of abnormal mitochondrial function. This differential effect on sperm and airway cilia has previously been recognized clinically in infertile men [64].

The dynamic behavior of mitochondria in a cell, their movement and morphology, is now recognized to be the result of intricate interactions between proteins on the outer surface of mitochondria and various components of the cytoskeleton, including actin filaments, microtubules and intermediate filaments [65]. It has previously been reported that in rat brain mitochondria, VDACs are binding sites for the microtubule-associated protein MAP2 [66]. Recently, novel proteins have been described from yeast that are localized either to cytoskeletal structures or to the mitochondria themselves [67–68]. Yeast bearing mutations in these proteins display an aberrant distribution of mitochondria at restrictive growth conditions. Mdm1 is one of these proteins involved in the distribution of mitochondria and their segregation during cell division. *In vitro* study has shown that Mdm1 protein can form structures reminiscent of intermediate filaments [69]. The structurally abnormal mitochondria in VDAC-deficient muscle and VDAC3-deficient sperm may result from the altered interactions between mitochondria and cytoskeletal elements, and/or alterations in proteins that play a role in fission/fusion of mitochondria, rather than being a consequence of a respiratory chain defect, energy production, or calcium signaling.

14.7 Mitochondrial Permeability Transition Pore and Synaptic Functions

Enhanced mitochondrial membrane permeability for proteins such a cytochrome *c* and SMAC/DIABLO is critical for the release of pro-apoptotic factors from mitochondria following induction of apoptosis. The exact mechanism for permeabilization of mitochondria has not been elucidated and several hypotheses have been proposed, including release through VDAC1 [12], oligomerized BAX [70] or ceramide channels [71]. It has been reported that VDAC is involved in the apoptotic transient increase of mitochondrial membrane permeability and is a component of the permeability transition pore (MPTP) complex [12]. The MPTP can be defined as a non-specific, voltage-dependent and cyclosporin A (CsA)-sensitive pore, permeant to any molecules below 1.5 kDa [72, 73]. It is believed that the MPTP is composed of VDAC in the mitochondrial outer membrane, the ANT in the inner membrane and cyclophilin D in the matrix compartment [73]. The MPTP opens under conditions of elevated matrix Ca^{2+}, especially when this is accompanied by oxidative stress and depleted adenine nucleotides. Hence, establishing the specific function, if any, of VDACs in the MPTP and the overall apoptotic process remains a central question.

Given the potential role of the MPTP in calcium homeostasis, it is interesting to note the involvement of VDACs in Ca^{2+} transport that was recently reported [74]. The authors demonstrated that purified VDAC reconstituted into lipid bilayers or liposomes is highly permeable to Ca^{2+}, possesses Ca^{2+}-binding sites and that Ruthinium Red (RuR) (an inhibitor of the Ca^{2+} uniporter) inhibits VDAC activity. The authors also showed a correlation between RuR inhibition of Ca^{2+} accumulation, MPTP opening and VDAC activity, all of which support the accumulating evidence for VDAC being a component of MPTP. More recently, a connection between mitochondrial hexokinase localization and the anti-apoptotic action of the serine/threonine protein kinase B (Akt/PKB) has been reported [75]. It has been proposed that Akt/PKB exerts its anti-apoptotic action by preventing closure of a MPTP component, VDAC. This interaction inhibits intracellular acidification, mitochondrial hyperpolarization and the decline in oxidative phosphorylation that normally precedes cytochrome *c* release during apoptosis. By increasing the coupling of glucose metabolism to oxidative phosphorylation via a hexokinase:VDAC interaction, it has been speculated that this pathway may contribute to regulating the MPTP opening.

Synaptic plasticity, a term used to describe alterations in the efficiency of synapses that are believed to underlay learning and memory formation, can be determined by standardized *in vitro* tests using extracellular recordings of stimulus-induced depolarizations. Paired-pulse facilitation (PPF) is widely regarded as a pre-synaptic phenomenon caused by residual Ca^{2+} present in the pre-synaptic terminal following a depolarization that, in turn, facilitates neurotransmitter release in response to a second depolarization [76, 77]. Long term potentiation (LTP), a long lasting increase in the amplitude of post-synaptic depolarizations, has been correlated with many long-lasting forms of learning and memory [78], and is believed to represent primarily a post-synaptic phenomenon.

VDAC3-deficient mice were observed to behave abnormally relative to wild-type littermates. This observation led us to hypothesize that VDACs, either through effects on ATP synthesis, calcium signaling, altered MPTP function or some combination of these processes, participate in synaptic plasticity. To determine the role that VDACs may play in learning and synaptic plasticity, Weeber et al. studied wild-type, and VDAC1-, VADC3- and VDAC1/3-deficient mice [30]. Using a battery of behavioral tests that quantify associative and spatial learning, it was shown that fear conditioning and spatial learning is disrupted in VDAC-deficient mice, with subtle differences between the different mutant strains in associative learning tasks (Figure 14.7). Importantly, there were no sensory or motor deficits identified in any of the mutant strains. This represents the first example of a genetically engineered mouse strain in which a mitochondrial defect leads to learning deficits. In the same report, it was demonstrated by electrophysiological recordings from the hippocampus that deficits in LTP and PPF occur in VDAC1- and VDAC3-deficient brain slices, respectively. As would be predicted, double-mutant mice exhibited deficits in both measures of synaptic plasticity. Thus, a connection was made between behavior, learning deficits and abnormal cell–cell communication. These results demonstrate that mitochondria play an important role in synaptic function, and,

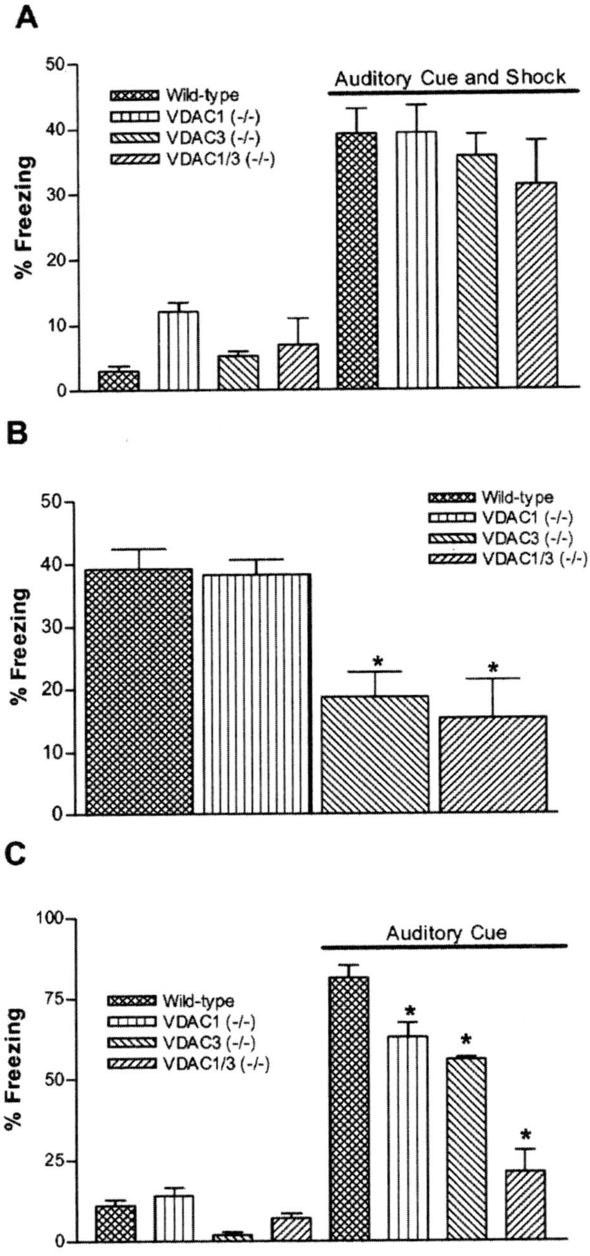

Figure 14.7 Associative learning impairments in VDAC-deficient mice. Initial training of VDAC mutant mice demonstrates normal freezing activity in response to the association of a mild foot shock and an auditory cue in the presence of a novel context (A). At 24 h following training, mice were reintroduced to the training context (B). Wild-type mice and VDAC1-deficient mice show similar freezing behavior. In comparison, VDAC3 and VDAC1/3 mice exhibit reduced contextual freezing behavior. When compared with control mice, all VDAC-deficient mice display a significantly reduced freezing behavior in response to the auditory cue in a novel context (C).

further, that each isoform has a unique function in brain synapses. Furthermore, inhibition of the MPTP by CsA in wild-type hippocampal slices reproduced the electrophysiological phenotype of VDAC-deficient mice. Hence, if VDACs are in fact a component of the MPTP, these results suggested for the first time that the MPTP has a dynamic, physiologic role in the central nervous system of mammals, and is not simply a pathologic endpoint to cellular death.

To further characterize the role of MPTP and VDAC in hippocampal synaptic plasticity and to address the issue of whether mitochondrial calcium buffering in the synapse is altered, Levy et al. [79] examined the effect of blocking the MPTP on synaptic transmission and plasticity. Using concentrations of CsA low enough not to effect calcineurin activity (a phosphatase of known importance in synaptic function), deficits in synaptic plasticity and an increase in baseline synaptic transmission was observed. The resting Ca^{2+} concentration in pre-synaptic terminals showed a transient increase immediately upon incubation with CsA that correlated with the changes in synaptic plasticity and transmission. These results suggest that the effect of MPTP blockade is via altered calcium handling. Although it has been reported that VDAC is present in the plasma membrane where it purportedly is concentrated in caveolae and caveolae-related domains [80], based upon subsequent studies of purified brain mitochondria the data from studies of VDAC-deficient brain tissues appear specific to mitochondria. Using an *in vitro* assay of Ca^{2+} transport in isolated brain mitochondria, Levy et al. demonstrated that CsA-exposed mitochondria buffer Ca^{2+} more rapidly, presumably trigger a more rapid loss of mitochondrial membrane potential, and therefore also release buffered calcium more rapidly (Figure 14.8). This phenomenon, known as calcium-induced calcium release, is believed to underlay the ability of mitochondria to buffer high levels of calcium, e. g. elevations associated with plasma membrane depolarization, and maintain sustained elevations in cytoplasmic calcium. Increased uptake of calcium following blockade of the MPTP suggests the MPTP acts as an efflux pathway for matrix calcium. Similarly, mitochondria lacking VDAC1 released Ca^{2+} more rapidly than wild-type mitochondria, however, the initial uptake of Ca^{2+} was neither as rapid nor as large (Figure 14.8). Based on these studies it was suggested that the MPTP and VDACs may contribute to the regulated uptake and release of cations from mitochondria to prevent Ca^{2+} overload and subsequent membrane depolarization. It was also demonstrated that it is not the total Ca^{2+} uptake but the rate of Ca^{2+} uptake that appears to play a role in triggering Ca^{2+} release. VDAC deficiency appears to be detrimental to the proper function of the MPTP because the absence of VDAC1 may result in an increased likelihood of mitochondrial depolarization and subsequent Ca^{2+} release.

Whether the alterations in the properties of the MPTP in VDAC-deficient brain is a direct result of the absence of VDACs or indirectly related to respiratory chain defects detected in VDAC-deficient brain samples (F. Scaglia and W. Craigen, unpublished results) needs further clarification. Brain mitochondria from VDAC1-deficient mice, in contrast to CsA-treated mitochondria, are impaired in mitochondrial Ca^{2+} uptake (Figure 14.8). Thus, the role of VDACs in Ca^{2+} flux may not only be as a component of the MPTP but also as a Ca^{2+} regulator. The overall re-

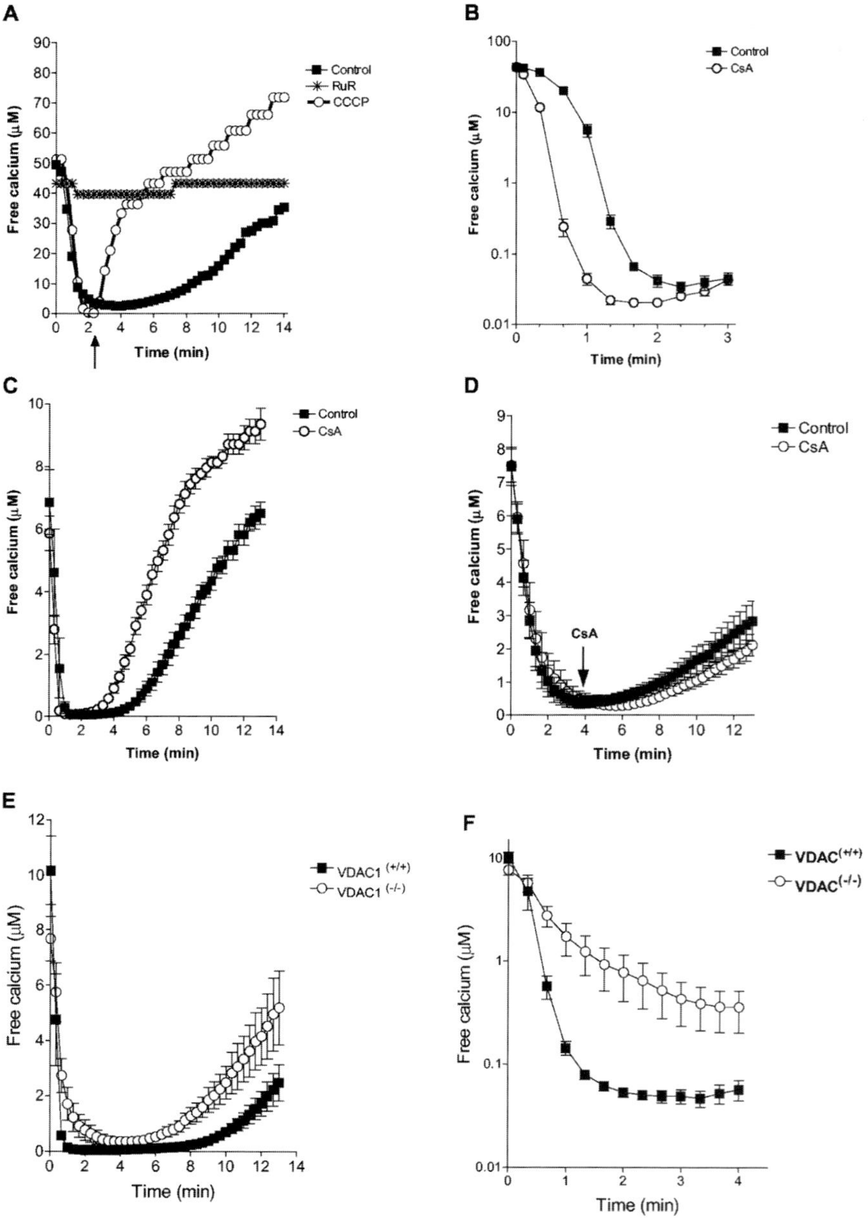

Figure 14.8 The MPTP is involved in mitochondrial Ca^{2+} handling. (A) When exposed to a bolus of free calcium, intact functional mitochondria rapidly absorb the calcium. Calcium overload eventually causes mitochondrial depolarization and subsequent calcium release. Calcium influx is mediated by the uniporter, which can be blocked by RuR, and is dependent on an intact mitochondrial proton gradient (CCCP added at the arrow after maximal calcium buffering). (B) Blocking the MPTP with CsA results in a significantly more rapid calcium influx. (C) Increased influx by CsA also leads to a more rapid mitochondrial depolarization and subsequent calcium release. (D) When CsA is added at the point of maximal buffering (added at the arrow), there is no change in calcium uptake or efflux. (E) VDAC1-deficient mitochondria depolarize and subsequently release calcium faster compared with wild-type controls resulting in significantly more extramitochondrial calcium during the efflux phase. (F) VDAC1-deficient mitochondria are impaired in their ability to buffer calcium as much as wild-type controls (from [79]).

sults highlighted by Weeber et al. and Levy et al. point to a dysregulation of resting Ca^{2+} in the presence of CsA or in the absence of VDACs. Changes to calcium homeostasis may in turn have an impact on signal transduction pathways important in synaptic transmission and subsequent plasticity. Determining the intramitochondrial Ca^{2+} content at rest and upon depolarization in intact cells will give further insight on the involvement of VDACs and MPTP in the dynamic phases of mitochondrial Ca^{2+}, and, indirectly, Ca^{2+} homeostasis in the cytosol.

14.8 VDACs and Apoptosis

The role of VDACs in apoptosis has remained a controversial topic for several years. Whether VDAC constitutes a release pathway into the cytosol or plays a role in compartmentalization of metabolites are the subjects of ongoing debate. Likewise, whether VDACs directly interact with pro- or anti-apoptotic proteins has not been clearly answered. The multidomain pro-apoptotic molecules BAK and BAX constitute essential components of the intrinsic death pathway, functioning at the level of both mitochondria and endoplasmic reticulum Ca^{2+} homeostasis [81]. Upon activation, BAX and BAK homo-oligomerize, resulting in the permeabilization of the mitochondrial outer membrane and, directly or indirectly, the release of several pro-apoptotic proteins, including cytochrome *c*, endonuclease G and SMAC/DIABLO, which initiates a caspase cascade leading to apoptosis. BAX and BAK function in parallel in that apoptosis can still occur in the absence of one or the other but not in the absence of both. Recently, Cheng et al. Described how VDAC2 is involved in regulating the activity of BAK [82]. By using a combination of protein cross-linking, affinity purification, liquid chromatography and tandem mass spectrometry, the authors showed that BAK interacts with VDAC2 and not VDAC1. It was demonstrated that endogenous BAK, but not BAX, efficiently co-precipitates with hemagglutinin-tagged VDAC2. The active conformation of BAK exhibits more proteolytic susceptibility than the inactive conformer. It was shown that the absence of VDAC2 increases the susceptibility of BAK to proteolysis, suggesting that VDAC2 regulates the conformation of BAK. By expressing VDAC1 or VDAC2 in BAK- or BAX-deficient embryonic fibroblasts, it was demonstrated that it is VDAC2 that inhibits apoptosis in cells deficient for BAX (but not BAK) upon treatment with the apoptosis inducers staurosporine or etoposide. It was also demonstrated that VDAC2 inhibits tBID-induced apoptosis of BAX-deficient cells, but not BAK-deficient cells.

VDAC2-deficient primary mouse embryonic fibroblasts were generated from chimeric embryos by microinjection of VDAC2-deficient ES cells into wild-type blastocysts. These cells have significantly more spontaneous apoptosis in culture, and are more sensitive to staurosporine and etoposide. VDAC2-deficient cells treated with staurosporine underwent a partial release of cytochrome *c* into the cytosol by 3 h and complete release by 5 h, while wild-type cells had not initiated any release of cytochrome *c* but that time. In concert, BAK undergoes oligomerization within 5 h following staurosporine treatment in the absence of VDAC2, but not in wild-type cells. This conformational activation of BAK coincides with complete release of cytochrome *c* and peak caspase activation. Following staurosporine treatment, VDAC2-deficient cells exhibited proteolytic activation of caspases 3 and 7 earlier than did wild-type cells. The responses of wild-type and VDAC2-deficient cells were similar when the extrinsic death pathway was activated by tumor necrosis factor-α.

It was concluded that in the absence of inhibition by VDAC2, BAK experiences an enhanced allosteric conformational activation resulting in the release of cytochrome *c*, caspase activation and mitochondrial dysfunction concordant with an increased susceptibility to apoptotic death. These results suggest that VDAC2 may have evolved for a specialized function distinct from that of the other VDAC isoforms and may be a novel target for modulating apoptosis. In contrast, the *in vivo* kinetics of apoptosis in the absence of VDAC1 or VDAC3 following a variety of induction protocols is not significantly different than that of wild-type animals (W. Decker, T Sheiko and W. Craigen, unpublished results). It remains to be established with certainty the nature of the release pathway for cytochrome *c*, but the most recent results in the literature suggest that a BAX channel [83] or a less well-characterized "mitochondrial apoptosis-induced channel" (MAC) [84] may constitute this pathway.

14.9 Conclusions

The mitochondrial outer membrane has historically been viewed as a constitutively permeable membrane, with the inner membrane being the site of regulated transport of metabolites involved in coupled respiration. In the past, *in vitro* studies have led to the hypothesis that VDACs may be involved in the regulation of a number of mitochondrial functions via regulated ion flux. It is becoming increasingly clear that the outer membrane potentially constitutes an additional site of regulation via control of metabolite flux, binding of various kinases, or through organelle or cytoskeletal interactions. However, extrapolation from the *in vitro* data to the role that VDACs may play in cell function has remained tenuous. For example, while VDACs are very sensitive to voltage gating, evidence for voltage gating *in vivo* is entirely lacking. The generation of VDAC-deficient mice and cell lines has provided data that in part bridges the gap that exists between the data obtained from reconstituted systems and *in vivo* VDAC function.

Upon isolation, mitochondria adopt a different configuration and loose the connections with cytosolic factors, including the cytoskeleton [54]. Thus, developing *in vivo* assessments of function are critical to understanding the roles of VDACs in the cellular context. For example, the data obtained with VDAC3-deficient sperm suggest that microtubule/VDAC interaction may play a significant role in mitochondrial function and sperm motility. Although the absence of VADC1 and VDAC3 leads to altered permeability of the outer membrane for ADP in various muscle types, the extent to which the absence of each isoform effects either subsarcolemmal or intermyofibrillar mitochondria is not known at present. Indeed, these two mitochondrial populations exhibit structural and biochemical differences [85]; hence, future studies will be needed to determine whether the distribution of VDAC isoforms is uniform amongst different mitochondrial populations or not. The data obtained from VDAC-deficient mice suggests that, although VDAC1 and VDAC3 may be part of the MPTP their physiologic function appears distinct, e. g. VDAC1 has a more prominent role in post-synaptic LTP whereas VDAC3 deficiency effects pre-synaptic PPF. Thus, only through studies in animal models does the complexity of VDAC function reveal itself.

Acknowledgments

This work was supported in part by the Muscular Dystrophy Association (K. A.), R01 GM55713 and R01 NS42319 (W. J.C.).

References

1 Colombini M. A candidate for the permeability pathway of the outer mitochondrial membrane. *Nature* **1979**, *279*, 643–645.

2 Colombini M. Anion channels in the mitochondrial outer membrane. *Curr Topics Membr* **1994**, *42*, 73–101.

3 Benz R. Porin from bacterial and mitochondrial outer membranes. *CRC Crit Rev Biochem* **1985**, *19*, 145–190.

4 Song J, Midson C, Blachly-Dyson E, Forte M, Colombini M. The topology of VDAC as probed by biotin modification. *J Biol Chem* **1998**, *273*, 24406–24413.

5 Casadio R, Jacoboni I, Messina A, De Pinto V. A 3D model of the voltage-dependent anion channel (VDAC). *FEBS Lett* **2002**, *520*, 1–7.

6 Mannella CA. Conformational changes in the mitochondrial channel protein, VDAC, and their functional implications. *J Struct Biol* **1998**, *121*, 207–218.

7 Song J, Colombini M. Indications of a common folding pattern for VDAC channels from all sources. *J Bioenerg Biomembr* **1996**, *28*, 153–161.

8 Brdiczka D, Beutner G, Ruck A, Dolder M, Wallimann T. The molecular structure of mitochondrial contact sites. Their role in regulation of energy metabolism and permeability transition. *BioFactors* **1998**, *8*, 235–242.

9 Cesar MDC, Wilson JE. Further studies on the coupling of mitochondrially bound hexokinase to intramitochondrially compartmented ATP, generated by oxidative phosphorylation. *Arch Biochem Biophys* **1998**, *350*, 109–117.

10 Wallimann T, Wyss M, Brdiczka D, Nicolay K, Eppenberger HM. Intracellular compartmentalization, structure and function of creatine kinase isoenzymes in tissues with high and fluctuating energy demands: the "phosphocreatine circuit" for cellular energy homeostasis. *Biochem J* **1992**, *281*, 21–40.

11 Vander Heiden M, Xian Li X, Gottleib E, Hill RB, Thompson CB, Colombini M. Bcl-x_L promotes the open configuration of the voltage-dependent anion channel and metabolite passage through the outer mitochondrial

membrane. *J Biol Chem* **2001**, *276*, 19414–10419.

12 Tsujimoto Y, Shimizu S. The voltage-dependent anion channel: an essential player in apoptosis. *Biochimie* **2002**, *84*, 187–193.

13 Blachly-Dyson E, Zambronicz EB, Yu WH, Adams V, MacCabe ER, Adelman J, Colombini M, Forte M. Cloning and functional expression in yeast of two human isoforms of the outer mitochondrial membrane channel, the voltage-dependent anion channel. *J Biol Chem* **1993**, *268*, 1835–1841.

14 Ha H, Hajek P, Bedwell DM, Burrows PD. A mitochondrial porin cDNA predicts the existence of multiple human porins. *J Biol Chem* **1993**, *268*, 12143–12149.

15 Sampson MJ, Lovell RS, Craigen WJ. Isolation, characterization, and mapping of two mouse mitochondrial voltage-dependent anion channel isoforms. *Genomics* **1996**, *33*, 283–288.

16 Sampson MJ, Lovell RS, Davison DB, Craigen WJ. A novel mouse mitochondrial voltage-dependent anion channel gene localizes to chromosome 8. *Genomics* **1996**, *36*, 192–196.

17 Bureau MH, Khrestchatiski M, Heeren MA, Zambrowicz EB, Kim H, Grizar TM, Colombini M, Tobin AJ, Olsen RW. Isolation and cloning of a voltage-dependent anion channel-like M_r 36,000 polypeptide from mammalian brain. *J Biol Chem* **1992**, *267*, 8679–8684.

18 Anflous K, Blondel O, Bernard A, Khrestchatiski M, Ventura-Clapier R. Characterization of rat porin isoforms: cloning of a cardiac type-3 variant encoding an additional methionine at its putative N-terminal region. *Biochim Biophys Acta* **1998**, *1399*, 47–50.

19 Dermietzel R, Hwang TK, Buettner R, Hofer A, Dotzler E, Kremer M, Deutzmann R, Thinnes FP, Fishman GI, Spray DC, Siemen D. Cloning and *in situ* localization of a brain-derived porin that constitutes a large-conductance anion channel in astrocytic plasma membranes. *Proc Natl Acad Sci USA* **1994**, *91*, 499–503.

20 Ryerse J, Blachly-Dyson E, Forte M, Nagel B. Cloning and molecular characterization of a voltage-dependent anion-selective channel (VDAC) from *Drosophila melanogaster. Biochim Biophys Acta* **1997**, *1327*, 204–212.

21 Mihara K, Sato R. Molecular cloning and sequencing of cDNA for yeast porin, an outer mitochondrial membrane protein: a search for targeting signal in the primary structure. *EMBO J* **1985**, *4*, 769–774.

22 Elkeles A, Devos KM, Graur D, Zizi M, Breiman A. Multiple cDNAs of wheat voltage-dependent anion channels (VDAC): isolation, differential expression, mapping and evolution. *Plant Mol Biol* **1995**, *29*, 109–124.

23 Al Bitar F, Roosens N, Smeyers M, Vauterin M, Boxtel JV, Jacobs M, Homble F. Sequence analysis, transcriptional and posttranscriptional regulation of the rice VDAC family. *Biochim Biophys Acta* **2003**, *1625*, 43–51.

24 Sampson MJ, Lovell RS, Craigen WJ. The murine voltage-dependent anion channel gene family. Conserved structure and function. *J Biol Chem* **1997**, *272*, 18966–18973.

25 Sampson MJ, Ross L, Decker WK, Craigen WJ. A novel isoform of the mitochondrial outer membrane protein VDAC3 via alternative splicing of a 3-base exon. *J Biol Chem* **1998**, *273*, 30482–30486.

26 Decker WK, Boules KR, Schatte EC, Towbin JA, Craigen WJ. Revised fine mapping of the human voltage-dependent anion channel loci by radiation hybrid analysis. *Mamm Genome* **1999**, *10*, 1041–1042.

27 Wu S, Sampson M, Decker WK, Craigen WJ. Each mammalian mitochondrial outer membrane porin protein is dispensable: effects on cellular respiration. *Biochim Biophys Acta* **1999**, *1452*, 68–78.

28 Sampson MJ, Decker WK, Beaudet AL, Ruitenbeek W, Armstrong D, Hicks MJ, Craigen WJ. Immotile sperm and infertility in mice lacking mitochondrial voltage-dependent anion channel

type 3. *J Biol Chem* **2001**, *276*, 39206–39212.

29 Anflous K, Armstrong DD, Craigen WJ. Altered mitochondrial sensitivity for ADP and maintenance of creatine stimulated respiration in oxidative striated muscles from VDAC1 deficient mice. *J Biol Chem* **2001**, *276*, 1954–1960.

30 Weeber EJ, Levy M, Sampson M, Anflous K. Armstrong DL, Brown SE, Sweatt JD, Craigen WJ. The role of mitochondrial porins and the permeability transition pore in learning and synaptic plasticity. *J Biol Chem* **2002**, *277*, 18891–18897.

31 Roos N, Benz R, Brdiczka D. Identification and characterization of the pore-forming protein in the outer membrane of rat liver mitochondria. *Biochim Biophys Acta* **1982**, *686*, 204–214.

32 Ludwig O, Krause J. Hay R, Benz R. Purification and characterization of the pore forming protein of yeast mitochondrial outer membrane. *Eur Biophys J* **1988**, *15*, 269–276.

33 Benz R, Brdiczka D. The cation-selective substate of the mitochondrial outer membrane pore: single-channel conductance and influence on intermembrane and peripheral kinases. *J Bioenerg Biomembr* **1992**, *24*, 33–39.

34 Blachly-Dyson E, Peng SZ, Colombini M, Forte M. Probing the structure of the mitochondrial channel, VDAC, by site-directed mutagenesis: a progress report. *J Bioenerg Biomembr* **1989**, *21*, 471–483.

35 Xu X, Decker W, Sampson MJ, Craigen WJ, Colombini M. Mouse VDAC isoforms expressed in yeast: channel properties and their roles in mitochondrial outer membrane permeability. *J Membr Biol* **1999**, *170*, 89–102.

36 Colombini M, Yeung CL, Tung J, Konig T. The mitochondrial outer membrane channel, VDAC, is regulated by a synthetic polyanion. *Biochim Biophys Acta* **1987**, *905*, 279–286.

37 Benz R, Wojtczak L, Bosch W, Brdiczka D. Inhibition of adenine transport through the mitochondrial porin by a synthetic polyanion. *FEBS Lett* **1988**, *231*, 75–80.

38 Gellerich FN, Bohnensack R, and Kunz W. Role of the mitochondrial outer membrane in dynamic compartmentation of adenine nucleotides. In: A Azzi, KA Nalecz, MJ Nalecz, L Wojtczak (Eds), *The Anion Carriers of the Mitochondrial Membranes*. Springer, Berlin, **1989**, pp. 349–359.

39 Liu MY, Colombini M. Regulation of mitochondrial respiration by controlling the permeability of the outer membrane through the mitochondrial channel, VDAC. *Biochim Biophys Acta* **1992**, *1098*, 255–260.

40 Lee AC, Zizi M, Colombini M. β-NADH decreases the permeability of the mitochondrial outer membrane to ADP by a factor of 6. *J Biol Chem* **1994**, *269*, 30974–30980.

41 Lee AC, Xu X, Colombini M. The role of pyridine dinucleotides in regulating the permeability of the mitochondrial outer membrane. *J Biol Chem* **1996**, *271*, 26724–26731.

42 Rostovtseva T, Colombini M. ATP flux is controlled by a voltage-gated channel from the mitochondrial outer membrane. *J Biol Chem* **1996**, *271*, 28006–28008.

43 Rostovtseva T, Colombini M. VDAC channels mediate and gate the flow of ATP: implications for the regulation of mitochondrial function. *Biophys J* **1997**, *72*, 1954–1562.

44 Vander Heiden MG, Chandel NS, Li XX, Schumacker PT, Colombini M, Thompson CB. Outer mitochondrial membrane permeability can regulate coupled respiration and cell survival. *Proc Natl Acad Sci USA* **2000**, *97*, 4666–4671.

45 Le Mallay V, Troppmair J, Benz R, Rapp UR. Negative regulation of mitochondrial VDAC channels by C-Raf kinase. *BMC Cell Biol* **2002**, *3*, 14.

46 Fiek C, Benz R, Roos N, Brdiczka D. Evidence for identity between the hexokinase-binding protein and the mitochondrial porin in the outer membrane of rat liver mitochondria. *Biochim Biophys Acta* **1982**, *688*, 429–440.

47 Linden M, Gellerfors P, Nelson BD. Pore protein and the hexokinase-binding protein from the outer membrane of rat liver mitochondria are identical. *FEBS Lett* **1982**, *141*, 189–12.

48 Brdiczka D, Wallimann T. The importance of the outer mitochondrial compartment in regulation of energy metabolism. *Mol Cell Biochem* **1994**, 133/*134*, 69–83.

49 Kottke M, Adams V, Wallimann T, Nalam VK, Brdiczka D. Location and regulation of octameric mitochondrial creatine kinase in the contact sites. *Biochim Biophys Acta* **1991**, *1061*, 215–225.

50 Savabi F. Interaction of creatine kinase and adenylate kinase systems in muscle cells. *Mol Cell Biochem* **1994**, *133/134*, 145–152.

51 Beutner G, Ruck A, Riede B, Welte W, Brdiczka D. Complexes between kinases, mitochondrial porin and adenylate translocator in rat brain resemble the permeability transition pore. *FEBS Lett* **1996**, *396*, 189–195.

52 Veksler VI, Kuznetsov AV, Sharov VG, Kapelko VI, Saks VA. Mitochondrial respiratory parameters in cardiac tissue: a novel method of assessment by using saponin-skinned fibers. *Biochim Biophys Acta* **1987**, *892*, 191–196.

53 Veksler VI, Kuznetsov AV, Anflous K, Mateo P, van Deursen J, Wieringa B, Ventura-Clapier R. Muscle creatine kinase-deficient mice. II. Cardiac and skeletal muscles exhibit tissue-specific adaptation of the mitochondrial function. *J Biol Chem* **1995**, *270*, 19921–19929.

54 Saks VA, Kuznetsov AV, Khuchua ZA, Vasilyeva EV, Belikova JO, Kesvatera T, Tiivel T. Control of cellular respiration by mitochondrial outer membrane and by creatine kinase in normal muscle and in pathology. *J Mol Cell Cardiol* **1995**, *27*, 625–645.

55 Kay L, Li Z, Mericskay M, Olivares J, Tranqui L, Fontaine E, Tiivel T, Sikk P, Kaambre T, Samuel JL, Rappaport L, Usson Y, Leverve X, Paulin D, Saks VA. Study of regulation of mitochondrial respiration *in vivo*. An analysis of influence of ADP diffusion and possible role of cytoskeleton. *Biochim Biophys Acta* **1997**, *1322*, 41–59.

56 Gellerich FN, Kapischke M, Kunz W, Neumann W, Kuznetsov A, Brdiczka D, Nicolay K. The influence of the cytosolic oncotic pressure on the permeability of the mitochondrial outer membrane for ADP: implications for the kinetic properties of mitochondrial creatine kinase and for ADP channeling into the intermembrane space. *Mol Cell Biochem* **1994**, *133*, 85–104.

57 Clark JF, Khuchua Z, Kuznetsov AV, Vassileva E, Boehm E, Radda GK, Saks VA. Actions of the creatine analogue beta-guanidinopropionic acid on rat heart mitochondria *Biochem J* **1994**, *300*, 211–216.

58 Walzel B, Speer O, Zanolla E, Eriksson O, Bernardi P, Wallimann T. Novel mitochondrial creatine transport activity. Implication for intracellular creatine compartments and bioenergetics. *J Biol Chem* **2002**, *277*, 37503–37511.

59 Taegtmeyer H. Switching metabolic genes to build a better heart. *Circulation* **2002**, *106*, 2043–2045.

60 Radford NB, Wan B, Richman A, Szczepaniak LS, Li J-L, Li K, Pfeiffer K, Schagger H, Garry DJ, Moreadith RW. Cardiac dysfunction in mice lacking cytochrome-*c* oxidase subunit VIaH. *Am J Physiol Heart Circ Physiol* **2002**, *282*, H726–H733.

61 Dessars B, Cochaux P. Genetics of male sterility. *Rev Med Brux* **1999**, *20*, A457–A462.

62 Martin-Du Pan RC, Bischof P, Campana A, Morabia A. Relationship between etiological factors and total motile sperm count in 350 infertile patients. *Arch Androl* **1997**, *39*, 197–210.

63 Merlino GT, Stahle C, Jhappan C, Linton R, Mahon KA, Willingham MC. Inactivation of a sperm motility gene by insertion of an epidermal growth factor receptor transgene whose product is overexpressed and compartmentalized during spermatogenesis. *Genes Dev* **1991**, *5*, 1395–1406.

64 Escudier E, Escalier D, Pinchon MC, Boucherat M, Bernaudin JF, Fleury-Feith J. Dissimilar expression of axonemal anomalies in respiratory cilia and sperm flagella in infertile men. *Am Rev Respir Dis* **1990**, *142*, 674–679.

65 Bereiter-Hahn J, Voth M. Dynamics of mitochondria in living cells: shape changes, dislocations, fusion and fission of mitochondria. *Microscop Res Tech* **1994**, *27*, 198–219.

66 Linden M, Karlsson G. Identification of porins as a binding site for MAP2. *Biochem Biophys Res Commun* **1996**, *218*, 833–836.

67 Berger KH, Yaffe MP. Mitochondrial distribution and inheritance. *Experientia* **1996**, *52*, 1111–1116.

68 Yaffe MP. The machinery of mitochondrial inheritance and behavior. Science **1999**, *283*, 1493–1497.

69 McConnell SJ, Yaffe MP. Nuclear and mitochondrial inheritance in yeast depends on novel cytoplasmic structures defined by the MDM1 protein. *J Cell Biol* **1992**, *118*, 385–395.

70 Antonsson B, Montessuit S, Lauper S, Eskes R, Martinou JC. Bax oligomerization is required for channel-forming activity in liposomes and to trigger cytochrome *c* release from mitochondria. *Biochem J* **2000**, *345*, 271–278.

71 Siskind LJ, Kolesnick RN, Colombini M. Ceramide channels increase the permeability of the mitochondrial outer membrane to small proteins. *J Biol Chem* **2002**, *277*, 26796–26803.

72 Bernardi P. Mitochondrial transport of cations: channels, exchangers and permeability transition. *Physiol Rev* **1999**, *79*, 1127–1155.

73 Halestrap AP, McStay GP, Clarke SJ. The permeability transition pore: another view. *Biochimie* **2002**, *84*, 153–166.

74 Gincel D, Zaid H, Shoshan-Barmatz V. Calcium binding and translocation by the voltage-dependent anion channel: a possible regulatory mechanism in mitochondrial function. *Biochem J* **2001**, *358*, 147–155.

75 Gottlob K, Majewski N, Kennedy S, Kandel E, Robey RB, Hay N. Inhibition of early apoptotic events by Akt/PKB is dependent on the first committed step of glycolysis and mitochondrial hexokinase. *Genes Dev* **2001**, *15*, 1406–1418.

76 Wu LG, Saggau P. Presynaptic calcium is increased during normal synaptic transmission and paired-pulse facilitation, but not in long-term potentiation in area CA1 of hippocampus. *J Neurosci* **1994**, *14*, 645–654.

77 Kamiya H, Zucker RS. Residual Ca^{2+} and short-term synaptic plasticity *Nature* **1994**, 371, 603–606.

78 Malenka RC, Nicoll RA. Long-term potentiation: a decade of progress? *Science* **1999**, *285*, 1870–1874.

79 Levy M, Faas GC, Saggau P. Craigen WJ, Sweatt JD. Mitochondrial regulation of synaptic plasticity in the hippocampus. *J Biol Chem* **2003**, *278*, 17727–17734.

80 Bathori G, Parolini I, Tombola F, Szabo I, Messina A, Oliva M, De Pinto V, Lisanti M, Sargiacomo M, Zoratti M. Porin is present in the plasma membrane where it is concentrated in caveolae and caveolae-related domains. *J Biol Chem* **1999**, *274*, 29607–29612.

81 Wei MC, Zong WX, Cheng EH, Lindsten T, Panoutsakopoulou V, Ross AJ, Roth KA, MacGregor GR, Thompson CB, Korsmeyer SJ. Proapoptotic BAX and BAK: a requisite gateway to mitochondrial dysfunction and death. *Science* **2001**, 292, 727–730.

82 Cheng EH, Sheiko TV, Fisher JK, Craigen WJ, Korsmeyer SJ. VDAC2 inhibits BAK activation and mitochondrial apoptosis. *Science* **2003**, *301*, 513–517.

83 Kuwana T, Mackey MR, Perkins G, Ellisman MH, Latterich M, Schneiter R, Green DR, Newmeyer DD. Bid, Bax, and lipids cooperate to form supramolecular openings in the outer mitochondrial membrane. *Cell* **2002**, *111*, 331–342.

84 Pavlov EV, Priault M, Pietkiewicz D, Cheng EH, Antonsson B, Manon S, Korsmeyer SJ, Mannella CA, Kinnally KW. A novel, high conductance channel of mitochondria linked to apoptosis in mammalian cells and Bax expression in yeast. *J Cell Biol* **2001**, *155*, 725–731.

85 Cogswell AM, Stevens RJ, Hood DA. Properties of skeletal muscle mitochondria isolated from subsarcolemmal and intermyofibrillar regions. *Am J Physiol* **1993**, *264 (Cell Physiol 33)*, C383–C389.

15
Gene Family Expression and Multitopological Localization of Eukaryotic Porin/Voltage Dependent Anion-selective Channel (VDAC): Intracellular Trafficking and Alternative Splicing

Vito De Pinto and Angela Messina

15.1
Introduction

Porin/voltage-dependent anion-selective channel (VDAC) is a small, abundant pore-forming protein first discovered in the mitochondrial outer membrane of eukaryotes [1, 2]. Soon after characterization of its physiological features it became clear that porin channel gating and selectivity properties are highly conserved [3–7], but, at the same time, that electrophysiological techniques are not a sufficient tool to discover all the cellular functions of this protein [8, 9]. By means of the most advanced molecular genetics technologies, several isoforms were discovered in almost all eukaryotic organism studied so far. As a consequence, an unexpectedly large body of literature on porin has appeared over the last 15 years covering a variety of different topics. Three main subjects have been surveyed: the molecular biology and genetics of porin/VDAC; the multitopological localization of porin/VDAC and its interaction with other subcellular components, and the involvement of porin/VDAC in the primary steps of apoptosis.

In this chapter we will review the most recent results dealing with the porin/VDAC gene families, and we will try to shed some light on another highly debated and controversial issue – the presence of porin in compartments different from mitochondria.

15.2
Molecular Biology of Porin/VDAC Gene Families

It became apparent only in recent years that eukaryotic cells have a set of genes encoding for porins and not just a single gene. This was indirectly noticed when human patients suffering from a generic mitochondrial encephalomyopathy were diagnosed with severe deficiency of VDAC in the skeletal muscle – these individuals could survive for some years, indicating the presence of other proteins in the mitochondria able to, at least partially, compensate for the porin deficiency [10, 11].

Bacterial and Eukaryotic Porins. Edited by Roland Benz

ISBN 3-527-30775-3

In higher eukaryotes like mouse and human it has been established that three porin/VDAC proteins are expressed [12–14]. They show sequence conservation and the prediction of putative secondary structure elements indicates that they could have similar structures [15]. Unfortunately, VDAC2 and VDAC3 isoforms have never been purified from animal tissues. Functional information has only been obtained from recombinant proteins [16]. These experiments showed that while VDAC1 and VDAC2 isoforms are able to form pores in lipid bilayers, VDAC3 has not evident pore-forming ability [16]. Each VDAC isoforms may, thus, have some distinct physiological role.

In lower eukaryotes like *Saccharomyces cerevisiae* [17] and *Drosophila melanogaster* [18], a different situation exists – a primary protein (VDAC1) is flanked by one or more isoforms of poorly known function.

In plants, a variable number of isoforms are used by the organism in more defined contexts [19, 20]. The idea that the various porin/VDAC isoforms might have different functional roles is a challenge for future years.

15.2.1
Porin/VDAC in the Yeast *S. cerevisiae*

VDAC from yeast was the first eukaryotic porin to be cloned and sequenced [21]. *S. cerevisiae* is one of the best studied model organism and a paradigm for lower eukaryotes.

In the yeast, two genes encoding for porin polypeptides have been discovered [17]. *POR1* encodes the primary porin expressed in this organism, a protein of 283 amino acids with a high sequence homology and pore-forming properties undistinguishable from the other eukaryotic porins. In yeast, porin is the most abundant protein of the mitochondrial outer membrane [22]. A yeast mutant lacking the outer membrane porin was constructed [22, 23] and retained all other major proteins of the mitochondrial outer membrane, but was deficient in mitochondrial cytochromes and initially did not grow on the non-fermentable carbon source, glycerol. However, it could slowly adapt to glycerol [23]. During this adaptation lag, cytochrome levels were raised again to normal levels and the accumulation of an 86-kDa polypeptide in the extramitochondrial cell fraction was observed. The 86-kDa polypeptide was later identified to be the major coat protein in virus-like particles (VLPs), a system of defense against other yeast strains [24]. The expression of VLPs does not provide an alternative channel pathway to the mitochondrion (M. Dihanich and R. Benz, unpublished data).

The discovery that deletion of the porin gene did not cause a lethal phenotype in yeast prompted the search for other channel-forming proteins in the mitochondrion. A comparison of detergent extracts of the mutant (strain HR 125-5A) membrane with the wild-type (strain HR 125-5B) revealed that the conductance in a black lipid bilayer membrane (BLM) induced by solubilized mutant membrane was only 20-fold smaller than the control [25]. Single-channel analysis of the mutant membrane showed a predominant pore with an estimated diameter of 1.2 nm [25]. Unfortunately, the molecular entity responsible for this conductance was not identified.

Another yeast strain lacking the porin gene was constructed by gene displacement and is able to grow on glycerol at 30 °C, but not at 37 °C [26]. The strain *Δpor1* M22-2 has been extensively used in many laboratories as a model system to evaluate the complementation pattern of porins from other sources.

Upon completion of the yeast genomic sequence, a second porin gene, *POR2*, was identified. The *POR2* gene was discovered by screening for genes which, when over-expressed, can correct the growth defect of *Δpor1* yeast [17]. The protein coded by *POR2*, called YVDAC2, has a 49 % amino acid sequence identity to the porin protein (also called YVDAC1). Subfractionation studies indicate that YVDAC2 is normally present in the outer membrane. However, it was not able to form channels in reconstituted systems. Deletion of the *POR2* gene alone has no detectable phenotype, while yeast strains *Δpor1Δpor2* are viable and show a similar temperature-sensitive growth defect in glycerol at 37 °C as *Δpor1* [17]. The physiological function of this second porin isoform in yeast is still uncertain.

15.2.2
Porin/VDAC Gene Family in *D. melanogaster*: Evidence of Alternative Splice Variants

The fly *D. melanogaster* is another important lower eukaryotic model system. Its genome was deciphered in 2000 and the third annotation of the genome sequence to a high quality is available [27]. Release 3 of the fly genome can be considered the most advanced and refined genomic picture of a metazoan available to date [28].

Porin was purified in our laboratory from *D. melanogaster* adult mitochondria [29] by the classical chromatographic procedure through hydroxyapatite/celite [30]. It showed very conserved functional features [29] and it is indeed about 60 % identical to the VDAC isoforms in mammals [31].

The corresponding cDNA (accession no. X-92408) was isolated by us [31] and later by others [32]. The porin gene was cloned and sequenced in our laboratory in 1998, and mapped by *in situ* hybridization on the second chromosome (2L32B3-4) [33]. A 4.5-kbp clone was entirely determined and contained the sequence corresponding to the cDNA divided in four exons (accession no. AJ000880). The beginning of the first exon was deduced by 5′-RACE-PCR experiments that elongated the cDNA used as a probe by 41 bp. In the same experiment, sequence analysis showed the expression of an apparently unrelated 5′-untranslated extension fused to the coding sequence. This second sequence surprisingly corresponded to another genomic region, enclosed between exon 1 and 2. The structure of the gene was then constituted by five exons: exon 1A indicated the most frequently used 5′ exon, while exon 1B was named the exon corresponding to the second 5′-untranslated region (UTR) (Figure 15.1). Exon 2 (320 bp) and exon 3 (228 bp) exclusively contain coding sequences. In particular, the first base in exon 2 precisely corresponds to the first base of the starting ATG codon. Exon 4 contains the remaining coding sequence, the stop triplet TAA and the 3′-UTR sequence [33] (Figure 15.1).

The novel feature of the *D. melanogaster* porin gene appeared to be the presence of alternative splice variants [33]. These alternative splice variants are expressed in

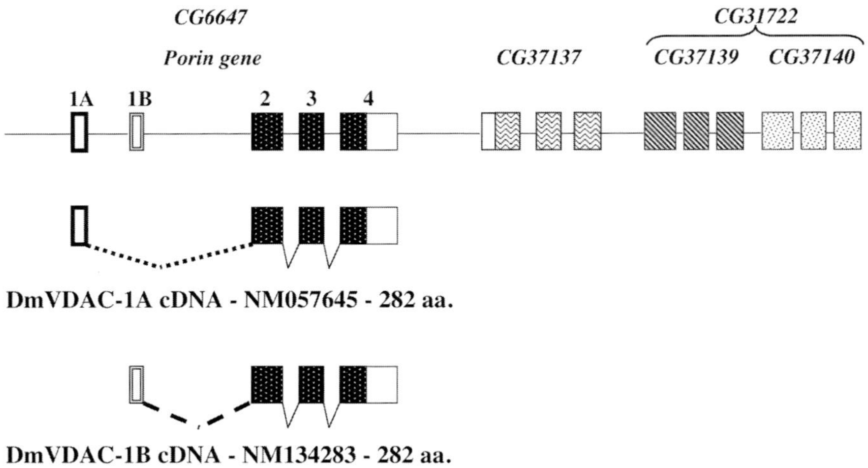

Figure 15.1 The *D. melanogaster* porin/VDAC gene cluster. The alternative splicing of exons 1A and 1B in the porin gene are also shown. The numbers refer to the annotation in the FlyBase GadFly Genome Database (http://flybase.bio.indiana.edu/) and the scaffold containing this region has accession no. AE003630. Not to scale.

the same tissues at the same time, as shown in [18]. They produce the same protein, since alternative transcripts vary in the location of a 5′ non-coding exon. Release 3 of the *D. melanogaster* genome revealed that in the fly about one-third of the alternatively transcribed genes generate only one protein product and there are more examples of alternative 5′-UTRs [28]. This can, thus, be considered a general mechanism suggesting the use of alternative promoters and the possibility of special regulation of the expression. However, the functional meaning of these alternative transcripts is unknown.

Information on these two mRNAs was obtained by performing disgenic crosses to remobilize *P* elements inserted in the *porin* transcription unit [18]. Two stocks carrying a *PlacW* insertion in the *porin* gene were identified and showed that the absence of the VDAC protein was lethal for the fly. The analysis of the revertants obtained by *P* element remobilization supported the notion that deletions encompassing exon 1A resulted in a lower expression or in a lack of the VDAC protein. On the other hand, removal of the region between exon 1A and exon 1B as well as deletion of exon 1B together with most of the downstream intron produced an apparently normal amount of the VDAC protein in homozygous viable lines [18]. The conclusion of these studies was that exon 1B is dispensable for the fly development and, since the 1B 5′-UTR does not add any protein pre-sequence, its functional role must be connected to the polynucleotide sequence itself.

The influence of the alternative 5′-UTRs 1A and 1B on coding sequences was also studied in transfected cells. 1A or 1B 5′-UTR were fused to reporter genes, like green fluorescent protein (GFP) or luciferase, and expressed in yeast or in fly cultured cells (R. Accardi, V. De Pinto, M. Tommasino et al., submitted).

In Δ*porin1* yeast cells the transformation by an exogenous construct carrying the 1A 5′-UTR resulted in the expression of the corresponding protein and in the expected cell phenotype. No porin protein at all was synthesized starting from a plasmid containing the 1B 5′-UTR. The same result was obtained when the GFP was fused downward the 1B 5′-UTR (Figure 15.2). A similar pattern was also detected when embryonal SL21 *D. melanogaster* cells were transfected. Again, 1B 5′-UTR had an inhibitory effect on the translation of a reporter gene fused downward. Surprisingly the over-expression of 1A VDAC in fly embryonal cells raised the 1A VDAC mRNA level, but also the 1B VDAC mRNA. In this experiment, however, the protein production remained quantitatively constant. The authors proposed that one of the functional meanings of such an alternative splicing is to keep VDAC protein production stable in the cell (R. Accardi, V. De Pinto, M. Tommasino et al., submitted).

The completion of the *D. melanogaster* genome produced another surprise. In the region adjacent to the genomic clone containing the *porin* locus, three additional putative genes with significant homology to porin/VDAC were discovered [18] (Figure 15.1). They were called *CG17137, CG17139* and *CG17140.* The genomic organization of the four genes was deduced *in silico* by alignment of the genomic sequences with expressed sequence tagged cDNA (EST) sequences. The additional transcription units are on the same strand of the *porin* gene and they also share a similar exon–intron organization. The 5′-UTR is present only in the genes *porin* and *CG17137.* The nucleotide distances separating the four *porin*-like

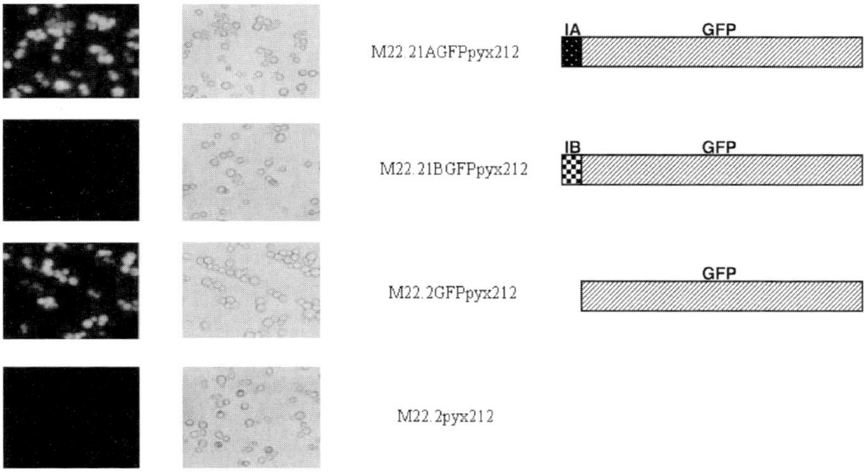

Figure 15.2 The transcription of exon 1B exerts an inhibitory effect upon the translation of a downstream GFP coding sequence. *D. melanogaster* exons 1A or 1B were fused to GFP, inserted in plasmids (pYX212) and used to transform yeast cells (M22.2) devoid of the porin gene. Upon transformation, the cells carrying the various constructs were observed under a fluorescence microscope. On the left are shown fluorescence microscopy images, in the central column light microscopy images and on the right the constructs used for transformation. From the top, cells carrying the construct 1A–GFP (M22.21A-GFPpyx212), the construct 1B–GFP (M22.21B-GFPpyx212), just the GFP (M22.2GFPpyx212) and the plasmid empty as a control.

genes are extremely reduced. In Release 3 of the fly genome, the putative genes *CG17139* and *CG17140* have been fused in a single transcription unit termed *CG31722* since ESTs exist that contain fused *CG17139* and *CG17140* sequences. *CG31722* appears to be specifically expressed in *D. melanogaster* testis cDNA. It is not clear whether *CG31722* produces a dicistronic mRNA or a transcript where a single coding sequence has the potential to produce the corresponding protein. Examples of tandemly duplicated genes transcribed dicistronically are reported in *D. melanogaster* [28].

Alignment of the porin/VDAC sequences deduced from the additional *D. melanogaster* genes shows pairwise sequence conservation. The protein sequence deduced from *CG17137* is 42 % identical to *D. melanogaster* VDAC, but it is only 26 % identical to the deduced amino acid sequence of *CG17140* or *CG31722* [18].

The expression of *CG17137* and *CG31722* is established at the transcription level, since the corresponding ESTs have been identified, but it is not clear whether these porin-like proteins are genuinely translated and what their role in the fly could be. In our laboratory, we have cloned and *in vitro* expressed the coding sequence corresponding to *CG17137*. The recombinant protein expressed *in vitro* from this gene was named DmPorin. DmPorin is able to form channels upon reconstitution in artificial lipid bilayer membranes. Its conductance is similar to native and recombinant DmPorin reconstituted in artificial membranes. This second isoform nevertheless shows interesting differences: DmPorin is clearly voltage independent and cation selective, while its counterpart isoform 1 is voltage dependent and anion selective (R. Aiello, R. Benz, V. De Pinto et al., J. Biol. Chem. Epub. Ms. 310572200).

This result supports the view that each VDAC gene may have a functional role also in *D. melanogaster*.

15.2.3
Porin/VDAC Gene Families in Mammals

15.2.3.1 Porin/VDAC Gene Family in Mouse: More Evidence of Alternative Splicing

An organic description of the gene family was reported in [14] and the result of specific gene knockouts is the subject of Chapter 14.

In brief, the mouse VDAC1 gene spans about 28 kbp; it is made up of nine exons (accession no. AF151092-097) and is mapped on the chromosome 11. The first exon contains the 5′-UTR and the start codon is located 4 bp into the second exon. The mouse VDAC2 gene is about 12 kbp long, divided in 10 exons (accession no. AF152220–227) and is mapped on the chromosome 14. The VDAC2 gene contains an additional exon in comparison with VDAC1 and 3: it expresses part of the 5′-UTR sequence. The start codon is located in the second exon. Another start codon in frame is in exon 3, but it is supposed that the most distal one is used by the ribosome [12]. Thus VDAC2 polypeptide should have an extension of 12 amino acids at the N-terminus relative to the VDAC1 protein. The mouse VDAC3 gene spans about 16 kbp on the chromosome and is again made up of nine exons (accession no. AF1515678-682) and is mapped on the chromosome 8. The start codon is located in the exon 2 [14].

Alternative polyadenylation sites were reported for the mouse VDAC2 gene. Two canonical and one aberrant polyadenylation sequences were found in the gene sequences. They originate different mRNAs with 3′-UTR sequences of different lengths and they were suggested to be implicated in different utilization of the translation product [14]. VDAC3 and 1 have single polyadenylation sites [14].

Evidence of a VDAC3 internal alternative splicing event was reported [34]. A 3-bp (ATG) mini-exon was shown to introduce a methionine 39 amino acids downstream the N-terminus of the mouse VDAC3 protein. This alternative splicing was confirmed by the presence of ESTs in the databases both for mouse and for human. In addition, the alternative ATG-containing VDAC3 isoform is specifically transcribed in the brain, heart and skeletal tissues [35]. A matter of interest was whether this alternative splicing might cause the production of a shortened form of VDAC3. The alternative isoform of mouse VDAC3, containing the internal ATG, was expressed in *Δpor1* yeast cells. This truncated form of VDAC3 was detectable by antibodies since a tag sequence was fused at the protein C-terminus. However, this experiment did not show the presence of a lower-molecular-weight band in Western blot, suggesting that the truncated form of the protein is not produced in the cell [36]. All efforts to observe channel-forming activity by the VDAC3 truncated form were unsuccessful [16]. Thus, it is unlikely that the shortened form has any physiological meaning. Also, the *in vivo* role of this alternatively spliced VDAC3 isoform remains unknown.

A second alternatively spliced porin/VDAC was claimed to be expressed in mouse. Buettner et al. reported that the VDAC1 gene could produce alternatively spliced transcripts [37]. In addition to the canonical one, they claimed that the utilization of the genomic sequence immediately 5′ upstream from the second exon could add an extension of 11 amino acids at the N-end of the protein [37]. Since this N-terminal extension resembles other plasma membrane targeting sequences, the authors speculated that this alternatively spliced product is responsible of the plasma membrane localization of porin. The mRNAs derived from the mouse VDAC1 gene were thus designed mt-VDAC1 (protein targeted to mitochondria) and pl-VDAC1 (protein targeted to plasma membrane) [37]. We will discuss the cellular consequences of this point in more detail in a later section, and we will instead focus here on the gene structure and function.

The transcription of pl-VDAC1 was verified by RACE-PCR and Northern blots with poly(A)$^+$ RNA purified from mouse brain. In whole brain the mt-VDAC1 was reported to be 5-fold more highly expressed than pl-VDAC1 [37]. Despite this relatively high abundance of the mRNA for pl-VDAC1 (the reader has to consider that porin/VDAC isoform 1 is a very abundant protein: hundreds of ESTs are reported in the sequence databanks!), only one EST containing the plasma membrane-targeting extension is reported in the databases (GenBank entry BU924179, expressed in mouse retina). In addition, the homologous human VDAC1 gene does not contain a similar 5′ extension of the exon 2. This is unexpected, since the gene structure is very conserved between human and mouse: the size of exons containing coding sequences is conserved and the exon–intron junctions are highly homologous [38]. This raises some doubt about the consistency of this proposed alternative splicing mechanism.

15.2.3.2 Porin/VDAC Genes in Human

Porin/VDAC isoform 1 gene

Human porin genes strictly resemble the mouse genes summarized above. They share the same number of exons for each gene isoform and the chromosomal localizations are synthenic [13, 38–40]. The main differences are expanded introns, possibly the result of genome evolution.

The porin/VDAC isoform 1 gene was mapped by *in situ* hybridization on the chromosome 5q31–32 [39]; the localization was then confirmed by radiation mapping [40]. The structure of the gene was determined in our laboratory before completion of the human genome project [38] (accession no. AJ250031-39). The human VDAC1 gene spans about 30 kbp and is organized in nine exons. About half of this region is covered by the intron 1 (14.1 kbp). This gene encodes the most abundant porin isoform, localized in the outer mitochondrial membrane and whose electrophysiological properties have been extensively studied [4]. The protein product of this gene has also been detected in the plasma membrane of B lymphocytes [41] and in other subcellular compartments [42], and isolated from caveolae [43].

Porin/VDAC isoform 2 gene

The porin/VDAC2 gene was mapped at chromosome 10q22 by various techniques [39, 40]. Its locus spans 16.4 kbp and its structure was determined to contain 10 exons (accession no. AF152220-227). Also in human, like in mouse, the VDAC2 gene originates three classes of mRNAs utilizing different polyadenylation sites [40]. Human VDAC2 cDNA (accession no. L06328) was first isolated by Blachly-Dyson et al. [12]. Its main distinguished feature is an extension of 11 amino acids predicted by the presence of a start codon in-frame with the ATG codon homologous to the start codon of any other porin-coding sequence. This porin isoform has never been isolated starting from animal tissues. Thus, the presence of the N-terminal 11-amino-acid extension has never been demonstrated by protein sequencing. However human VDAC2 has been expressed in heterologous systems [12, 16] where it showed pore-forming activity similar in many respects to the VDAC1 isoform [16].

Soon after the discovery of VDAC2, a new porin-encoding cDNA was fortuitously isolated during cloning of a different human gene [44]. This new protein sequence (termed Humpor in this article, accession no. L08666) showed an N-terminal sequence of 26 amino acids different from the 11-amino-acid N-terminal extension discussed above. Synthesis of two proteins of different lengths was shown by translation of this cDNA in a reticulocyte lysate system [44]. This means that both start sites can be potentially used. Transfection experiments in COS7 cells performed by another group confirmed that both start sites could be apparently used at roughly similar quantitative levels [45]. Interestingly, when a similar plasmid carrying this cDNA sequence was expressed in yeast, the first ATG codon was exclusively used for translation initiation, demonstrating that the ribosome recognition of the second start codon is specific for high eukaryotes [45]. ESTs cDNA containing the sequence encoding Humpor are present in the databases. Humpor is enriched in heart and B lymphocytes [45].

Porin/VDAC isoform 3 gene

The porin/VDAC3 cDNA was discovered by Rahmani et al. in the process of studying the interaction of the HBX protein, encoded by the hepatitis B virus genome, with cellular proteins using the yeast two-hybrid method (accession no. U90943). The corresponding gene was mapped on chromosome 8p11.2 [13]; it spans 13.3 kbp and is made up of nine exons (accession no. AF151678-682). Human VDAC3 mRNA is highly enriched in testis, but it is also expressed in many other tissues tested [13].

The protein encoded by this gene is the least known and the most intriguing among the various isoforms. From the scarce information available for the human and mouse porin/VDAC3 (they show only 5 amino acid exchanges upon a sequence of 283) this is the only porin unable to complement the incapacity of the yeast strain *Δpor1* M22-2 to grow on glycerol at 37 °C [13]. In reconstituted systems mammalian porin isoform 3 seldom forms channels and these channels are not closed under the application of high voltages, a typical feature of this class of proteins [16]. Finally, a leucine zipper domain was shown in this protein between amino acids 150 and 171 [14] – a domain of unexplained function for this protein.

Alternative splicing events in human genes

We have investigated the human genome and the related EST cDNA sequences libraries with BLAST tools [46]. The query probes were, respectively, the human cDNA sequences encoding for the three human isoforms. Results obtained from the genome search were matched with those from EST libraries to get the most likely picture of the porin/VDAC human genes expression (Messina et al., in preparation).

Search with the cDNA encoding human VDAC1 (accession no. L06132) and VDAC2 (accession no. L06328) upon EST collections, and cross-analysis between the major human sequence databanks (GenBank and Ensembl), showed the presence of more exons than reported for both genes. Human VDAC1 gene showed two additional exons. Exon 1bis (Figure 15.3) is 157 bp long and is located a few bases after exon 1. It appears to be a true exon since it has the canonical splicing consensus sequence at the 3′ extremity. Furthermore, ESTs exist containing this exon fused to the remaining exons 2–9 (i.e. ENSESTT00002576024, 022 or 020). This alternative splice variant of VDAC1 does not change the coding frame, since exon 1, skipped in this transcript, contains the 5′-UTR like exon 1bis. Another likely alternative exon is located 52 bp before exon 1 (Figure 15.3). We call it exon 0; it is 429 bp long and has the 3′ splicing site. Its use is very similar to exon 1bis: it is present in another splice variant of the transcript containing another alternative 5′-UTR. Several ESTs containing this sequence are present in the databases (i.e. BE618448 or BG469465) and also a whole mRNA sequence is registered in GenBank as AK095989. The protein sequence is again not altered by transcript modification.

The human VDAC2 gene shows a similar pattern (Figure 15.4). An alternative exon (exon 2bis, 177 bp long) is located between exons 2 and 3. Cross-analysis with ESTs and mRNA reveals that this exon codes the N-terminal sequence

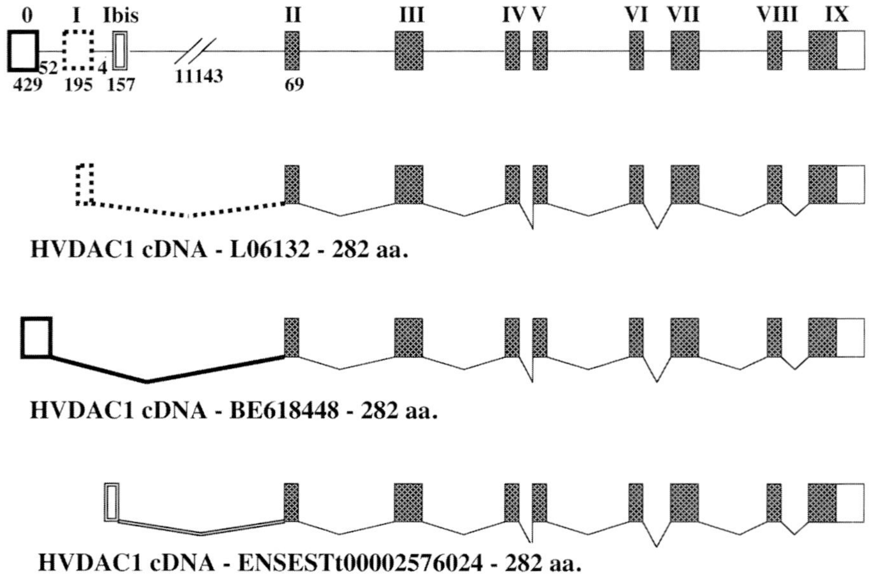

Figure 15.3 Alternative splicing events in the human porin/VDAC1 gene. The structure of the gene as deduced by a cross-analysis between EST libraries and the human genome is shown at the top. Three 5′ alternative splicing events are shown below. The solid boxes represent exons coding amino acid sequences. The empty boxes are transcribed, but not translated regions. Not to scale.

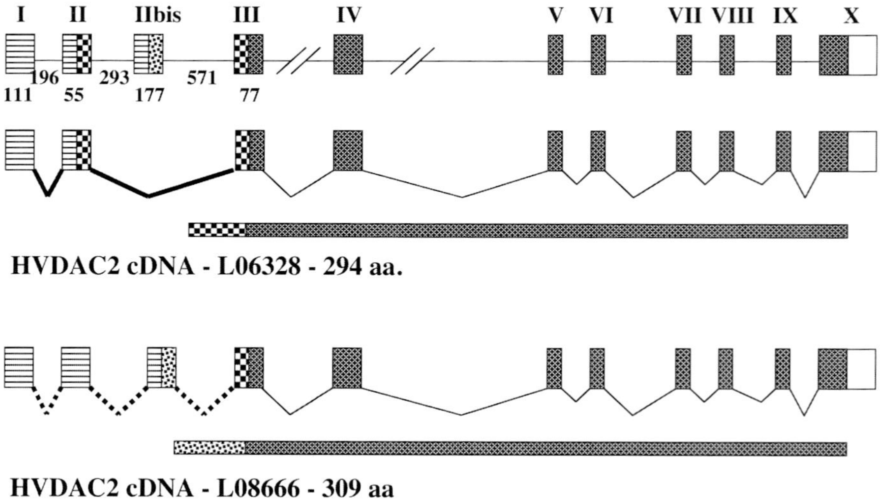

Figure 15.4 Alternative splicing events in the human porin/VDAC2 gene. The structure of the gene as deduced by a cross-analysis between EST libraries and the human genome is shown at the top. Two 5′ alternative splicing events are shown below. The solid boxes represent exons coding amino acid sequences. The empty boxes are transcribed, but not translated regions. Not to scale.

reported in Humpor [44]. Thus, this alternative splicing could have potential effects upon proteins since VDAC2 containing a different N-terminal sequence would be expressed.

As shown for *D. melanogaster*, the human porin/VDAC gene family seems to use alternative splicing at the 5′ extremity of the genes. Alternatively spliced mRNAs might exert an influence in pre-translational processes, even if they are still poorly known. Also for VDAC2, where different N-terminal sequences have been reported [12, 44] as cDNA translation, there is not yet any biochemical demonstration of their whole translation in the cell. Thus, it cannot be ruled out that also for VDAC2 the alternative splicing events at the 5′ extremity of the gene might have the physiological aim to produce transcripts with different 5′-UTRs. In any case, the functional meaning of such alternative splice variants has to be understood.

In the human VDAC3 gene, the only alternative splicing event reported in the literature and confirmed by the genome sequence corresponds to the single ATG internal exon described in Section 15.2.3.2.3.

Regulatory elements in porin genes

Analysis by publicly available software of putative promoters of porin genes has been reported [14, 33, 38]. The putative promoters of the VDAC1 gene from mouse and human showed 45 % identity upon a sequence of 1243 bp [38]. Putative transcription starts were localized at –29 and –55 with respect to the beginning of the cDNA reported [38]. Other putative transcription starting sites were found by several software approaches [NNPP (neural network promoter prediction: www.fruitfly.org/seq_tools/promoter.html) and Genomatix (www.genomatix.de/cgi-bin/eldorado/main.pl)] at about 600 bp before the start of the cDNA reported in [12]. Interestingly, these promoters are compatible with the alternatively spliced mRNAs indicated in this review (Figure 15.3). A sterol-repressing element (SRE-1) is present in the promoter region of human and mouse VDAC1 [38, 14] and mouse [14] VDAC2 genes. SREs regulate the synthesis of proteins that are involved in cholesterol traffic. Porin binds preferentially to cholesterol [47] and was reported to be part of the peripheral benzodiazepine receptor, which appears to be involved in the transport pathway for cholesterol between the outer and inner mitochondrial membranes [48–49]. Furthermore, a nuclear respiratory factor (NRF)-2 element was shown in the human VDAC1 putative promotor. NRFs are transcription factors involved in the regulated expression of proteins belonging to the respiratory complexes. This could mean that porin may also be regulated together with main components of the respiratory chain. A final interesting elements found by bioinformatic analysis was an SRY element. SRY is a protein involved in sex determination and in testis development [50]. This could explain the absence or the reduction of VDAC1 in sperm [51].

Cloned putative promoter regions from mouse VDAC1–3 genes were used to direct the expression of reporter proteins [14]. These experiments showed the highest levels of activity in the transient transfection assays for the VDAC2 cloned putative promotor and quite low activities for the other candidates. A possible explanation for the very low levels of expression induced especially by the VDAC1 putative promoter is that the regions used for the experiment did not contain the true promoter.

15.2.4
Porin/VDAC Gene Families in Plants

In plants, pore-forming channel or porin-like proteins have been reported in various subcellular organelles. Mitochondria and chloroplasts contain specific proteins for metabolite passage through external membranes. In the chloroplast outer envelope, the presence of a general diffusion channel was reported in 1984 [52]. More recently, three channel-forming proteins termed OEP16, OEP21 and OEP24 were described: they show differences in substrate specificity and gating behavior [53–55]. Outer envelope proteins (OEPs) are reminiscent of mitochondrial porins also because secondary sequence analysis showed that they should have amphipathic β-barrel structures [56]. A specific role for OEP16 in nitrogen metabolism has been proposed, while OEP21 has an ATP-binding site and in the presence of nucleotides dramatically changes its functional properties. OEP24 forms the least specific channel and it appears to be the most similar to VDACs: it can functionally complement the porin-deficient yeast strain *Δpor1* [57]. Chloroplast porins have been recently reviewed in [56].

Previous works reported the purification of porins from mitochondria of maize [58], wheat [59] and pea [60]. Later, porins from non-green plastids were shown, revealing conserved features with the mitochondrial porins [61]. Finally, evidence of a porin/VDAC gene family has been presented in potato [62]. Plant mitochondrial porins share with the homologous proteins from animals and fungi a very conserved pore-forming activity and a similar structure, consisting of about 280 amino acids where several amphipathic β-strands can be easily identified [19, 20]. It has been shown that these isoforms are expressed ubiquitously, but at different levels, in the plant's organs [19, 20]. The structure of the genes has not been studied yet and, thus, the regulatory regions determining the expression features in plants are not known.

Porin/VDAC family genes have been characterized in maize, wheat and rice, all monocotyledonous. In rice, the expression of three porin/VDAC cDNAs was studied by Northern blotting [20]. These isoforms were similarly highly expressed during seedling development and the transcription decreased afterwards. They were transcribed in all the organs of the plant, but the quantitative levels were variable. For example, isoform 2 (osvdac2) was highly expressed in the flower after pollination, while isoforms 1 and 3 were expressed in large amounts especially in the root and in the flower before pollination. The protein expression level as detected by Western blotting was not affected by plant stress conditions [20].

In wheat, a family of three porin/VDAC genes was found [19]. Isoform 1 (Tavdac1) was expressed at the highest levels in mature embryos and decreased in the following days. This isoform transcript is the most abundant one. In flowers, isoform 2 (Tavdac2) is poorly expressed in comparison with isoforms 1 and 3 (Tavdac3). Interestingly, isoform 3 is expressed more in anthers than in pistils. The authors hypothesized that the different isoforms might perform different functions or just reflect mitochondrial heterogeneity in the different plant tissues [19]. Functional characterization of the wheat porin/VDACs was performed upon expression in *Δpor1* yeast cells [63]. All three plant porin isoforms were able to complement the defective phenotype shown by the yeast porin-deficient mutant (see Section 15.2.1), but isoform 3 showed a slower adaptation curve. These three isoforms were purified from the mutant yeast cells and studied in planar lipid bilayers. They showed differences in the single-channel activity, and different and various closed states were exhibited. Furthermore, isoforms 2 and 3, different from isoform 1, did not show a symmetric response in the voltage dependence curves. An asymmetric pattern of insertion in the membrane for the three isoforms was proposed [63].

15.2.5 Molecular Evolution of Porin/VDAC

Phylogenetic trees were obtained for porin sequences available in the public databases [14, 20, 64]. These analyses indicated an early separation among fungi, plants and animals. In vertebrates there are three clades corresponding to VDAC1–3 isoforms. There is a general agreement among different groups in assuming that VDAC3 could be considered the oldest protein. The divergence between VDAC3 and VDAC1/2 was estimated at 365 ± 60 million years ago, while the divergence between VDAC1 and VDAC2 was estimated at 289 ± 63 million years ago [64]. VDAC1, apparently the most successful isoform, would be the most recent porin (Figure 15.5A).

From an analysis extended to plant porin sequences, it has been proposed that the duplications that gave rise to the multigene families of VDACs probably occurred independently in animals and plants after the divergence between the two kingdoms [20]. Furthermore, in plants, VDAC duplications probably happened after the divergence between monocotyledons and dicotyledons [20]. This conclusion means that the numbering of plant isoforms is unrelated to he homologues in animals (Figure 15.5B).

The porin/VDAC genes organization in *D. melanogaster* [33], mouse [14] and human [38, 40] has been thoroughly elucidated. Analysis of the *D. melanogaster porin* gene shows a possible primordial gene organization with three (or four, see above Section 15.2.2) exons. In addition, duplication events might have originated other genes in the fly bearing the same exon–intron organization. Comparison of the *D. melanogaster* and human VDAC genes shows that three coding exons in the fly produced eight coding exons in human. Intriguingly, exon 3 of *D. melanogaster porin* gene is perfectly conserved also in mouse and human genes, where it is the longest coding exon [38]. It has been suggested that this exon might encode a functional important domain: predictive algorithms indeed localized putative large hydrophilic loops connecting transmembrane β-strands in this exon [33]. Large hydrophilic loops are exposed protein moieties and are, thus, able to interact with external molecules. The second exon of *D. melanogaster porin* gene produced four

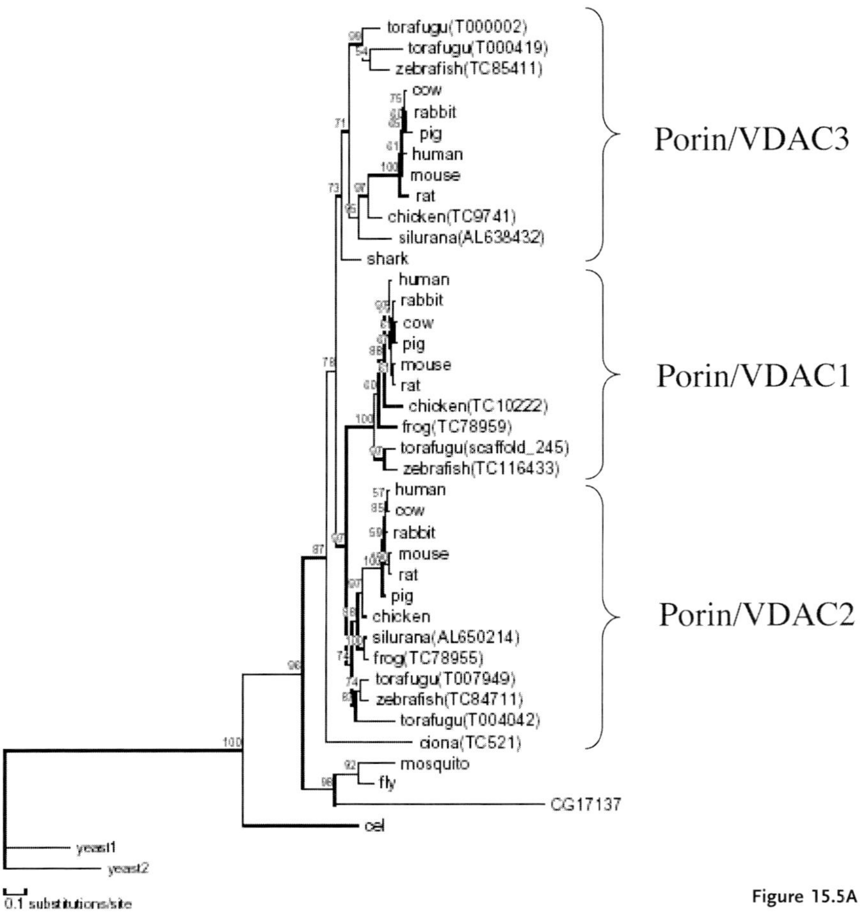

Figure 15.5A

Figure 15.5 Phylogenetic trees of the porin/VDAC genes. (A) The maximum-likelihood tree inferred from VDAC sequences of animals. Numbers along the branches show the percentage occurrence of nodes in 100 replicates of simulation. The tree was constructed using PROTML. Kindly provided by C. Lanave and C. Saccone (Bari). (B) Phylogenetic tree for VDAC sequences from animals, plants and fungi. The tree was constructed using the UPGMA method. This is a rooted method. The position of the branching point between two sequences is a direct measure of their similarity. By courtesy of Dr F. Homblé (Brussels).

exons in mammals, while the remaining three exons were derived from interruptions in the last fly exon. In mouse and human, the coding sequence of all porin/VDAC isoforms is thus distributed among eight exons. This organization is very well conserved between the same isoforms in mouse and human, indicating that these genes are orthologous, and among the three isoforms of the same species. The respective coding exons have exactly the same size in the three genes both in mouse [14] and in human [38]. Thus, the gene duplication should have arisen after such an exon–intron organization was established. Since it is very difficult

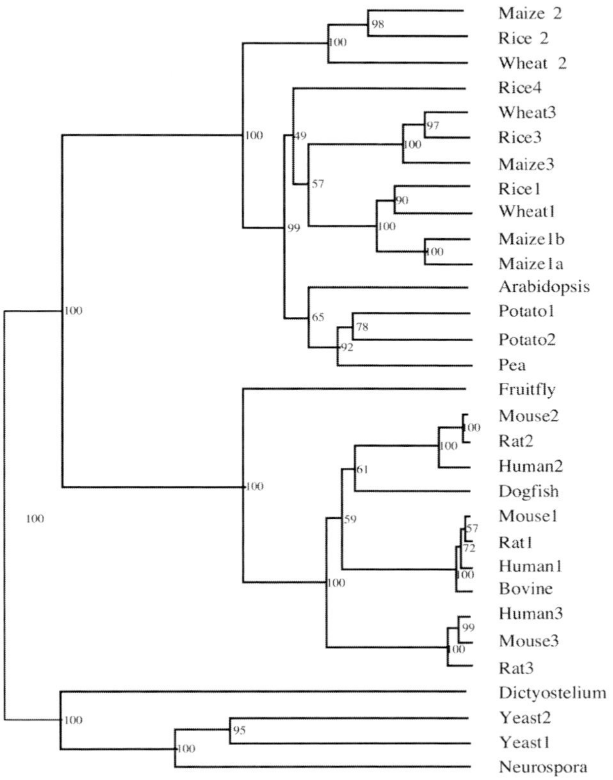

Figure 15.5B

to imagine that each *D. melanogaster* VDAC gene might have evolved in the same way, the mammal genes should have a different evolutionary meaning.

15.3 Multitopological Localization of VDAC in the Cell

The extramitochondrial localization of porin was shown for the first time by Thinnes et al. [41, 65]. A review by Bathori et al. [66] reports the results that appeared in the literature on this topic until 1999 and the reader is referred to it for further details.

15.3.1 Porin in the Plasma Membrane

Thinnes et al. fortuitously co-purified porin together with human transplantation antigens. Intrigued by this protein, they sequenced it by Edman degradation. In fact this was the first high eukaryotic porin/VDAC amino acid sequence reported [65]. The same group raised antibodies against the protein, named "porin 31HL", purified from a crude membrane preparation from human lymphocytes. Later, monoclonal antibodies against the N-terminal end of the protein were developed by the same group [67]. Application of the antibodies to an Epstein–Barr virus-transformed lymphocyte line was used to generate secondary immunofluorescence images. The images obtained with this techniques showed a fluorescent staining limited to the cell surface [41] (Figure 15.6).

This finding was subsequently confirmed by a number of studies employing several cell lines or tissues (normal and cultured lymphocytes [68, 69], epithelial cells [70], astrocytes [71], and the post-synaptic membrane fraction from brain [72]) and flow cytometry as well as electron microscopy immunogold labeling and immunofluorescence techniques. Immunofluorescence showed a punctuate staining distribution on the cell surface, compatible with a localization in membrane subdomains of the protein. In addition, the experiments with monoclonal antibodies indicate that the N-terminal end of the protein is exposed on the cell surface.

The weak point of this line of evidence is that all immunotopological evidence was obtained by the same set of anti-VDAC antibodies and by histochemical techniques alone. The reliability of this evidence was thus questioned by Yu et al. [45], who investigated the subcellular distribution of epitope-tagged porins. They introduced at the C-terminus of the porin human isoforms VDAC1, VDAC2 and VDAC2′ (i. e. Humpor discussed in Section 15.2.3.2.2) tag sequences encoding known epitopes. Tag-modified porins cDNAs were used to transfect cultured

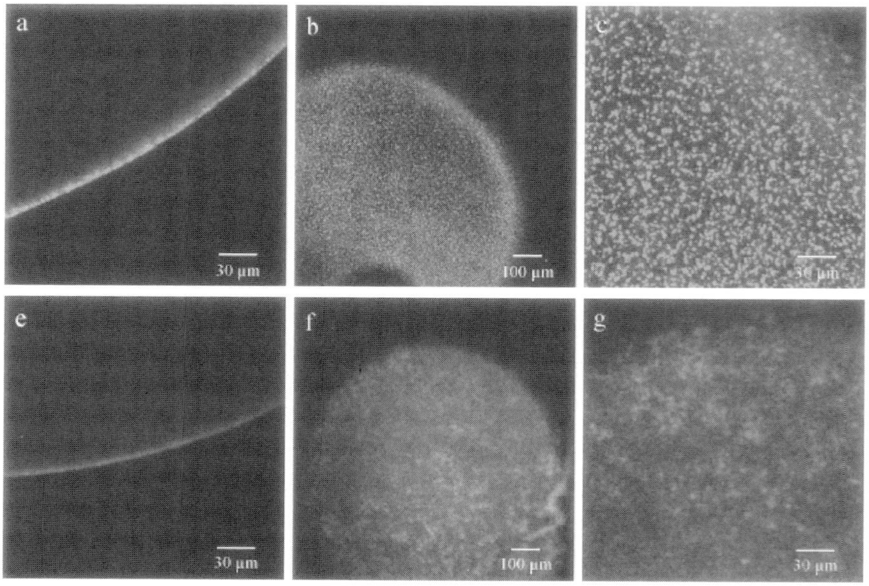

Figure 15.6 Visualization of human VDAC1 distribution at the surface of *X. laevis* oocytes using a humanVDAC1–EGFP fused protein. Confocal laser microscopy of hVDAC1/EGFP (a–c)- and EGFP1 (e–g)-expressing *Xenopus* oocytes 7 days after mRNA injection. (a and e) Border of the oocytes in the equatorial plane; (b, c, f and g) oocyte surfaces (magnification: a, c, e and g, ×400; b and f, ×75). From [78].

cells, and the intracellular target of each porin isoform was studied by subcellular fractionation, light level immunocytochemistry and immunoelectron microscopy. Experiments with antibodies against the tagged proteins showed that all three isoforms were exclusively located in fractions or subcellular regions that contained mitochondrial marker proteins. The authors concluded that the data supporting a non-mitochondrial localization of VDAC1 were probably artifacts, perhaps due to unspecific immunoreactions and redistribution of porin among subcellular fraction in procedures involving the use of detergents [46].

New biochemical evidence supporting the presence of porin in the plasma membrane came from other groups. Lisanti et al. [73] reported the presence of VDAC1 in a catalogue of proteins identified in caveolae. Caveolae are domains of the plasma membrane where specific lipid and proteins are subcompartimentalized. A family of proteins termed caveolins organizes these membranous domains that have specific functions in the trafficking between the plasma membrane and the rest of the cell [74]. On the other hand, in our laboratory the purification and functional characterization of porin in plasma membrane was accomplished [43]. Our group took advantage of the work of Jakob et al. [75]. These investigators labeled intact cells with the membrane-impermeable reagent NH-SS-biotin and, based on the assumption that NH-SS-biotin could react only with protein segments exposed to the external surface of the plasma membrane, identified VDAC1 as a biotinylated protein [75]. Starting from intact CEM cells labeled with NH-SS-biotin, caveolae were isolated and subject to a purification procedure consisting in two chromatographic steps: a first step in HTP/celite was used to isolate porin from other membrane proteins [30]; next, an affinity purification step on streptavidin–agarose was used to separate VDAC1 protein labeled with NH-SS-biotin from unlabeled VDAC1, i. e. not unequivocally present in the plasma membrane [75]. The procedure resulted in a single protein band on SDS–PAGE that was identified with porin/VDAC1 on the basis of its electrophoretic mobility, reactivity with polyclonal and monoclonal antibodies, and experiments in planar bilayers that displayed typical porin electrophysiological activity after reconstitution [43].

15.3.1.1 **Targeting to the Plasma Membrane**

The finding of porin/VDAC in the plasma membrane poses new problems, and in particular considerations about how this protein may reach its target and what is its function there.

In a recent paper, Buettner and coworkers explained the targeting of VDAC1 to the plasma membrane with an alternative splicing event that would add a targeting pre-sequence to the VDAC1-coding sequence [38] (see Section 15.2.3.1). This is a possibility that has been demonstrated in other cases, like for transcription factor A which can be targeted to the mitochondrion or to the nucleus depending of the pre-sequence fused to it [76]. Evidence for the involvement of the "pl-" (plasma membrane) pre-sequence has been recently reported [77]. These investigators supported the plasmalemmal localization of VDAC1 in C1300 mouse neuroblastoma cells with three kinds of experiment. RT-PCR of C1300 mRNA using primers specific to the leader sequence of pl-VDAC showed a band with the expected molecular weight and sequence. Immunofluorescence confocal microscopy of fixed cells showed co-localization of VDAC with the cholera toxin β subunit, a marker of plasma membrane lipid rafts. At the end transfection of C1300 with an antisense oligonucleotide directed against the pl-specific leader sequence reduced the cellular immunostaining on the plasma membrane [77].

Reports by the group of Thinnes, on the other hand, have always outlined the possibility that VDAC1 might be targeted to the plasma membrane without such a specific sequence. In particular, they have demonstrated that, upon transfection in *Xenopus laevis* of a cDNA containing the human VDAC1-coding sequence carrying a FLAG epitope at the N-terminus, but lacking the plasma membrane targeting peptide, the modified porin could be detected on the surface of the oocytes [78].

Understanding the targeting mechanism of porin into the plasma membrane requires preliminary comprehension of the targeting mechanism into the mitochondria.

The post-translational import pathway of porin/VDAC has been elucidated in some detail. It has been proposed that the precursor protein binds the surface receptor Tom 20 via hydrophobic interactions. Krimmer et al. reported also binding to Tom 5, 7 and 22 [79]. In particular, Tom 20 would be capable of catalyzing direct insertion of the protein into the lipid bilayer or might play a direct role in guiding the conformational change that allows the trans-bilayer β-barrel channel to form [80, 81]. This import mechanism would take into account the lack of any signal sequence in the porin/VDAC [80–82]. Also, a specific chaperone of VDAC is not known. Most of the insertion would happen via a spontaneous integration of the porin protein recruited by Tom 20 into the lipid bilayer [80, 83]. Previous experiments showed indeed that purified detergent-solubilized VDAC is able to integrate into lipid bilayers as a functional entity spontaneously and in a cooperative way [83]. Other investigators showed that even the water-soluble porin/VDAC1, i. e. porin purified from cells or tissues and delipidated, regains its pore-forming activity in planar bilayers after incubation with lipids [84]. More impressive, the same polypeptide expressed in *Escherichia coli* and purified under denaturing conditions can refold in the presence of lipids showing functional features undistinguishable from those of the native protein [85]. This is similar to the targeting process of bacterial porins. In bacteria, outer membrane proteins are secreted in the periplasmic space in an unfolded form, and they subsequently insert and fold spontaneously in the outer membrane. The capacity of porins to fold spontaneously in the presence of lipids and in the absence of any accessory proteins has been proved for many years [86–87].

These results indicate that there is a specific contribution of the particular structure of porin/VDAC1 in the membrane insertion process. Bacterial porins and possibly VDACs have the potential to spontaneously insert in a phospholipid bilayer. The most important question thus remains: what would confer exclusive membrane specificity to the process?

Another hypothesis taking into account the absence of targeting sequences in VDAC and the potential of its spontaneous insertion in any phospholipid bilayer shifts the targeting decision at a transcription level. When we noticed the presence of alternative splicing events not affecting the VDAC amino acid sequence, but only the untranslated mRNA regions, we suggested that this process could be responsible for the different targeting in the cell [33, 66]. Different pools of ribosomes exist and it has been shown that translationally active ribosomes accumulate on the surface of yeast mitochondria [88–89]. In yeast, for example, the transcript for VDAC1 is translated in the cytoplasm, while the VDAC2 mRNA is reported close to mitochondria (Claude Jacq, Paris, France, personal communication). Thus, mRNA specifically directed to mitochondria could carry a population of VDAC molecules directly in the organelle by a co-translational mechanism [90, 91]. Another mRNA devoid of this signal could produce a population of VDAC molecules free to diffuse in the cytoplasm and, thus, also able to insert into the plasma membrane. Since alternative splicing events at the 5′-UTR producing different transcripts have indeed been found for VDAC1 not only in *Drosophila melanogaster* [33], but also in human (see Section 15.2.3.2.4 and Figure 15.3), they could carry specific targeting signals in the mRNA untranslated regions [92, 93].

Taken together, the information available in the literature does not allow us to exclude any of the above-mentioned targeting mechanism.

15.3.1.2 **Porin/VDAC Function in the Plasma Membrane**

The presence of a putative unspecific pore-forming protein in the plasma membrane is considered a potentially lethal occurrence for the cell. In spite of this, there are several proposals in the literature claiming functional roles of VDAC in the plasma membrane.

Consistent with the physiological function of porin/VDAC1 in the mitochondrion, it has been suggested that this protein might also form channels in the plasma membrane. Porin-like activities in the plasma membrane have been searched by patch-clamping (for a review, see [66]). Large conductance values, compatible with porin features, have been observed in excised patches. Since the porin-like conductance in the plasma membrane showed similarities with the Maxi Cl^- channel properties, the molecular identity of these two channels was proposed [71, 77, 94]. Unfortunately the same activities were not recorded in the cell-attached mode and this was explained by the need to maintain the porin channel strictly closed by some unknown mechanism or some modulator molecule until the cell would order its opening (e. g. [71, 95]). Recently, the Maxi Cl^- channel conductance both under cell-attached and whole-cell patch-clamp conditions was shown in cells treated with anti-estrogens [96]. Interestingly, when C1300 cells were transfected with pl-VDAC antisense oligonucleotides the magnitude of the anti-estrogen-activated Maxi Cl^- currents was greatly reduced [77]. This could thus correlate at a molecular level VDAC1 and the Maxi Cl^- channel in the plasma membrane.

In a recent communication, it has been proposed that the plasma membrane VDAC1 can function as a redox enzyme, capable of reducing extracellular ferricyanide in the presence of intracellular NADH [97]. Once again, investigators isolated porin by chance. They sought to identify the plasma membrane enzyme responsible for NADH:ferricyanide reduction. The presence of a NADH:ferricyanide reductase in the plasma membrane is suggested by the finding that oxidation of cytosolic NADH causes concomitant reduction of ferricyanide, an artificial impermeable electron acceptor [98, 99]. This enzymatic activity is referred to as plasma membrane NADH-oxidoreductase (PMOR). Starting from a purified Namalwa cell plasma membrane, Triton X-100-solubilized proteins were subject to several chromatographic procedures. The enrichment of NADH:ferricyanide activity led to the isolation of a 35-kDa band. It was identified by MALDI-TOF analysis as human VDAC1 [97]. To confirm that VDAC1 can function as an NADH:ferricyanide reductase, anti-VDAC1 antibodies were used to immunoprecipitate the protein. The immunopellet was shown to have NADH:ferricyanide reductase activity. Finally, VDAC1 was purified from plasma membrane and from rat liver mitochondria. These purified samples exhibited NADH:ferricyanide reductase activity. Although it is unlikely that VDAC1 would function as a reductase in mitochondria, these data show that in the presence of NADH, VDAC1 can directly catalyze the reduction of ferricyanide. This novel finding poses new questions: how the same polypeptide may function in different ways in two different membranes? Likely explanations are that (i) the porin polypeptide may assume a different conformation when targeted to the plasma membrane and (ii) porin/VDAC1 is controlled by some effector. In this last case, subdomain localization, e. g. in caveolae, might be a strategic way to prevent any harmful function of the VDAC protein [97]

Other possible roles for VDAC1 in the plasma membrane come from more recent publications. VDAC was identified in the plasma membrane of human endothelial cells as the receptor for plasminogen Kringle 5 in these cells [100] and, from the specific binding of a neuroactive steroid analog, it was proposed to be part of the $GABA_A$ receptor in rat brain membranes [101].

Porin/VDAC1 has also been detected, in addition to mitochondria, in the sarcoplasmic reticulum of skeletal muscle [102–106] – its function there is still unknown.

15.3.2
Interactions of Porin with other Cellular Structures

The first evidence that porin could bind other proteins comes from its identification as the hexokinase receptor on the mitochondria surface [107, 108]. Other kinases such as glucokinase, glycerol kinase, mitochondrial creatine kinase and c-Raf kinase have been found to associate with VDAC [109–111]. The association of cytosolic kinases with VDAC is thought to give a preferential access to the phosphorylated substrates produced in the mitochondrion. In this respect, porin/VDAC might be involved in the coupling of the glucose metabolism with oxidative phosphorylation.

Also connected with a regulatory function at the border between mitochondrial and cellular metabolisms is the presence of porin/VDAC1 in contact sites – structures where the outer and the inner mitochondrial membranes are fused [112]. Contact sites were shown to have a functional meaning and are a privileged location for interactions of porin/VDAC1 with other proteins (this is the subject of another contribution in this book, see Chapter 16).

Furthermore, porin/VDAC1 appears to interact with cytoskeletal proteins. By affinity chromatography, microtubule-associate protein (MAP) 2 was shown to bind a porin dimer [113]. Human recombinant gelsolin, a Ca^{2+}-dependent protein that modulates actin assembly and disassembly, was demonstrated to bind directly to porin/VDAC reconstituted in liposomes and to block its activity [114]. In a two-hybrid screening, by using human VDAC1 as a bait protein, it was shown that porin/VDAC1 interacts with the dynein light chain Tctex-1 and the heat-shock protein peptide-binding protein 74 [115]. These proteins were expressed as recombinant binding partners and were able to alter the electrophysiological properties of human VDAC1, thus also reflecting a functional role in the cell [115]. Finally, actin in the monomeric form (G-actin), but not in the fibrous form, was able to reduce by as much as 85 % the VDAC-mediated membrane conductance at elevated membrane potentials in reconstituted systems [116]. These reports raise the idea that porin/VDAC1 might act as a mediator of the interactions between mitochondria and cytoskeletal structures. Thus, some domain of porin/VDAC might have a role as a docking structure to support the motion and the distribution of mitochondria to high-energy-consuming districts in specialized cells [117, 118]. Interestingly, most of the interactions with cytoskeleton protein reported in these works affected the pore-forming activity of VDAC1 [114–116].

Another recent role attributed to porin/VDAC was as an actor in the Ca^{2+} trafficking in the cell. Subdomains of the endoplasmic/sarcoplasmic reticulum [(mitochondria-associated membranes (MAMs)] [119] look juxtaposed to mitochondria [120]. These domains have been implicated in the phospholipid transfer to the mitochondria. In these subdomains (MAMs) the enrichment of Ryenodine receptors together with that of the Ca^{2+} ATPase pump from the sarcoendoplasmic reticulum was evidenced [121]. The recent report of a tight regulation of Ca^{2+} flow through VDAC [122] and the evidence that over-expression of VDAC facilitates Ca^{2+} signal propagation to mitochondria [123] indicate that the OMM, and thus VDAC, may have an important role in the physiological control of Ca^{2+} signaling between IP_3R/RyR (IP_3 and cyanodine receptor) and mitochondria [124].

If the porin/VDAC1 interaction is mediated by a close relationship with sarcoplasmic structures, as it seems to be the case in MAMs, the reported localization of porin/VDAC1 in the sarcoplasmic reticulum of skeletal muscle [102–106] could be explained as the result of a co-purification of porin/VDAC1 with sarcoplasmic membranes or with the histochemical staining of very close membranes where the exclusive organellar localization of the fluorescence would be undistinguishable. If this is the case, the subcellular localization of porin/VDAC1 would be reduced essentially to the mitochondrial outer membrane and the plasma membrane.

15.4
Conclusions

In this work we have reviewed a section of the very large and growing literature concerned with the pore-forming protein VDAC or porin. As usual in this kind of work, many areas of research await clarification and this hinders our ability to draw a definitive picture even about the limited topics we have focused on – the expression of porin genes families and the multitopological localization of this protein. In particular, the function of the VDAC2 and VDAC3 isoforms in mammalian cell is not appreciated: the genetic mechanisms underlying the necessary coordination of VDAC isoform gene expression have not yet been elucidated and the *cis*-acting regions of the corresponding genes have been until now poorly studied. The presence of porin/VDAC1 in the plasma membrane looks to be reasonably established, despite its apparent oddity, and the contribution reporting a novel function of porin/VDAC1 in the plasma membrane [97] opens new research fields whose consequence have still to be considered. The increasing number of papers indicating the interaction of porin with other cellular proteins or macromolecular structures witnesses a growing interest in the cellular coordination of mitochondrial metabolism with other functions. In this case, the role of porin/VDACs, by far the most abundant protein(s) on the mitochondrion surface, should overwhelm the simple pore-forming function. Thus, knowledge of the VDAC(s) structure at atomic resolution is urgently required to understand how and what other functions could be associated with this still elusive protein.

Acknowledgments

The authors acknowledge University of Catania, COFIN 2003058409_004 and FIRB-MIUR RBNE01ARR4 for the financial support.

References

1 Schein, S. J., Colombini, M., Finkelstein, A. Reconstitution in planar lipid bilayers of a voltage-dependent anion-selective channel obtained from paramecium mitochondria. *J Membr Biol* **1976**, *30*, 99–120.

2 Colombini, M. A candidate for the permeability pathway of the outer mitochondrial membrane. *Nature* **1979**, *279*, 643–645.

3 Benz, R. Porin from bacterial and mitochondrial outer membranes. *CRC Crit Rev Biochem* **1985**, *19*, 145–190.

4 De Pinto, V., Ludwig, O., Krause, J., Benz, R., Palmieri, F. Porin pores of mitochondrial outer membranes from high and low eukaryotic cells: biochemical and biophysical characterization. *Biochim Biophys Acta* **1987**, *894*, 109–119.

5 Sorgato, M. C., Moran, O. Channels in mitochondrial membranes: knowns, unknowns, and prospects for the future. *Crit Rev Biochem Mol Biol* **1993**, *18*, 127–171.

6 Benz, R. Permeation of hydrophilic solutes through mitochondrial outer membranes: a review on mitochondrial porins. *Biochim Biophys Acta* **1994**, *1197*, 167–196.

7 Colombini, M., Blachly-Dyson, E., Forte, M. VDAC, a channel in the outer mitochondrial membrane. *Ion Channels* **1996**, *4*, 169–202.

8 Blachly-Dyson, E., Forte, M. VDAC channels. *IUBMB Life* **2001**, *52*, 113–118.

9 De Pinto, V., Messina, A., Accardi, R., Aiello, R., Guarino, F., Tomasello, M. F., Tommasino, M., Tasco, G., Casadio, R., Benz, R., De Giorgi, F., Ichas, F., Baker, M., Lawen, A. New functions of an old protein: the eukaryotic porin or voltage dependent anion selective channel (VDAC). *It J Biochem* **2003**, *52*, 17–24.

10 Huizing, M., Ruitenbeek, W., Thinnes F. P., De Pinto, V., Wendel, U., Trijbels, F. J., Smit L. M., ter Laak, H. J., van den Heuvel, L. P. Deficiency of the voltage-dependent anion channel: a novel cause of mitochondriopathy. *Pediatr Res* **1996**, *39*, 760–765.

11 De Pinto, V., Messina, A., Schmid, A., Simonetti, S., Carnevale, F., Benz, R. Characterization of channel-forming activity in muscle biopsy from a porin-deficient human patient. *J Bioenerg Biomembr* **2000**, *32*, 459–467.

12 Blachly-Dyson, E., Zambronicz, E. B., Yu, W. H., Adams, V., McCabe, E. R., Adelman, J., Colombini, M., Forte, M. Cloning and functional expression in yeast of two human isoforms of the outer mitochondrial membrane channel, the voltage-dependent anion channel. *J Biol Chem* **1993**, *268*, 1835–1841.

13 Rahmani, Z., Maunoury, C., Siddiqui, A. Isolation of a novel human voltage-dependent anion channel gene. *Eur J Hum Genet* **1998**, *6*, 337–340.

14 Sampson, M. J., Lovell, R. S., Craigen, W. J. The murine voltage-dependent anion channel gene family. Conserved structure and function. *J Biol Chem* **1997**, *272*, 18966–18973.

15 Casadio, R., Jacoboni, I., Messina, A., De Pinto, V. A 3D model of the voltage-dependent anion channel (VDAC). *FEBS Lett* **2002**, *520*, 1–7.

16 Xu, X., Decker, W., Sampson, M. J., Craigen, W. J., Colombini, M. Mouse VDAC isoforms expressed in yeast: channel properties and their roles in mitochondrial outer membrane permeability. *J Membr Biol* **1999**, *170*, 89–102.

17 Blachly-Dyson, E., Song, J., Wolfgang, W. J., Colombini, M., Forte, M. Multicopy suppressors of phenotypes resulting from the absence of yeast VDAC encode a VDAC-like protein. *Mol Cell Biol* **1997**, *17*, 5727–5738.

18 Oliva, M., De Pinto, V., Barsanti, P., Caggese, C. A genetic analysis of the porin gene encoding a voltage-dependent anion channel protein in *Drosophila melanogaster*. *Mol Genet Genomics* **2002**, *267*, 746–756.

19 Elkeles, A., Devos, K. M., Graur, D., Zizi, M., Breiman, A. Multiple cDNAs of wheat voltage-dependent anion channels (VDAC): isolation, differential expression, mapping and evolution. *Plant Mol Biol* **1995**, *29*, 109–124.

20 Al Bitar, F., Roosens, N., Smeyers, M., Vauterin, M., Boxtel, J. V., Jacobs, M., Homble, F. Sequence analysis, transcriptional and posttranscriptional regulation of the rice VDAC family. *Biochim Biophys Acta* **2003**, *1625*, 43–51.

21 Mihara, K., Sato, R. Molecular cloning and sequencing of cDNA for yeast porin, an outer mitochondrial membrane protein: a search for targeting signal in the primary structure. *EMBO J* **1985**, *4*, 769–774.

22 Dihanich, M., Suda, K., Schatz, G. A yeast mutant lacking mitochondrial porin is respiratory-deficient, but can recover respiration with simultaneous accumulation of an 86-kd extramitochondrial protein. *EMBO J* **1987**, *6*, 723–728.

23 Michejda, J., Guo, X. J., Lauquin, G. J. The respiration of cells and mitochon-

dria of porin deficient yeast mutants is coupled. *Biochem Biophys Res Commun* **1990**, *171*, 354–361.

24 Dihanich, M., van Tuinen, E., Lambris, J. D., Marshallsay, B. Accumulation of viruslike particles in a yeast mutant lacking a mitochondrial pore protein. *Mol Cell Biol* **1989**, *9*, 1100–1108.

25 Dihanich, M., Schmid, A., Oppliger, W., Benz, R. Identification of a new pore in the mitochondrial outer membrane of a porin-deficient yeast mutant. *Eur J Biochem* **1989**, *181*, 703–708.

26 Blachly-Dyson, E., Peng, S. Z., Colombini, M., Forte, M. Selectivity changes in site-directed mutants of the VDAC ion channel: structural implications *Science* **1990**, *247*, 1233–1236.

27 Celniker, S. E., Wheeler, D. A., Kronmiller, B., Carlson, J. W., Halpern, A., Patel, S., Adams, M., Champe, M., Dugan, S. P., Frise, E., et al. [Multi-author contribution]. Finishing a whole-genome shotgun: release 3 of the *Drosophila melanogaster* euchromatic genome sequence. *Genome Biol* **2002**, *3*, 0079.

28 Misra, S., Crosby, M. A., Mungall, C. J., Matthews, B. B., Campbell, K. S., Hradecky, P., Huang, Y., Kaminker, J. S., Millburn, G. H., Prochnik, S. E., et al. [Multi-author contribution]. Annotation of the *Drosophila melanogaster* euchromatic genome: a systematic review. *Genome Biol* **2002**, *3*, 0083.

29 De Pinto, V., Benz, R., Caggese, C., Palmieri, F. Characterization of the mitochondrial porin from *Drosophila melanogaster. Biochim Biophys Acta* **1989**, *987*, 1–7.

30 De Pinto, V., Prezioso, G., Palmieri, F. A simple and rapid method for the purification of the mitochondrial porin from mammalian tissues. *Biochim Biophys Acta* **1987**, *905*, 499–502.

31 Messina, A., Neri, M., Perosa, F., Caggese, C., Marino, M., Caizzi, R., De Pinto, V. Cloning and chromosomal localization of a cDNA encoding a mitochondrial porin from *Drosophila melanogaster. FEBS Lett* **1996**, *384*, 9–13.

32 Ryerse, J., Blachly-Dyson, E., Forte, M., Nagel B. Cloning and molecular characterization of a voltage-dependent anion-selective channel (VDAC) from *Drosophila melanogaster. Biochim Biophys Acta* **1997**, *1327*, 204–212.

33 Oliva, M., Messina, A., Ragone, G., Caggese, C., De Pinto, V. Sequence and expression pattern of the *Drosophila melanogaster* mitochondrial porin gene: evidence of a conserved protein domain between fly and mouse. *FEBS Lett* **1998**, *430*, 327–332.

34 Anflous, K., Blondel, O., Bernard, A., Khrestchatiski, M., Ventura-Clapier, R. Characterization of rat porin isoforms: cloning of a cardiac type-3 variant encoding an additional methionine at its putative N-terminal region. *Biochim Biophys Acta* **1998**, *1399*, 47–50.

35 Sampson, M. J., Ross, L., Decker, W. K., Craigen, W. J. A novel isoform of the mitochondrial outer membrane protein VDAC3 via alternative splicing of a 3-base exon. *J Biol Chem* **1998**, *273*, 30482–30486.

36 Decker, W. K., Craigen, W. J. The tissue-specific, alternatively spliced single ATG exon of the type 3 Voltage-Dependent Anion Channel gene does not create a truncated protein isoform *in vivo. Mol Genet Metab* **2000**, *70*, 69–74.

37 Buettner, R., Papoutsoglou, G., Scemes, E., Spray, D. C., Dermietzel, R. Evidence for secretory pathway localization of a voltage-dependent anion channel isoform. *Proc Natl Acad Sci USA* **2000**, *97*, 3201–3206.

38 Messina, A., Guarino, F., Oliva, M., van den Heuvel, L. P., Smeitink, J., De Pinto, V. Characterization of the human porin isoform 1 (HVDAC1 gene by amplification on the whole human genome: a tool for porin deficiency analysis. *Biochem Biophys Res Commun* **2000**, *270*, 787–792.

39 Messina, A., Oliva, M., Rosato, C., Huizing, M., Ruitenbeek, W., van den Heuvel, L. P., Forte, M., Rocchi, M. De Pinto, V. Mapping of the human Voltage-Dependent Anion Channel isoforms 1 and 2 reconsidered. *Biochem Biophys Res Commun* **1999**, *255*, 707–710.

40 Decker, W. K., Boules, K. R., Schatte, E. C., Towbin, J. A., Craigen, W. J. Revised

fine mapping of the human voltage-dependent anion channel loci by radiation hybrid analysis. *Mamm Genome* **1999**, *10*, 1041–1042.
41 Thinnes, F. P., Götz, H., Kayser, H., Benz, R., Schmidt, W. E., Kratzin, H. D., Hilschmann, N., Kratzin, H. D., Hilschmann, N. Zur Kenntnis der Porine des Menschen I. Reinigung eines Porins aus menschlichen B-Lymphozyten (Porin 31HL) und sein topochemischer Nachweis auf dem Plasmalemm der Herkunftszelle. *Biol Chem Hoppe-Seyler* **1989**, *370*, 1253–1264.
42 Lewis, T. M., Dulhunty, A. F., Junankar, P. R., Stanhope, C. Ultrastructure of sarcoballs on the surface of skinned amphibian skeletal muscle fibres. *J. Muscle Res Cell Motil* **1992**, *13*, 640–653.
43 Bathori, G., Parolini, I., Tombola, F., Szabò, I., Messina, A., Oliva, M., De Pinto, V., Lisanti, M., Sargiacomo, M., Zoratti, M. Porin is present in caveolae and caveolae-related domain. *J Biol Chem* **1999**, *274*, 29607–29612.
44 Ha, H., Hajek, P., Bedwell, D. M., Burrows, P. D. A mitochondrial porin cDNA predicts the existence of multiple human porins. *J Biol Chem* **1993**, *268*, 12143–12149.
45 Yu, W. H., Wolfgang, W., Forte, M. Subcellular localization of human voltage-dependent anion channel isoforms. *J Biol Chem* **1995**, *270*, 13998–14006.
46 Altschul, S. F., Gish, W., Miller, W., Myers, E. W., Lipman, D. J. Basic local alignment search tool. *J Mol Biol* **1990**, *215*, 403–410.
47 De Pinto, V., Benz, R., Palmieri, F. Interaction of non-classical detergents with the mitochondrial porin. *Eur J Biochem* **1989**, *183*, 179–187.
48 McEnery, M. W., Snowman, A. M., Trifiletti, R. R., Snyder, S. H. Isolation of the mitochondrial benzodiazepine receptor: association with the voltage-dependent anion channel and the adenine nucleotide carrier. *Proc Natl Acad Sci USA* **1992**, *89*, 3170–3174.
49 Papadopoulos, V., Amri, H., Boujrad, N., Cascio, C., Culty, M., Garnier, M., Hardwick, M., Li, H., Vidic, B., Brown, A. S., Reversa, J. L., Bernassau, J. M., Drieu, K. Peripheral benzodiazepine receptor in cholesterol transport and steroidogenesis. *Steroids* **1997**, *62*, 21–28.
50 Koopman, P. *Sry* and *Sox9*: mammalian testis-determining genes. *Cell Mol Life Sci* **1999**, *55*, 839–856.
51 Hinsch, K. D., Asmarinah, Hinsch, E., Konrad, L. VDAC2 (porin-2) expression pattern and localization in the bovine testis. *Biochim Biophys Acta* **2001**, *1518*, 329–333.
52 Flügge, U. I., Benz, R. Pore-forming activity in the outer membrane of the chloroplast envelope. *FEBS Lett* **1984**, *169*, 85–89.
53 Pohlmeyer, K., Soll, J., Steinkamp, T., Hinnah, S., Wagner, R. Isolation and characterization of an amino acid-selective channel protein present in the chloroplastic outer envelope membrane. *Proc Natl Acad Sci USA* **1997**, *94*, 9504–9509.
54 Pohlmeyer, K., Soll., J., Grimm, R., Hill, K., Wagner, R. A high-conductance solute channel in the chloroplastic outer envelope membrane. *Plant Cell* **1998**, *10*, 1207–1216.
55 Bölter, B., Soll, J., Hill, K., Hemmler, R., Wagner R. A rectifying ATP-regulated solute channel in the chloroplastic outer envelope from pea. *EMBO J* **1999**, *18*, 5505–5516.
56 Bölter, B., Soll, J. Ion channels in the outer membranes of chloroplasts and mitochondria: open doors or regulated gates? *EMBO J* **2001**, *20*, 935–940.
57 Rohl, T., Motzkus, M., Soll, J. The outer envelope protein OEP24 from pea chloroplasts can functionally replace the mitochondrial VDAC in yeast. *FEBS Lett* **1999**, *460*, 491–494.
58 Aljamal, J. A., Genchi, G., De Pinto, V., Stefanizzi, L., De Santis, A., Benz, R., Palmieri F. Purification and characterization of porin from corn (*Zea mays* L.) mitochondria. *Plant Physiol* **1993**, *102*, 615–621.
59 Blumenthal, A., Kahn, K., Beja, O., Galun, L., Colombini, M., Breiman, A. Purification and characterization of the voltage-dependent anion selective

channel (VDAC) protein from wheat mitochondrial membranes. *Plant Physiol* **1993**, *101*, 579–587.

60 Schmid, A., Kromer, S., Heldt, H. W., Benz, R. Identification of two general diffusion channels in the outer membrane of pea mitochondria. *Biochim Biophys Acta* **1992**, *1112*, 174–180.

61 Fischer, K., Weber, A., Brink, S., Arbinger, B., Schunemann, D., Borchert, S., Heldt, H. W., Popp, B., Benz, R., Link, T. A., Eckerskorn, C., Flügge, U. J. Porins from plants. Molecular cloning and functional characterization of two new members of the porin family. *J Biol Chem* **1994**, *269*, 25754– 25760.

62 Heins, L., Mentzel, H., Schmid, A., Benz, R., Schmitz, U. K. Biochemical, molecular, and functional characterization of porin isoforms from potato mitochondria. *J Biol Chem* **1994**, *269*, 26402–26410.

63 Elkeles, A., Breiman, A., Zizi, M. Functional differences among wheat voltage-dependent anion channel (VDAC) isoforms expressed in yeast. Indication for the presence of a novel VDAC-modulating protein? *J Biol Chem* **1997**, *272*, 6252– 6260.

64 Saccone, C., Caggese, C., D'Erchia, A. M., Lanave, C., Oliva, M., Pesole, G. Molecular clock and gene evolution. *J Mol Evol* **2003**, *57*, 1–9.

65 Kayser, H., Kratzin, H. D., Thinnes, F. P., Götz, H., Schmidt, W. E., Eckart, K., Hilschmann, N. Zur Kenntnis der Porine des Menschen II. Charakterisierung und Primarstruktur eines 31-kDa-Porins aus menschlichen BLymphozyten (Porin 31HL). *Biol Chem Hoppe-Seyler* **1989**, *370*, 1265–1278.

66 Báthori, G., Parolini, I., Szabó, I., Tombola, F., Messina, A., Oliva, M., Sargiacomo, M., De Pinto, V., Zoratti, M. Extramitochondrial porin: facts and hypotheses. *J Bioenergetics Biomembr* **2000**, *32*, 79–89.

67 Babel, D., Götz, W., Götz, H., Thinnes, F. P., Jürgens, L., König, U., Hilschmann, N. Studies on human porin. VI. Production and characterization of eight monoclonal mouse antibodies against the human VDAC "Porin 31HL" and their application for histotopological studies in human skeletal muscle. *Biol Chem Hoppe-Seyler* **1991**, *372*, 1027–1034.

68 König, U., Götz, H., Walter, G., Babel, D., Hohmeier, H. E., Thinnes, F. P., Hilschmann, N. Studies on human porin. V. The expression of "porin31HL" in the plasmalemma is not by cell transformation. *Biol Chem Hoppe-Seyler* **1991**, *372*, 565–572.

69 Cole, T., Awni, L. A., Nyakatura, E., Götz, H., Walter, G., Thinnes, F. P., Hilschmann, N. Studies on human porin. VIII. Expression of "Porin 31HL" channels in the plasmalemma of the acute-lymphoblastic-leukemia cell line KM3 as revealed by light- and electron-microscopy. *Biol Chem Hoppe-Seyler* **1992**, *373*, 891–896.

70 Puchelle, E., Jacquot, J., Fuchey, C., Burlet, H., Klossek, J.-M., Gilain, L., Triglia, J.-M., Thinnes, F. P., Hilschmann, N. Studies on human porin. IX. Immunolocalization of porin and CFTR channels in human surface respiratory epithelium. *Biol Chem Hoppe-Seyler* **1993**, *374*, 297–304.

71 Dermietzel, R., Hwang, T. K., Buettner, R., Hofer, A., Dotzler, E., Kremer, M., Deutzmann, R, Thinnes, F. P., Fishman, G. I., Spray, D. C., Siemen, D. Cloning and *in situ* localisation of a brain-derived porin that constitutes a large-conductance anion channel in astrocytic plasmamembranes. *Proc Natl Acad Sci USA* **1994**, *91*, 499–503.

72 Moon, J. I., Jung, Y. W., Ko, B. H., De Pinto, V., Jin, I., Moon I. S. Presence of a voltage-dependent anion channel 1 in the rat postsynaptic density fraction. *Neuroreport* **1999**, *10*, 443–447.

73 Lisanti, M. P., Scherer, P. E., Vidugiriene, J., Tang, Z., Hermanowski Vosatka, A., Tu, Y. H., Cook, R. F., Sargiacomo, M. Characterization of caveolin-rich membrane domains isolated from an endothelial-rich source: implications for human disease. *J Cell Biol* **1994**, *126*, 111–126.

74 Razani, B., Woodman, S. E., Lisanti, M. P. Caveolae: from cell biology to animal physiology. *Pharmacol Rev* **2002**, *54*, 431–467.

75 Jakob, C., Götz, H., Hellmann, T., Hellmann, K. P., Reymann, S., Flörke, H., Thinnes, F. P., Hilschmann, N. Studies on human porin. XIII. The type-1 VDAC "porin 31HL" biotinylated at the plasmalemma of trypan blue excluding human B lymphocytes. *FEBS Lett* **1995**, *368*, 5–9.
76 Larsson, N. G., Garman, J. D., Oldfors, A., Barsh, G. S., Clayton, D. A. A single mouse gene encodes the mitochondrial transcription factor A and a testis-specific nuclear HMG-box protein. *Nature Genet* **1996**, *13*, 296–302.
77 Bahamonde, M. I., Fernandez-Fernandez, J. M., Guix, F. X., Vazquez, E., Valverde, M. A. Plasma membrane Voltage-Dependent Anion Channel mediates antiestrogen activated Maxi Cl^- currents in C1300 neuroblastoma cells. *J Biol Chem* **2003**, *278*, 33284–33289.
78 Steinacker, P., Awni, L. A., Becker, S., Cole, T., Reymann, S., Hesse, D., Kratzin, H. D., Morys-Wortmann, C., Schwarzer, C., Thinnes, F. P., Hilschmann, N. The plasma membrane of *Xenopus laevis* oocytes contains voltage-dependent anion-selective porin channels. *Int J Biochem Cell Biol* **2000**, *32*, 225–234.
79 Krimmer, T., Rapaport, D., Ryan, M. T., Meisinger, C., Kassenbrock, C. K., Blachly-Dyson, E., Forte, M., Douglas, M. G., Neupert, W., Nargang, F. E., Pfanner, N. Biogenesis of porin of the outer mitochondrial membrane involves an import pathway via receptors and the general import pore of the TOM complex. *J Cell Biol* **2001**, *152*, 289–300.
80 Schleiff, E., Turnbull, J. L. Characterization of the N-terminal targeting signal binding domain of the mitochondrial outer membrane receptor, Tom 20. *Biochemistry* **1998**, *37*, 13052–13058.
81 Schleiff, E., Silvius, J. R., Shore, G. C. Direct membrane insertion of voltage-dependent anion-selective channel protein by mitochondrial Tom20. *J Cell Biol* **1999**, *145*, 973–978.
82 Schleiff, E. Signals and receptors: the translocation machinery on the mitochondrial surface. *J Bioenerg Biomembr* **2000**, *32*, 55–66.
83 Xu, X., Colombini, M. Self-catalyzed insertion of proteins into phospholipid membranes. *J Biol Chem* **1996**, *271*, 23675–23682.
84 Popp, B., Schmid, A., Benz, R. Role of sterols in the functional reconstitution of water-soluble mitochondrial porins from different organisms. *Biochemistry* **1995**, *34*, 3352–3361.
85 Popp, B., Court, D. A., Benz, R., Neupert, W., Lill, R. The role of the N and C termini of recombinant Neurospora mitochondrial porin in channel formation and voltage-dependent gating. *J Biol Chem* **1996**, *271*, 13593–13599.
86 Surrey T., Jahnig F. Refolding and oriented insertion of a membrane protein into a lipid bilayer. *Proc Natl Acad Sci USA* **1992**, *89*, 7457–7461.
87 Tamm, L. K., Arora, A., Kleinschmidt, J. H. Structure and assembly of β-barrel proteins. *J Biol Chem* **2001**, *276*, 32399–32402.
88 Kellems, R. E., Allison, V. F., Butow, R. A. Cytoplasmic type 80S ribosomes associated with yeast mitochondria. II. Evidence for the association of cytoplasmic ribosomes with the outer mitochondrial membrane *in situ*. *J Biol Chem* **1974**, *249*, 3297–3303.
89 Kellems, R. E., Allison, V. F., Butow, R. A. Cytoplasmic type 80S ribosomes associated with yeast mitochondria. IV. Attachment of ribosomes to the outer membrane of isolated mitochondria. *J Cell Biol* **1975**, *65*, 1–14.
90 Verner, K. Co-translational protein import into mitochondria: an alternative view. *Trends Biochem Sci* **1993**, *18*, 366–371.
91 Lithgow, T. Targeting of proteins to mitochondria. *FEBS Lett* **2000**, *476*, 22–26.
92 Corral-Debrinski, M., Blugeon, C., Jacq, C. In yeast, the 3′ untranslated region or the presequence of ATM1 is required for the exclusive localization of its mRNA to the vicinity of mitochondria. *Mol Cell Biol* **2000**, *20*, 7881–7892.
93 Marc, P., Margeot, A., Devaux, F., Blugeon, C., Corral-Debrinski, M.,

Jacq, C. Genome-wide analysis of mRNAs targeted to yeast mitochondria. *EMBO Rep* **2002**, *31*, 159–164.
94 Janisch, U., Skofitsch, G., Thinnes, F. T., Graier, W. F., Groschner, K. Evidence for the involvement of plasma membrane porin in agonist-induced activation of vascular endothelial cells. *Naunyn-Schmiedeberg's Arch, Pharmacol* **1993**, *347*, R72.
95 Mitchell, C. H., Wang, L., Jacob, T. J. C. A large-conductance chloride channel in pigmented ciliary epithelial cells activated by GTPgammaS. *J Membr Biol* **1997**, *158*, 167–175.
96 Diaz, M., Bahamonde, M. I., Lock, H., Muñoz, F. J., Hardy, S. P., Posas, F., Valverde, M. A. Okadaic acid-sensitive activation of Maxi Cl^- channels by triphenylethylene antioestrogens in C1300 mouse neuroblastoma cells. *J Physiol* **2001**, *536*, 79–88.
97 Baker, M. A., Ly, J. D., De Pinto, V., Lawen, A. VDAC1 is a transplasma membrane NADH:ferricyanide reductase. *J Biol Chem* **2004**, *279*, 4811–4819.
98 Baker, M. A., Lawen, A. Plasma membrane NADH-oxidoreductase system: a critical review of the structural and functional data. *Antioxid Redox Signal* **2000**, *2*, 197–212.
99 Navas, P., Sun, I. L., Morré, D. J., Crane, F. L. Decrease of NADH in HeLa cells in the presence of transferrin or ferricyanide. *Biochem Biophys Res Commun* **1986**, *135*, 110–115.
100 Gonzalez-Gronow, M., Kalfa, T., Johnson, C. E., Gawdi, G., Pizzo, S. V. The voltage-dependent anion channel (VDAC) is a receptor for plasminogen kringle 5 on human endothelial cells. *J Biol Chem* **2003**, *278*, 27312–27318.
101 Darbandi-Tonkabon, R., Hastings, W. R., Zeng, C.-M., Akk, G., Manion, B. D., Bracamontes, J. R., Steinbach, J. H., Mennerick, S. J., Covey, D. F., Evers, A. S. Photoaffinity labeling with a neuroactive steroid analogue. 6-Azi-pregnanolone labels Voltage-Dependent Anion Channel-1 in rat brain. *J Biol Chem* **2003**, *278*, 13196–13206.
102 Lewis, T. M., Roberts, M. L., Bretag, A. H. Immunolabelling of VDAC, the mitochondrial voltage-dependent anion channel, on sarcoplasmic-reticulum from amphibian skeletal muscle. *Neurosci Lett* **1994**, *181*, 83–86.
103 Jürgens, L., Kleineke, J., Brdiczka, D., Thinnes, F. P., Hilschmann, N. Localization of type-1 porin channel (VDAC) in the sarcoplasmic reticulum. *Biol Chem Hoppe-Seyler* **1995**, *376*, 685–689.
104 Shoshan-Barmatz, V., Hadad, N., Feng, W., Shafir, I., Orr, I., Varsanyi, M., Heilmeyer L. M. G. VDAC:porin is present in sarcoplasmic reticulum from skeletal muscle. *FEBS Lett* **1996**, *386*, 205–210.
105 Shafir, I., Feng, W., Shoshan-Barmatz, V. Dicyclohexylcarbodiimide interaction with the voltage-dependent anion channel from sarcoplasmic reticulum. *Eur J Biochem* **1998**, *253*, 627–638.
106 Massa, R., Marlier, L. N., Martorana, A., Cicconi, S., Pierucci, D., Giacomini, P., De Pinto, V., Castellani, L. Intracellular localization and isoform expression of the voltage-dependent anion channel (VDAC) in normal and dystrophic skeletal muscle. *J Muscle Res Cell Motil* **2000**, *21*, 433–442.
107 Fiek, C., Benz, R., Roos, N., Brdiczka, D. Evidence for identity between the hexokinase-binding protein and the mitochondrial porin in the outer membrane of rat liver mitochondria. *Biochim Biophys Acta* **1982**, *688*, 429–440.
108 Linden, M., Gellerfors, P., Nelson, B. D. Pore protein and the hexokinase-binding protein from the outer membrane of rat liver mitochondria are identical. *FEBS Lett* **1982**, *141*, 189–192.
109 Adams, V., Griffin, L., Towbin, J., Gelb, B., Worley, K., McCabe, E. R. B. Porin interaction with hexokinase and glycerol kinase: metabolic microcompartmentation at the outer mitochondrial-membrane. *Biochem Med Metab Biol* **1991**, *45*, 271–291.
110 Savabi F. Interaction of creatine kinase and adenylate kinase systems in muscle cells. *Mol Cell Biochem* **1994**, *133/134*, 145–152.

111 Le Mallay, V., Troppmair, J., Benz, R., Rapp, U. R. Negative regulation of mitochondrial VDAC channels by C-Raf kinase. *BMC Cell Biol* **2002**, *3*, 14.
112 Brdiczka, D. Contact sites between mitochondrial envelope membranes. Structure and function in energy- and protein-transfer. *Biochim Biophys Acta* **1991**, *1071*, 291–312.
113 Linden, M., Karlsson, G. Identification of porin as a binding site for MAP2, *Biochem Biophys Res Commun* **1996**, *218*, 833–836.
114 Kusano, H., Shimizu, S., Koya, R. C., Fujita, H., Kamada, S., Kuzumaki, N., Tsujimoto, Y. Human gelsolin prevents apoptosis by inhibiting apoptotic mitochondrial changes via closing VDAC. *Oncogene* **2000**, *19*, 4807–4814.
115 Schwarzer, C., Barnikol-Watanabe, S., Thinnes, F. P., Hilschmann, N. Voltage-dependent anion-selective channel (VDAC) interacts with the dynein light chain Tctex1 and the heat-shock protein PBP74, *Int J Biochem Cell Biol* **2002**, *34*, 1059–1070.
116 Xu, X., Forbes, J. G., Colombini, M. Actin modulates the gating of *Neurospora crassa* VDAC. *J Membr Biol* **2001**, *180*, 73–81.
117 Leterrier, J. F., Rusakov, D. A., Nelson, B. D., Linden, M. Interactions between brain mitochondria and cytoskeleton: evidence for specialized outer membrane domains involved in the association of cytoskeleton-associated proteins to mitochondria *in situ* and *in vitro*. *Microsc Res Tech* **1994**, *27*, 233–261.
118 Wagner, O. I., Lifshitz, J., Janmey, P. A., Linden, M., McIntosh, T. K., Leterrier J.-F. Mechanisms of mitochondria–neurofilament interactions. *J Neurosc* **2003**, *23*, 9046–9058.
119 Cui, Z., Vance, J. E., Chen, M. H., Voelker, D. R., Vance, D. E. Cloning and expression of a novel phosphatidylethanolamine *N*-methyltransferase. A specific biochemical and cytological marker for a unique membrane fraction in rat liver. *J Biol Chem* **1993**, *268*, 16655–16663.
120 Hajnóczky, G., Csordás, G., Madesh, M., Pacher, P. The machinery of local Ca^{2+} signalling between sarco-endoplasmic reticulum and mitochondria. *J Physiol* **2000**, *529*, 69–81.
121 Csordás, G., Hajnóczky, G. Sorting of calcium signals at the junctions of endoplasmic reticulum and mitochondria. *Cell Calcium* **2001**, *29*, 249–262.
122 Gincel, D., Zaid, H., Shoshan-Barmatz, V. Calcium binding and translocation by the voltage-dependent anion channel: a possible regulatory mechanism in mitochondrial function. *Biochem J* **2001**, *358*, 147–155.
123 Rapizzi, E., Pinton, P., Szabadkai, G., Wieckowski, M. R., Vandecasteele, G., Baird, G., Tuft, R. A., Fogarty, K. E., Rizzuto, R. Recombinant expression of the voltage-dependent anion channel enhances the transfer of Ca^{2+} microdomains to mitochondria. *J Cell Biol* **2002**, *159*, 613–624.
124 Hajnóczky, G., Csordás, G., Yi, M. Old players in a new role: mitochondria-associated membranes, VDAC, and ryanodine receptors as contributors to calcium signal propagation from endoplasmic reticulum to the mitochondria. *Cell Calcium* **2002**, *32*, 363–377.

16
Function of the Outer Mitochondrial Membrane Pore (Voltage-dependent Anion Channel) in Intracellular Signaling

Mikhail Vyssokikh and Dieter Brdiczka

16.1
Introduction

Eukaryotic porins are membrane proteins that form aqueous channels in the cell membrane and the mitochondrial outer membrane. In contrast to bacterial porins that have a similar function, the structure of eukaryotic porins is not known at a useful resolution. The mitochondrial outer membrane pore was first characterized by Marco Colombini [1] as a voltage-dependent anion channel (VDAC). The properties of VDACs have been widely investigated in reconstituted systems by several groups [1, 2] exploring the main differences compared to bacterial porins, i. e. sensitivity to voltage already at 30 mV and to the polarity of the applied voltage. In general, bacterial and mitochondrial outer membranes have segregating functions, and pore proteins in the membranes control limited exchange. However, in bacteria, the outer membrane has more protective functions, whereas in mitochondria, communicative functions are more important.

The VDAC plays an important role in the coordination of the communication between mitochondria and cytosol. A substantial aspect of this management is a transient formation of complexes with other proteins. This will be the topic of this chapter.

16.2
Structure and Isotypes of VDAC

Although the VDAC amino acid sequence is very different compared to the bacterial porins, it is assumed that the mitochondrial outer membrane pore may form a β-barrel composed of 16 β-strands analogous to the know structure in bacteria [3]. Three different isotypes of VDAC are expressed in cells [4]. Not much is known in mammals about tissue-specific distribution, intracellular localization and function of the different isotypes. All experiments mentioned here relate to the VDAC isotype 1. This isotype might be the main species in the outer mitochondrial mem-

Bacterial and Eukaryotic Porins. Edited by Roland Benz

ISBN 3-527-30775-3

brane. This is deduced from experiments with yeast expressing two VDAC isotypes. When type 1 VDAC was deleted, type II could not replace it, but instead TOM 40, a pore for mitochondrial precursor peptides, was overexpressed [5]. Furthermore, Anflous *et al.* [6] followed their expression in different tissues by cloning rat heart VDAC isotypes. It was observed that VDAC type 1 was the most abundant.

16.3 The Influence of Phospholipids on VDAC Structure

Sterols have been found in isolated and purified mitochondrial VDAC such as ergosterol in the case of *Neurospora crassa* [7] and cholesterol in VDAC from bovine heart [8]. Sterols influence the channel properties of reconstituted VDAC. Popp et al. [9] found them to be essential for proper insertion of soluble precursor VDAC into bilayer membranes and also observed a 10 times increase of conductance after addition of cholesterol to native VDAC. It was postulated by the authors that sterols may increase the hydrophobicity of the surface of the VDAC β-barrel structure exposed to the phospholipid matrix of the membrane. The sterols may thus be required to obtain the high-conducting VDAC structure. Complete saturation of VDAC with sterols may even cause a reduction of voltage-dependent conductance regulation. The outer mitochondrial membrane contains 10% cholesterol, while almost nothing is found in the inner membrane [10]. However, the cholesterol in the outer membrane appears to be clustered. Ardail et al. [11] were able to isolate two contact site fractions differing significantly in cholesterol content (9.4 versus 0.4%). Both fractions bound hexokinase and, thus, contained the specific binding protein VDAC. Treatment of mitochondria with digitonin that interacts with cholesterol removes only parts of the outer membrane, leaving

Table 16.1 Lipid composition of hexokinase and creatine kinase complexes from contact sites of rat kidney mitochondria.

Fraction	*Cholesterol*	*Cardiolipin*	*Cardiolipin/cholesterol*
Kidney			
creatine kinase from cristae	1	36	36
creatine kinase (periph.)	10	26	2.6
hexokinase complex	2	22	11
Liver			
OMCS	9.4	20.2	2.15
IMCS	0.4	21.5	53.7

In the periphery, mitochondrial creatine kinase interacts with VDAC in the outer membrane and cardiolipin in the peripheral inner membrane. The enzyme is also localized in the cristae, where it interacts with two inner membranes. For comparison, the data from Ardail et al. [11] are shown at the bottom (OMCS = outer membrane contact sites, IMCS = inner membrane contact site). Data are shown as percentage of total lipid mass.

small vesicles attached at contact sites were hexokinase is bound [12]. This also suggests inhomogeneous distribution of cholesterol in the outer membrane. As with Ardail et al., we were able to characterize different types of contact sites on the basis of their composition by different functionally interacting proteins, as will be described below. We determined the lipid composition of the different contact-forming complexes (Table 16.1) and found a 5-fold difference of cholesterol content between two contact site-generating complexes. On the basis of this observation we assume the existence of structures like lipid rafts [13] in the outer membrane as a kind of outer membrane-signaling phospholipid domain. The localization of VDAC inside or outside the cholesterol-containing lipid raft-like areas might explain the different properties and structure of VDAC observed in the two complexes, as will be described below.

16.4
VDAC Conductance and Ion Selectivity

Analysis of the isolated VDAC after reconstitution in artificial bilayer membranes resulted in the calculation of the pore diameter of 4 nm at a voltage smaller than 30 mV. In this high conductance state (4 nS at 1 M KCl in the bathing fluid) the pore is anion selective. Above 30 mV, the diameter of the pore is reduced to 2 nm. The conductance decreases to 2 nS and ion selectivity changes to cation selectivity [1, 2]. The voltage-dependent conductance variations are linked to large structural modifications of VDAC. So far, no information about the nature of these changes is available. It has been postulated that a positively charged loop moves out of the channel [14]. However, it is also possible that negative charges move into the mouth of the channel, as has been observed for bacterial porins [15]. Macromolecules such as dextran, that cannot penetrate the pore, have an osmotic effect on the aqueous interior of the channel [16] and thereby increase voltage sensitivity. In the presence of dextran the low-conductance, cation-selective state is adopted already at 10 mV [17]. It has been postulated that the pore acquires this state in the contact sites where the inner membrane potential extends to the outer membrane by capacitive coupling [18]. Recently, a new idea of generation of a potential at the outer membrane was proposed on a basis of metabolic cycling and translocation of ATP, ADP and P_i across the inner membrane, and different resistance in the contact sites [19].

16.5
Physiological Significance of the Voltage Dependence

VDAC in the low-conductance, cation-selective state is not permeable for ATP, ADP and other negatively charged small molecules. This has been demonstrated by Colombini et al. investigating VDAC reconstituted in artificial membranes [20]. In isolated mitochondria, enzymes in the inter-membrane space such as adenylate kinase, had no access to external adenine nucleotides in the presence of König's polyanion [21] that induces the low-conductance state of VDAC [22]. If we convey these results to the physiological situation where more contact sites are formed and macromolecules are present, we have to assume that most VDAC are in the low-conductance, cation-selective state. Indeed, *in situ* mitochondria of permeable cardiomyocytes or skinned red muscle fibers had a 10 times higher apparent K_m for ADP compared to isolated mitochondria from the same tissue. The high K_m could be reduced to *in vitro* values by generating porous outer membranes with digitonin or by the pro-apoptotic protein Bax, suggesting that *in situ* the outer membrane was not freely permeable for ADP [23]. In addition to pure voltage-dependent regulation, specific proteins in the inter-membrane space or attached to the outer surface of mitochondria have been described that physiologically may as well be involved in regulation of the pore permeability [24, 25].

16.6
Porins as Specific Binding Sites

In bacteria, porins are receptors for various phages. VDAC at the mitochondrial periphery acts in a similar way as a binding site for enzymes such as hexokinase [26] and glycerol kinase [27]. A shift in trans-membrane topology in the membrane also changes the surface-exposed loops between the β-strands. Considering the voltage dependence of the trans-membrane topology, this means that VDAC in the contact sites might have a role to reflect functional changes of the inner membrane potential at the mitochondrial surface. Alternatively, such a signal at the surface of the outer membrane may be generated by interaction of VDAC with the adenine nucleotide translocator (ANT; see below) if we assume that the ANT would interact exclusively with a certain VDAC state.

16.7
VDAC senses Inner Membrane Functions in the Contact Site

In support of this idea, it has been observed by immune electron microscopy and binding studies that hexokinase binds with higher capacity to the outer membrane pore in the contact sites compared to pores beyond contacts [28.]. Freeze-fracturing analysis of isolated liver mitochondria revealed that contact sites could be induced by dextran or atractyloside and suppressed by glycerol or dinitrophenol [29, 30]. Thus, hexokinase binding to isolated outer membrane or mitochondria with induced or suppressed contact sites was studied. Hexokinase showed sigmoidal-type binding to mitochondria with contact sites, while binding to pure outer membrane or mitochondria with suppressed contact sites led to hyperbolic binding curves [29]. This suggested a cooperative binding of hexokinase to the pore in the contact site area. In agreement with this, hexokinase, by binding, was activated and bound hexokinase was found to form tetramers [31]. In a recent investigation,

Hashimoto and Wilson were analyzing different epitopes at the surface of bound hexokinase by specific monoclonal antibodies. The authors observed a variation of the bound hexokinase structure that correlated with functional changes of oxidative phosphorylation in the inner membrane [32]. This direct reflection of inner membrane functions at the mitochondrial periphery can be explained by interaction between VDAC in the outer and ANT in the inner mitochondrial membranes [33–35].

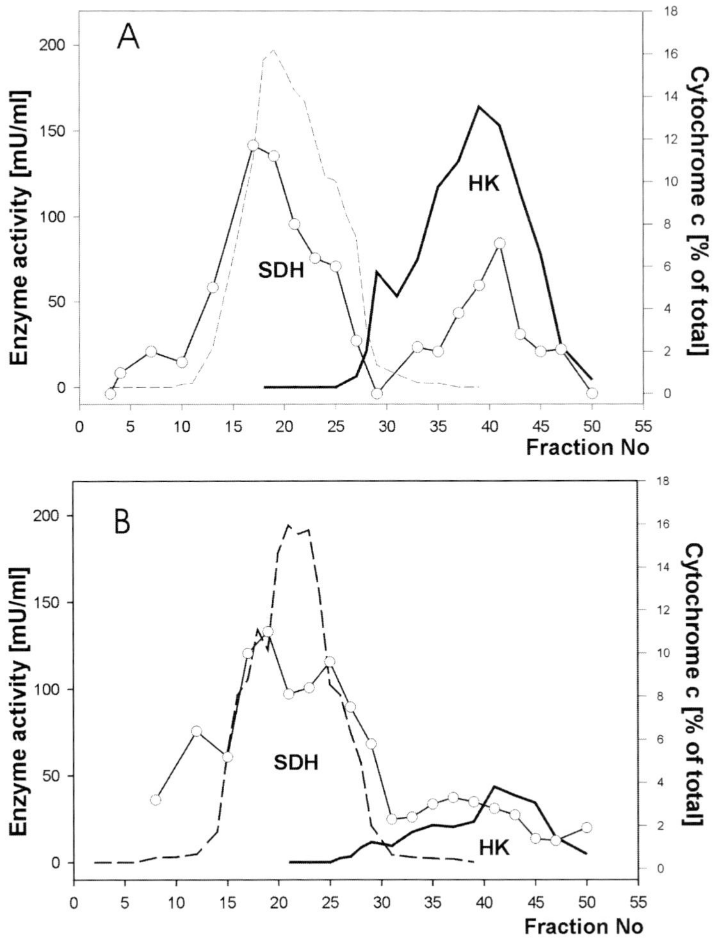

Figure 16.1 Variation of cytochrome *c* and hexokinase in the contact site fraction of kidney mitochondria by perturbing ANT structure. After osmotic shock subfractions of rat kidney mitochondria were separated in a sucrose density gradient (54.3 %, left; 30.8 %, right). The distribution of inner membrane (SDH) and contact site (HK) fractions is shown. Cytochrome *c* concentration (empty circles) in the different fractions was determined by specific antibodies and varied significantly in 500 μM atractyloside-treated (A), compared to 250 μM bongkrekate-treated (B) mitochondria.

Because of the high capacity of contact sites for hexokinase binding, the enzyme was used as a marker for isolation of this membrane fraction from osmotically disrupted mitochondria by density gradient centrifugation [36, 37]. In a recent application, hexokinase was used to indicate structural changes of VDAC correlated to the functional state of ANT. The contact site isolation method in kidney mitochondria revealed that hexokinase activity in the contacts was decreasing when the ANT was shifted to the m-conformation by pre-treatment with bongkrekic acid, whereas it was increasing after induction of the c-conformation by pre-incubation with atractyloside (Figure 16.1).

We do not know how the structure of the VDAC changes and whether the alterations are induced by the membrane potential or by the interaction with the ANT. However, we can summarize the observations of voltage-dependent conductance changes and studies of hexokinase binding as showing that VDAC alters its trans-membrane topology correlated to functional mitochondrial states. Because it affects the affinity and capacity of hexokinase binding, we can define a tensed and relaxed state of the pore in analogy to the regulation of allosteric enzymes. The tensed state would be the low-conductance, cationically selective pore in a complex with the ANT, whereas the relaxed state would be characterized by high conductance and anion selectivity without complex formation (Figure 16.2).

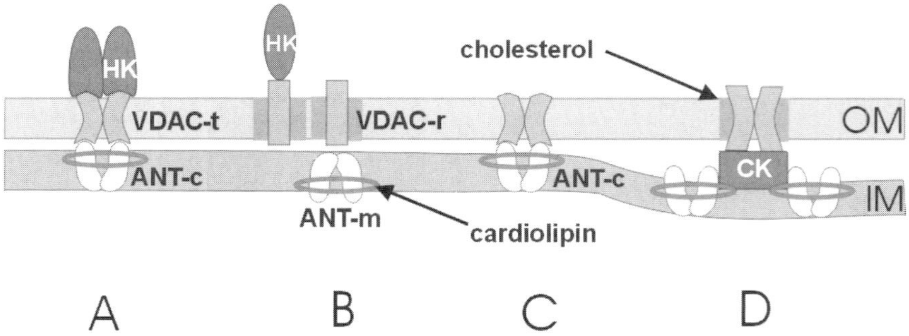

Figure 16.2 Different configuration of the VDAC–ANT complex connected to either hexokinase or creatine kinase. Each of the three components of the two kinase complexes can exist in different configurations that support or inhibit complex formation. The ANT (ANT-c, induced by atractyloside) in its c-conformation faces the cytosol and interacts with the outer membrane pore (VDAC), whereas ANT in the m-conformation (ANT-m induced by bongkrekate) orientates to the matrix and does not interact with VDAC. The ring represents cardiolipin encircling the ANT. Cardiolipin might change from the outer to inner leaflet of the membrane correlated to the ANT structure. The active ANT is a dimer. VDAC-t (tensed) is a dimer when it interacts with the ANT-c in a cholesterol-free lipid matrix. VDAC-r (relaxed) is in a cholesterol-rich lipid matrix (depicted as bars). It is not bound to the ANT and may also exist as a monomer. This structure of VDAC has low affinity for hexokinase and is present in the complex with the octamer of creatine kinase (CK). (A) Hexokinase (HK) associates to oligomers preferentially in the contact sites formed by the VDAC–ANT-c complex. The oligomerization of hexokinase leads to activation of the enzyme. (B) Monomers of hexokinase bind to VDAC beyond the contacts with less affinity and activity. (C) VDAC–ANT complexes may also exist free without bound kinases. (D) Creatine kinase (CK) binds to VDAC as octamer. The octamer preferentially associates with cardiolipin that is bound to the ANT.

16.8
Cytochrome *c* is a Component of the Contact Sites

It was observed that not only hexokinase, but also cytochrome *c* is a component of the contact sites, and its concentration increased in the presence of atractyloside and decreased upon treatment with bongkrekic acid (Figure 16.1). Thus, as observed for hexokinase, the structure of the ANT was responsible for the cytochrome *c* distribution at the surface of the inner membrane.

16.9
Isolation and Characterization of VDAC–ANT Complexes

Complexes between VDAC and ANT were generated *in vitro* with isolated porin and ANT proteins [33]. However, they were also isolated from Triton X-100-dissolved brain or kidney membranes by binding the attached hexokinase to an anion exchanger. It was found that the bound hexokinase complex contained VDAC and ANT isotype I [35]. A second complex composed of porin, creatine kinase and ANT could also be isolated from the same membrane extracts [34, 38]. Based on the separation and purification of the different complexes, two types of VDAC–ANT aggregates could be defined: one complex where VDAC and ANT interacted directly and VDAC achieved higher affinity for hexokinase (Figure 16.2A), and a second complex where VDAC interacted indirectly with the ANT through a mitochondrial creatine kinase octamer. In this case, VDAC exposed a different structure at the surface that had low affinity for hexokinase (Figure 16.2D), as hexokinase activity was always absent in the creatine kinase complexes. From cross-linking studies in isolated yeast mitochondria we know that VDAC was a dimer in the intact outer membrane [39].

Determination of the molecular weight of the isolated complexes suggested that tetramers of the active kinases were present in both cases [34]. Hexokinase is active as a monomer of 100 kDa, whereas the functionally active mitochondrial creatine kinase unit is a dimer of 85 kDa. Four dimers of mitochondrial creatine kinase form a cubic structure with identical top and bottom faces that are able to connect two membranes [40]. It is known from *in vitro* studies that the octamer of mitochondrial creatine kinase interacts directly with the outer membrane pore [41], while the interaction with the ANT may be indirect through the cardiolipin [42] that is known to be tightly bound to the ANT [43] (Figure 16.2D).

16.10
Reconstitution of VDAC–ANT Complexes

The hexokinase–porin–ANT complex and the porin–creatine kinase–ANT complex were isolated and functionally reconstituted in vesicles [34]. ATP loaded into the vesicles did not leak out, although VDAC was a component of the complexes. However, the kinases in the complexes had access to the internal ATP through the ANT. This was shown by either glucose-6-phosphate formation from external glucose via hexokinase or creatine phosphate generation by creatine kinase from external creatine. Both reactions were inhibited by blocking the antiporter activity of the ANT by atractyloside. The results suggested that the kinases, after reconstitution, were functionally coupled to the ANT. The ANT was controlling the exchange of adenine nucleotides through VDAC and was working as an antiporter [44]. However, the functional state of the ANT in the reconstituted complexes could be converted into an unspecific uniporter (resembling the permeability transition pore) by addition of Ca^{2+}, as has been demonstrated also for several related mitochondrial transport systems by Dierks et al. [45].

16.11
Importance of Metabolic Channeling in Regulation of Energy Metabolism

The tight functional coupling of peripheral kinases to the ANT is important because of two reasons: (i) it regulates the activity of oxidative phosphorylation by ADP and (ii) it increases the free energy (ΔG) in the ATP system. To clarify the first point we may imagine the situation after eating a carbohydrate-rich meal: a high glucose concentration enters the resting muscle cell that must be phosphorylated and converted to UDP-glucose to accomplish incorporation into glycogen. The ATP for this process is more efficiently provided by oxidation of pyruvate in the mitochondria than by lactate formation. Hexokinase bound to the VDAC–ANT complex induces the oxidative pathway by converting intra-mitochondrial ATP into ADP and by that activating oxidative phosphorylation. To describe the second advantage of metabolic channeling, we have to consider that all ATP energy-consuming processes in the cell (including ion pumps) depend on the transfer of the γ-phosphoryl group of ATP according to the reaction: ATP + X = X-P + ADP. In a subsequent step, X-P dissociates when, for example, the ion is pumped. This means the efficiency of ATP to drive these phosphorylation reactions depends on the level of ADP and P_i during the reaction according to the equation of $\Delta G = \Delta G^{o\prime} + RT \ln [ADP] [P_i]/[ATP]$. Thus, the power ($\Delta G$) of ATP and the flux through the ATP-dependent reaction would increase if ADP and P_i would be low or be continuously withdrawn from the reaction. In the complex between mitochondrial creatine kinase and the ANT these requirements are perfectly fulfilled. The equilibrium of the reaction will not be reached as the ATP just exported is directly utilized and the ADP produced is immediately taken up into the matrix by the ANT. P_i is excluded from the reaction because of complex formation. Thus, metabolic channeling through functional coupling between ANT and creatine kinase pushes the balance of the reaction between creatine and ATP far to the side of phosphocreatine production. A further advantage of this coupling to the ANT is that the higher free energy of ATP at the surface of the inner membrane is preserved as a higher phosphocreatine/creatine quotient. ATP is electrogenically exported by the ANT at the expense of the membrane potential. It has been calculated that this process increases the free energy of ATP by an additional 12 kJ mol^{-1} [46]. Thus, the phosphocreatine/creatine ratio could be increased by this quantity through coupling between mitochondrial creatine kinase and the ANT.

It has been criticized that the functional interaction between ANT and creatine kinase may be physiologically irrelevant because of the approximately 10-fold difference in ATP turnover of the two partners. However, one has to consider that at least four of the slower ANT dimers may be attached to one of the fast creatine kinase octamers (Figure 16.2D).

16.12
The VDAC–ANT Complex as Permeability Transition Pore

Mitochondria contain a yet-unidentified structure that forms a large unspecific pore under conditions of high matrix Ca^{2+}, P_i or oxidative stress [47]. Opening of this so-called permeability transition pore was fully reversible upon withdrawing Ca^{2+} [48]. Electrophysiological measurements revealed properties of the pore related to VDAC suggesting that it may be a component of the whole structure [49, 50]. The permeability transition pore could be specifically blocked by cyclosporin A, which binds to cyclophilin D, a peptidylprolyl *cis–trans* isomerase present in the mitochondrial matrix [48, 51]. By employing a cyclophilin D affinity matrix, a complex between VDAC and ANT was isolated by Crompton et al. after extracting heart mitochondrial membranes with the detergent CHAPS [52]. This complex was reconstituted in vesicles, and pore opening by Ca^{2+} and Pi addition was registered in cyclosporin A-sensitive manner. Comparable results were obtained with the isolated hexokinase–VDAC–ANT complex. Cyclophilin was co-purified through isolation of the complex [38]. The complex was reconstituted in phospholipid vesicles that were loaded with malate. The internal malate was released correlated to increasing Ca^{2+} concentrations. The process was inhibited by cyclosporin A, suggesting that the reconstituted hexokinase–porin–ANT complex had properties of the permeability transition pore (Figure 16.3A).

Two questions arise: (i) whether the ANT is the only structure that causes permeability transition and (ii) whether the ANT forms the pore only as a complex with VDAC. Concerning the first question, it is possible that all related antiporters such as P_i/OH- or the Glu/Asp exchanger can form a permeability transition pore depending on the actual conditions. As to the second question, it is possible to claim that the purified ANT also forms unspecific pores in the presence of Ca^{2+}. This was shown by reconstitution of ANT in bilayer membranes [53] or in vesicles [54]. However, these pores were not sensitive to cyclosporin A as cyclophilin D was missing. Cyclophilin D is thus a second important regulatory component of the permeability transition pore [51, 55]. The idea that VDAC may also have regulatory functions of the permeability transition was supported by the analysis of the reconstituted VDAC–creatine kinase–ANT complex. In contrast to the reconstituted hexokinase complex there was no release of enclosed malate from the vesicles by addition of Ca^{2+} (Figure 16.3B). As shown schematically in Figure 16.3(B), VDAC is

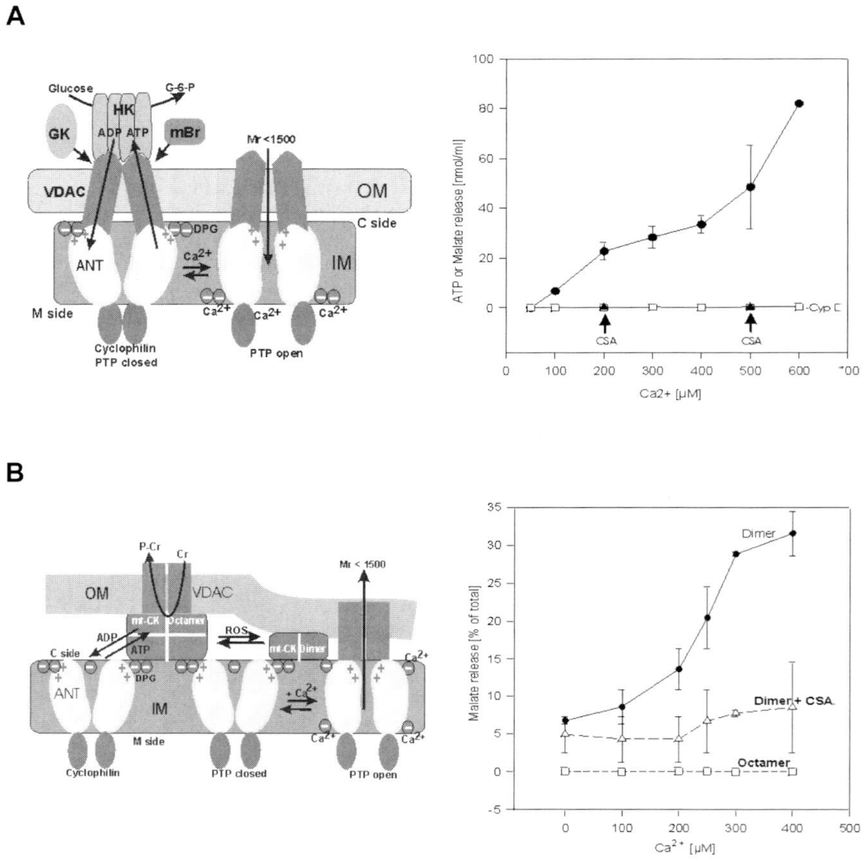

Figure 16.3

linked to the octamer of creatine kinase that hinders interaction with the ANT. However, direct interaction between VDAC and ANT was possible after dissociation of the creatine kinase octamer. In this case malate was liberated dependent on Ca^{2+} addition (Figure 16.3B). The release was inhibited by cyclosporin A. The results suggested that VDAC such as cyclophilin D might play a role in regulation of the permeability transition pore. A possible role of VDAC might be to keep the ANT in the c-conformation, exposing the ATP/ADP-binding site to the cytosolic surface of the inner membrane according to the re-orientating carrier model proposed by Klingenberg [56]. In this conformation that is induced by atractyloside, the conversion of the ANT into an unspecific pore by Ca^{2+} is facilitated and leads to permeability transition. However, if VDAC and the ANT interact with the octamer of mitochondrial creatine kinase, conversion of ANT to the permeability transition pore-like state would be suppressed (Figure 16.3B). In fact this has been observed in a transgenic mouse model expressing mitochondrial creatine kinase in liver mitochondria [57]. The permeability transition pore was opened in iso-

◀ **Figure 16.3** The VDAC–ANT complex as permeability transition pore. (A, left panel) The scheme depicts the presumable arrangement of the hexokinase porin ANT complex in the contact sites of mitochondria. Hexokinase (HK) binds as a tetramer to the VDAC–ANT complex. Glycerol kinase (GK) and the mitochondrial benzodiazipin receptor (mBr) bind to the same pore configuration. (A, right panel) The complex was isolated from a Triton X-100 extract of brain membranes and was reconstituted in phospholipid vesicles that were loaded with malate or ATP. Cyclophilin was co-purified with the complex and could be extracted from the complex with phosphate buffer, pH 4.5. The enclosed malate was released from the vesicles by Ca^{2+} between 100 and 600 μM. The malate release was inhibited either by cyclophilin extraction (–Cyp D) or by incubation of the vesicles with 0.5 μM cyclosporin A (CSA) after addition of 200 or 500 μM Ca^{2+}. According to Brustovetsky and Klingenberg [53], Ca^{2+} interacts with the cardiolipin (DPG) around the ANT, thus changing the structure of the ANT from a ATP/ADP antiporter to an unspecific uniporter state. The uniporter state of the ANT may act as a permeability transition pore (PTP) in intact mitochondria. (B, left panel) The scheme shows the possible arrangement of a complex between VDAC, the octamer of mitochondrial creatine kinase and the ANT in the contact sites of mitochondria. When the octamer dissociates, VDAC and ANT can directly interact. (B, right panel) The complex was isolated from a Triton X-100 extract of brain membranes and was reconstituted in egg yolk liposomes that were loaded with malate. The malate was not released by increasing Ca^{2+} concentrations (Octamer). When the octamer was dissociated into dimer by 20 min incubation of the liposomes with 5 mM $MgCl_2$, 20 mM creatine, 50 mM KNO_3 and 4 mM ADP, VDAC–ANT complexes could form. In these complexes Ca^{2+} could shift the ANT structure to the unspecific permeability transition pore (PTP) and the enclosed malate was liberated (Dimer). The malate permeability of the PTP was inhibited by pre-incubation of the vesicles with 100 nM cyclosporin A (dimer + CSA). The octamer–dimer equilibrium of the mitochondrial creatine kinase can be shifted to dimer by reactive oxygen species (ROS) [74].

lated mouse liver mitochondria by atractyloside and Ca^{2+}. In contrast, in the liver mitochondria from transgenic mice containing creatine kinase, the permeability transition was inhibited by substrates that stabilized the octamer of creatine kinase such as creatine and cyclocreatine.

It should be mentioned here that mitochondrial creatine kinase, in addition to ts important role of organization in the peripheral mitochondrial compartment, is also localized in the cristae [58] where it has certainly different functions and is not linked to VDAC.

16.13 The VDAC–ANT Complex as a Target for Bax-dependent Cytochrome *c* Release

There are several signal- and tissue-specific pathways to induce apoptosis. One of them is a release of cytochrome *c* from mitochondria [59–61] that activates caspases [60] by binding to a cytoplasmic protein Apaf-1 in the presence of dATP [62]. The mechanism by which external Bax releases cytochrome *c* is still controversial, and may also depend on the actual localization of cytochrome *c* in addition to the binding and incorporation of Bax at the mitochondrial surface.

VDAC and ANT interact preferentially when adopting a specific molecular conformation. The ANT has to be in the c-conformation as atractyloside induces the

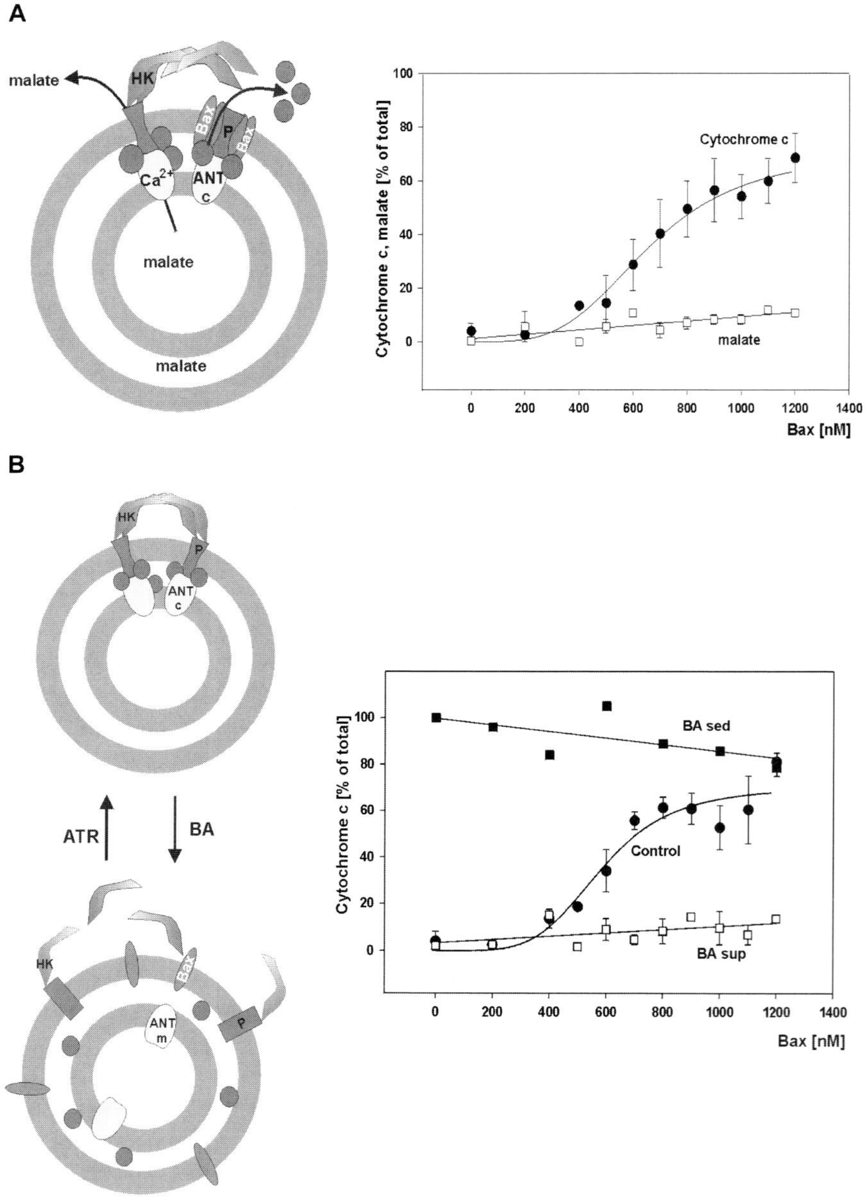
A
malate
HK
Bax
P
Bax
Ca^{2+}
ANT
c
malate
malate
Cytochrome c, malate [% of total]
Cytochrome c
malate
Bax [nM]
B
HK
P
ANT
c
ATR
BA
HK
Bax
ANT
m
P
Cytochrome c [% of total]
BA sed
Control
BA sup
Bax [nM]

Figure 16.4

◀ **Figure 16.4** Bax-dependent release of cytochrome *c* from the hexokinase VDAC–ANT complex. (A) Schematic representation of the reconstituted hexokinase (HK)–VDAC (P)–ANT complex in multilayered vesicles. Cytochrome *c* (circles) that was still attached to the complexes could be released by Bax, as shown in the right panel. However, malate, loaded into the vesicles, was not liberated by Bax, but by 200–500 μM Ca^{2+} through opening the ANT as a permeability transition pore (Figure 16.3A). This suggested that Bax was specifically interacting with the cytochrome *c* in the porin–ANT complex. The sigmoidal curve of cytochrome *c* release might indicate that Bax is accomplishing this process as an oligomer. (B) Schematic representation of the association/dissociation of the hexokinase–VDAC (P)–ANT complex by atractyloside (ATR) or bongkrekate (BA) after reconstitution in multilayer vesicles. Treatment of the complexes with bongkrekate changed the ANT into the m-conformation (ANT-m). This led to dissociation of the complex and a correlated structural change of VDAC, with a decreased affinity for hexokinase and Bax. Cytochrome *c* might have redistributed as well. As shown in the right panel, Bax was unable to release cytochrome *c* after the structural change of the complex by bongkrekate. All cytochrome *c* remained in the sediment (BA sed) in contrast to the control, without bongkrekate treatment, and nothing appeared in the supernatant (BA sup).

complex with the outer membrane pore [35], while bongkrekate shifts the ANT to the m-conformation and by that represses this interaction. The VDAC in the complex with the ANT appears to be in the low-conductance, cationically selective, tensed state. This can be deduced from the observation that heart muscle mitochondria *in situ* show a reduced response to external ADP [23]. The structural change of the outer membrane pore achieved by interaction with the ANT is recognized at the mitochondrial surface and leads to higher affinity of hexokinase. Recent observations suggest that this tensed pore structure is the preferred target of Bax molecules. Pastorino et al. [63] found that hexokinase and Bax compete for the same binding site, and Capano and Crompton [64] co-precipitated VDAC and ANT when Bax was immune precipitated from extracts of cardiomyocytes.

16.14 The VDAC–ANT Complexes contain Cytochrome *c*

As described above it was observed that the contact sites contained cytochrome *c*. As the VDAC–ANT complexes are derived from the contact sites it was not surprising that the complexes contained a significant amount of cytochrome *c* [65]. This was found although the complex was eluted from anion exchanger column by 200 mM KCl suggesting that the cytochrome *c* was not bound by ionic interaction. To investigate whether Bax would be able to interact with the cytochrome *c* within the VDAC–ANT complexes, the latter were reconstituted in phospholipid vesicles as described above (Figure 16.4A). After reconstitution, the vesicles were loaded by malate. Bax liberated the endogenous cytochrome *c*, but did not release the internal malate (Figure 16.4A). The Bax-dependent liberation of endogenous cytochrome *c* was abolished when the VDAC–ANT complex was dissociated by bongkrekate (Figure 16.4B).

16.15 The Importance of the Kinases in Regulation of Apoptosis

Release of cytochrome *c* from mitochondria proceeds in two steps. The first involves the release of a small fraction of cytochrome *c* from the compartment between the outer and peripheral inner membranes and the contact sites. It is induced by Bax at low concentrations [66] in the early phase [64]. The second step includes release of additional fractions of cytochrome *c* located at the surface of the cristae membranes and results from opening of the permeability transition pore, followed by mitochondrial swelling and membrane disruption [67]. The massive cytochrome *c* release may also be a consequence of VDAC closure preceding membrane disruption [68]. The observed failure of mitochondria to exchange adenine nucleotides with the cytosol may be a consequence of the transformation of VDAC into the tensed state. The two described kinases bound to VDAC are involved in the regulation of both processes of cytochrome *c* liberation.

16.16 Suppression of Bax-dependent Cytochrome *c* Release and Permeability Transition by Hexokinase

Hexokinase binds to the outer membrane pore in its tensed state, when VDAC interacts with the ANT. In this state VDAC is not freely permeable for adenine nucleotides. This has been already observed in mitochondria in the presence of dextran that increases hexokinase binding and contacts between VDAC and ANT [29]. In these mitochondria, external pyruvate kinase had less access to the ADP produced by hexokinase [69]. Although VDAC in the tensed state does not allow ATP transfer, hexokinase utilizes mitochondrial ATP, suggesting a structural modification of VDAC through binding of the enzyme.

On the other hand, the activity of mitochondrial-bound hexokinase was found to be important for protein kinase B-linked suppression of cytochrome *c* release and apoptosis [70]. The experiments suggested that following withdraw of growth hormone, Bax liberated the small fraction of cytochrome *c* from the contact sites. In complete agreement with this, the Bax-dependent release of cytochrome *c* bound in the VDAC–ANT complexes described above was inhibited by activity of hexokinase in the presence of glucose and ATP [65]. This was explained by stabilization of hexokinase binding to the VDAC that is also the main target of Bax. In addition to this effect on Bax-dependent cytochrome *c* release, activation of hexokinase inhibited Ca^{2+}-dependent opening of the permeability transition pore by ADP production [38].

16.17
Suppression of Permeability Transition by Mitochondrial Creatine Kinase

Another kinase which binds to VDAC is mitochondrial creatine kinase. This enzyme interacts with VDAC exclusively in the octameric state, while the dimer has only weak affinity to porin [41, 42]. Thus, the association–dissociation equilibrium between octamer and dimer determines the formation of octamer–VDAC complexes. Ligation of the coronary artery in guinea pigs led to decrease of octamer from 85 to 60 % in the infarcted area of the heart [71]. Because of the dissociation of the creatine kinase octamer, the possibility of direct VDAC–ANT interaction was increased. As suggested above, the latter complexes may be the prerequisite of Ca^{2+}-induced permeability transition, followed by mitochondrial swelling, membrane disruption, cytochrome *c* release and apoptosis. Indeed, apoptotic cells encircle the area of necrotic cells in the infarcted heart.

Heart, kidney and brain mitochondria contain two ANT isotypes. ANT isotype I was localized in the peripheral inner membrane having higher affinity to VDAC and cyclophilin, whereas isotype II was present in the cristae membranes [35]. When ANT I was overexpressed in cells, apoptosis was induced [72]. However, co-expression of cyclophilin D or creatine kinase together with ANT I abolished apoptosis induction. Moreover, it has been observed that patients suffering from dilated cardiomyopathy had significantly higher ANT I mRNA levels [73]. This phenomenon, such as in the tissue culture experiments mentioned above, may be responsible for the increased apoptosis rate observed in case of cardiomyopathy [74]. Considering that the ANT I in the peripheral inner membrane would form a permeability transition pore, we assume that proteins interacting with ANT I such as cyclophilin and the octamer of mitochondrial creatine kinase would reduce the probability of pore opening and apoptosis induction. However, the chance of pore opening and permeability transition would increase if the ratio between the ANT I and its regulatory proteins is changing either by induction of ANT I or by dissociation of the octamer of mitochondrial creatine kinase.

16.18
Conclusion

The first electron microscopic observation of contact sites as dynamic structures [75] reflected morphological perturbations in mitochondrial membranes induced by association or dissociation of the described kinase complexes. The different components of the complexes interact in various ways depending on their actual structures that are regulated by the mitochondrial membrane potential and the energy metabolism of the cell. Physiologically, the complexes are transient structures that are subjected to regulation. During isolation the complexes become stable because alternatively interacting partners are removed. Thus, the reconstituted complexes represent certain functional states that the different components have adopted. We have shown how perturbation of a single component such as the

ANT changes the function of the whole complex and is reflected at the mitochondrial surface by VDAC. This means that physiological signals will change the complex architecture if they affect the properties of individual components such as VDAC conductance or hexokinase/creatine kinases activity with significant regulatory consequences for the cell.

References

1 Colombini M. A candidate for the permeability pathway of the outer mitochondrial membrane. *Nature* **1979**, *279*, 643–645.
2 Benz R. Permeation of hydrophilic solutes through mitochondrial outer membranes: review on mitochondrial porins. *Biochim Biophys Acta Rev Biomembr* **1994**, *1197*, 167–196.
3 Casadio R, Jacoboni I, Messina A, De Pinto V. A 3D model of the voltage-dependent anion channel (VDAC). *FEBS Lett* **2002**, *520*, 1–7.
4 Sampson MJ, Lovell RS, Craigen WJ. The murine voltage-dependent anion channel gene family. Conserved structure and function. *J Biol Chem* **1997**, *272*, 18966–18973.
5 Kmita H, Budzinska M. Involvement of the TOM complex in external NADH transport into yeast mitochondria depleted of mitochondrial porin1 *Biochim Biophys Acta* **2000**, *1509*, 86–94.
6 Anflous K, Blondel O, Bernard A, Khrestchatisky M, Ventura-Clapier R. Characterization of rat porin isoforms: cloning of cardiac style-3 variant encoding an additional methionine at its putative N-terminal region. *Biochim Biophys Acta* **1998**, *1399*, 47–50.
7 Freitag H, Neupert W, Benz R. Purification and characterization of a pore protein of the outer mitochondrial membrane from *Neurospora crassa*. *Eur J Biochem* **1982**, *162*, 629–636.
8 DePinto V, Benz R, Palmieri F. Interaction of non-classical detergents with the mitochondrial porin. A new purification procedure and characterization of the pore-forming unit. *Eur J Biochem* **1989**, *183*, 179–187.
9 Popp B, Schmid A, Benz R. Role of sterols in the functional reconstitution of water-soluble mitochondrial porins from different organisms. *Biochemistry* **1995**, *34*, 3352–3361.
10 Daum G. Lipids of mitochondria. *Biochim Biophys Acta* **1985**, *822*, 1–42.
11 Ardail D, Privat JP, Egret-Charlier M, Levrat C, Lerme F, Louisot P Mitochondrial contact sites. Lipid composition and dynamics. *J Biol Chem* **1990**, *265*, 18797–18802.
12 Brdiczka D, Wallimann T. The importance of the outer mitochondrial compartment in regulation of energy metabolism. *Mol Cell Biochem* **1994**, 133/134, 69–83.
13 Simons K, Ikonen E. Functional rafts in cell membranes. *Nature* **1997**, *387*, 569–572.
14 Song J, Midson C, Blachly-Dyson E, Forte M, Colombini M. The sensor regions of VDAC are translocated from within the membrane to the surface during the gating processes. *Biophys J* **1998**, *74*, 2926–2944.
15 Welte W, Nestel U, Wacker T, Diederichs K. Structure and function of the porin channel. *Kidney Int* **1995**, *48*, 930–940.
16 Zimmerberg J, Parsegian VA. Polymer inaccessible volume changes during opening and closing of a voltage-dependent ionic channel. *Nature* **1986**, *323*, 36–39.
17 Gellerich FN, Wagner M, Kapischke M, Wicker U, Brdiczka D. Effect of macromolecules on the regulation of

the mitochondrial outer membrane pore and the activity of adenylate kinase in the inter-membrane space. *Biochim Biophys Acta* **1993**, *1142*, 217–227.

18 Benz R, Kottke M, Brdiczka D. The cationically selective state of the mitochondrial outer membrane pore: a study with intact mitochondria and reconstituted mitochondrial porin. *Biochim Biophys Acta* **1990**, *1022*, 311–318.

19 Lemeshko VV. Model of the outer membrane potential generation by the inner membrane of mitochondria. *Biophys J* **2002**, *82*, 684–692.

20 Rostovtseva T, Colombini M. VDAC channels mediate and gate the flow of ATP: implications for the regulation of mitochondrial function. *Biophys J* **1997**, *72*, 1954–1962.

21 König T, Kocsis B, Mészaros L, Nahm K, Zoltán S, Horváth I. Interaction of a synthetic polyanion with rat liver mitochondria. *Biochim Biophys Acta* **1977**, *462*, 380–389.

22 Benz R, Wojtczak L, Bosch W, Brdiczka D. Inhibition of adenine nucleotide transport through the mitochondrial porin by a synthetic polyanion. *FEBS Lett* **1988**, *210*, 75–80.

23 Saks V, Kuznetsov A, Khuchua Z, Vasilyeva E, Belikova J, Kesvatera T, Tiivel T. Control of cellular respiration *in vivo* by mitochondrial outer membrane and by creatine kinase. A new speculative hypothesis: possible involvement of mitochondrial–cytoskeleton interactions. *J. Mol Cell Cardiol* **1995**, *27*, 625–645.

24 Liu MY, Colombini M. A soluble mitochondrial protein increases the voltage dependence of the mitochondrial channel, VDAC. *J Bioenerg Biomembr* **1992**, *24*, 41–46.

25 Saks V, Kaambre T, Sikk P, Eimre M, Orlova E, Paju K, Piirsoo A, Appaix F, Kay L, Regnitz-Zagrosek V, Fleck E, Seppet E. Intracellular energetic units in red muscle cells. *Biochem J* **2001**, *356*, 643–657.

26 Fiek Ch, Benz R, Roos N, Brdiczka D. Evidence for identity between the hexokinase-binding protein and the mitochondrial porin in the outer membrane of rat liver mitochondria. *Biochim Biophys Acta* **1982**, *688*, 429–440.

27 Östlund A.K, Göhring U, Krause J, Brdiczka D. The binding of glycerol kinase to the outer membrane of rat liver mitochondria: its importance in metabolic regulation. *Biochem Med* **1983**, *30*, 231–245.

28 Weiler U, Riesinger I, Knoll G, Brdiczka D. The regulation of mitochondrial-bound hexokinases in the liver. *Biochem Med* **1985**, *33*, 223–235.

29 Wicker U, Bücheler K, Gellerich FN, Wagner M, Kapischke M, Brdiczka D. Effect of macromolecules on the structure of the mitochondrial intermembrane space and the regulation of hexokinase. *Biochim Biophys Acta* **1993**, *1142*, 228–239.

30 Bücheler K, Adams V, Brdiczka D. Localization of the ATP/ADP translocator in the inner membrane and regulation of contact sites between mitochondrial envelope membranes by ADP. A study on freeze fractured isolated liver mitochondria. *Biochim Biophys Acta* **1991**, *1061*, 215–225.

31 Xie G, Wilson JE. Tetrameric structure of mitochondrially bound rat brain hexokinase: a cross-linking study. *Arch Biochem Biophys* **1990**, *276*, 285–293.

32 Hashimoto M, Wilson JE. Membrane potential-dependent conformational changes in mitochondrially bound hexokinase in brain. *Arch Biochem Biophys* **2000**, *884*, 163–173.

33 Bühler S, Michels J, Wendt S, Rück A, Brdiczka D, Welte W, Przybylski M. Mass spectrometric mapping of ion channel proteins (Porins) and identification of their supramolecular membrane assembly. *Proteins* **1998**, *2*, 63–73.

34 Beutner G. Rück A, Riede, B, Welte W, Brdiczka D. Complexes between kinases, mitochondrial porin and adenylate translocator in rat brain resemble the permeability transition pore. *FEBS Lett* **1996**, *396*, 189–195.

35 Vyssokikh MY, Katz A, Rück A, Wuensch C, Dörner A, Zorov DB, Brdiczka D. Adenine nucleotide translocator isoforms 1 and 2 are differently

distributed in the mitochondrial inner membrane and have distinct affinities to cyclophilin D. *Biochem J* **2001**, *358*, 349–358.

36 Ohlendieck K, Riesinger I, Adams V, Krause J, Brdiczka D. Enrichment and biochemical characterization of boundary membrane contact sites in rat-liver mitochondria. *Biochim Biophys Acta* **1986**, *860*, 672–689.

37 Adams V. Bosch W, Schlegel J, Wallimann T, Brdiczka D. Further characterization of contact sites from mitochondria of different tissues: topology of peripheral kinases *Biochim Biophys Acta* **1989**, *981*, 213–225.

38 Beutner G, Rück A, Riede B, Brdiczka D. Complexes between porin, hexokinase, mitochondrial creatine kinase and adenylate translocator display properties of the permeability transition pore. Implication for regulation of permeability transition by the kinases. *Biochim Biophys Acta* **1998**, *1368*, 7–18.

39 Krause J, Hay R, Kowollik C, Brdiczka D. Cross-linking analysis of yeast mitochondrial outer membrane. *Biochim Biophys Acta* **1986**, *860*, 690–698.

40 Rojo M, Hovius R, Demel RA, Nicolay K, Wallimann T. Mitochondrial creatine kinase mediates contact formation between mitochondrial membranes. *J Biol Chem* **1991**, *266*, 20290–20295.

41 Brdiczka D, Kaldis Ph, Wallimann Th. *In vitro* complex formation between octamer of creatine kinase and porin. *J Biol Chem* **1994**, *269*, 27640–27644.

42 Schlattner U, Dolder M, Wallimann T, Tokarska-Schlattner M. Mitochondrial creatine kinase and mitochondrial outer membrane porin show direct interaction that is modulated by calcium. *J Biol Chem* **2001**, *276*, 48027–48030.

43 Beyer K, Klingenberg M. ADP/ATP carrier protein from beef heart mitochondria has high amounts of tightly bound cardiolipin, as revealed by ^{31}P nuclear magnetic resonance *Biochemistry* **1985**, *24*, 3821–3826.

44 Krämer R, Palmieri F. Metabolite carriers in mitochondria. In: Ernster L (Ed.), *Molecular Mechanisms in Bioenergetics*. Elsevier, Amsterdam, **1992**, pp. 359–384.

45 Dierks T Salentin A, Heberger C, Krämer R. The mitochondrial aspartate/glutamate and ADP/ATP carrier switch from obligate counterexchange to unidirectional transport after modification by SH-reagents. *Biochim Biophys Acta* **1990**, *1028*, 268–280.

46 Heldt H.W, Klingenberg M, Milovancev M. Differences between the ATP–ADP ratios in the mitochondrial matrix and in the extramitochondrial space. *Eur J Biochem* **1972**, *30*, 434–440.

47 Harworth RA, Hunter PR. Allosteric inhibition of the Ca^{2+}-activated hydrophilic channel of the mitochondrial inner membrane by nucleotides. *J Membr Biol* **1980**, *57*, 231–236.

48 Crompton M, Costi A. Kinetic evidence for a heart mitochondrial pore activated by Ca^{2+} and oxidative stress. *Eur J Biochem* **1988**, *178*, 448–501.

49 Zorati M, Szabao I. The mitochondrial permeability transition. *Biochim Biophys Acta* **1995**, *1241*, 139–176.

50 Kinally KW, Zorov DB, Antonenko YN, Snyder SH, McEnery MW, Tedeschi H. Mitochondrial diazepine receptor linked to inner membrane ion channels by nanomolar actions of ligand. *Proc Natl Acad Sci USA* **1993**, *90*, 1374–1378.

51 Halestrap AP, Davidson AM. Inhibition of Ca^{2+}-induced large amplitude swelling of mitochondria by cyclosporin A is probably caused by binding to a matrix peptidylprolyl *cis–trans*-isomerase and preventing it interacting with the adenine nucleotide translocase. *Biochem J* **1990**, *268*, 153–160.

52 Crompton M, Virji S, Ward JM. Cyclophilin-D binds strongly to complexes of the voltage-dependent anion channel and the adenine nucleotide translocase to form the permeability transition pore. *Eur J Biochem* **1998**, *258*, 729–735.

53 Brustovetsky N, Klingenberg M. The mitochondrial ADP/ATP carrier can be reversibly converted into a large channel by Ca^{2+}. *Biochemistry* **1996**, *35*, 8483–8488.

54 Rück A. Dolder M, Wallimann T, Brdiczka D. Reconstituted adenine

nucleotide translocase forms a channel for small molecules comparable to the mitochondrial permeability transition pore. *FEBS Lett* **1998**, *426*, 97–101.
55 Crompton M. Ellinger H, Costi. A. Inhibition by cyclosporin A of a Ca^{2+}-dependent pore in heart mitochondria activated by inorganic phosphate and oxidative stress. *Biochem J* **1888**, *255*, 257–360.
56 Klingenberg M, Grebe K, Falkner G. Interaction between the binding of ^{35}S-atractyloside and bongkrekic acid at mitochondrial membranes. *FEBS Lett* **1971**, *16*, 301–303.
57 O'Gorman E, Beutner G, Dolder M, Koretsky AP, Brdiczka D, Wallimann T. The role of creatine kinase in inhibition of mitochondrial permeability transition *FEBS Lett* **1997**, *414*, 253–257.
58 Kottke M, Wallimann Th, Brdiczka D. Dual localization of mitochondrial creatine kinase in brain mitochondria. *Biochem Med Metabol Biol* **1994**, *51*, 105–117.
59 Yang JLX, Bhalla K, Kim CN, Ibrado AM, Cai J, Peng TI, Jones DP, Wang X. Prevention of apoptosis by Bcl-2: release of cytochrome *c* from mitochondria blocked. *Science* **1997**, *275*, 1129–1132.
60 Liu X, Kim CN, Yang J, Jemmerson R, Wang X. Induction of apoptotic program in cell-free extracts: requirement for dATP and cytochrome *c*. *Cell* **1996**, *86*, 147–157.
61 Kluck RM, Green DR, Newmeyer DD. The release of cytochrome *c* from mitochondria: a primary site for Bcl-regulation of apoptosis. *Science* **1997**, *275*, 1132–1136.
62 Saleh A S. S, Acharya S, Fishel R, Alnemri ES. Cytochrome *c* and dATP-mediated oligomerization of Apaf-1 is a prerequisite for procaspase-9 activation. *J Biol Chem* **1999**, *274*, 17941–17945.
63 Pastorino JG, Shulga N, Hoek JB. Mitochondrial binding of hexokinase II inhibits Bax-induced cytochrome *c* release and apoptosis. *J Biol Chem* **2002**, *277*, 7610–7618.
64 Capano M, Crompton M. Biphasic translocation of BAX to mitochondria. *Biochem J* **2002**, *367*, 169–178.
65 Vyssokikh MY, Zorova L, Zorov D, Heimlich G, Jürgensmeier JM, Brdiczka D. Bax releases cytochrome *c* preferentially from a complex between porin and adenine nucleotide translocator. Hexokinase activity suppresses this effect. *Mol Biol Rep* **2002**, *29*, 93–96.
66 Pastorino JG, Tafani, M, Rothman, RJ, Marcineviciute A, Hoek JB, Farber JL. Functional consequences of sustained or transient activation by Bax of the mitochondrial permeability transition pore. *J Biol Chem* **1999**, *274*, 31734–31739.
67 Halestrap AP, McStay GP, Clarke SJ. The permeability transition pore complex: another view. *Biochimie* **2002**, *84*, 153–166.
68 Vander Heiden MG, Chandel NS, Williamson EK, Schumacker PT, Thompson CB. Bcl-x_L regulates the membrane potential and volume homeostasis of mitochondria. *Cell* **1997**, *91*, 627–637.
69 Gellerich FN, Laterveer FD, Zierz S, Nicolay K. The quantification of ADP diffusion gradients across the outer membrane of heart mitochondria in the presence of macromolecules. *Biochim Biophys Acta* **2002**, *1554*, 48–56.
70 Gottlob K, Majewski N, Kennedy S, Kandel E, Robey RB, Hay N. Inhibition of early apoptotic events by Akt/PKB is dependent on the first committed step of glycolysis and mitochondrial hexokinase. *Genes Dev* **2001**, *15*, 1406–1418.
71 Soboll S Brdiczka D, Jahnke D, Schmidt A, Schlattner U, Wendt S, Wyss M, Wallimann T. Octamer–dimer transitions of mitochondrial creatine kinase in heart disease. *J Mol Cell Cardiol* **1999**, *31*, 857–866.
72 Bauer MK, Schubert A, Rocks O, Grimm S. Adenine nucleotide translocase-1, a component of the permeability transition pore, can dominantly induce apoptosis. *J Cell Biol* **1999**, *147*, 1493–1502.
73 Dörner A. Schulze K. Rauch U, Schultheiss HP. Adenine nucleotide translocator in dilated cardiomyopathy:

pathophysiological alterations in expression and function. *Mol Cell Biochem* **1997**, *174*, 261–269.

74 Dolder M, Wendt S, Wallimann T. Mitochondrial creatine kinase in contact sites: interaction with porin and adenine nucleotide translocase, role in permeability transition and sensitivity to oxidative damage [Review]. *Biol Signals Recept* **2001**, *10*, 93–111.

75 Knoll G, Brdiczka D. Changes in freeze-fracture mitochondrial membranes correlated to their energetic state. *Biochim Biophys Acta* **1983**, *733*, 102–110.

Index

Bacterial and Eukaryotic Porins. Edited by Roland Benz

ISBN 3-527-30775-3

F

G

H